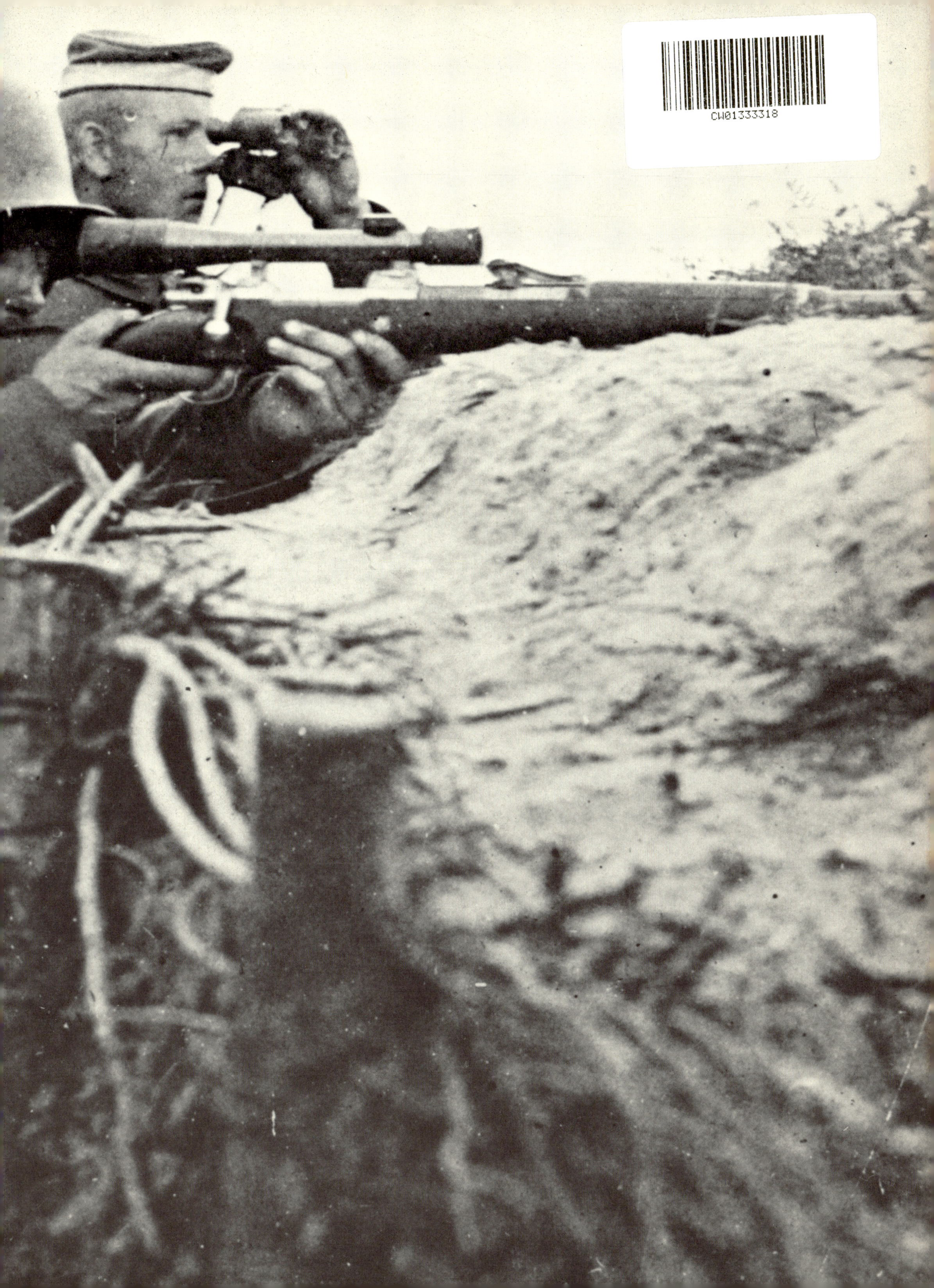

The German Sniper

1914 - 1945

The German Sniper

1914 - 1945

by Peter R. Senich

Paladin Press — Boulder, Colorado
Arms and Armour Press — London and Melbourne

The German Sniper 1914 - 1945
Copyright © 1982 by Peter Senich

All rights reserved. Except for use in a review, no part of this publication may be reproduced, stored in a retrieval system, or transmitted in any form by any means electrical, mechanical, or otherwise, without first seeking the written permission of the copyright owner and of the publisher.

Published by Paladin Press, a division of Paladin Enterprises, Inc., P.O. Box 1307, Boulder, Colorado U.S.A. 80306.

Published in the United Kingdom by Arms and Armour Press, Lionel Leventhal Ltd., 2-6 Hampstead High St., London NW3 1QQ England; and at 4-12 Tattersalls La., Melbourne, Victoria 3000 Australia.

ISBN (U.S.A.) 0-87364-223-6.
ISBN (U.K.) 085368536-3

Printed in the United States of America.

First Edition

Library of Congress Cataloging in Publication Data

Senich, Peter R.
  The German Sniper 1914-1945.

  1. Shooting, Military. 2. Rifles. 3. Telescopic sights. 4. Germany. Heer. Infanterie—Drill and tactics. I. Title.
UD335.G3S463   355.5'47   81-18828
ISBN 0-87364-223-6   AACR2

*Dedicated to my Mom and Dad.
God never made better ones.*

"The real glory of war is surviving."
Samuel Fuller

Contents

      Preface ix
      Credits and Acknowledgments xi
      German Production Marks xiii

I	Sharpshooter—The Beginning 1
II	Scharfschutzen—Gewehr 98 25
III	Rifles For Sniping—Arming The Wehrmacht 79
IV	Combat Experiences—The Eastern Front 113
V	Camouflage—A Special Skill 127
VI	The Sharpshooter Award 147
VII	Karabiner 98k—Short Side Rail System 159
VIII	Karabiner 98k—Zielfernrohr 41 185
IX	Karabiner 98k—Turret Mount System 215
X	Karabiner 98k—Long Side Rail System 245
XI	Karabiner 98k—Claw Mount System 269
XII	Scharfschutzen—Waffen-SS 281
XIII	Selbstladegewehr 41 301
XIV	Selbstladegewehr 43 323
XV	Telescopic Sights—The Assault Rifles 369
XVI	Sniper—The Soviet Approach 401
XVII	Sniping Weapons Miscellany 419

Preface

In late 1914 the action of World War I settled into static or trench warfare along the Western front. The combatants who were separated by just a few meters in some cases, found the usual man-sized target was reduced to that of a head peering over a parapet.

This change and the increased necessity for firing in marginal light at dawn and dusk prompted the Imperial German High Command to field competent marksmen armed with telescopic-sighted rifles. Their purpose was to harass, impede, and destroy movement of enemy personnel opposing them.

Systematic application of special telescopic-sighted rifles had proved to be of immense value during the Great War. This practice, however, was all but discontinued by the time the Wehrmacht emerged as part of the new order under Adolf Hitler's Third Reich in the mid-1930s.

In World War II scores of highly trained Soviet sharpshooters took an alarming toll of German combat personnel during the early months of the Russian campaign. The German High Command had no means to counteract the tactics they had introduced to modern warfare, a bitter realization and one they viewed with particular disdain. It spurred them to institute remedial measures calculated to offset the tremendous advantage held by the snipers in the Red Army.

Without question, the effect of Soviet sharpshooting was the predominant influence determining the direction of the overall German sniping program from late 1941 until the end of the war in 1945.

Despite the lasting global notoriety of German snipers, lack of historical data on the equipment and circumstances of the Scharfschutzen (sharpshooter), as he was officially referenced, has made this unique specialist an enigmatic segment of the German Armed Forces, chronicles.

Following the capitulation of the Imperial German Army in 1918 and Hitler's Wehrmacht in 1945, untold quantities of significant documents and records were either destroyed or so completely dispersed that a concise determination of this fascinating aspect of German military activity has required extensive and painstaking research. Unfortunately, this loss of documentation has accounted for many of the misconceptions that have been perpetuated through the years.

Interest in military sniping continues to grow, and observation of a few German sniping rifles appears to be sufficient qualification to write about them. Too many contemporary offerings dealing with this subject constitute little more than an exercise in imagination rather than a presentation of events as they actually transpired.

In the interest of factuality, with few exceptions, the material for this work was not drawn from secondary sources. Regardless of the original intent, secondary sources, books about books, are not accurate because they tend to replicate errors made by the writer they quote.

In contrast to this prevalent practice, the policy of utilizing only that material having accredited origins was responsible for what may appear to be incomplete segments. The material presented herein consists of primary information gleaned from original reports, documents, letters, archive microfilm, direct observations, and knowledgeable individuals who served within or in close proximity to the German Ordnance system as it then existed.

This book provides a comprehensive presentation of German sniping practice as carried forth in World War I and World War II. In addition, for the first time serious military students and martial arms collectors can trace the chronological development of a surprising number of telescopic rifle sights, mounts, and weapons employed by the German military for sharpshooting purposes from 1914 through 1945.

During preparation for this work, I was fortunate to be able to contact an assortment of wonderfully cooperative people who provided me with significant information pertinent to the subject of German sniping.

Were it not for the many friends made through the course of this endeavor, and their contributions over and above what any author has a right to expect, this book would not have been brought to this point. Although these individuals deserve all due credit for helping me locate and piece together the myriad of informational odds and ends, any errors in putting these pieces in their final form properly falls to me.

Credits and Acknowledgments

The following organizations rendered invaluable assistance in the form of original documents, technical information, microfilm, photographs, and other archival material. For this help I wish to thank the staffs of:

Library of Congress
National Archives
Berlin Document Center
U.S. Army Ordnance Museum
Springfield Armory Museum
West Point Museum
Public Archives of Canada
Imperial War Museum
Bayerische Armeemuseum
Bayerische Nationalmuseum
Militargeschichtliches Forschungsamt
Bibliothek fur Zeitgeschichte
Heeresgeschichtliches Museum
Bundesarchiv
Bundesarchiv—Militararchiv
Zentralbibliothek der Bundeswehr

QAD - Pattern Room, Enfield Lock
Deutsche Gesellschaft fur Wehrtechnik
Kungl. Armemuseum
U.S. Army Photographic Agency
U.S. Army Infantry Museum
Army Ordnance Association
National Rifle Association
Smithsonian Institution
Bundesministerium fur Landesverteidigung
Armamentarium
Musee Royal De L'Armee et D'Histoire—Militaire
Josef Schneider & Co.
Mauser—Jagdwaffen GMBH
Carl Walther Sportwaffenfabrik
Steyr—Daimler—Puch AG.
Hensoldt Optische Werk AG.
Wehrgeschichtliches Museum

In addition, the following gentlemen provided me with collection photographs, vital data, and in many cases, their encouragement:

Al Peirce
Monty Lutz
John Fitzpatrick
Ludwig Olson
Fred Hicks
Dick Boyd
Jost-Burkhard Anderhub
Franz Achleithner

Chris Burton
William Stonley
James Stonley
Robert F. MacKinnon
Thomas B. Nelson
Daniel Musgrave
David Gill
Tim Kader

Robert Taylor
Dennis Fossey
J.D. Truby
Robert Simpson
Morton Gray
Allan Cors
Hugh Brock
Don Collova
Ed Macauley
Roger Cleverly
Hans Widhofner
William Adams
Ian V. Hogg
Howard Kyle
H. Gaertner
Johann Hanke
George Petersen
Peter Stahl
Daniel W. Kent
Dolf Goldsmith
John Minnery
Harlan Reinfeld
Roy Shetter
Bill Xanten

Russell Ostrem
Warren Odegard
Wolfgang Friedrich
Lt. Col. F.B. Conway (Ret)
Paul M. Senich
Jim Flurchick
John Minnery
Osprey Publications—Michael Roffe
Bill Harris
Elroy Sanford
John Bannister
John Cross
Jean Huon
Glen de Ruiter
Raymond Crockett
Nat Rosenblatt
Thomas Shannon
Mike Marine
Niel Broky
Robert Fisch
Bill Hickey
Robert Jensen
Colin Stevens
John Leach

A very special expression of appreciation to:

Brad Murphy, for sharing his vast knowledge relevant to German sniping hardware;

Donald G. Thomas, a true professional in the strictest sense and indefatigable enforcer of technical accuracy;

Ed Owen, who comes close to having studied the K98k weapon system on a par with German Ordnance;

David Heinmuller, for the years of invaluable contributions;

Sam Newland, the major contributor of weapon photographs used in this work;

and to Major Paul M. Seibold, for his clinical accuracy in translating an overwhelming assemblage of complex German Ordnance documents.

Last and most important perhaps: to my wife and family for their patience and understanding.

Peter R. Senich

German Production Marks

This list delineates the German manufacturer's production marks (codes) relative to the hardware covered in this volume.

Telescopic Sights

- bek– Hensoldt Werk fur Optik und Mechanik, Herborn
- bmj– Hensoldt & Sohne, M., Optische Werk A.-G., Wetzlar
- bzz– J.G. Farbenindustrie A.-G., Camerawerk, Munchen
- cad– Kahles, Karl, Optiker, Wien
- cag– Swarovski, D., Glasfabrik und Tyrolit-Schleifmittel-Werk, Wattens/Tirol
- clb– Wohler, Dr. F.A., Optische Fabrik, Kassel
- cxn– Busch A.-G., Emil, Optische Jndustrie, Rathenow
- ddv– Oculus, Spezialfabrik ophthalmologischer Jnstrumente, Berlin
- ddx– Voigtlander & Sohn A.-G., Braunschweig
- dkl– Schneider & Co., Josef, Optische Werke, Kreuznach
- dow– Opticotechna G.m.b.H., Prerau, Czechoslovakia
- dym– Runge & Kaulfuss, Fabrik f. Feinmecha- und Optik, Rathenow
- eso– Optische Werke G. Rodenstock, Munchen
- fvs– Spindler & Hoyer, mechan. u. optische Werkstatte K.-G., Gottingen
- fzg– Feinmechanik e. G.m.b.H., Kassel
- gkp– Ruf & Co., Nachfolger der optischen Werke, Kassel
- hap– Kohl A.-G., Max, Physikalische Apparate/Laboratoriums Einrichtungen Chemniz
- jve– Optisches Werk Ernst Ludwig, Weixdorf
- kay– Ford–Werke A.-G., Werk Berlin
- kov– Etablissement Barbier, Bernard et Turenne, Paris
- mow– Seidenweberei Berga C.W. Crous & Co., Berg/Elster
- rln- Zeiss, Carl, Militarabteilung, Jena

Small Arms

- ac– Walther, Carl, Waffenfabrik, Zella-Mehlis
- ar– Mauser Werke A.-G., Werk Borsigwalde, Berlin-Borsigwalde, Eichborndamm
- bcd– Gustloff-Werke, Werk Weimar, Weimar
- bnz– Steyr-Daimler-Puch A.-G., Werk Steyr
- byf– Mauser Werk A.G., Oberndorf a.N.
- ce– Sauer & Sohn, J.P., Gewehrfabrik, Suhl
- dot– Waffenwerke Brunn A.-G., Werk Brunn, Brno, Czechoslovakia
- duv– Berlin-Lubecker Maschinenfabriken, Werk Lubeck
- fzs– Krieghoff, Heinrich, Waffenfabrik, Suhl
- qve– Unconfirmed

Small Arms (Early)

S/27

    Erma B. Geipel, G.m.b.H., Erfurt

27

S/42 Mauser-Werk A.-G., Oberndorf a.N.

42

S/237 Berlin-Lubecker Maschinenfabriken, Werk Lubeck

237

S/243 —Mauser Werke A.-G., Werk Borsigwalde, Berlin-Borsigwalde, Eichborndamm

Miscellaneous

cww— Weiss, Carl, Lederwarenfabrick, Braunschweig

gnn— Pryo-Werk G.m.b.H., Spec.-fabrik fur Pyrometerbau, Hannover

jvb— Wessel & Muller, Beschlagfabrik, Luckenwalde

ofh— Metall—u. Kunstharzwerk G.m.b.H., Komotau

zs— Unconfirmed

blc— Zeiss, Carl, Militarabteilung, Jena—Early code found on Zeiss military optics and very few military contract telescopic rifle sights.

kxv— Jackenroll, A., Optische Anstalt G.m.b.H., Berlin—Even though this code was assigned to Jackenroll, their military contract sights (Ajack) were not known to bear this mark.

CHAPTER I

Sharpshooter — The Beginning

In the years preceding the First World War the German optical industry had attained a sophisticated capability for producing a myriad of optical devices that were recognized for their excellence throughout the world. Included were "telescope sights for rifles."

The German military, in their continuing quest for accurately directed rifle fire, had conducted evaluations of telescopic sights to determine their military potential.

Although the sights were ultimately rejected for service use, two significant conclusions were drawn from their early testing. The theory of optically increasing the visual range of the rifleman in order to provide proportionate gains in long range hits proved false; increased vision was of little value where the bullet trajectory was so long that any hits were attributed strictly to chance. It was decided, however, that the rifle scopes utilized by hunters of that period did possess light-gathering capabilities which permitted effective sighting at diverse ranges under light conditions that made target definition virtually impossible with standard sights and the ordinary eye.

During the eventful August of 1914, Imperial German forces swept into Belgium with the heart of France as their objective. Telescopic sighted rifles were not a part of their armament.

The winter of 1914-1915 saw a significant change in the character of the fighting; the open field war of move and countermove came to an abrupt end with each side consolidating its position opposite each other in mile after mile of parallel fortifications stretching approximately 500 miles from the North Sea to the Swiss frontier.

An excellent summation of trench warfare is set forth by John Ellis in his contemporary work, *Eye-Deep in Hell*.

Both sides were forced to dig deep holes in the ground and concentrate upon breaking up any attacks launched by their adversaries.
The armies remained in these holes for the next four years, millions of men trapped in a desolate strip of territory, living and dying in a wilderness of trenches, dugouts, craters, shattered villages and forests of lifeless tree-stumps, a desert in the midst of civilization, that became more featureless with each passing day.

Trench warfare was not new, but by the end of hostilities the enormity and futility of this facet of the Great War would both stagger and totally compromise the sensibilities of modern man.

The Imperial German High Command, grasping the importance of this development, commenced preparation for position warfare with characteristic Teutonic aplomb.

Recognition of the increased necessity for firing at dawn and dusk prompted the systematic procurement, distribution, and combat application of telescopic-sighted weapons for select riflemen well-versed in the use of optical sights as hunters or participants in prewar shooting organizations.

Sighting a telescope equipped "emergency issue" commercial sporting rifle. A German sniper is shown during early action on the western front.

The advantage was obvious: the rifle scopes permitted German marksmen effective sighting in marginal light under adverse conditions which precluded the use of conventional open sights. The ability for the telescopic sights to function in subdued light was the principal reason for their adoption by the German army at this time.

Long after the Germans introduced the deadly sharpshooter (Scharfschutzen) to the western front, the appalling increase in Allied casualties was attributed to "stray bullets." The French and the British were the ones who finally realized, with great indignation, what the Germans had contrived and associated the unusual number of fatal head and heart wounds to this new equipment and technique.

Until then, an impersonal aura bordering on anonymity gave most combat personnel a sense of shelter or security, much the same as being lost or hidden in a crowd. The thought of being singled out and shot at a forgetful moment of exposure proved to be particularly unnerving. The effect on Allied troop morale was devastating.

A British officer, commenting on this dire situation in early 1915, stated: "Only those who have been in a trench opposite Hun snipers that had the mastery, know what a hell life can be made under these circumstances."

The armed forces of Belgium and France, having borne the brunt of initial German thrusts, were further perplexed by the thoroughness of the German marksmen facing them, and they fielded snipers of their own. At no time, however, did their efficiency and organization remotely compare with German efforts and success in this regard.

In contrast to their allies, snipers fielded by the British and Canadian Expeditionary Forces (the latter closely aligned with evolving British practice) emerged as the principal antagonists of the German Scharfschutzen to the extent that by war's end they were considered the standard of comparison.

When the German sniping menace became manifest in late 1914, the British army had no provision for sniping equipment in their stores and were forced to improvise until suitable weapons could be procured. In exactly the same manner as the Germans had obtained their original sniping arms, from civilian sources, they hurriedly collected and pressed into service hunting or target rifles originally intended for sporting use in England. Infrequently captured sniping weapons of the Germans were also used.

Because of the variety of civilian arms and resultant difficulty in providing correct ammunition for them at the front, any benefit derived from their use was more psychological to the British troops observing them rather than damaging to the Imperial German forces opposing them.

When the War Office decided that the only method capable of neutralizing the effectiveness of German sniping rested with "out sniping them," the British sniping effort bordered on the threshold of reality. Through concentrated efforts on the home front the Short Magazine Lee Enfield, Mark III (SMLE) rifle, the standard British service arm fitted with telescopic sight found its way to the trenches by the spring of 1915. The British discovered, however, that efficient sniping entailed considerably more effort and organization than just delivering weapons to the front.

Owing to haphazard methods of placing their sharpshooters and equipment in the trenches, the British permitted German marksmen the luxury of maintaining their sniping superiority through 1915 as well. With no official guidelines for systematic distribution, British sniping rifles were treated as trench stores to be meted out by quartermaster-sergeants having little or no regard for their significance or capability.

To add to the confusion, only a few among those receiving sniping issue were either competent riflemen or had used a telescopic sight before. It was later officially estimated that six out of every ten British sniping rifles issued in this manner were rendered useless shortly afterward. Reports indicated that battalions fielding untrained snipers lost men to the Germans at alarming rates. There was no circumventing the conclusion that unorganized sniping had proved of little value.

By the end of 1915, schools for Sniping-Observation and Scouting (S.O.S.) were established through the diligent efforts of a few British officers and men sensitive to the dire need for fielding trained sharpshooters. Results were gradual, but casualties directly attributed to Ger-

INSTRUCTIONS FOR USE OF S.m.K. CARTRIDGES
AND RIFLES WITH TELESCOPIC SIGHTS
1915

With the manufacturing of the K bullet being difficult and expensive, this cartridge must be used only for precision shooting when a great penetration is sought. The S.m.K. cartridge is distributed only to marksmen supplied with the Model 98 rifle and telescopic sight. These men must use these cartridges wisely and not give them away.

If necessary, these cartridges can be fired from a machine gun against a fortified implacement, armored shelter (pillbox) or armored aircraft.

The rifles with telescopic sight in the German Army are of two kinds:

a) The standard rifle Model 98 on which a telescope has been installed. The first order (late 1914) was for 15,000 rifles.

b) Hunting rifles with telescope. All of these in Germany have been requisitioned. These rifles have less strength than the rifle 98 and can fire the 88 cartridge only.

The weapons with telescopic sight are very accurate up to 300 meters. They must only be issued to qualified marksmen who can assure results when firing from trench to trench, and especially at dusk or during clear nights when ordinary weapons are not satisfactory.

The marksman must shoot with discretion and the rifles must not be fired for salute or suppressive fire. Marksmen are not limited to the location of their unit, and are free to move anywhere they can see a valuable target. Sentries have the duty to signal the marksman, such targets they themselves can determine.

The marksman will use his telescope to watch the enemy front, recording his observations on a note book, as well as his cartridge consumption and probable results of his shots.

Marksmen are exempted from additional duty.

They will wear a special badge of two crossed oak leaves above the upper badge of the cap.

Translation of an extremely rare document outlining German sniper duties during their early days of trench dominance.

man snipers were reduced significantly and eliminated altogether in certain sectors. By mid-1917, sniper-training schools operated throughout the British force.

Initial efforts concentrated on neutralizing the tremendous advantage held by the Germans and assumed a countersniping posture from the onset. Because there existed no established standards from which valuable reference could be drawn, the British were compelled to imitate German methods.

From a standpoint of historical reference, if the British had not carefully recorded German sniping practice along with their own remedial methods and innovations during this hectic period, significant details concerning the Imperial German sniping effort would have been relegated to certain obscurity.

In contrast to the early and futile efforts of their British adversaries, the German High Command had wisely restricted the issuance of telescopic-sighted rifles to qualified marksmen having prior experience with rifle scopes. This decision constituted the principal reason for their immediate success along the western front.

Also, as the war progressed and the need for additional snipers increased, only the most competent riflemen were selected and given thorough instruction in the use of their special equipment and in the fine art of staying alive while plying their chosen trade. It was considered an honor to be a sharpshooter. According to various accounts, many of the more proficient German snipers attained a surprising measure of notoriety among their comrades in arms. Unlike the highly publicized aviators who were held in high esteem by the German press, the average line troops could easily relate to "their sharpshooters," who shared the same plight in the trenches as they.

Sharpshooting equipment abounded in some sectors and was virtually nonexistent in others. Telescopic-sighted rifles were fielded on an independent basis by the various German states who maintained their own armies within the German Corps system: Prussia, Saxony, Bavaria, and Wurttemberg.

Sharpshooting rifles were generally issued at the company level, with ultimate responsibility for their care and maintenance delegated to the individual marksman and to the competent supporting ordnance personnel who handled major adjustments and repairs.

With few exceptions, the expert German sharpshooters were exempt from additional duties and were given free rein to move about their sectors in quest for suitable targets. While specific tactics were largely a matter of individual preference, two basic rules were followed: to fire no more than one or two shots from the same point, and to remain as mobile as conditions permitted. In addition to excellent marksmanship, the German specialists were quite adept at concealing themselves and their positions. Detection was an infrequent occurrence, especially during the early months of the war when they possessed a decided edge in the art of sniping.

German snipers were known to work alone or, as circumstances dictated, with an observer who selected suitable targets and reported results. A particularly significant mode of German sniping found enterprising marksmen venturing forth into "no man's land" before dawn to set up a sniping hide for a given day. While to some this area was considered infinitely less hazardous than the trenches, no one could question the resolve needed for such precarious duty. Nevertheless, this practice proved effective enough to offset the inherent danger. The greatest detriment rested with the sniper having to lie virtually motionless through an entire day while Allied riflemen and machine gun crews constantly scanned the area for sniper activity. Evidence strongly suggests, however, that the vast majority of German sharpshooters preferred to excercise prudence and to operate from the relative safety of their own lines.

Even though the uniform of the German soldier blended favorably with the surroundings prevalent on the western front, the use of camouflage was added to the skills demanded of the successful sniper. This was especially necessary in the later stages of the war when more and more sniping activity was conducted in the open, and not being detected was the only protection from swift retaliation.

Two methods of camouflage were practiced by the German forces. One applied to individuals, particularly those functioning as snipers or observers, and the other to positions and equipment. Camouflage uniforms did not see

During early combat, steel shields were carried by the sniper and placed on top of the parapet for protection from return fire.

The protective steel plates were incorporated directly into the trench works as the war progressed.

general issue, but special suits or robes fashioned from canvas or burlap sacking were available for sniper use. Found to be virtually indispensable for disguising the outline of a man's figure, the outfits were painted to conform to the area of operation. They were effective perhaps but not very popular because the unique suits were quite cumbersome and exceedingly uncomfortable during the warmer seasons.

In the trenches, improvised hoods or face veils made of gauze or burlap were frequently used for altering the head form, and the rifle was either "pattern-painted" or wrapped with strips of cloth or burlap colored to match the immediate surroundings.

In the beginning of the war the British had looked upon various sections of German trench works with particular disdain, sardonically noting the "un-military fashion" in which great piles of twisted barbed wire, corrugated metal sheets, boards, and trash of every imaginable description had been haphazardly scattered about. There was a method to this apparent German madness; in many sectors, particular emphasis had been placed on the role of the German sharpshooters. From amidst this debris, small narrow openings (loopholes) had been cunningly placed through which a telescopic-sighted rifle could be fired with virtual impunity. As a further ruse, some of the openings were left obvious but rarely used. Then, when least expected, a quick shot would find an unsuspecting mark. The application of loopholes in field fortifications was not a German innovation, but their use in this specific manner sprang from German ingenuity at this time.

The Germans were also reported to have been the first to employ "sniper plates" or shields, as they were variously cited, fashioned from steel plates of varying thickness. Well-placed and concealed in their trench works, the heavy plates, with loopholes cut into them, afforded the sniper adequate protection, especially since return fire had to pass directly through the small opening to render any harm.

It was small wonder that Allied riflemen, at this stage armed with rifles having only conventional sights, were hard pressed to place accurately their shots through the armored German loopholes even at comparatively short ranges. Although there were a number of marksmen proficient enough to hit the small apertures, they were so few in number that their efforts were of minor value.

Without the benefit of satisfactory armor-piercing small arms ammunition early in the war, the imaginative British used special high-velocity and large-bore rifles originally intended for elephant hunting on the African continent to penetrate or, as then stated, "bash-in" the German plates. It will suffice to mention that where such rifles were brought to bear, the relatively unhindered existence of the German specialists was somewhat less secure.

The Germans, on the other hand, had provided armor-piercing ammunition for sniper use in early 1915 and during the latter stages were known to employ their 13mm Mauser-system, antitank rifle (T-Gewehr). Originally developed to combat Allied tanks, the rifle was used for countersniping fortified British positions. Allied machine guns and their crews were considered prime targets, and when one was sighted, a solitary sniper could put it out of action with a single, well-aimed hit either on the weapon itself or on the crew members.

As the war progressed, it was determined that well-prepared, carefully camouflaged positions established in the open behind the front lines were best for sniping and observation purposes. To minimize chance of detection, posts were frequently abandoned and reestablished elsewhere; if discovered, they were easily demolished by a brief mortar or artillery barrage. The Allies would periodically shell or machine gun every ruin or dilapidated structure near the front lines as a deterrent to German use for sniping or observation.

Compared to the fighting in France and Belgium, the war on the eastern front was totally different. A highly trained German army faced a horde of Russian soldiers along a shifting front more than a 1000 miles long between the Baltic Sea and the Black Sea. Although the Russians vastly outnumbered the Germans, they were simply no match for them. Most of the ill-equipped Russian troops were poor peasants who had no wish to die for their tsar. With little industry to provide equipment and a thoroughly inadequate supply corps, the Russian army was in constant need of almost everything necessary to wage a successful war.

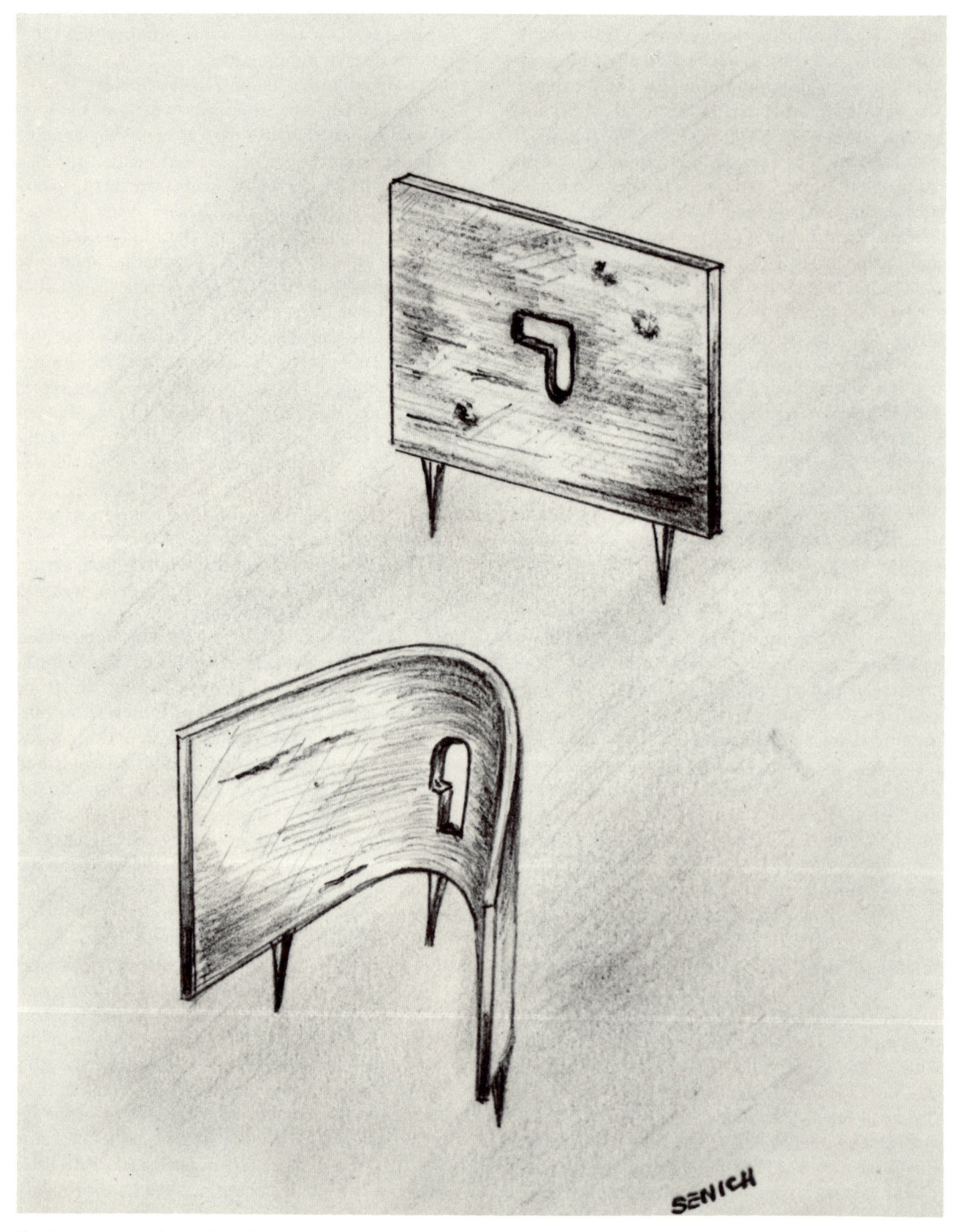

Typical steel sniper shields employed by both factions during the Great War. A notch was provided for sighting with offset scopes.

Even though German sharpshooters were reported to have seen considerable action on the eastern front, few documents detailing their activities and results have survived.

Inasmuch as the Russian army possessed neither the sniping capability nor the means to effectively combat snipers, the effect of uncontested German sharpshooting could have only been awesome.

The Allies had learned their lesson well, but in the years following the Great War, the demonstrated value of organized sniping slipped into a state of lethargy in all major military powers with one exception, Russia's revolutionary Red Army. When the German blitzkreig thundered into Russia in 1941, scores of trained Red Army snipers rendered countless numbers of German officers and combat personnel K.I.A. (Killed in Action).

Ironically, the German army found itself confronted with the same plight they had placed before their adversaries in 1914—that of being on the receiving end of optically-directed rifle fire with no means to counteract it.

German sharpshooter and observer at work in the trenches. Equipped with binoculars, an observer's primary function was the selection of suitable targets.

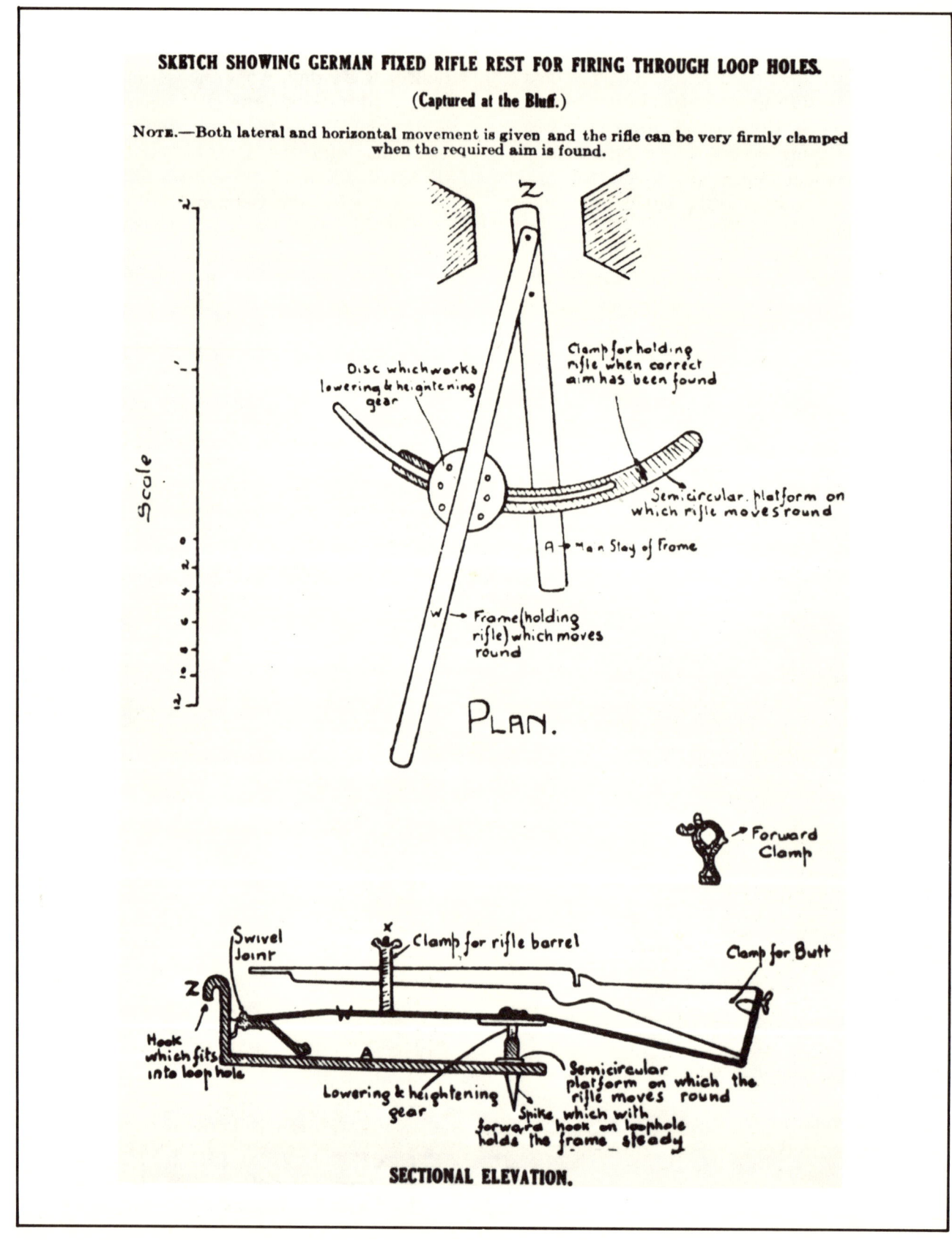

An excerpt from a World War I U.S. Army training pamphlet illustrates a captured "German fixed rifle rest for firing through loopholes."

A sketch made in 1918 by AEF combat artist George Harding depicts a German marksman plying his trade from a ruined building.

British sniper team in action against Imperial German forces. France, 1915.

Despite their cumbersome nature, camouflage suits made from painted canvas were to see extensive use on the western front. The tree trunk, fashioned from steel, served to conceal and protect the marksman.

Although British issue in this case, there was little difference between the camouflage sniper suits employed by both factions.

A duplication of German trench works used for training British snipers to pinpoint concealed loopholes.

Hoods of gauze or burlap were used to alter the head form for sniping activity.

A 1916 photograph with personnel of the 88th Infantry Regiment, Mainz, holding a Scharfschutzen-Gewehr 98.

Even though the French army fielded a number of sharpshooters during the Great War, their efforts did not pose a serious threat to German forces.

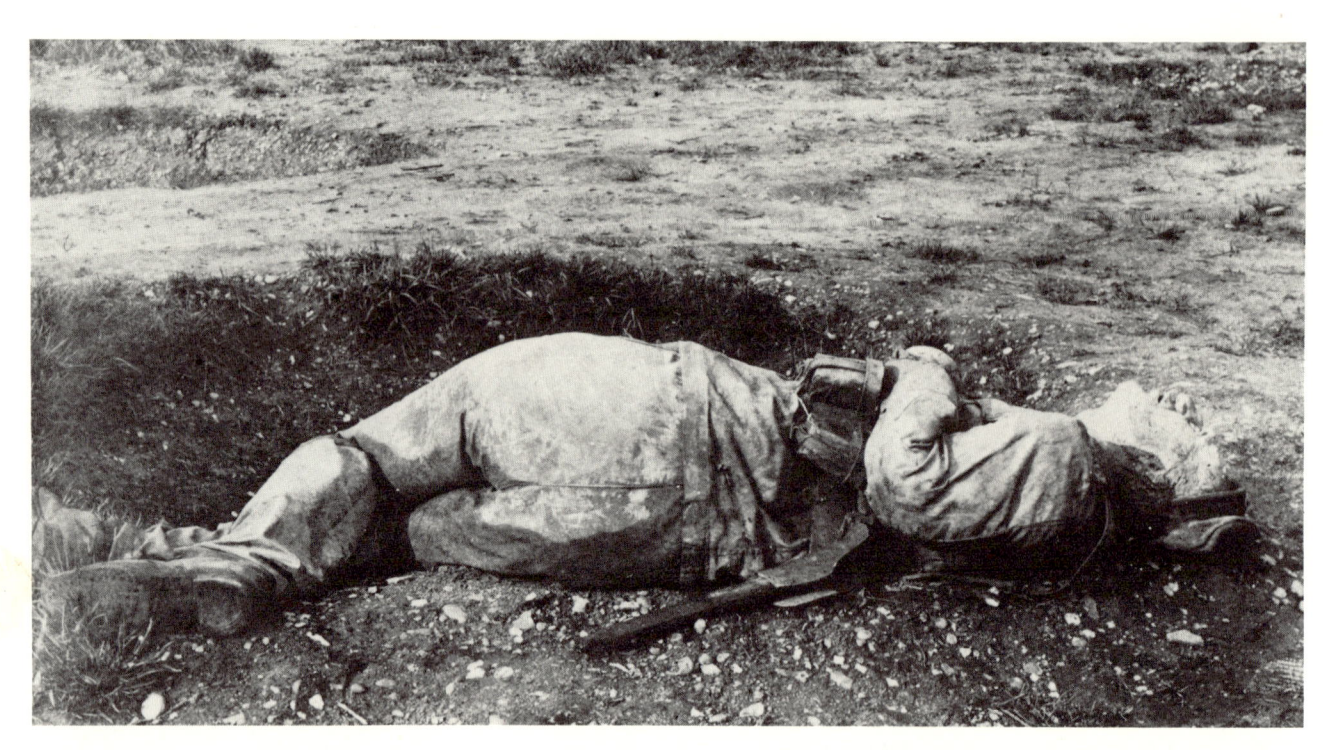

Dummy forms placed in "no man's land" were used for concealment purposes by both factions. In this case the form was dressed to resemble a fallen German soldier; German snipers made similar use of forms outfitted in captured British garb.

British efforts to pinpoint the locations of well-camouflaged German sharpshooters included the use of cleverly detailed heads fashioned from paper-mache. The head forms, used on an individual basis, were placed on a stick and slowly raised over the parapet to draw sniper fire.

Civilian personnel (French) responsible for making the dummy head forms are shown with molds used for that purpose.

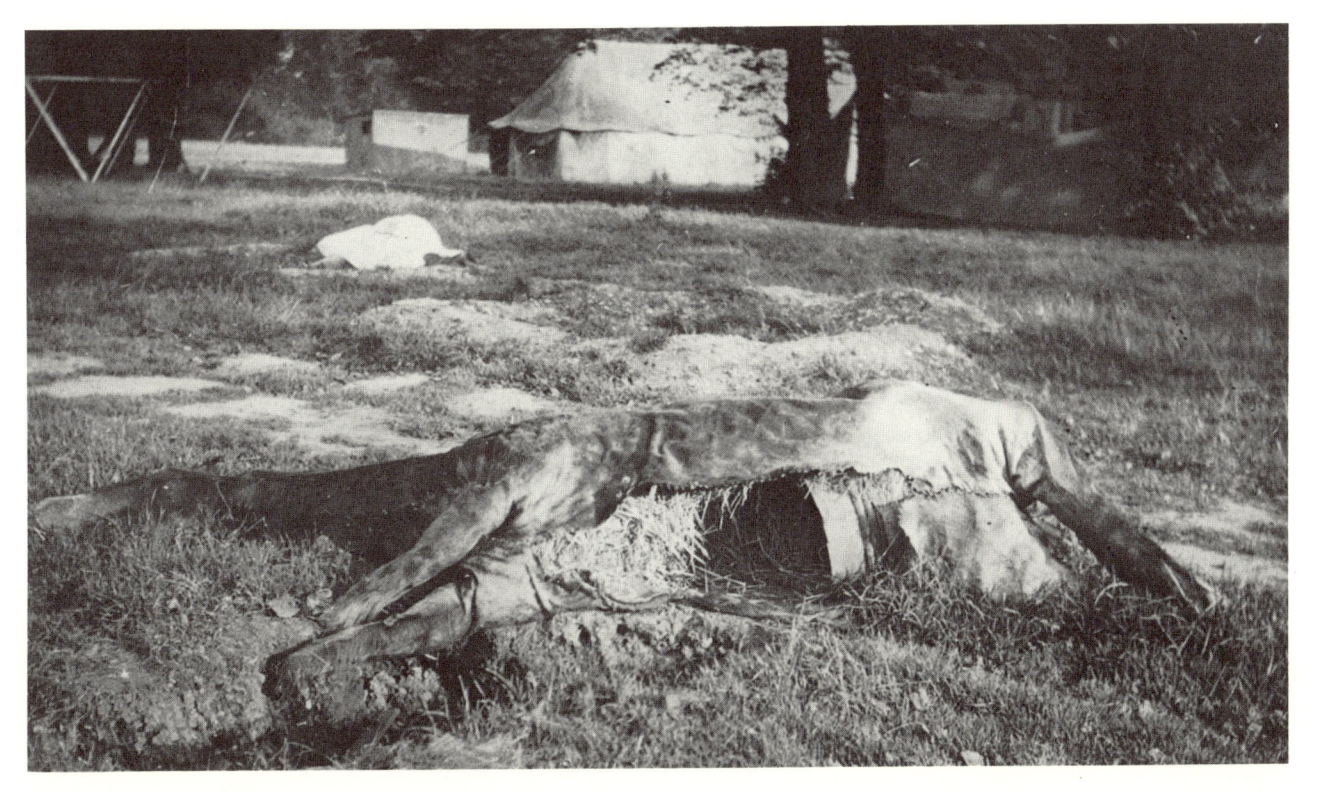

A unique ruse employed by enterprising British snipers saw the use of horse hides suitably arranged to resemble the form of a dead horse. Sniping was then conducted from behind the dummy carcass.

In most cases, the repair and maintenance of German sharpshooting equipment beyond the limits of basic field support were conducted by ordnance personnel in shops such as this.

The remains of a ruined building, a typical position used both for German sniping and observation. It was virtually impossible for the Allies to prevent sporadic use of such positions while being subjected to periodic shelling and machinegun fire.

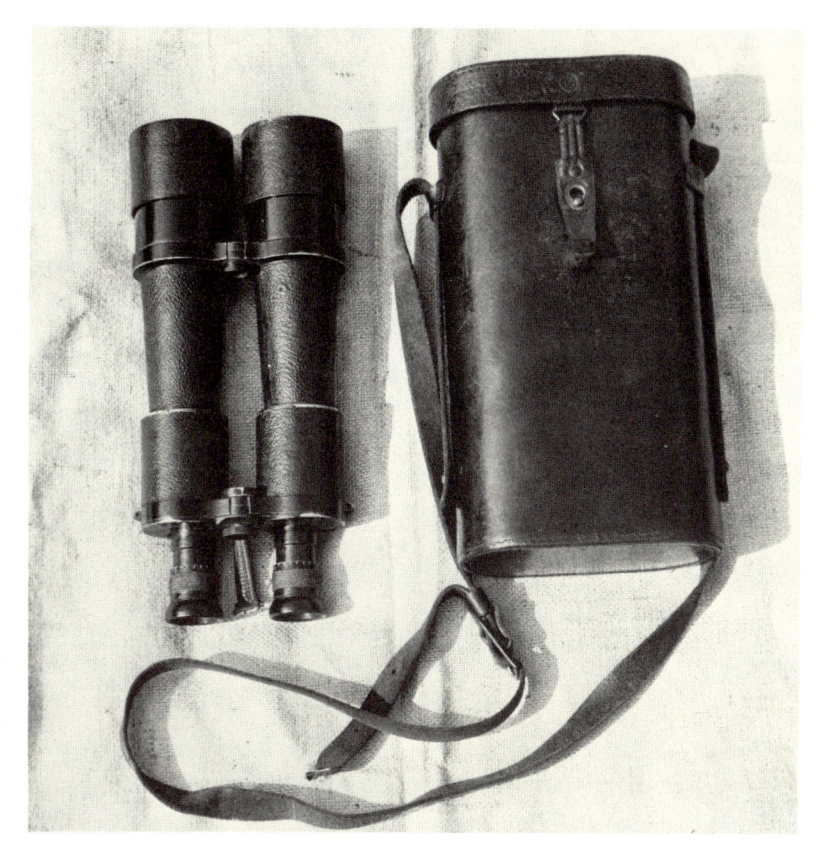

Typical 10 x 50 binoculars (Carl Zeiss, 1917) employed by German snipers during the Great War.

Below: Although binoculars made by Zeiss, Goerz et al were basic army issue, they were an essential item for German sharpshooters and their observers. In this case the "Fernglas 08" binoculars (approx. 6 x 30) were made by C. P. Goerz and bear a 1916 date.

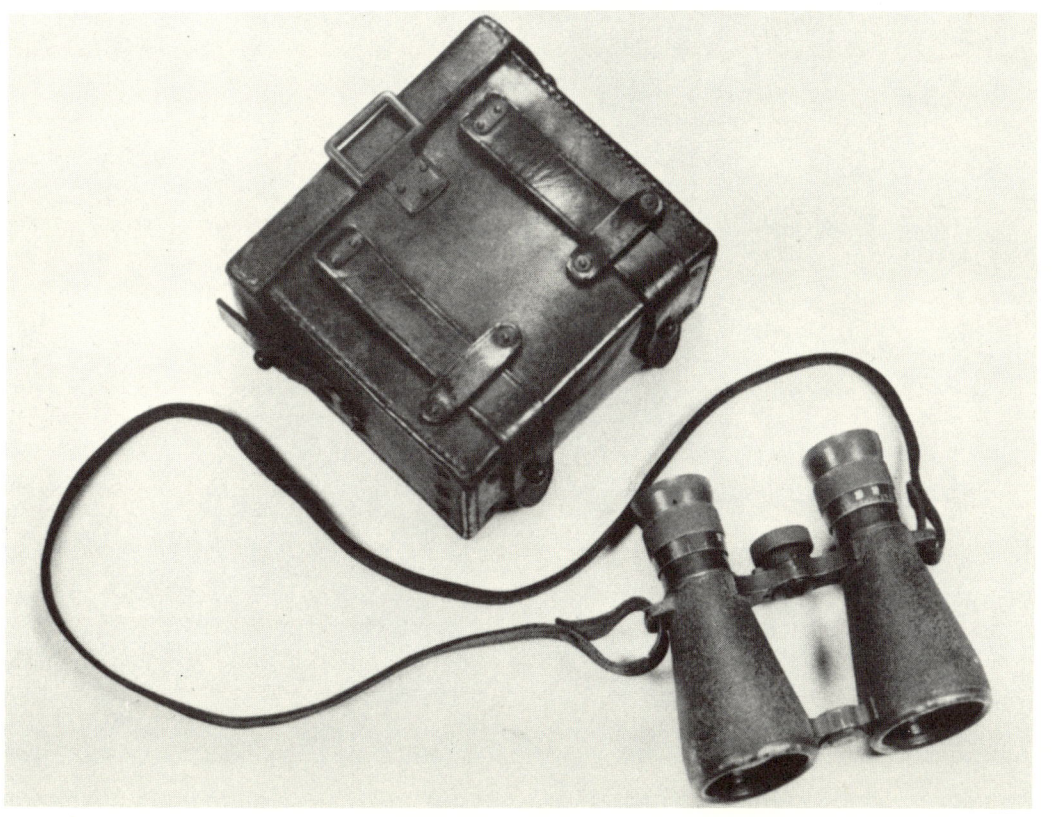

While only unusual circumstances would account for the close proximity of three sharpshooters under combat conditions, the German marksmen are shown sighting their weapons from the cover of a stone wall.

Under smoke cover, silhouette figures fashioned from millboard were pulled into a standing position using special wire apparatus. Employed by the British on one or both flanks, the figures were exposed a few minutes before an attack (zero hour). Their purpose, to draw concealed sniper and machinegun fire to reveal German positions.

SHARPSHOOTER — THE BEGINNING

British marksman at rest during a lull in combat. The rifle, a select S.M.L.E., Mk III, is fitted with an offset-mount 3.5-power Periscope Prism Company telescopic sight. Note the leather carrying case and lens caps (lower left). British sharpshooters were to emerge as the principal antagonists of German snipers during World War I.

Heavy steel German face armor. Variously cited as intended for sniper use, it is difficult to envision a highly mobile German sharpshooter handicapping himself with so awkward a device.

Side view of German face armor. Note the padding and remnants of leather used to hold the shield in place.

A British rifleman (circa World War I) demonstrates captured German face armor.

An interesting view of a Scharfschutzen-Gewehr 98 in use. Note the leather objective and ocular shields in place on the scope and the carrying case slung from the sniper's belt. There are no grasping grooves in the fore-end of the stock in this case.

While utilized primarily for observation purposes, German snipers also operated from similar positions located well behind the front lines.

CHAPTER II

Scharfschutzen — Gewehr 98

When the use of telescopic-sighted rifles became a matter of priority in late 1914, initial efforts to field sharpshooting equipment included the acquisition of sporting arms from civilian sources throughout Germany.

German military authorities had no illusions regarding the durability of standard hunting rifles in a combat environment, but at that point, their sole purpose centered on gaining the advantage of optically-directed rifle fire in the shortest possible time.

In consideration of the variety of small arms ammunition for prewar German sporting arms, it was decided that only weapons with "Mauser-systems" firing either the Patrone-88 or the S-Patrone would be accepted for use at the front. This measure was calculated to simplify the distribution of correct ammunition for sharpshooter use in the "Jagdgewehre mit Handelsublichem Zielfernrohre" (hunting rifles with commercially available telescopic sights).

Although the anticipated availability of sporting arms firing the Patrone-88 prompted their inclusion, this ammunition had been considered obsolete long before 1914. Both the Patrone-88 and the S-Patrone (Spitzgeschoss Patrone or pointed-bullet cartridge), the latter serving as the standard cartridge for German infantry during World War I, were 7.9mm ammunition. But they differed in principal dimensions to the extent that it was unsafe to attempt firing the improved S-Munition in rifles chambered for the obsolete Patrone-88.

With the bulk of the hunting arms belonging to the more affluent German citizens, what had first appeared to be a relatively simple task proved fruitless because most individuals, while sympathetic, preferred to retain their weapons. As a result, the quantity of arms collected for sniper use at the front fell far short of anticipated needs. With no other recourse, the German military procured telescopic-sighted sporting carbines and rifles from various commercial arms establishments on a commission basis.

While specific details concerning the original "emergency issue" sniping arms remain unknown, many of the Jagdgewehre (hunting rifles) employed for early German sharpshooter use were sporting arms manufactured at the original Mauser factory (Waffenfabrik Mauser) at Oberndorf. Whether these particular weapons were donated by patriotic citizens, or as some experts contend, obtained in part by direct agreement with the highly respected Mauser firm remains unconfirmed.

To reduce the chance of accidents at the front, a small plate bearing the silhouette of the Patrone-88 with its distinctive, long, round-nosed bullet and the admonition, "NUR FUR PATRONE-88, KEINE S-MUNITION VERWENDEN" (only for Patrone-88, unsuitable for S-Munition, was attached to the left side of the rifle stock. Of further interest were those emergency issue rifles bearing a small steel or brass plate on which was stated that the weapon had been "donated" to the Fatherland for the war effort. The plate was located on either side of the stock.

In spite of their emergency service, the civilian sporting arms performed as well as could

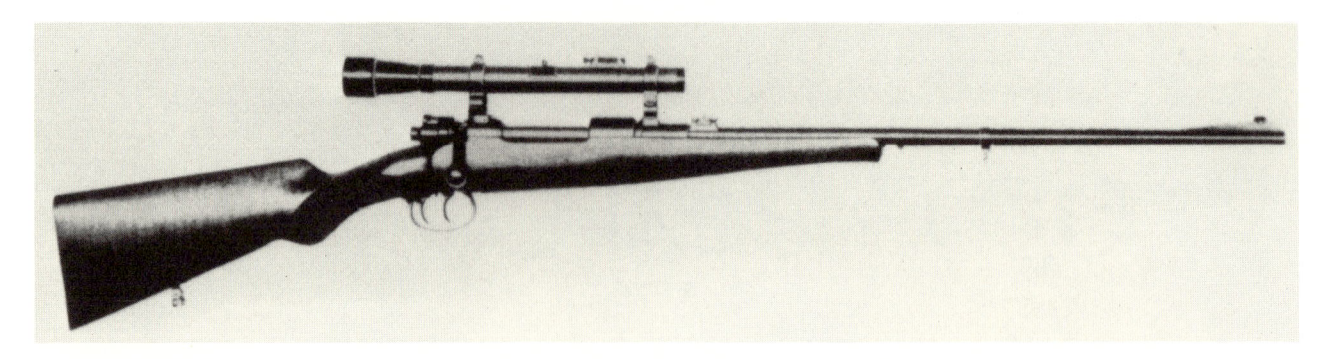

Pre-war telescopic-sighted Mauser sporting rifle typical of those pressed into German military service for sharpshooting purposes. Note the double-set trigger assembly.

Single-claw rear base assembly as utilized with an offset mount Gewehr 98 sniping rifle (Waffenfabrik Mauser).

be expected. They were never intended to withstand the rigors of sustained combat, and following a brief period in the trenches, reports indicated even arms of the highest commercial quality were deemed too fragile or lacking sufficient accuracy.

Possibly of greatest concern were numerous reports of barrel failures resulting from the use of S.m.K. (Spitzgeschoss mit Stahlkern or pointed-bullet with metal core) armor-piercing ammunition, which had been placed at the disposal of German sharpshooters early in 1915. In many cases the thin wall sporter barrels were not equal to the task.

The ultimate number of Jagdgewehre pressed into military service remains conjectural, but on an overall basis they were apparently of sufficient quantity to cause much consternation during the winter of 1914-1915.

Concurrent with the hasty acquisition of sporting arms was the fitting of telescopic sights to a number of standard Model 98 infantry rifles (Gewehr 98). This was done by civilian firms as directed by the German military. Whether from civilian or military origins, the quantity of sharpshooting rifles reaching the front lines by late 1914 is believed to have been extremely limited. A precedent had been established, and as duly noted by German authorities the hastily improvised Zielfernrohr-Gewehr 98 (telescopic-sighted Rifle 98) proved satisfactory for use in the trenches (unlike its civilian counterparts). Thus, with the standard infantry arm serving as the nucleus for subsequent German sniping issue, by the spring of 1915 modified versions of the Gew. 98 found their way to the front lines in ever increasing numbers.

The original Imperial German military sniping arms were brought to fruition during the latter part of 1914 when various commercial telescopic sights and mountings were adapted to the Gewehr 98.

In order to meet requirements, rifles were selected from newly manufactured pieces on an as needed basis, and variants based on Gew. 98 arms manufactured by private contractors have been reported. The vast majority of Imperial German sniping rifles fielded during the Great War appear to have been based on weapons produced by the government arsenals at Spandau, Danzig, Erfurt, and Amberg between 1915 and 1918.

Officially known as the Scharfschutzen-Gewehr 98 (sharpshooter-Rifle 98), the special rifles were carefully fitted with 3-power or 4-power commercial hunting type telescopic sights and provided with a turned-down bolt handle to clear the sight. In most cases a clearance cut was made on the right side of the stock above the trigger guard to facilitate grasping the modified bolt handle.

Even though rifle scopes and related hardware (rings, mounts, and bases) varied greatly, as the illustrations in this chapter show, telescopic sights were mounted either directly over the receiver and in line with the bore or offset slightly to the left side of the receiver. The latter method permitted clip-loading with the sight in place. The standard "claw-mount" locking system, developed by commercial firms in prewar Germany for scope mounting on sporting arms, saw extensive but by no means exclusive use with Imperial German sniping rifles dating from this period. Besides providing a positive mounting, the claw-mount system allowed the sight to be readily detached and removed from the rifle.

After the prospective sniping weapons were selected at the various arsenals, the telescopic sights were painstakingly fitted by civilian gunsmiths or by competent military armorers charged with this specific responsibility. Despite the weapons' dire need at the front, the methods employed for sight mounting were diverse and extremely tedious by today's standards.

Since the various components were hand fitted, each sharpshooting rifle had its own personality, which reflected the technique and craftsmanship of a particular gunsmith or armorer. As a result, scope mounting varied to the extent that sights were not always fully interchangeable from one weapon to another even if they had been assembled with identical components by the same individual. Therefore, when originally mated, the telescopic sight on each Scharfschutzen-Gewehr 98 was either stamped or engraved with the rifle serial number, a measure intended to keep the scope and rifle paired.

Top view of Mauser Gewehr 98 rear base. In this case the locking lever turns under and is spring-indexed in both front and rear positions. Note the large central screw (5) which serves to lock the other four in place.

Top view of Mauser Gewehr 98 front base assembly. The mounting screws were recessed and covered with soft solder, an unusual practice. Both front and rear bases are offset to the left of the receiver.

Judging from the multiplicity of base and mount variations present on surviving examples of both "top" and "offset" mount sniping arms produced during the war, the adoption of a specific telescope mounting system for universal German military use was obviously a low priority. Fielding satisfactory sniping rifles in quantity remained the prime consideration.

It is generally recognized that the top and offset mounting practice originated within the ordnance systems of Bavaria and Prussia respectively during the latter part of 1914. But to categorize an Imperial German sniping rifle as either Bavarian or Prussian in origin based solely on the position of the sight is erroneous.

Although various contemporary offerings echo an "acute rarity" of Imperial German sniper rifles, in reality, a surprising number of original, authentic pieces are retained within various military museums and private martial arms collections throughout the world. In addition to the emergency issue sporting arms pressed into early service, which certainly qualify as rare, a Scharfschutzen-Gewehr 98 mounting an original issue matching serial number telescopic sight complete with carrying case also deserves this distinction.

Even though the Scharfschutzen-Gewehr 98 proved relatively trouble-free in the trenches, experience soon taught the German marksman that regardless of how carefully his issue had been chosen or the components fitted, the special rifle required certain attention to sustain its sniping efficiency.

As it was then stated: "The life of the barrel depends more often on proper cleaning than from use. To the expert rifle shot, the rifle is a living thing, particularly the bore."

In addition to judicious and mandatory maintenance of his weapon on an overall basis, the marksman was further admonished not to tamper with the telescopic sight beyond those basic adjustments necessary for daily operation. Directives had stipulated that "Only authorized, qualified personnel should be allowed to adjust and zero telescopic sights, a measure intended to prevent sights from being rendered useless unnecessarily once they were distributed for combat use.

Due to changes in the ammunition in use, in the shooting characteristics of the rifle, and in the sight itself, it was absolutely necessary to rezero a telescopic-sighted rifle frequently for elevation and deflection (windage) to insure optimum accuracy. Confronted with the realities that existed in the trenches, after a brief period the directives relevant to scope adjustments were quietly laid to rest at the official level as well.

Even though the telescopic sights and mounts were generally as rugged as the technology of that era permitted, a great deal of attention was required by the sniper to keep his issue in correct order. Often waist-deep in water or crawling over treacherous terrain, the sniper had a difficult time keeping the lenses clean and dry. A telescope in the rain was virtually useless, and even in a light fog or mist the water droplets would cloud the glass, while only a few specks of dirt or mud could scratch the lenses enough to render the sight almost useless.

Additional care to prevent sun or bright light reflecting from the objective lens was a continual concern. To obviate this possibility, improvised tubes fashioned from heavy cardboard or sheet metal were slipped over the front portion of the tube.

The German sniper had no alternative but to develop a protective attitude for his equipment, for if it was ever knocked out of alignment or otherwise rendered questionable, the nature of the circumstances he operated under seldom afforded an opportunity for repair or sighting-in. A simple mistake or mental lapse however brief brought certain death.

Through the course of the war, all types of telescopic sights, regardless of their configuration or manufacturer, were mounted on both top and offset variations of the Scharfschutzen-Gewehr 98. As such, they saw unilateral combat use in the unified armies of Prussia, Saxony, Bavaria, and Wurttenburg, which comprised the Imperial Germany Army during this era.

Recognition of the excellence of German rifle scopes prompted the United States Army, the first major power to adopt sniping equipment (1908), to evaluate a 3-power Goerz telescopic sight as part of their progressive efforts to enhance their sharpshooting capability.

Details of the Goerz evaluations are set

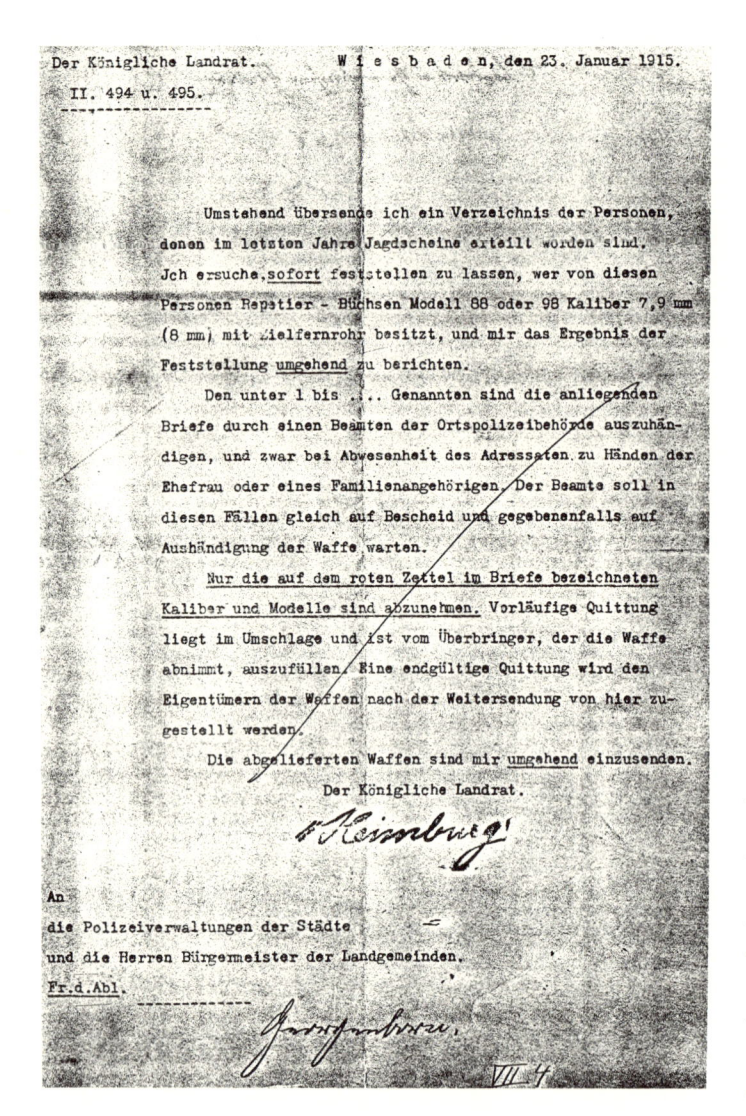

An official WW1 German document dated 23 January 1915 from the Wiesbaden District Magistrate to Police authorities and mayors of the rural communities ordering the confiscation of all 7.9 mm Model 88 or 98 telescopic sighted hunting rifles from private citizens. To simplify matters, a list of those having a hunting license from the previous year was included with this directive.

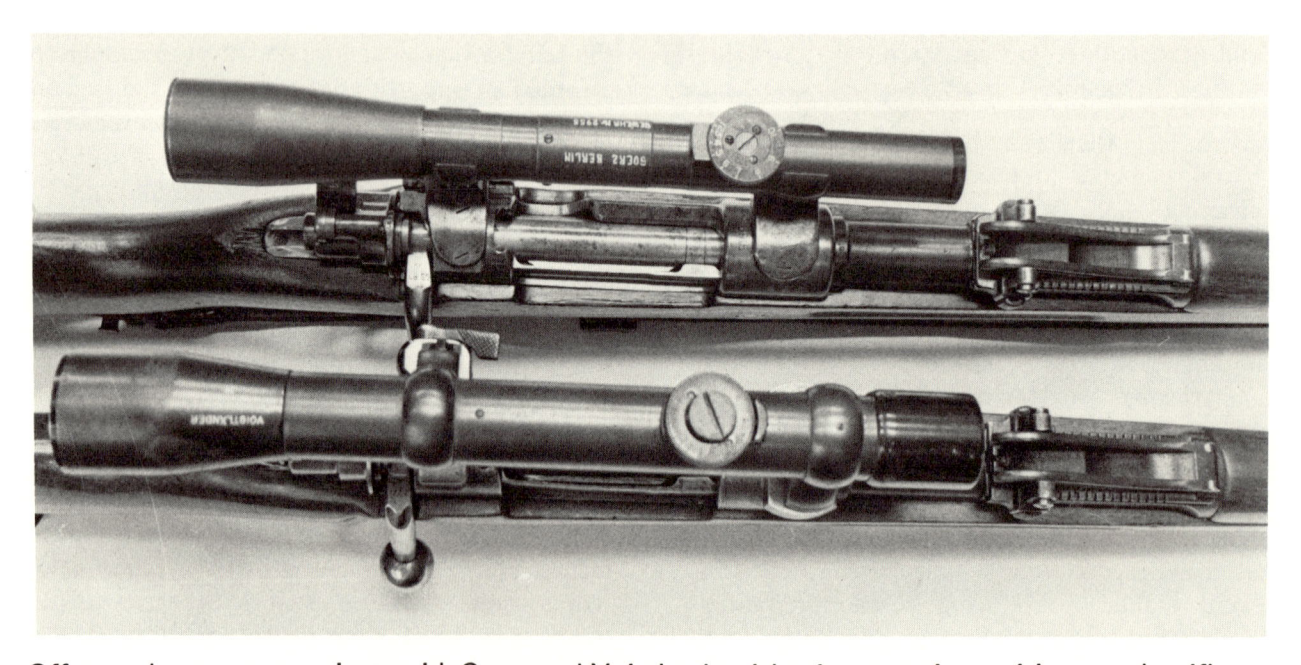

Offset and top mount variants with Goerz and Voigtlander sights in respective positions on the rifles.

forth in a U.S. Ordnance report dated 18 December 1915 which states in part:

C.P. Goerz Musket Sight—This sight mounted on a U.S. Model 1903 Rifle was obtained from the C.P. Goerz Company, of Berlin, in August 1913, and after a preliminary test and relocating of the sight on the rifle, was submitted to the School of Musketry for a competitive test of all available types of telescopic-sights.

Conclusions of the School of Musketry— The report of the School of Musketry states that in its opinion this sight (C.P. Goerz) possesses all the essential requirements of such an instrument, viz., power, definition, field, ready adjustments, simplicity, strength, rigidity and convenience to the user, and that for military use it is superior to the Warner & Swasey type.

Recommendations—The recommendations of the School of Musketry were:

(A) That the Goerz Telescopic-Sight be adopted for issue replacing the Warner & Swasey Sight.

(B) That they be issued at the rate of two to each organization of the mobile army armed with the rifle.

Acting accordingly, in February 1916 the U.S. Chief of Ordnance communicated with representatives of the C. P. Goerz Company concerning the conditions under which that firm would permit the manufacture of their sight in the United States. The communication was acknowledged, but no further reply was received. Considering circumstances at that time, the war in Europe, this came as no surprise to American military authorities.

Further testimony to the high regard for German rifle scope design was the introduction by the British of their Pattern '18 Telescopic-Sight late in the war. It was a device that bore a marked resemblance to a typical German scope in every major characteristic including the claw-mount system.

Of all the rifle scopes employed with the Gew. 98, the unique prismatic variants were perhaps the least known. A limited-use item at best, one of the few specific references to its German combat application is found in U.S. Ordnance papers, which also cite the evaluation of a captured Zeiss prismatic scope. Marketed by Zeiss, Goerz, and Hensoldt prior to the war, the 2-power and 3-power prismatic sights were noted for their extremely clear definition and extensive field of view.

From the beginning, incessant demands for sniping equipment had precluded the use of a standard type rifle scope. As a result, sights procured for German military service during the Great War were extremely diverse in origin and configuration. Any semblance of mass production did not exist, and sights were painstakingly hand-crafted to the extent that no single manufacturer could satisfy all military requirements.

While sights produced by the major optical firms saw extensive use, a substantial number of rifle scopes were procured from smaller establishments that ceased to exist in the economic chaos that engulfed Germany following the war. Three-power and 4-power hunting scopes manufactured by Gerard, Goerz, and Zeiss were reportedly mounted to the first military Zielfernrohr-Gewehr 98 fielded late in 1914. The use of sights originally intended for the commercial market continued until the availability of prewar devices was expended. Although specific details involving telescope procurement have not been divulged, it is known that the German military had contracted with various optical firms to produce "telescope sights suitable for military use."

It remains almost impossible to distinguish an early pure commercial scope from a variant made during the war expressly for fitting to the Scharfschutzen-Gewehr 98. Perhaps the most recognizable military variants were those finished with a textured olive-green coating as opposed to those finished with a standard commercial blueing used on the vast majority of rifle scopes fielded during this period. Excluding the myriad of lesser known sights produced by obscure German firms, the most commonly used rifle scopes were furnished by Gerard, Oigee, Goerz, Zeiss, Fuess, Busch, Hensoldt, and Voigtlander.

Although German telescopic sights differed in principal characteristics such as dimensions, reticle patterns, and method of focus, in all of them elevation adjustments were made within the scope tube. Elevation settings were readily obtained by turning a graduated drum located directly on top of the telescope until the desired range was opposite a stationary reading line. Rifle scopes from this era have been noted with range scales marked in divisions expressing values of 100 meter increments from 1-3, 1-4, 1-8, 1-9, 1-10, 1-12 and 200-400-600.

Adjustments for windage (right or left) necessitated external movement of the telescope tube by use of a small tool resembling a contemporary skate key which turned a screw-driven dovetail incorporated into the rear mount assembly. Inasmuch as the forward end of the telescope was held by a rigid mount and could not pivot correspondingly, lateral adjustments required the utmost care to prevent excessive stress on the entire sight assembly.

Reticle patterns, or "graticules," as they were then referenced, did not appear to have been standardized. A wide variety were employed for sharpshooter use throughout the war, particularly during the early stages when commercial hunting scopes were prevailing issue.

The most common reticle pattern perhaps consisted of a vertical post with horizontal side bars extending from the outside edge of the sight picture leaving an opening in the center on either side of the post. The gap (space) between the horizontal bars conformed to a predetermined value, that is "x" cm at a distance of "x" meters, and thus provided an effective means for estimating the distance of the target. While identical at first glance, these values varied accordingly in rifle scopes of different manufacture.

Even though other types of reticle designs could also be used for range estimation, documents cite the vertical post and side bar pattern (Normalabsehen) as having proved the best in marginal light, a factor which unquestionably accounts for its prevalent use. This reticle pattern was but one of the numerous standard types popular in prewar Germany and was not developed expressly for military sniper use as some contend.

As originally fielded, in addition to the standard items furnished with the Gewehr 98, sling, muzzle cover, and cleaning kit, the Scharfschutzen-Gewehr 98 was issued with a carrying case for holding the telescope assembly when removed from the rifle and with basic tools (lens brush, windage tool) necessary for cleaning and adjusting the sight.

Scope cases were fashioned from leather and formed metal as well as from micarta and cardboard of varying thickness with a canvas or heavy material covering. Although shoulder straps have been noted, belt loops were more commonly used for carrying purposes. Case lids were generally stamped with the name and location of the manufacturer, sight magnification (power), and the Gew .98 serial number.

In addition to the "Lederschutzdeckel" (leather protection lids) used for covering the objective and ocular ends of the sight when not in use, many of the telescopic sights were also issued with leather eye shields.

Soft rubber cups dating from this period were intended as a protective measure to prevent damage to the objective end of the telescope tube when the sight was attached or removed rather than as a safeguard for the marksman's eye.

The Germans placed particular emphasis on night sniping, due to the level of effectiveness their marksmen had demonstrated with innovative techniques during the early months of the war. Much to the chagrin of Allied line troops who could not even rely on darkness for relief from the German sharpshooters, in some cases and in some sectors German specialists were actually more efficient in the dark with their conventional rifle scopes and the special night sighting devices developed for use with the Gewehr 98.

Since ordinary rifle sights proved worthless in the dark, efforts to improve this situation prompted early experiments with phosphorous paste applied to both the front and rear sight in conjunction with a thin strip of white paper or tape runing the length of the handguard. Even though special sights with luminous elements were eventually fielded for use with the Gew. .98, from all indications they were not very popular because the ever-present mud in the trenches would completely obliterate the luminous forms thereby defeating their purpose.

In addition to the conventional medium power telescopic sights, which proved invaluable especially on clear nights, the most significant German sighting apparatus developed primarily

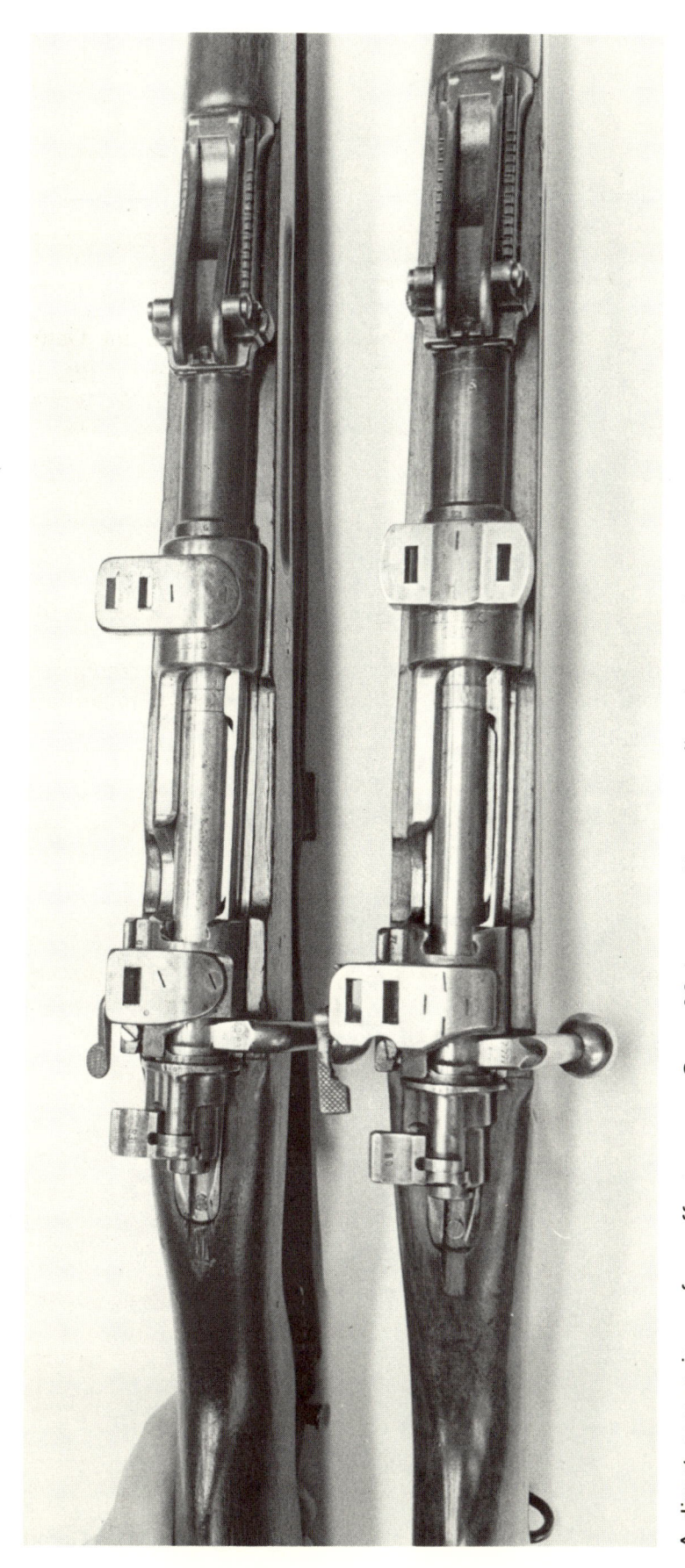

A direct comparison of an offset mount Gew. 98 (scope offset slightly to the left of the receiver) to a top mount variant (scope directly over the receiver). The locking levers on both rear base assemblies are shown in the open or unlocked position.

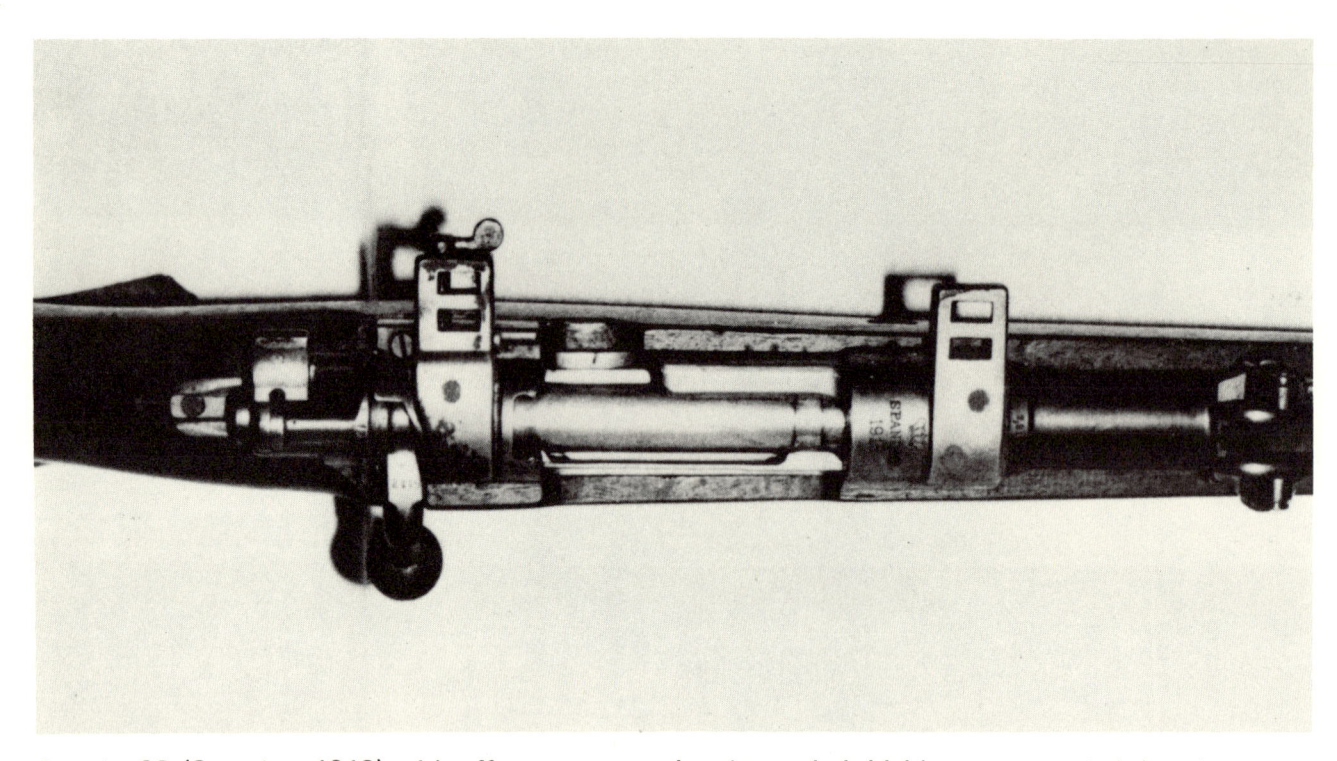

Gewehr 98 (Spandau, 1918) with offset-mount receiver bases. It is highly recommended that the reader carefully study and compare the examples of receiver bases and scope mountings illustrated in this chapter to fully comprehend the differences that existed in World War I Imperial German sniping hardware.

An offset mount Scharfschutzen-Gewehr 98 with a 3-power telescopic sight by Gerard.

for shooting in subdued light consisted of a short metal tube containing a unique bifocal lens similar to that found in contemporary eyeglasses. Readily mounted directly over the rear sight assembly of any standard Gew. 98, the 2.5-power bifocal or "twilight sight," as it was referenced, was utilized in combination with a special device having a white pyramid form on its rear face which snapped onto the conventional front sight.

When viewed through the main scope assembly, sighting the mark entailed aligning the top or apex of the pyramid on the target and gently squeezing off a round. The most desirable feature of this system rested with its relatively uncomplicated operation which permitted satisfactory results when employed by any competent rifleman. Manufactured by the Carl Zeiss firm (Jena), the innovative units were issued with a small rectangular leather case for carrying or storing the sighting assembly when removed from the rifle.

Perhaps the most enigmatic sniping variants to see combat use during the Great War were the telescopic-sighted Model 95 (Mannlicher) rifles employed by Austro-Hungarian forces.

A multinational force, the army of Austro-Hungary was a major participant in World War I and a principal German ally. Unfortunately, when this vast army broke up and vanished into history in November 1918, many of the finite details concerning their weapon developments and subsequent field use ceased to exist or were relegated to obscurity where they haved remained. Therefore, a discussion of Austro-Hungarian sniping arms remains confined to the study of the surviving weapons.

With the "straight-pull" Model 95 serving as the base weapon, telescopic-sighted versions of both the rifle (Repetier-Gewehr M95) and the carbine (Repetier-Carabiner M95) have been noted mounting a variety of 3-power and 4-power rifle scopes manufactured in Germany. Fitted with special offset receiver bases, the scope rested slightly to the left of the receiver, thereby permitting clip loading of the 8mm rimmed cartridges and use of the standard sights.

The rear scope mounting utilized a small stud (post), which was inserted into the base and locked in place with a spring-loaded latch positioned directly beneath the sight. The front mount, fitted with a single claw, hooked into the receiver base which consisted of an integral dovetail assembly. Contrary to the standard German practice of rear mount windage, lateral scope adjustments were effected by moving the front mount assembly as required.

While the value of adapting a conventional service rifle for sharpshooting purposes is quite obvious, the inherent, excessive muzzle blast and flash characteristic of the short-barreled military carbines dating from this era raises serious doubts as to the practicality of a telescopic-sighted weapon of this type.

Nevertheless, as authentic surviving examples of the Austrian Carabiner M95 with its 482mm barrel bear witness, a requirement for this equipment evidently existed in the Austro-Hungarian army at some point in the war. Of parallel interest is evidence which strongly suggests that the Germans also fielded a weapon of this type in the form of their Karabiner 98a, in reality a short rifle with a 599mm barrel. While the extent of such conversions remains unsubstantiated and subject to debate, at least one bonafide weapon of this type, the Karabiner 98a, mounting a 3-power Gerard scope and original flash hider (a cylindrical sheet metal tube), has been noted.

When the Armistice brought an end to hostilities in 1918, as dictated by the Treaty of Versailles, Germany was methodically stripped of the capability to wage an offensive war.

The size of the postwar German army (Reichsheer) was restricted to a standing force of 100,000 men equipped with only certain types of defensive armament. An unspecified number of sharpshooting weapons based on the Karabiner 98b, the improved version of the Gewehr 98, were reportedly fielded at the company level.

In the final days when the end of the war was imminent, countless German snipers thought it prudent to discard or destroy their Scharfschutzen-Gewehr 98. The fate of the vast majority of the sharpshooting rifles that were surrendered to the Allies when the Imperial German Army capitulated remains unknown.

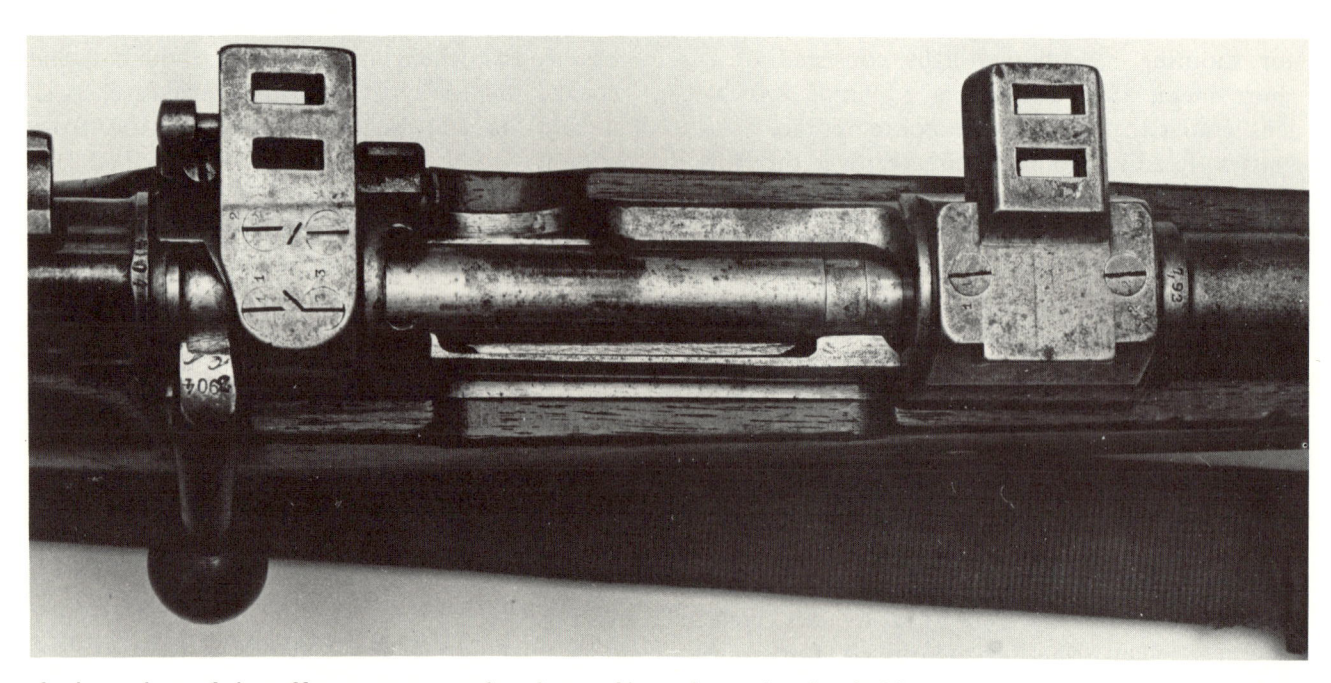

A close view of the offset mount receiver bases. Note the spring-loaded button located on the back of the rear base, which was pressed to accept the double-claw rear scope mount and released to lock the sight in place.

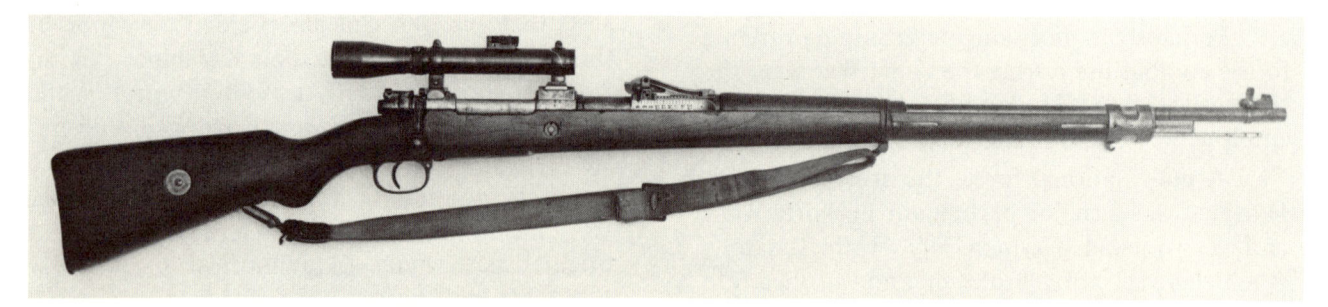

Right view of offset mount Scharfschutzen-Gewehr 98 with 3-power Gerard scope. Although sharp-shooting rifles were fielded with stocks that have neither grasping grooves nor a recess for the turned down bolt handle, the stock shown was not issued with this weapon.

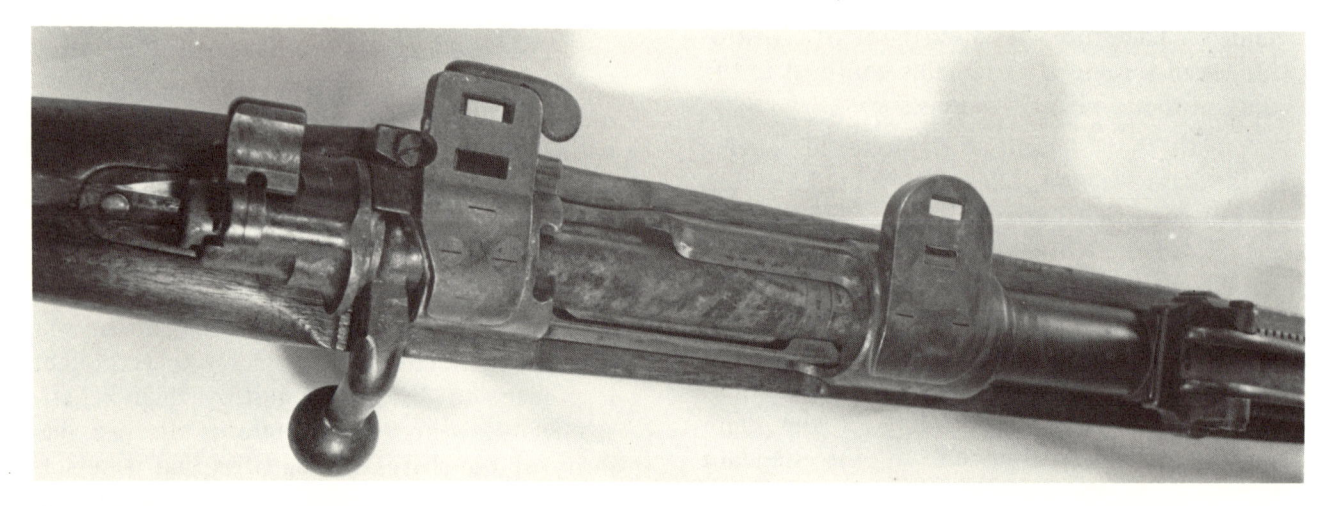

A variant offset mount receiver base Gewehr 98. The rear base locking lever is rotated forward to a locked position. Both screws and solder were used to attach scope bases to the receivers.

SCHARFSCHUTZEN — GEWEHR 98

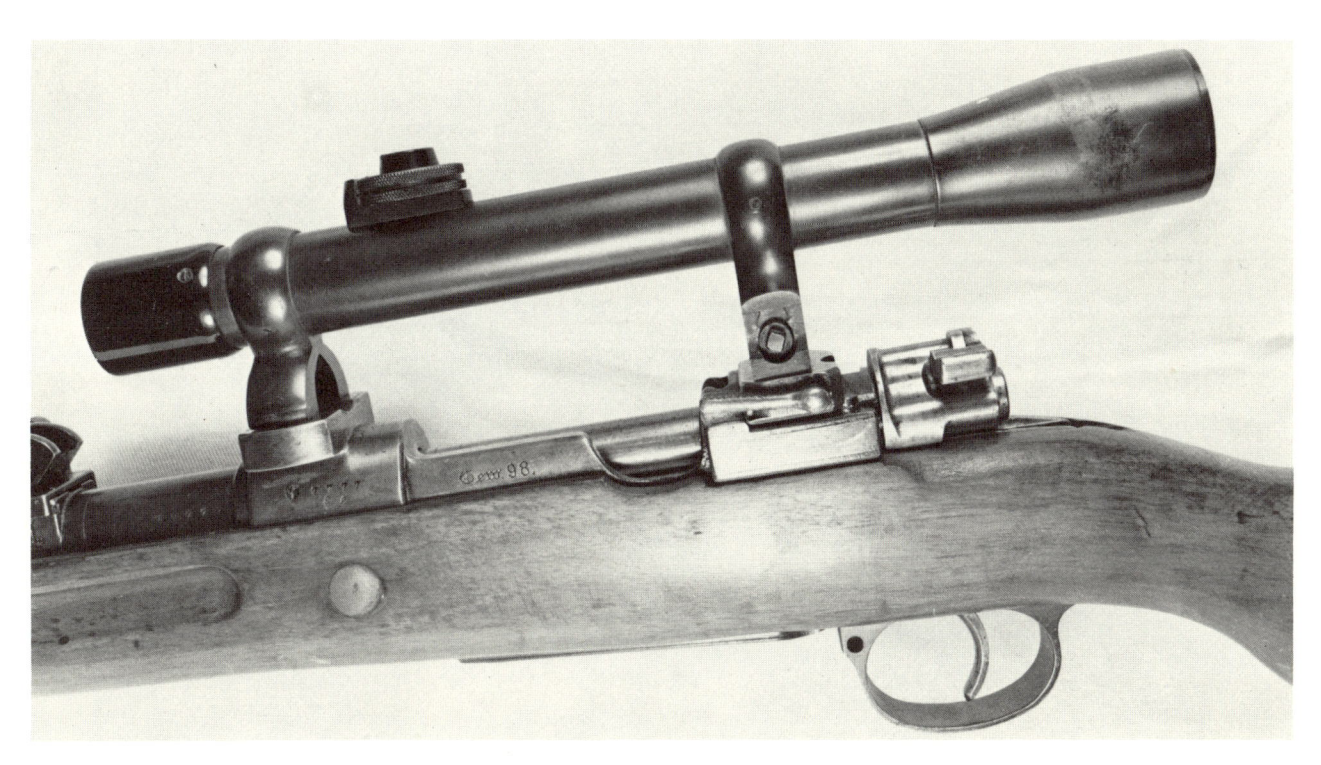

Gewehr 98 top mount sharpshooting rifle with a 3-power Voigtlander sight.

Even though the telescope was mounted directly over the bore, the configuration of the mounts permitted use of the conventional sights with the scope in place.

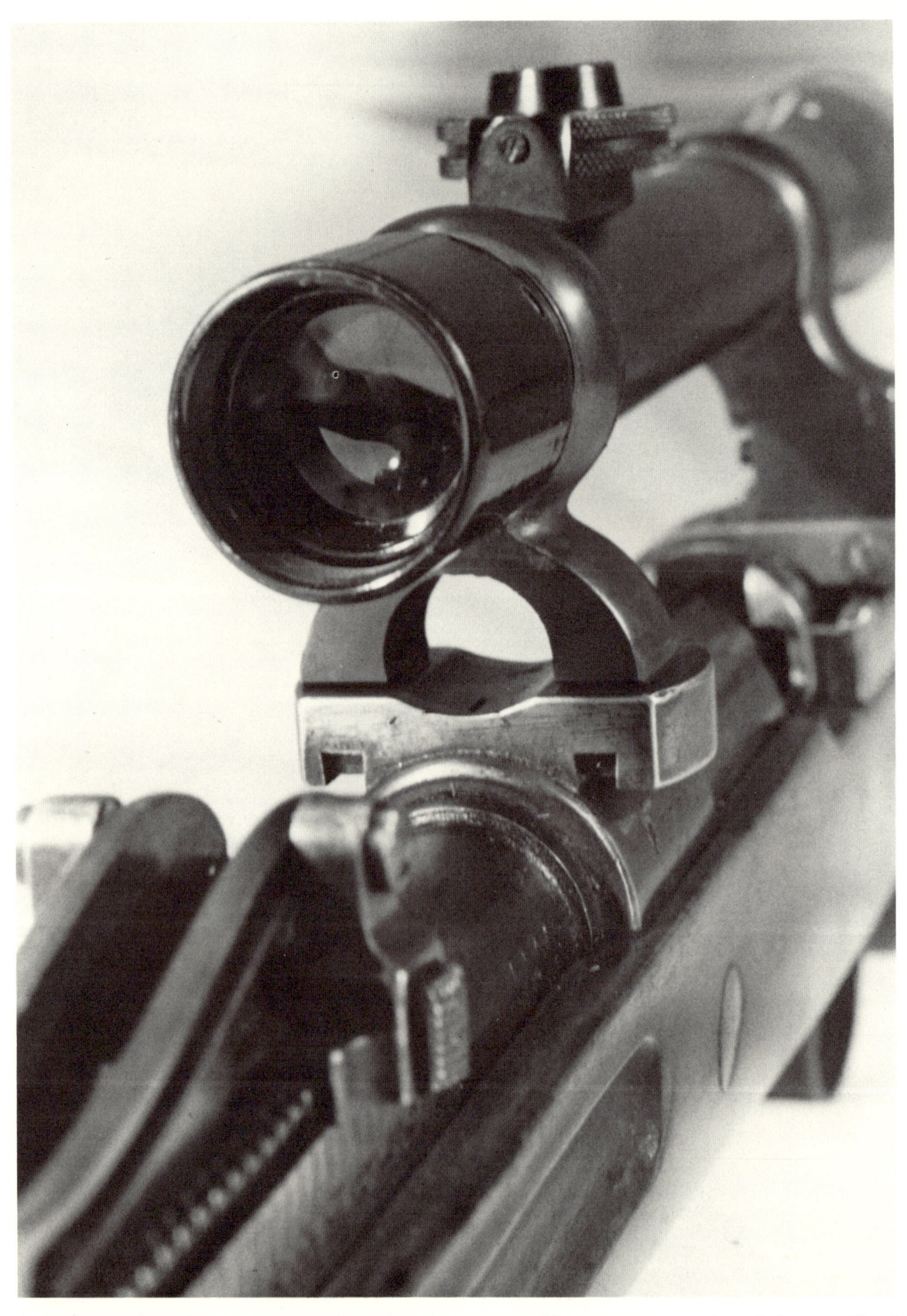

Close view of the front mount and receiver base assembly is representative of those utilized with the top mount variations.

The rear scope ring and mount, an integral unit, made use of a worm-driven dovetail for windage adjustments.

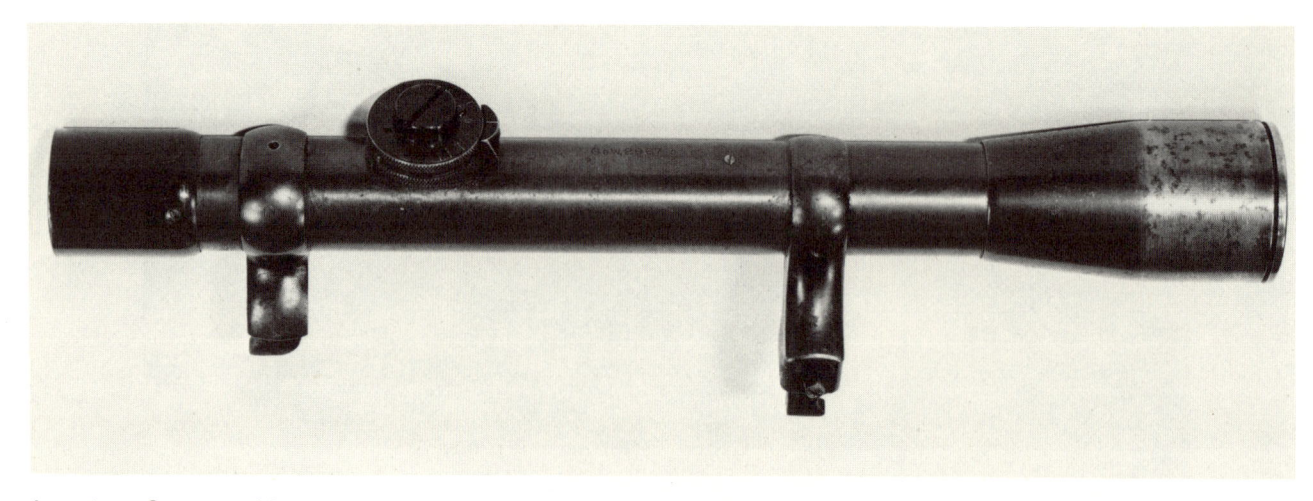

A variant 3-power Voigtlander sight (top mount). Compare the elevation adjustment drum saddle to that on the preceding scope.

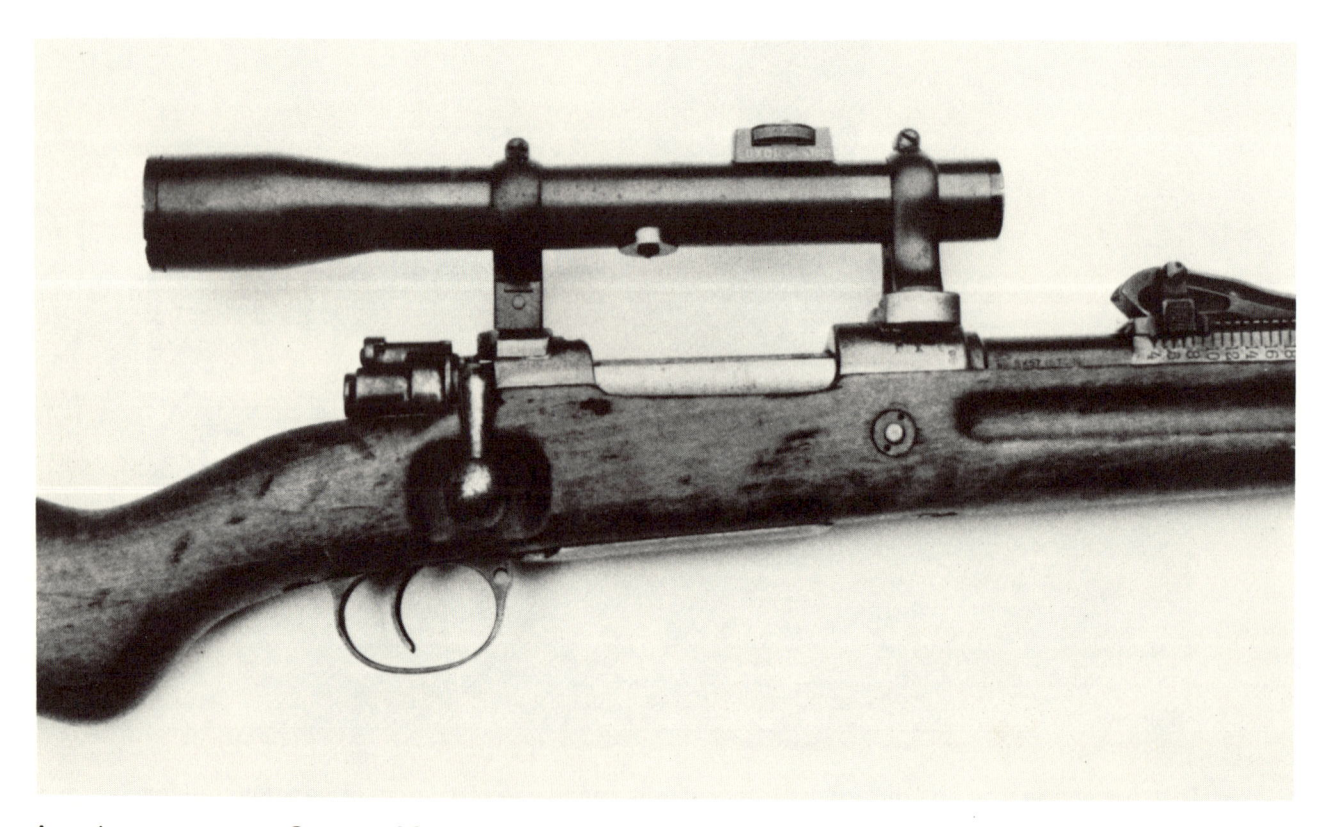

A variant top mount Gewehr 98 with an Oigee telescopic sight.

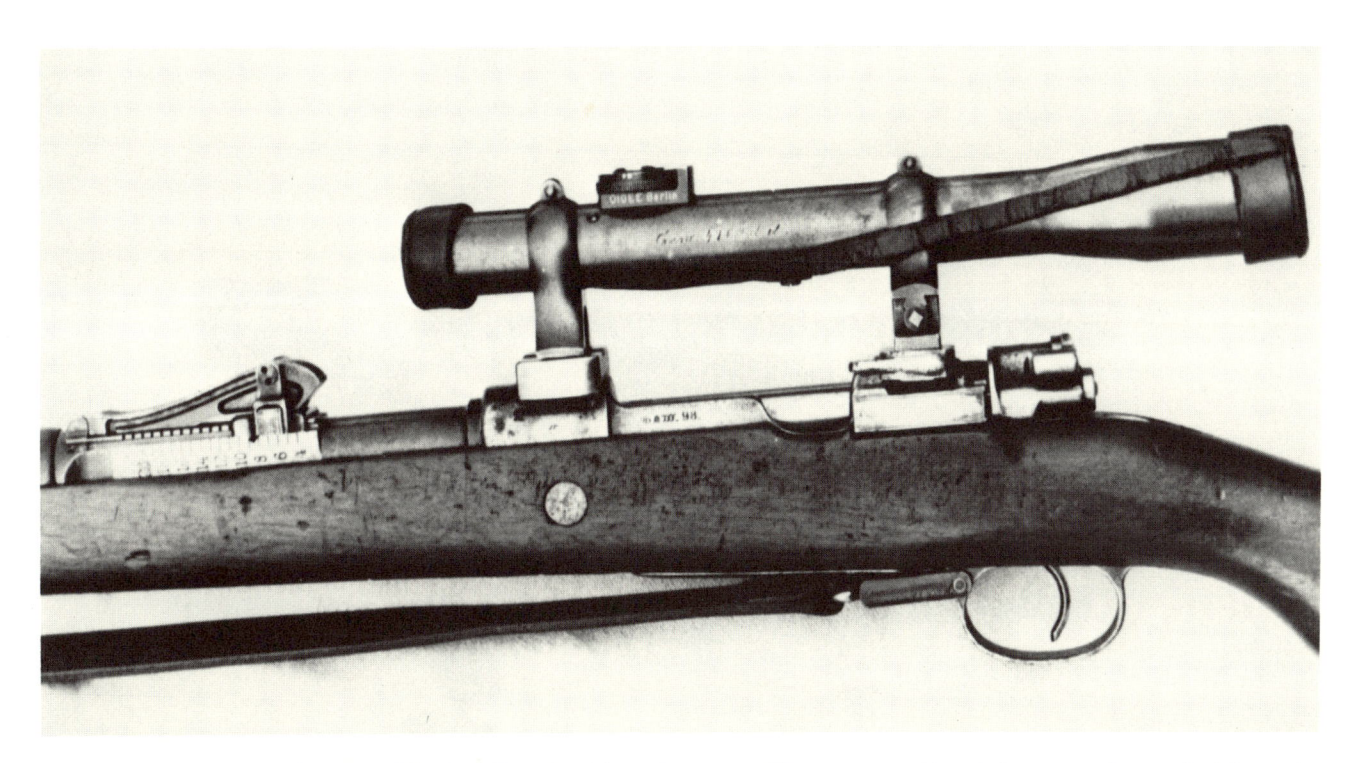

Another top-mount Gewehr 98 also fitted with a 3-power Oigee scope. Careful comparison of this and the preceding illustration reveals the subtle differences characteristic of World War I German sniping equipment. Note the positioning of the front bases.

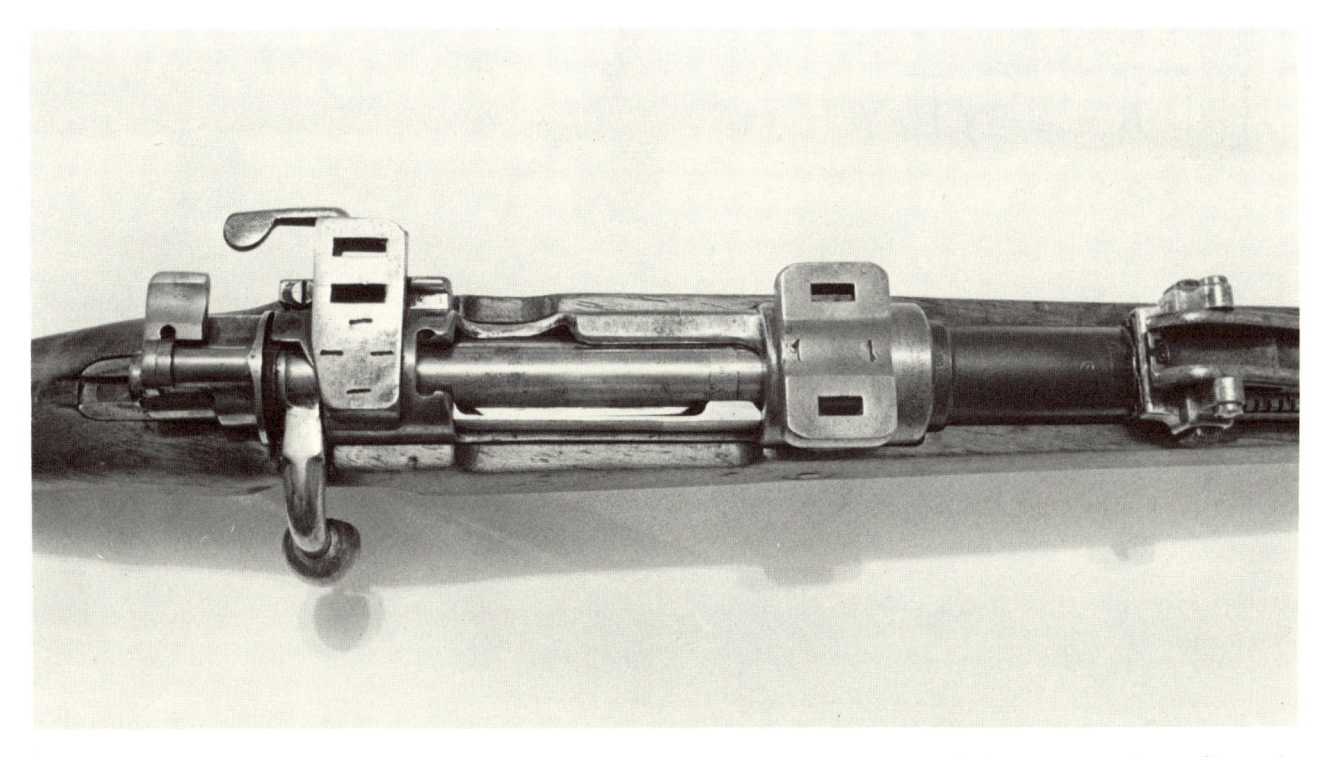

A variant top mount receiver base Gewehr 98. With few exceptions, sniper rifle base mounting reflected typical German craftsmanship. Note the careful installation and finishing of the base mounting screws.

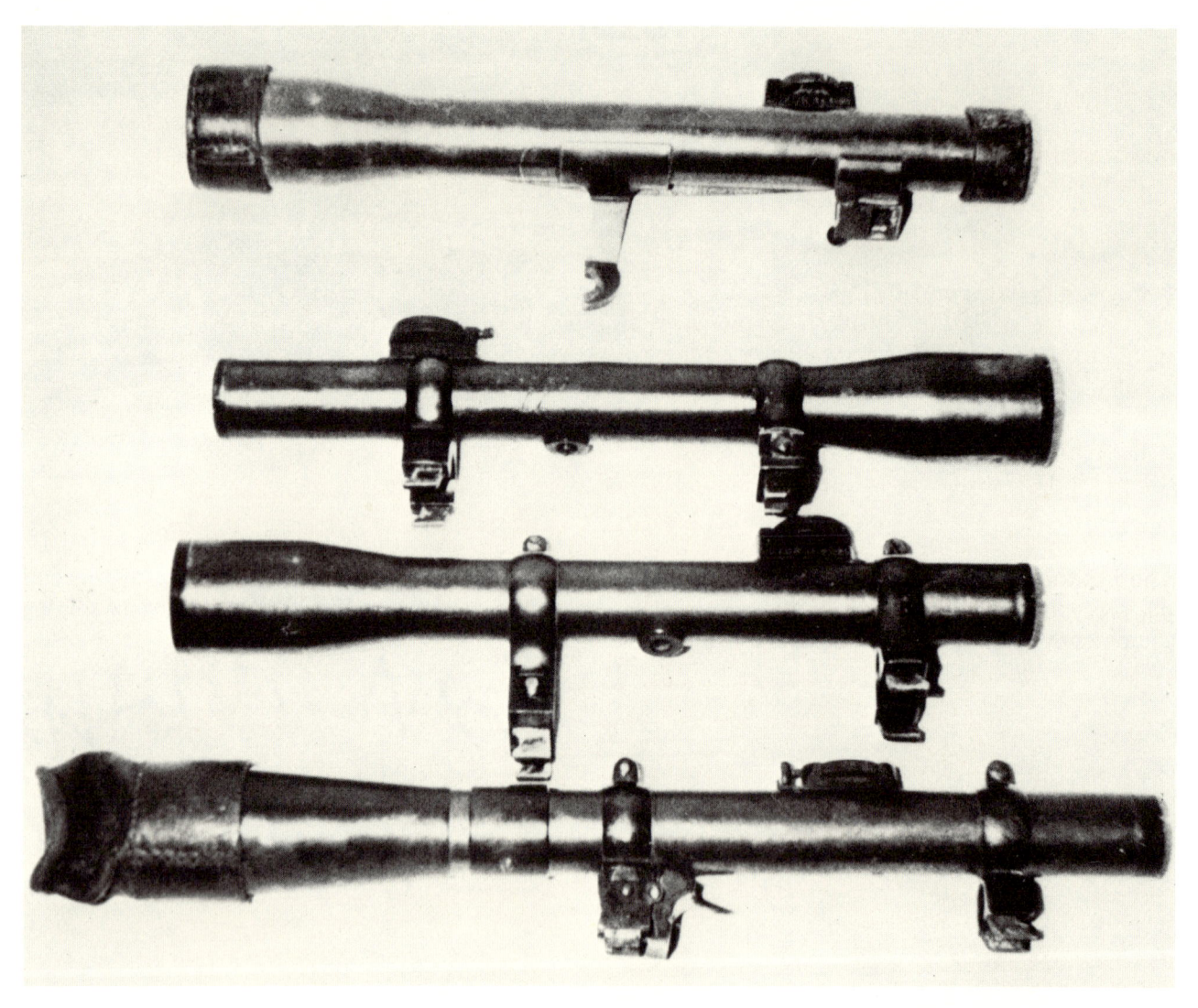

An illustration of various World War I German telescopic rifle sights serves to emphasize the variance in scope mounting practice exercised during this period. Note the in-line mounts on the bottom sight.

SCHARFSCHUTZEN — GEWEHR 98

Close view of top mount Gewehr 98 with a 3-power Dr. Walter Gerard sight. The device behind the elevation range drum served to focus the scope.

Typical World War I German telescopic sight fielded for use with the Scharfschutzen-Gewehr 98. In this case, a 3-power Oigee device made in Berlin.

Right view of Oigee scope which was originally paired with a top-mount Gewehr 98 sniping rifle.

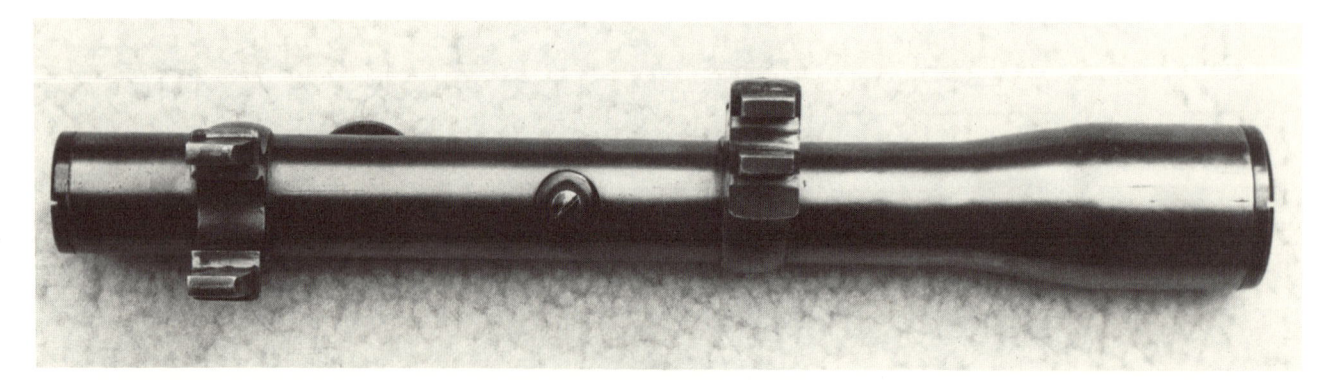

Underview of the Oigee scope showing the positioning of the claw mounts. The small circular device was used for focus adjustment.

SCHARFSCHUTZEN — GEWEHR 98

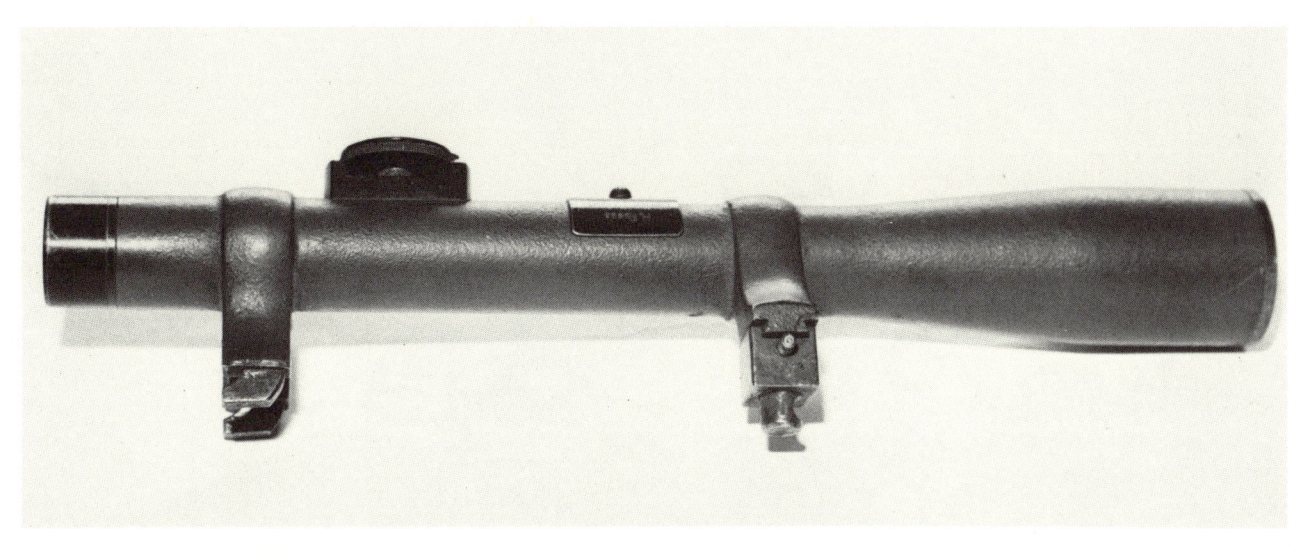

R. Fuess telescopic sight as produced for military service. The green exterior finish (textured) is similar to that which appears on various World War I German binoculars, etc.

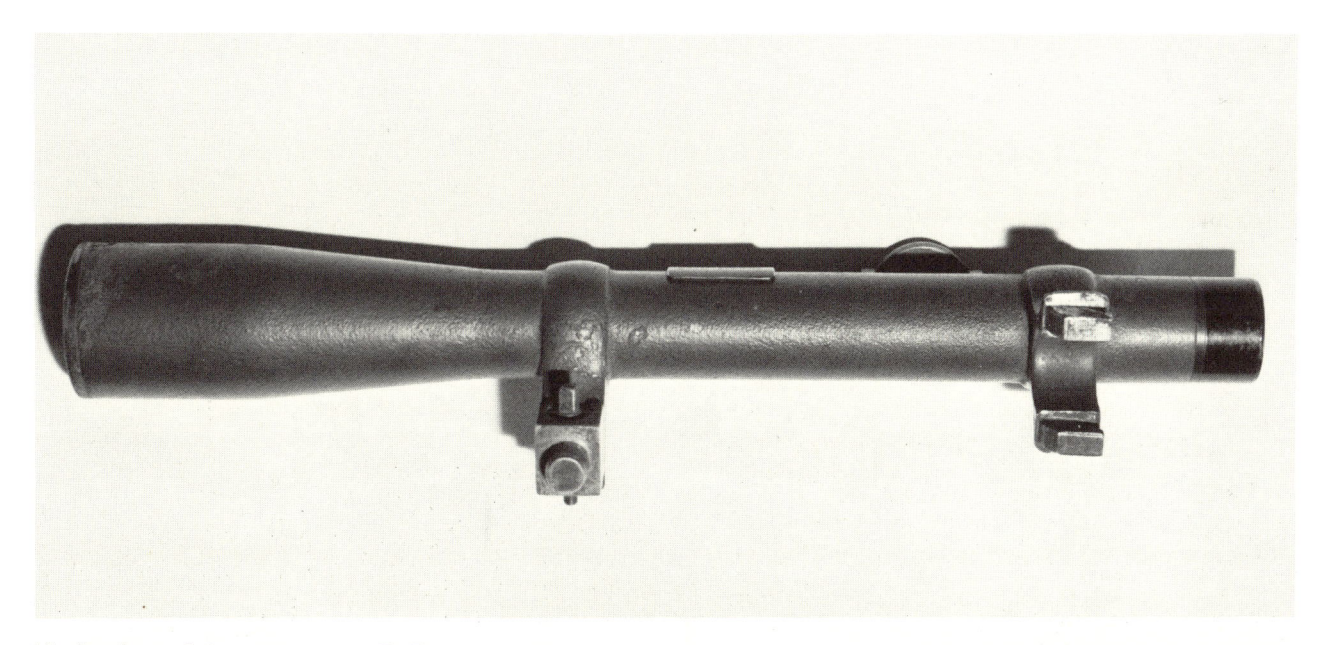

Underview of the top mount R. Fuess sniping scope showing the front claws and single-post (stud) rear mounting.

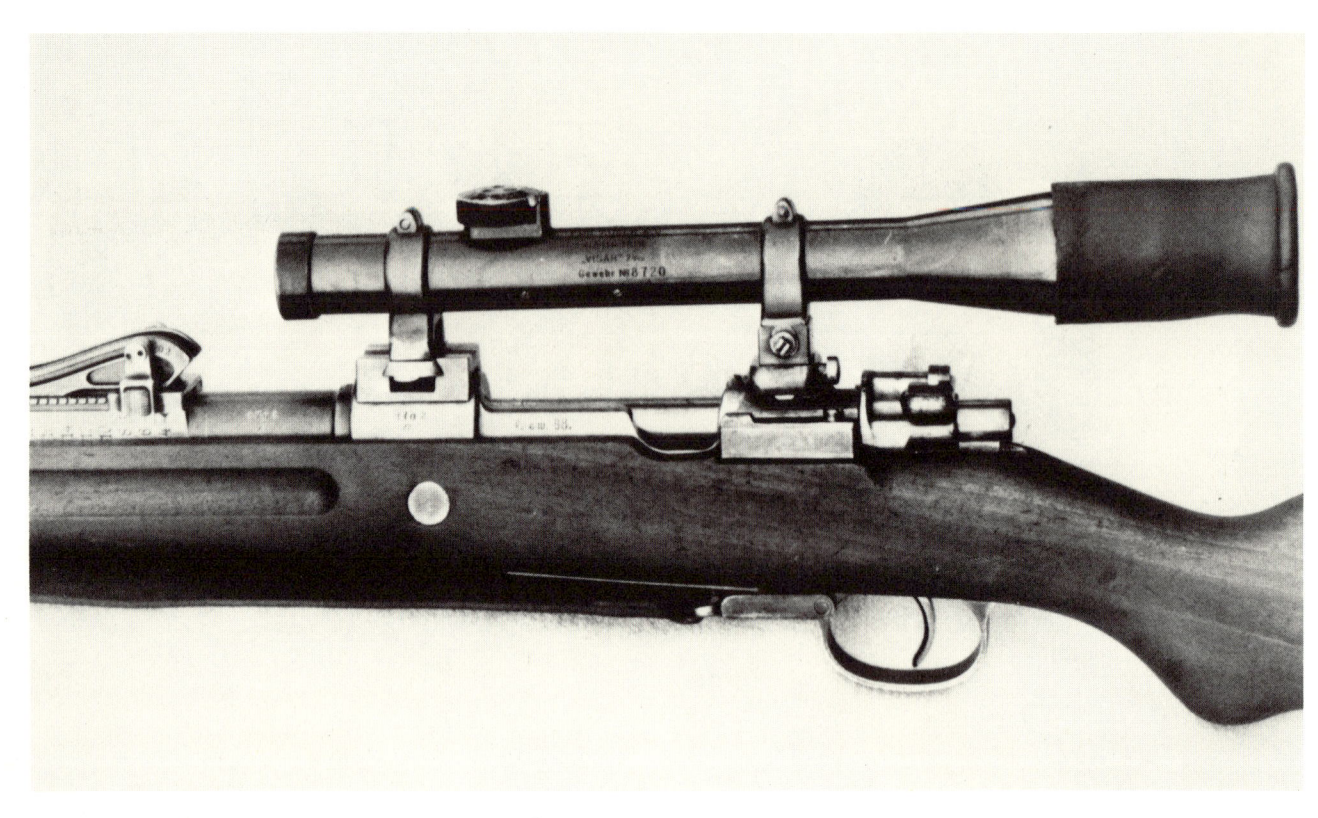

Scharfschutzen-Gewehr 98 mounting a 2¾-power rifle scope by Busch with receiver bases uncommon among the top mount variations. The rubber eye-guard is not an original item.

The Gewehr 98 with Busch sight removed. Note the push buttom release on the rear base.

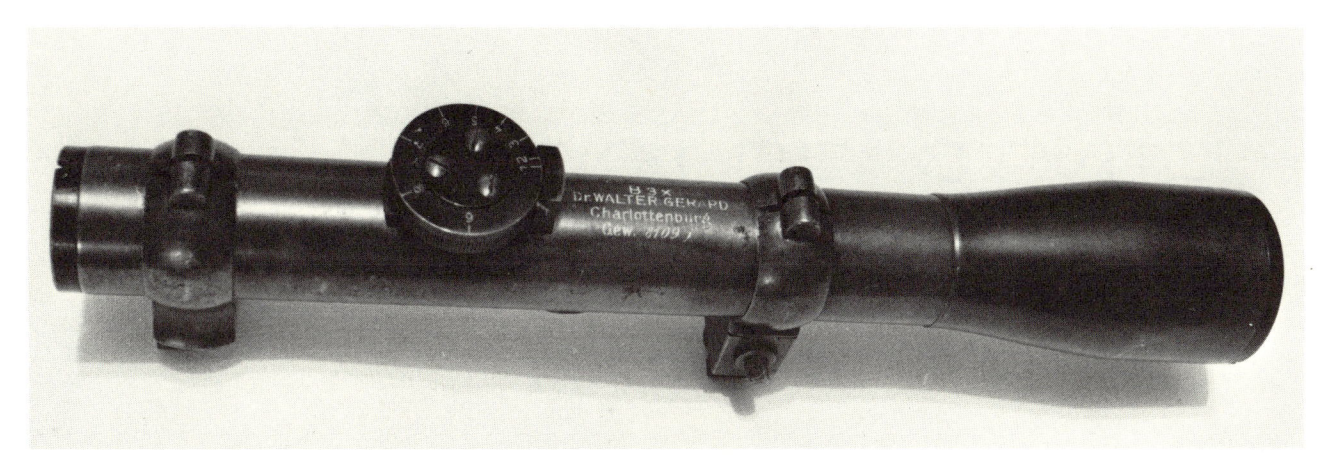

Top view of Gerard scope with in-line mounts. Even though the practice of mounting rifle scopes directly over the bore (top mount) saw extensive use during World War I, the use of mounts in this particular manner (in line) was not common.

A 3-power Dr. Walter Gerard sniper sight with in-line claw-mounts, that is, both mounts placed directly in line with the axis of the bore. As such, the conventional rifle sights could not be used with the scope in place. Note the rifle serial number on both mounts.

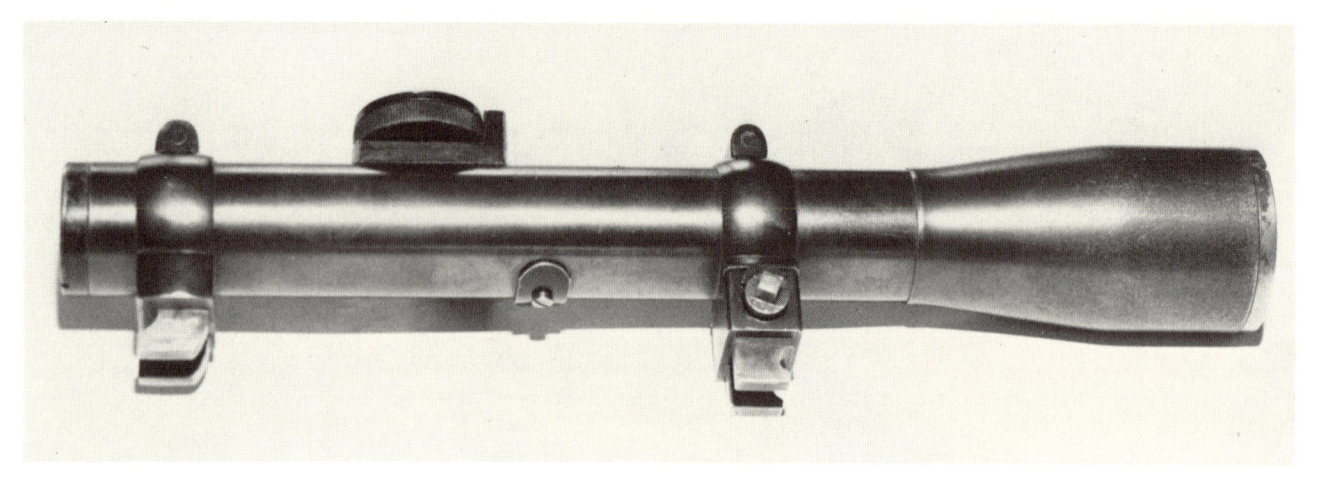

Another double-claw, in-line mount Gerard scope is shown for comparison. The sight and mounts bear a corresponding Gewehr 98 number (179).

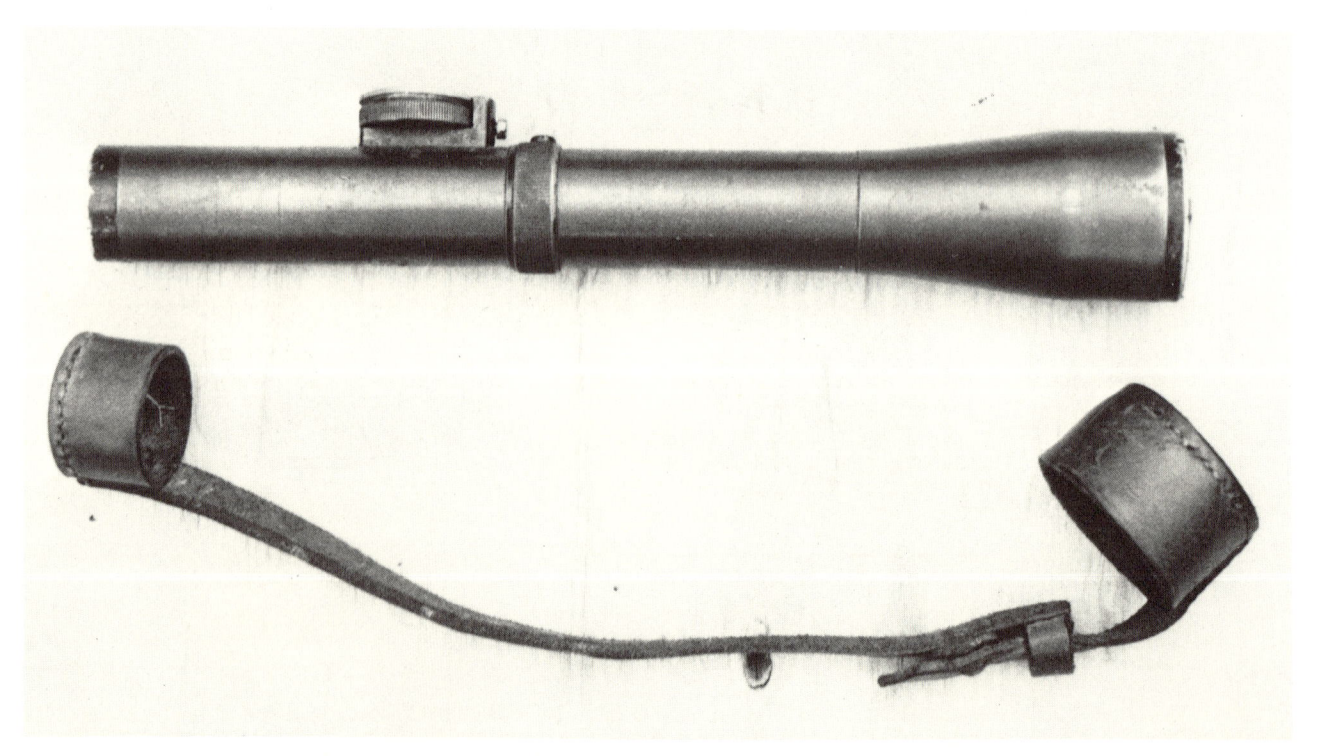

A variant Dr. Walter Gerard telescopic sight. Note the use of a full focus ring.

Close view of markings as they appear on the variant Gerard sight.

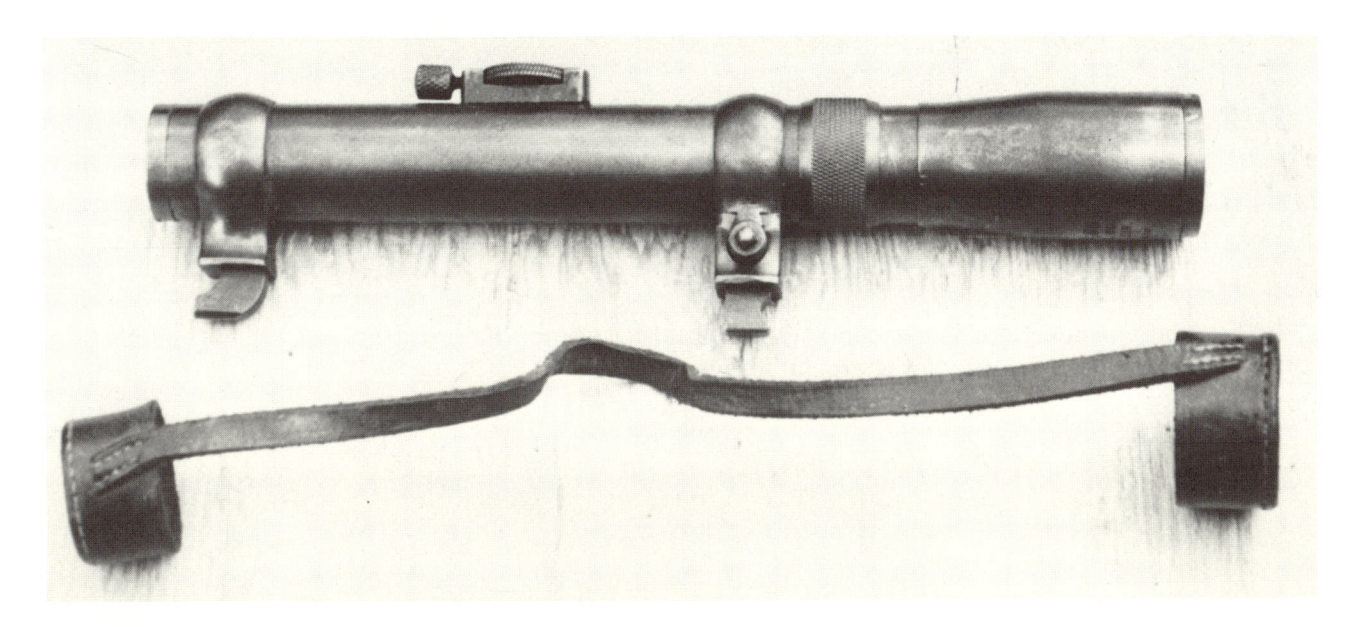

Single-claw, in-line mounting with Otto Bock telescopic sight. Note the focus adjustment ring.

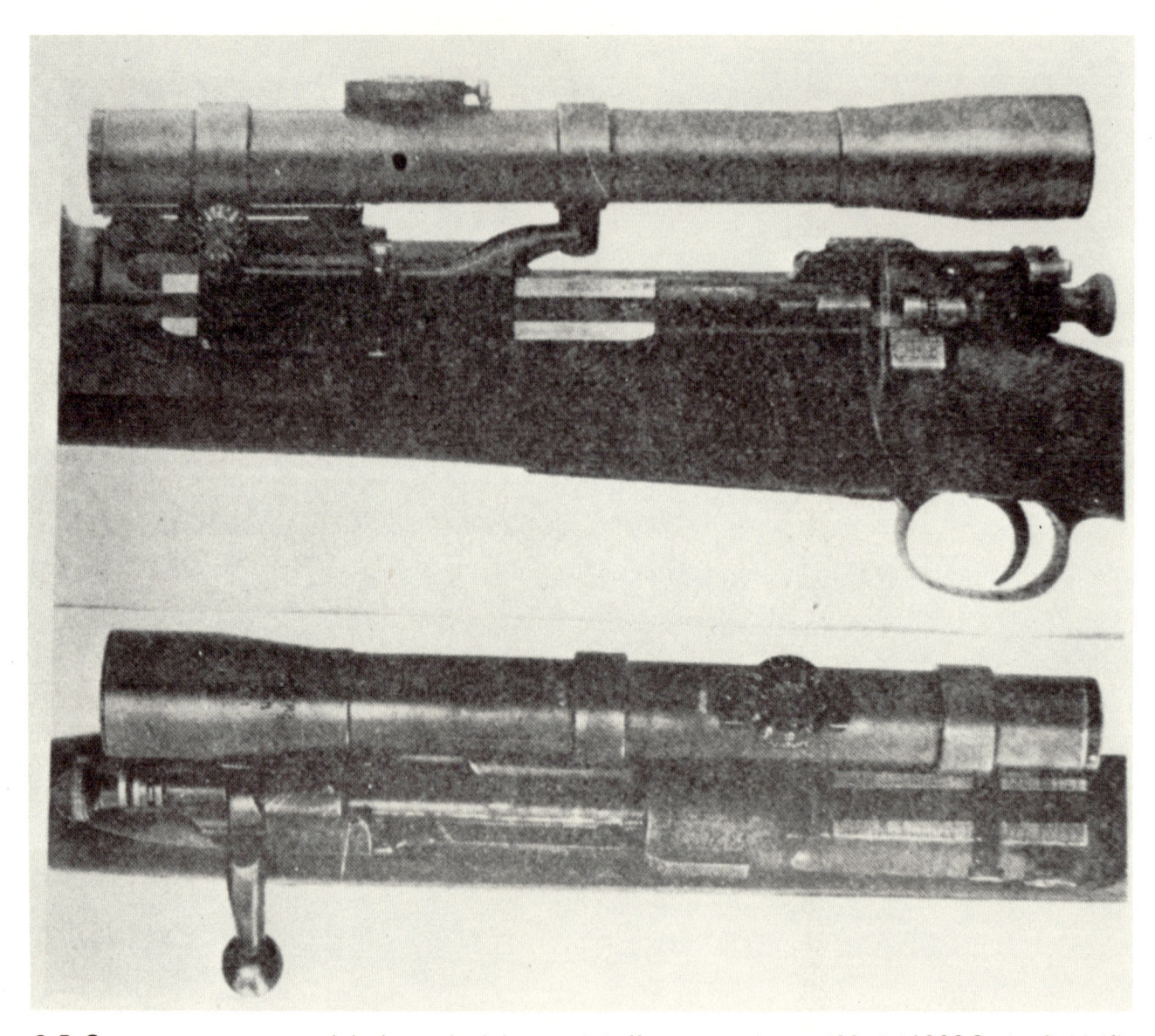

C. P. Goerz prewar commercial telescopic sight, special offset-mounting and Model 1903 Springfield rifle as evaluated by the U.S. Army for sharpshooter use in 1915.

SCHARFSCHUTZEN — GEWEHR 98

A comparatively rare World War I German sharpshooter rifle, dubbed the "semi-turret" variation (contemporary reference), is believed to represent late war efforts to field a standard type telescopic sight and mounting for the Gewehr 98.

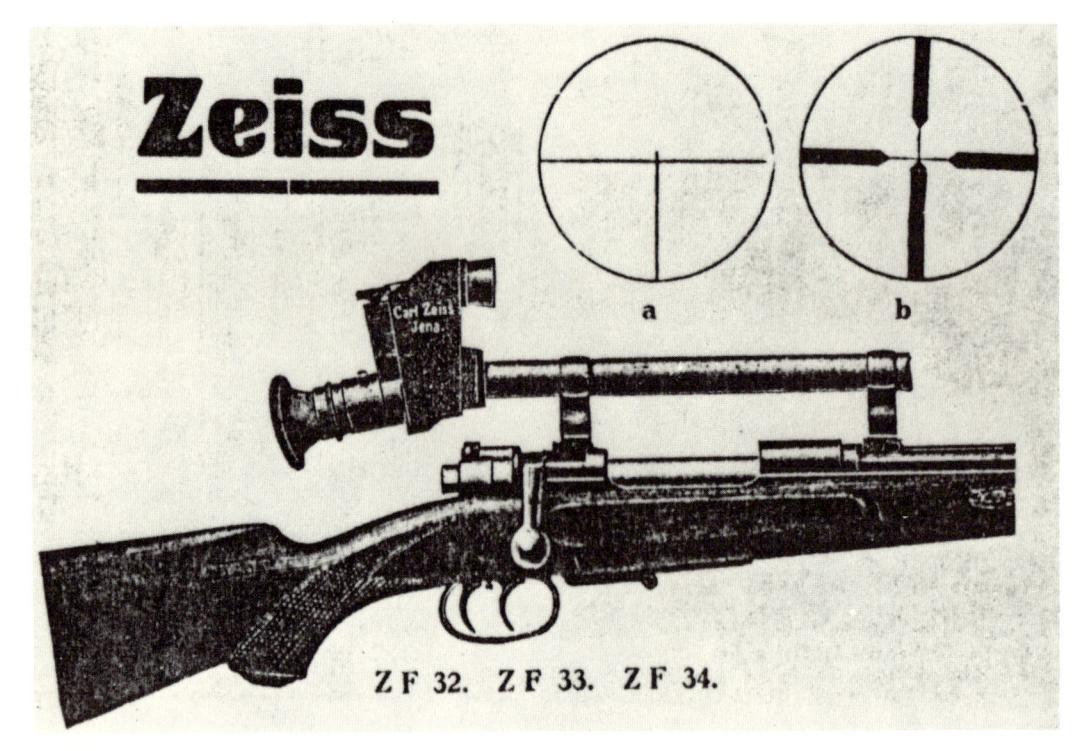

Zeiss prismatic rifle scope as depicted in prewar German sales literature. Although prismatic sights were reportedly fielded during World War I, the extent of their combat use remains obscure.

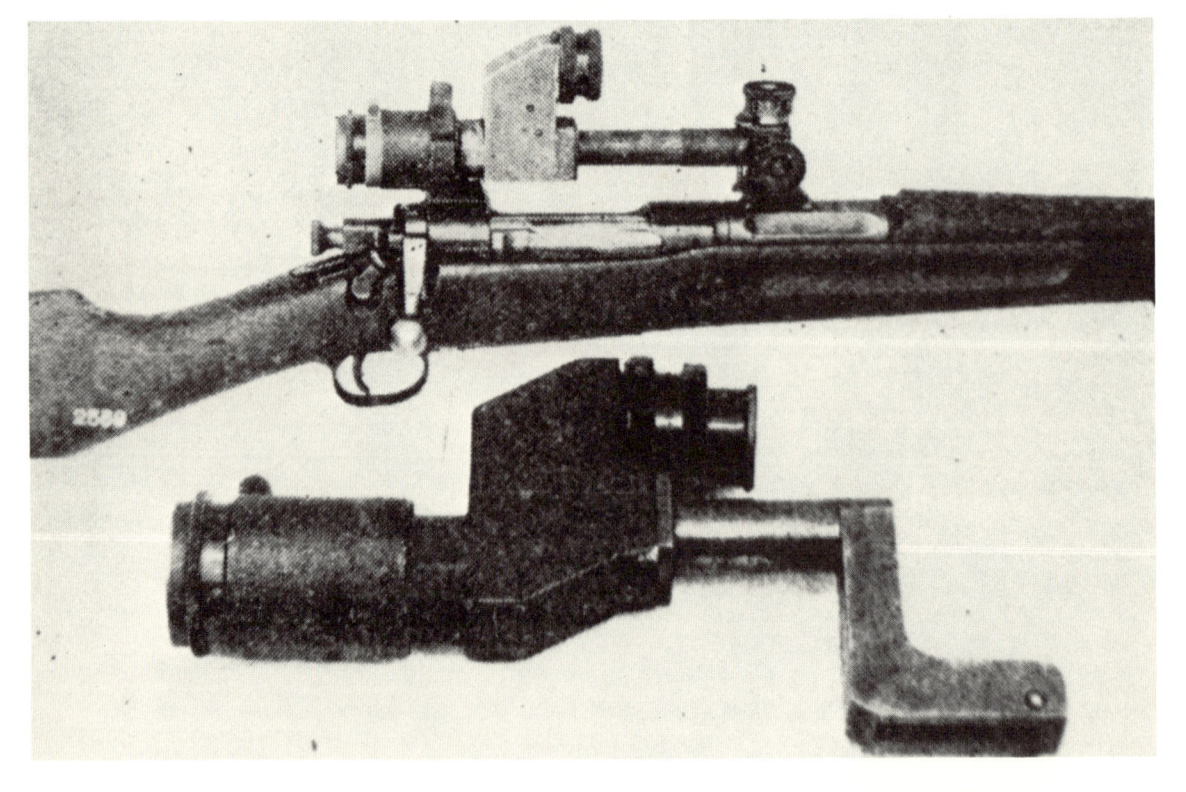

Zeiss prismatic rifle sight tested by the U.S. Army shortly after World War I. The top illustration shows the Zeiss device mounted to an M1903 Springfield rifle for testing purposes. The lower view depicts the sight as originally "captured."

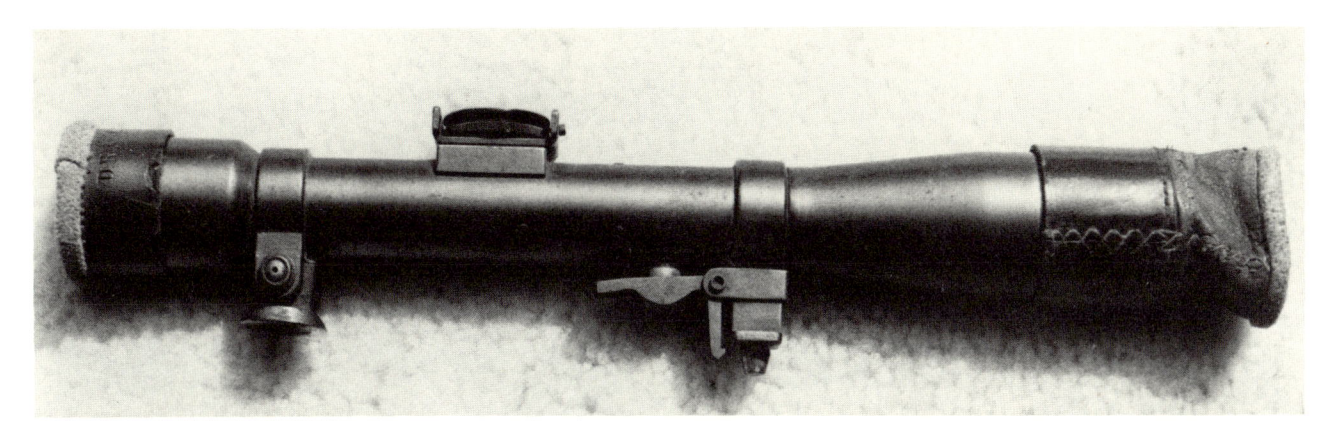

Side view of C. P. Goerz semi-turret sight showing the in-line scope mounting (both mounts directly over the bore axis). The front mount assembly was turned into a corresponding recess in the front base; the sight was then locked in place with a spring-loaded latch (lever) located directly beneath the scope, an integral member of the single-stud (post) rear mounting.

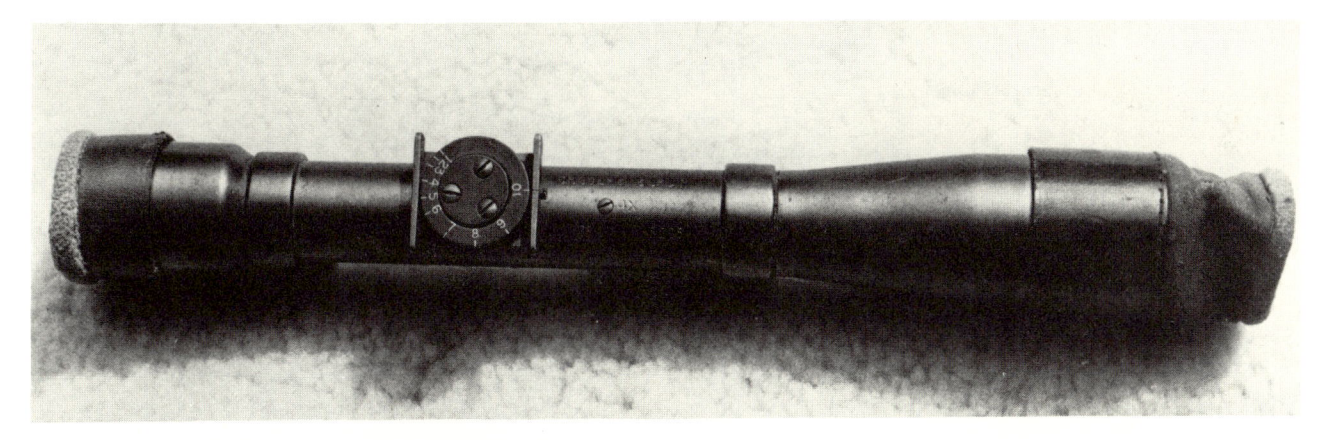

Top view of late war 4-power Goerz rifle scope. Note the protective "ears" on either side of the elevation range drum, a feature unique to German sights of this era.

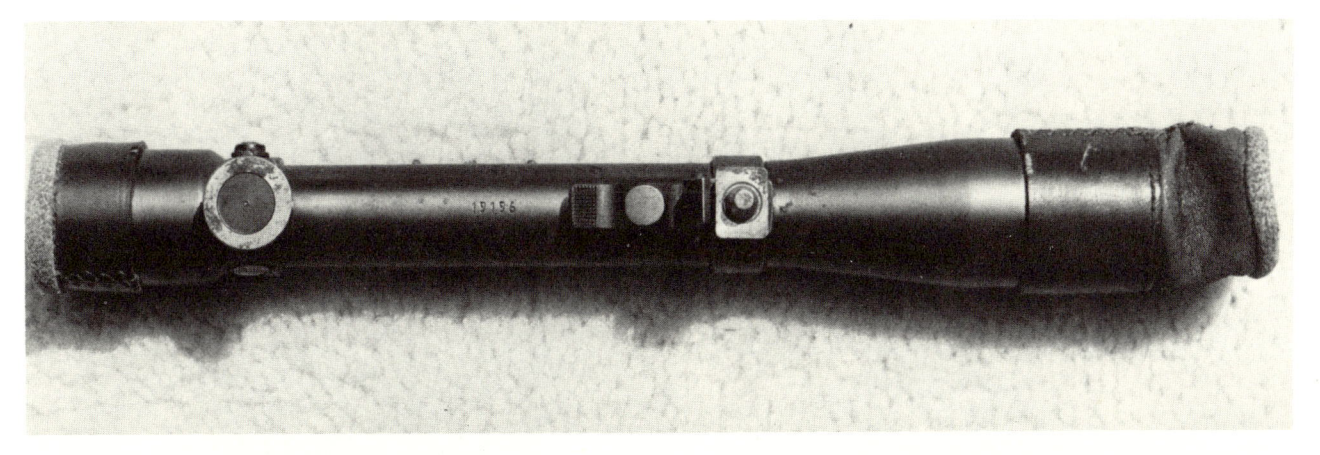

Receiver base view of the C. P. Goerz sight showing its unusual mounting system. This design provided a positive means of engaging the receiver bases, unlike that of a standard claw-mount scope, which was simply tilted upwards ("tipped-up"), engaged, and lowered into place. The overall configuration of the Goerz mounting required the front mount to swing upward when the circular section was engaged with the receiver base during installation or removal of the scope.

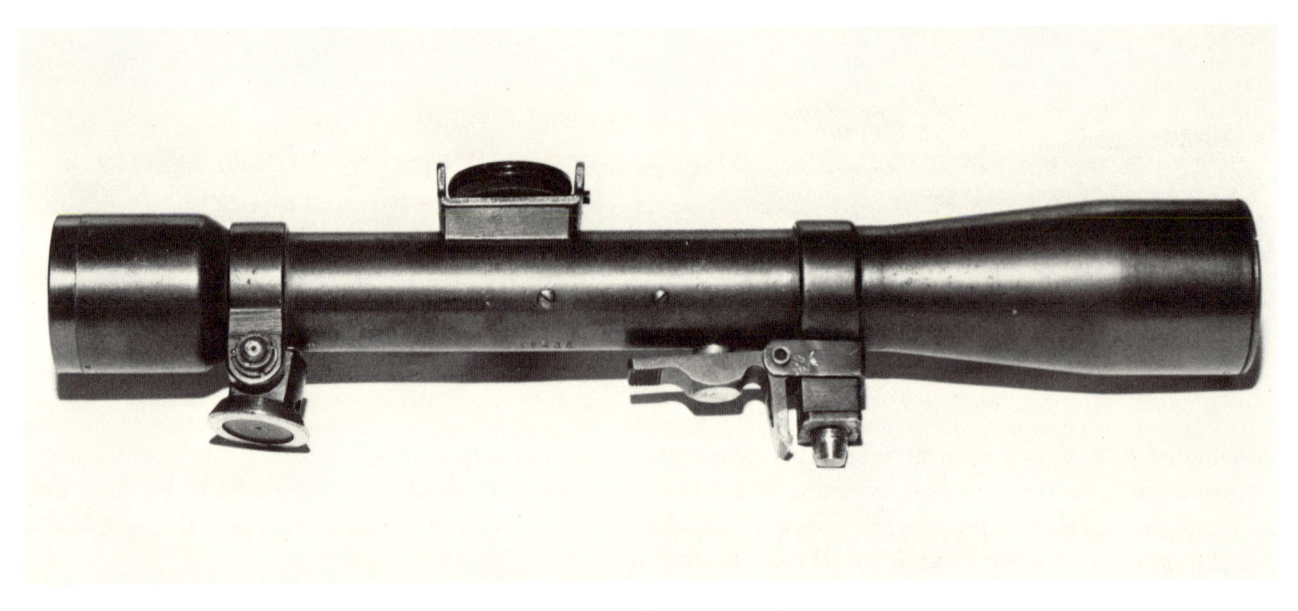

An additional late war C. P. Goerz sight is shown for comparison purposes.

Leather ocular and objective shields furnished with the 4-power C. P. Goerz semi-turret rifle scope.

SCHARFSCHUTZEN — GEWEHR 98

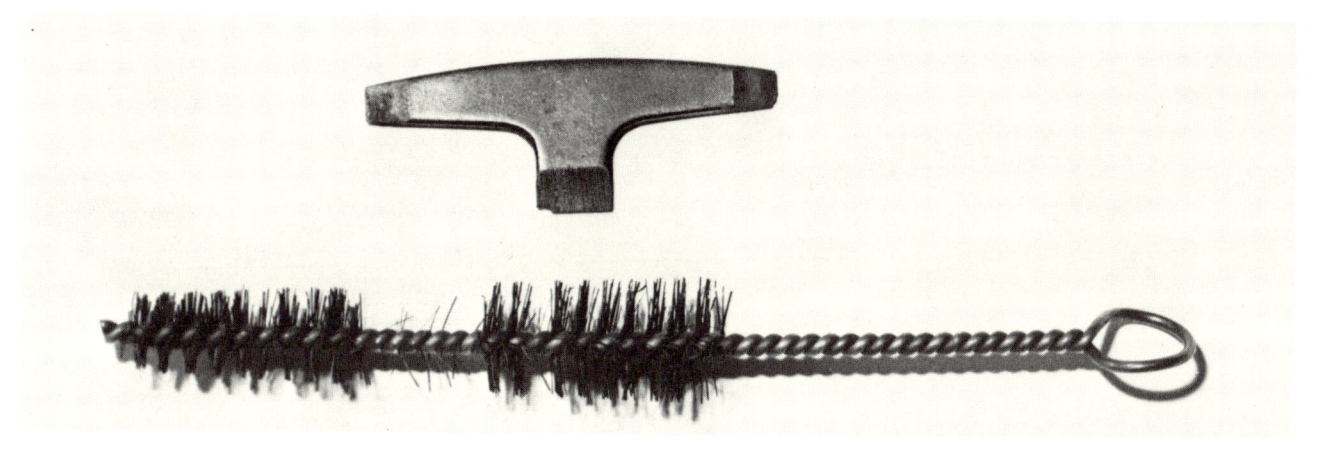

Late war C. P. Goerz adjustment tool and cleaning brush. A pocket inside the case held the brush.

Telescope carrying case for the C. P. Goerz "semi-turret" sighting system.

Late war C. P. Goerz scope case lid (underside) showing a small pouch and manner in which the adjustment tool was held in place. The canvas-covered metal case has a leather lid with Goerz markings.

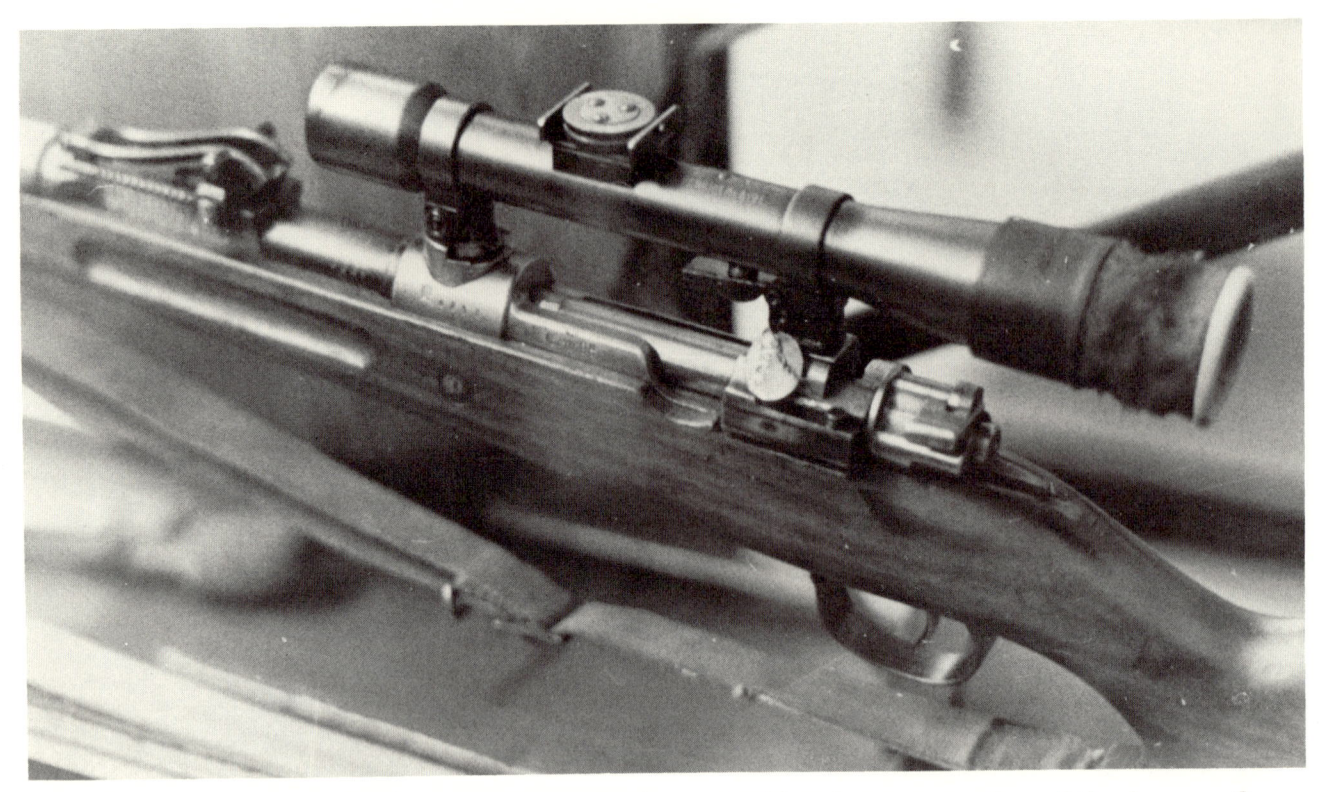

Although complete rifles of this type (semi-turret) are considered rare, a number of the 4-power Goerz sights intended for this variant have been noted on both sides of the Atlantic Ocean with all major sight characteristics identical in every case.

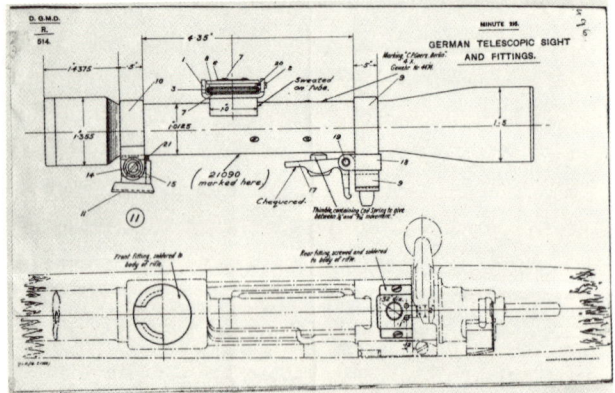

Left: Postwar British Ordnance drawing dating from 1919 depicts the German "semi-turret" system as used with the Gewehr 98.

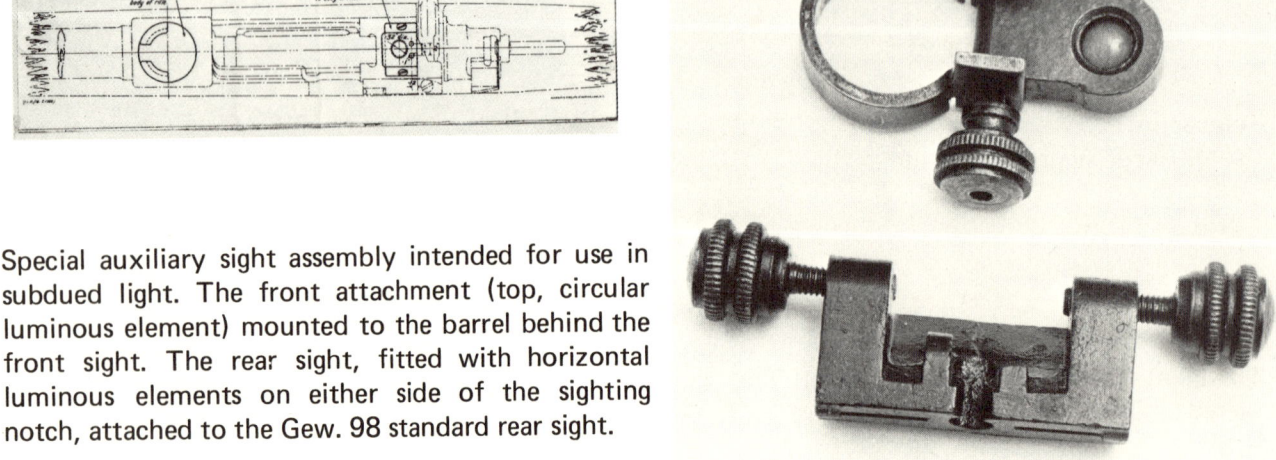

Special auxiliary sight assembly intended for use in subdued light. The front attachment (top, circular luminous element) mounted to the barrel behind the front sight. The rear sight, fitted with horizontal luminous elements on either side of the sighting notch, attached to the Gew. 98 standard rear sight.

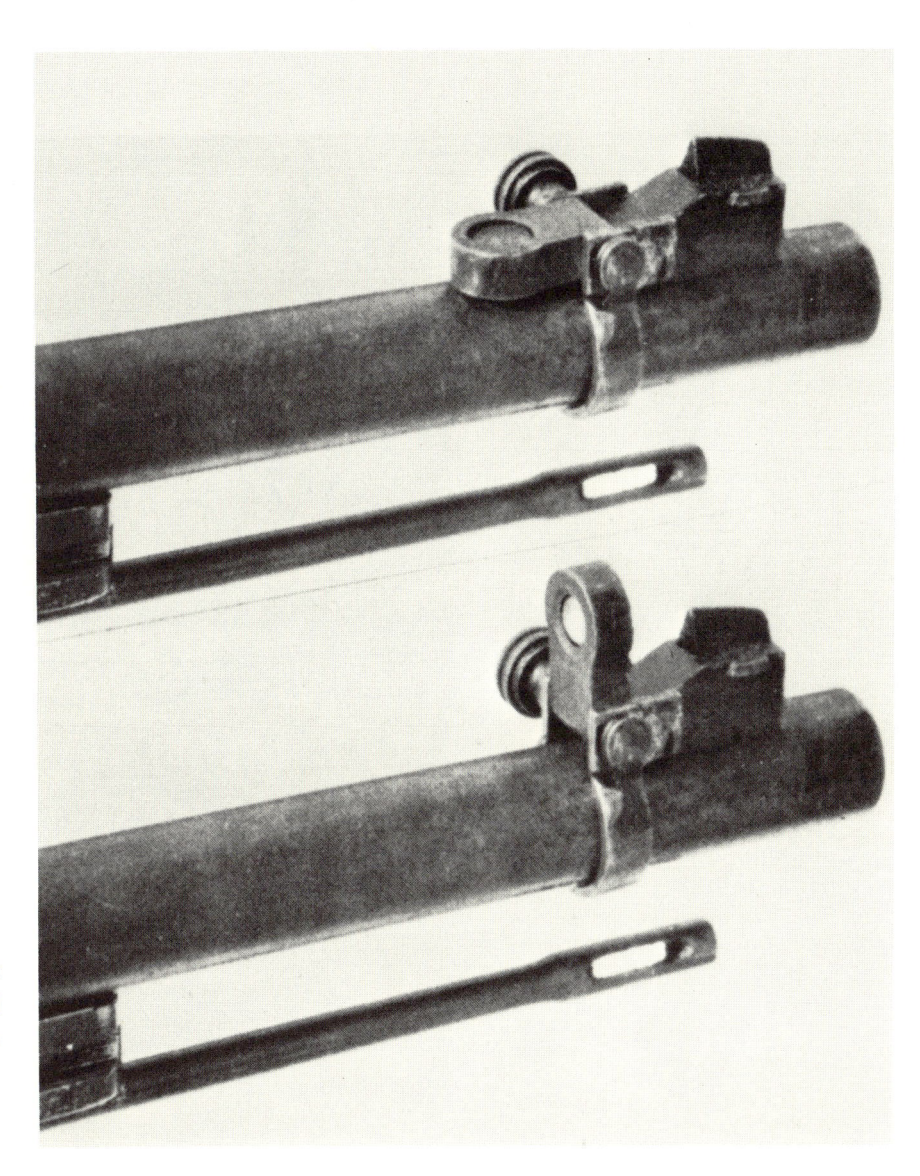

Gewehr 98 luminous front sight attachment was folded against the barrel when not in use.

A Gewehr 98 with extension magazine (25 rounds) and unique 2.5-power Zeiss sighting system developed primarily for use in marginal light. Even though the special magazine increased the cartridge capacity, it made rifle handling extremely difficult and was not popular with the German line troops.

In addition to being readily affixed to the rear sight assembly of a standard Gewehr 98, the uncomplicated operation of bifocal sights permitted satisfactory results when utilized by any competent rifleman. The sight is fitted with a compressed-fiber eye shield.

While sights of this type were essentially the same, in some cases a protective coating similar to clear lacquer was applied over the blue finish to retard the formation of rust during sustained field use. The leather eye shield was held by a small retaining ring.

SCHARFSCHUTZEN — GEWEHR 98

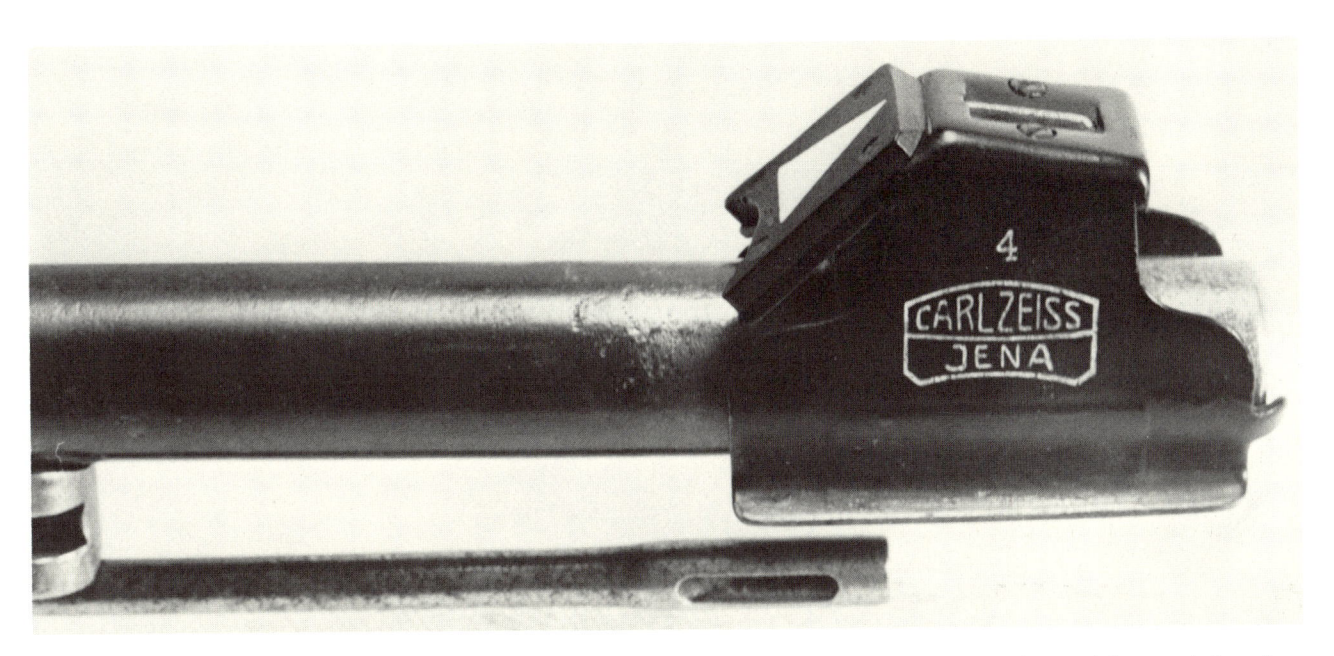

The front sight attachment for the Zeiss scope merely clamped over the muzzle and front sight of a standard Gewehr 98.

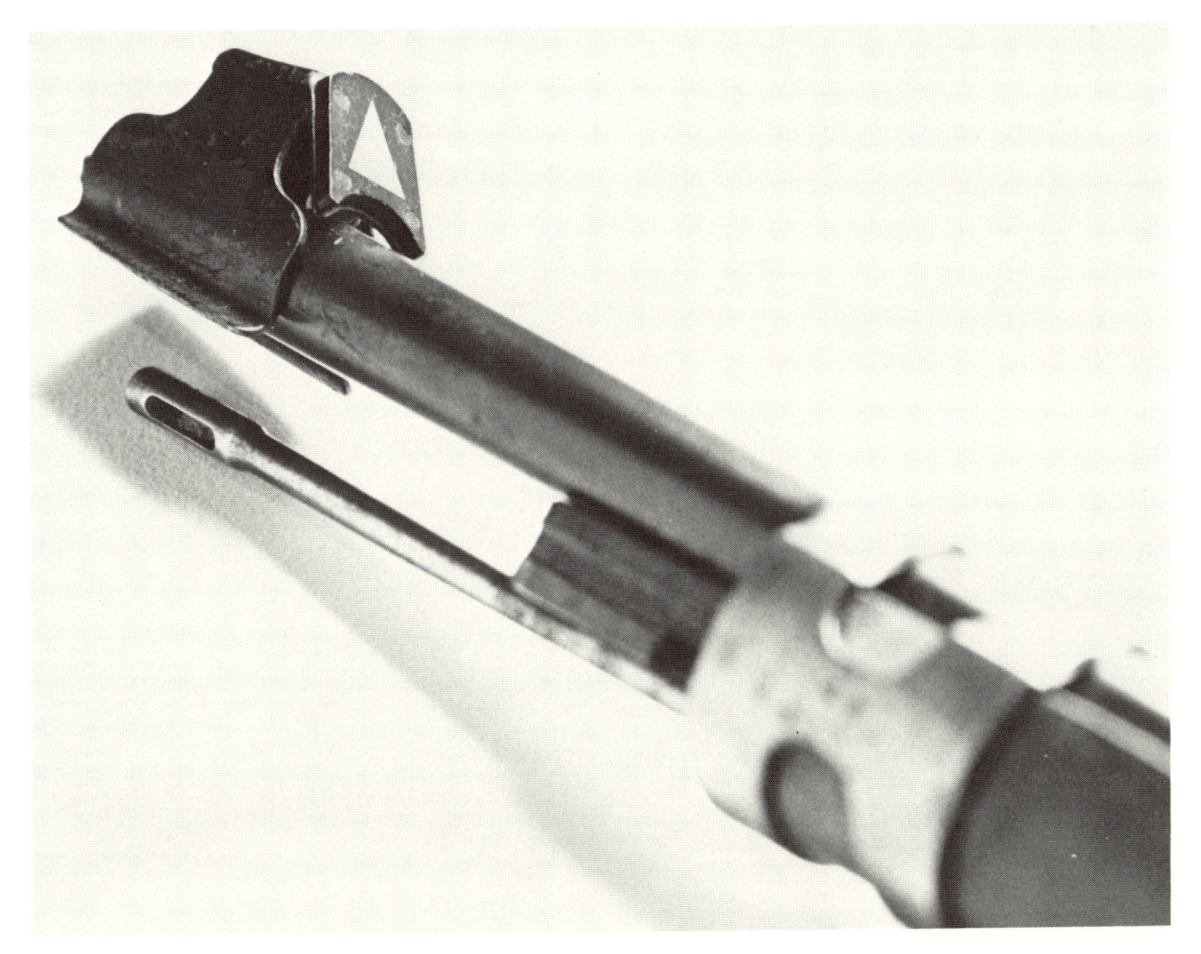

There was no reticle in the Zeiss bifocal sight; the white pyramid on the front attachment served as the aiming point for the rifleman when viewed through the scope.

Alternate view of another Zeiss sight showing typical markings (eye shield missing). Unlike conventional sniping scopes fielded during this period, bifocal sights were not numbered to a particular rifle.

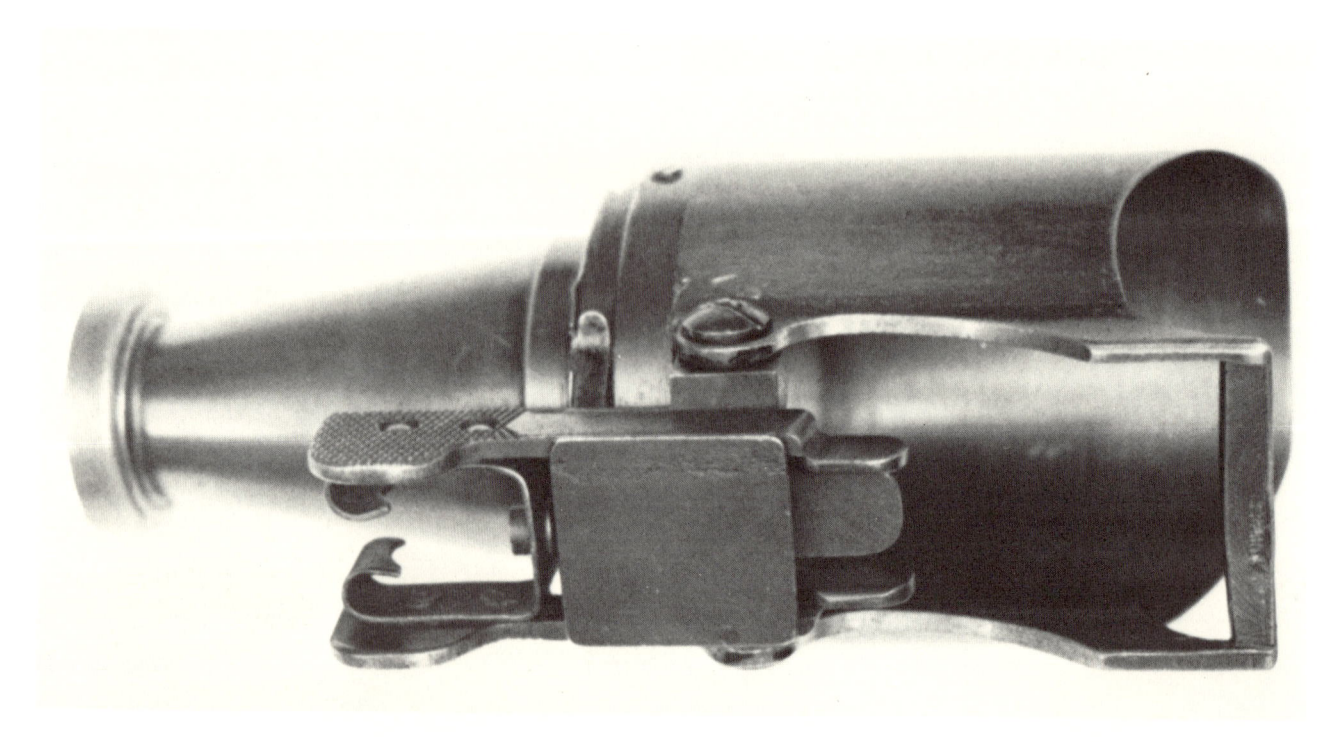

A simple spring-loaded clamping device allowed the bifocal sight to be readily attached or removed from the Gew. 98.

A small rectangular leather case was provided for holding the Zeiss bifocal sight assembly when removed from the rifle.

Internal view of the carrying case with Zeiss sighting apparatus in place.

Telescopic-sighted Austro-Hungarian Model 95 (Mannlicher) rifle as fielded for sniper use during the Great War.

Left view of Austro-Hungarian M95 sniping rifle.

SCHARFSCHUTZEN — GEWEHR 98

Close view of M95 sniper rifle showing the 3-power Oigee "Luxor" scope in place on the receiver. As in German practice, rifle serial numbers were placed on the scope as originally issued.

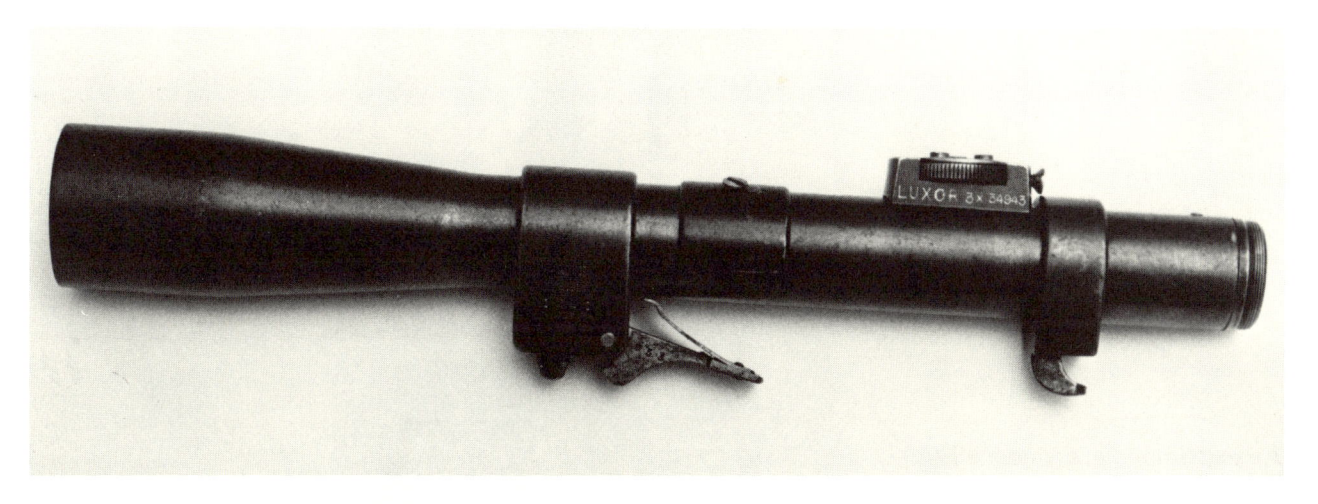

The telescopic sight attached to the receiver bases by means of a single-claw front mount and a vertical stud (post) with spring-loaded latch (lever) rear mount assembly.

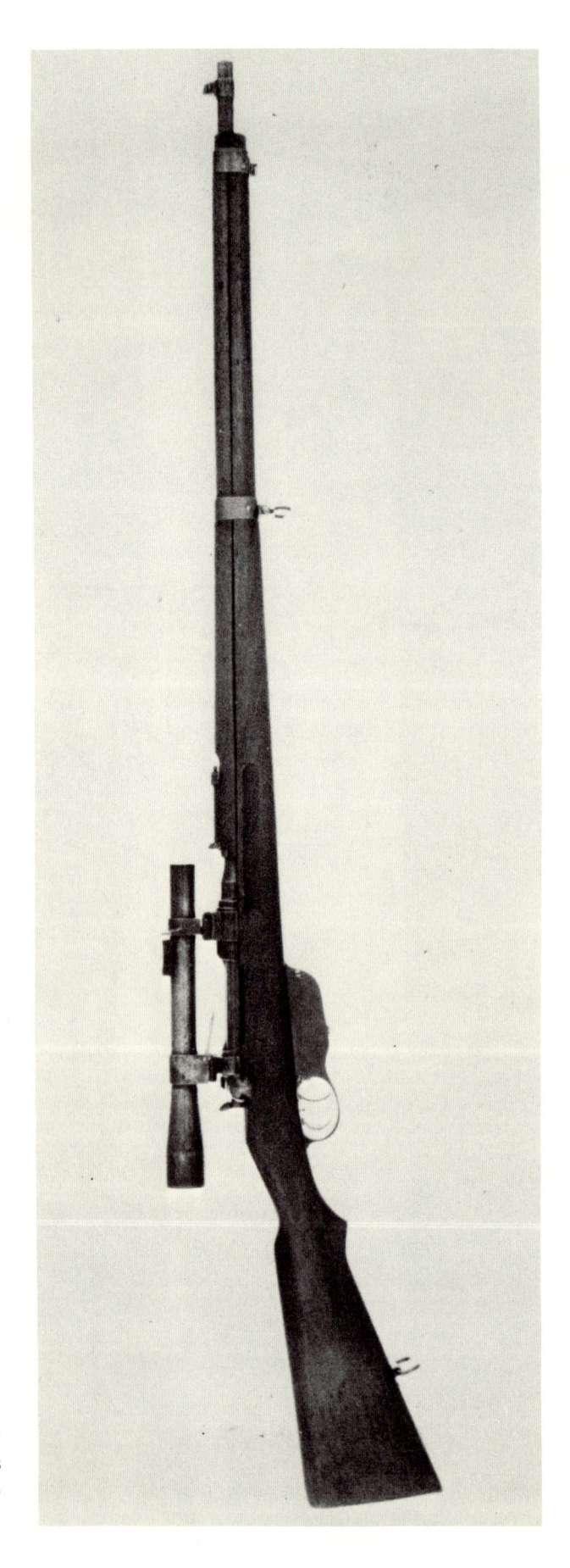

An extremely rare Model 95 Austrian sharpshooter rifle with its original double-set trigger assembly.

SCHARFSCHUTZEN — GEWEHR 98

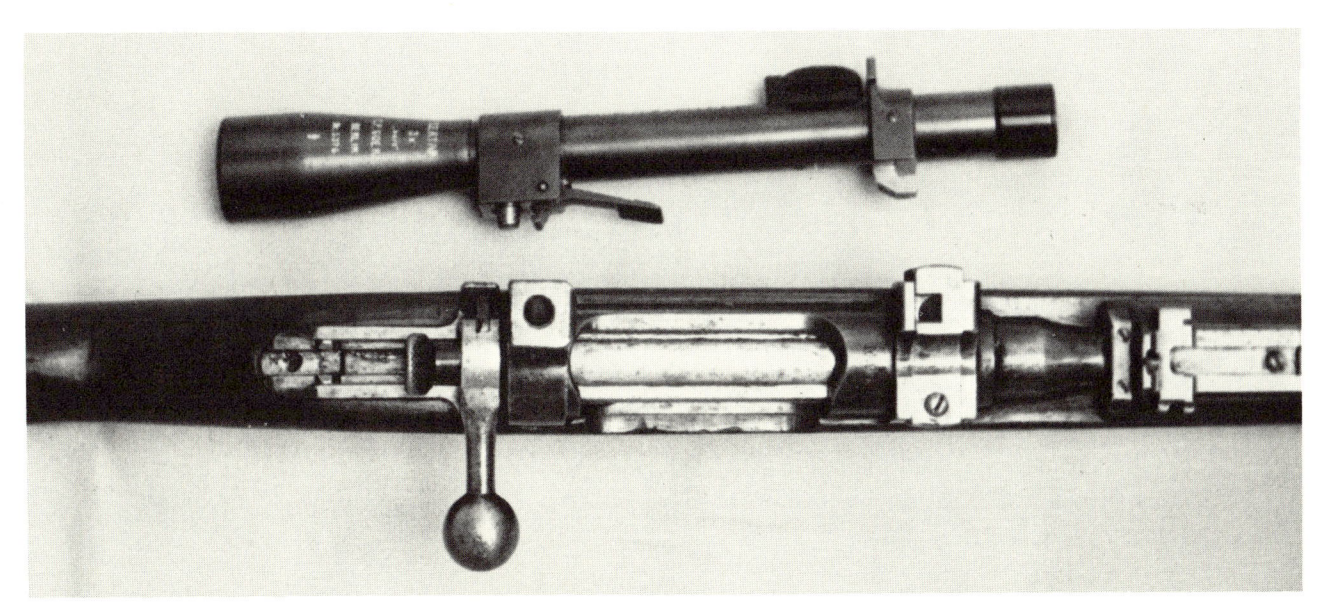

Top view of M95 sniper rifle showing the manner in which the scope bases were positioned on the receiver. Windage adjustments were made by loosening the screw on the front base and gently tapping the dovetail assembly as required. Note the variant spring-loaded latch (lever) on the C. P. Goerz rear scope mount.

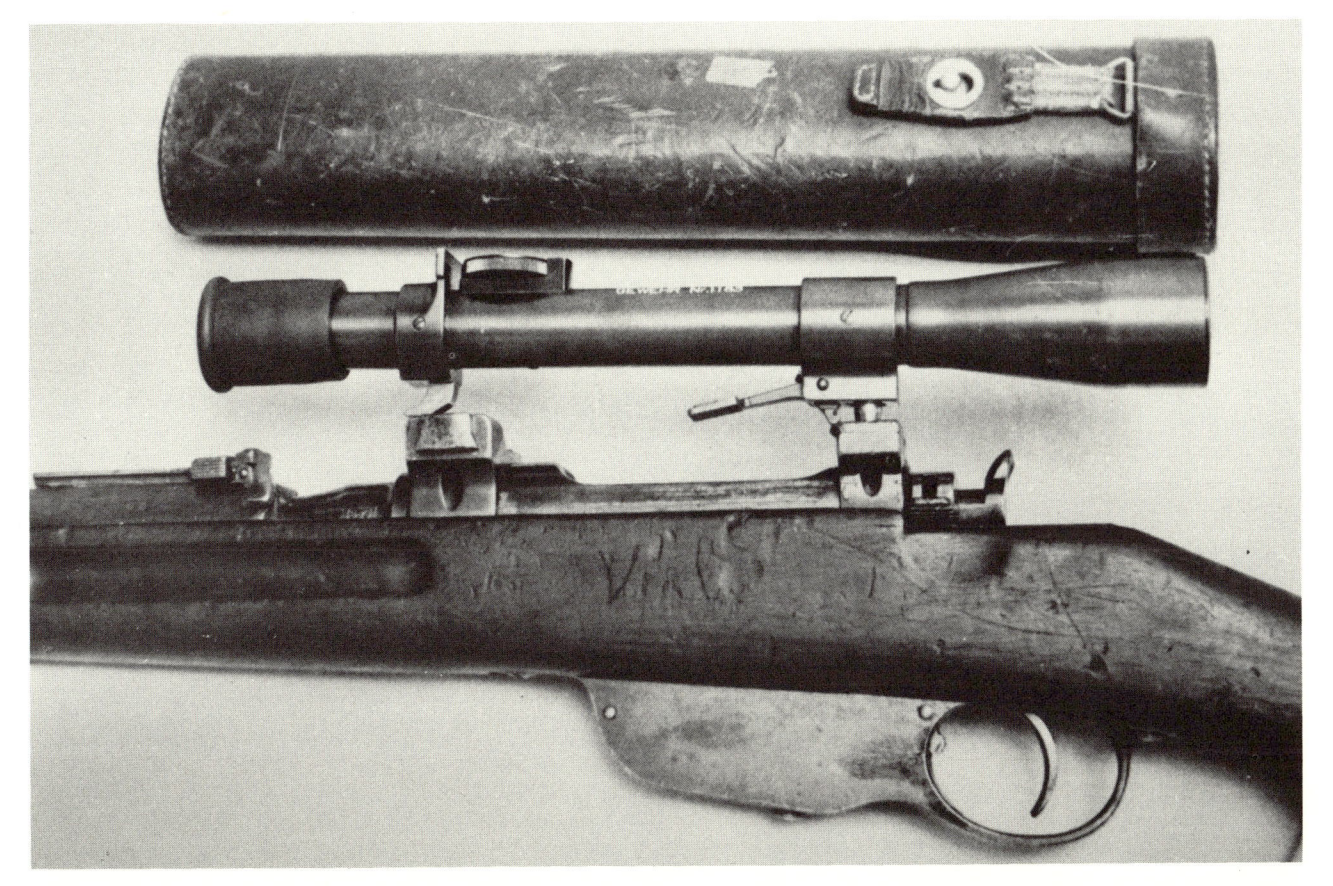

Model 95 sniper rifle, C. P. Goerz scope, and original carrying case. Cups made of soft rubber were first intended to be placed on the end of the scope to prevent damage to the tube when the sight was "tipped-up" for installation or removal.

A unique, original World War I Austro-Hungarian Model 95 carbine in sharpshooting trim.

Left view of telescopic-sighted Model 95 carbine.

Close view of M95 carbine with its 3-power C. P. Goerz scope. Note the difference between the rear mount locking latch (lever) assembly and that on the Model 95 sniper rifle.

Left view of the Goerz scope illustrates the unique method of preventing loss of the leather lens caps. Bases were attached to both rifle and carbine receivers with solder and screws.

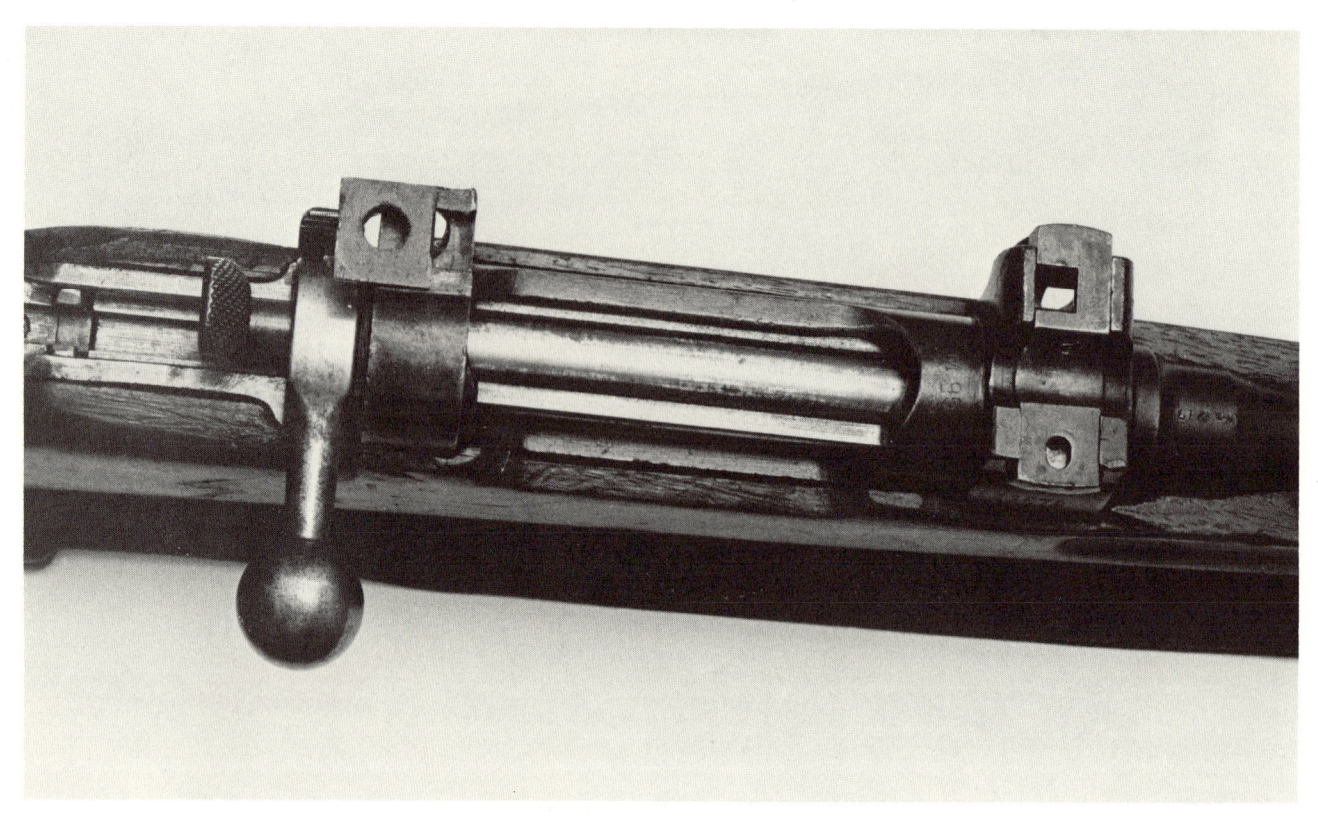

Top view of M98 carbine with telescope removed. The rear base top plate is loose and has slide back slightly, while the front base dovetail locking screw is missing.

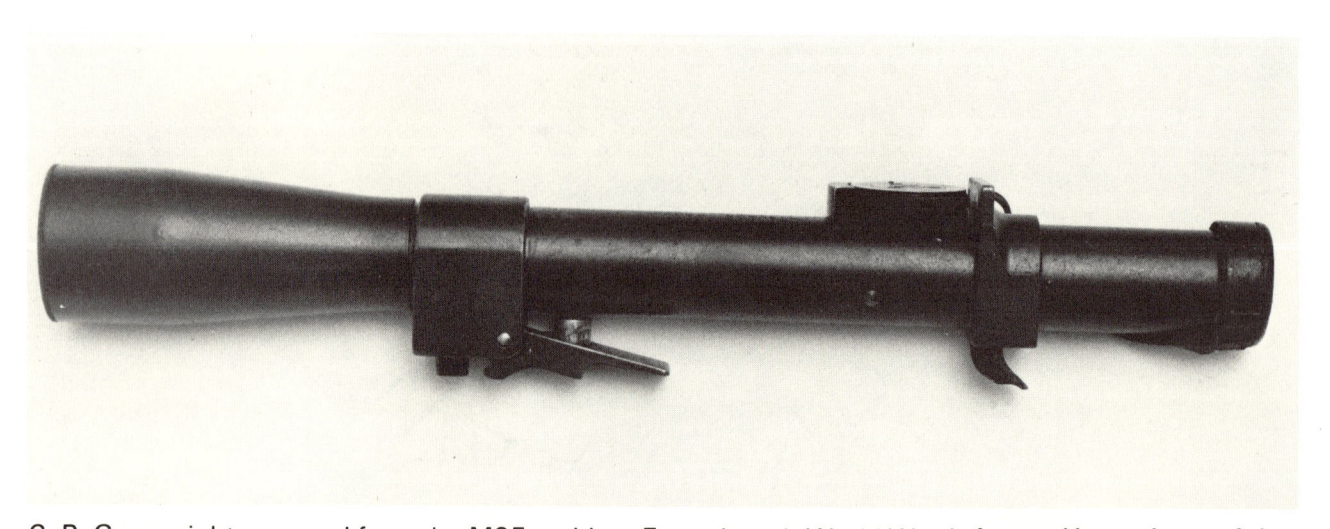

C. P. Goerz sight removed from the M95 carbine. Even though World War I Austro-Hungarian sniping arms were essentially the same, subtle differences in scope mountings have been noted on various pieces.

SCHARFSCHUTZEN — GEWEHR 98

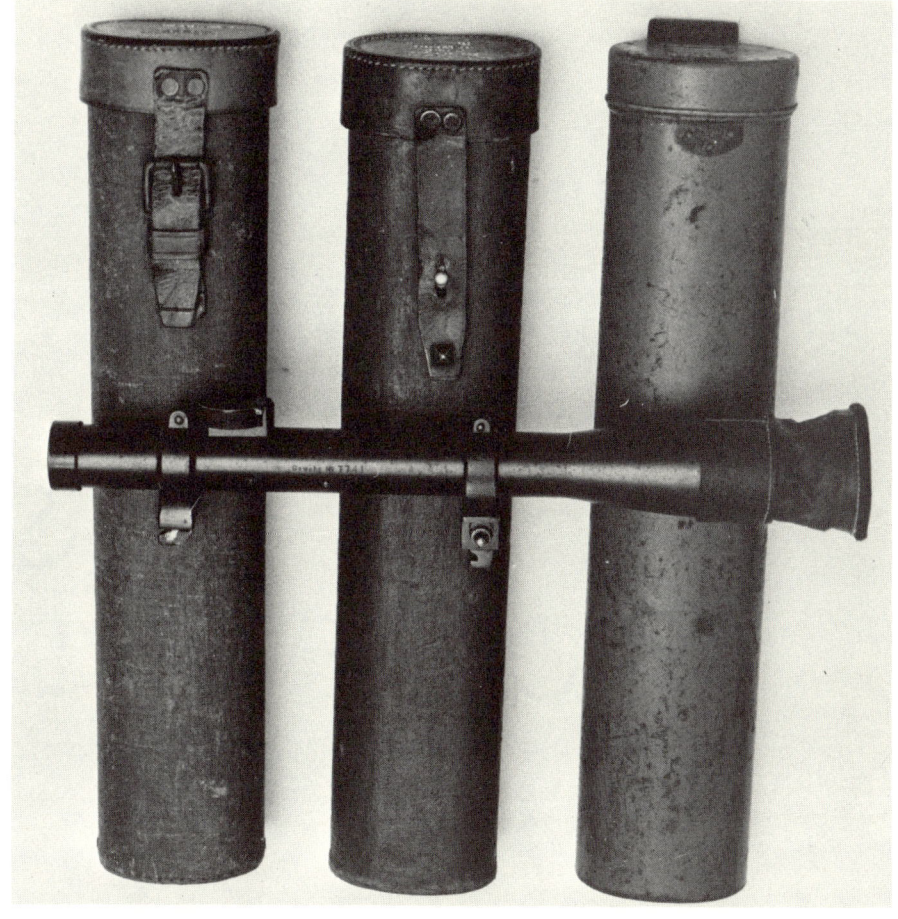

Typical World War I German scope cases issued for carrying the sight assembly when removed from the rifle. Most were fitted with small compartments under the lid or in the case for holding basic tools used for cleaning or adjusting the sight. The sight, made by Busch, was originally issued with the formed metal case on the right.

Comparative view of scope case lid markings. Buckles, studs, or friction latches were used to keep the case lids closed while leather belt loops were usually provided for carrying purposes.

C. P. Goerz scope case lid with markings typical of those found on German cases dating from World War I. Note the rifle number, Gewehr No. 2955.

A variant windage instrument tool (skate key) typical of those issued with various German sniper scopes.

Voigtlander & Sohn scope carrying case lid markings. The canvas covered cardboard case has a leather lid.

The underside of the Voigtlander case lid showing a small compartment for the scope tools.

Telescopic sight windage adjustment tool. (variation)

Dr. Walter Gerard scope carrying case showing the lid markings. The canvas covered metal case has a leather lid.

Underside of the Gerard scope case lid showing provisions for holding the windage adjustment tool and lens brush.

Protective leather eye shield typical of those issued with World War I German sniping scopes.

Protective leather eye shield in place on a Gerard scope.

An uncommon telescope carrying case without manufacturer's markings.

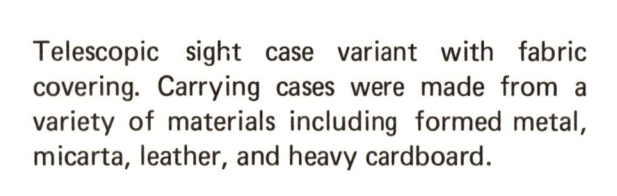

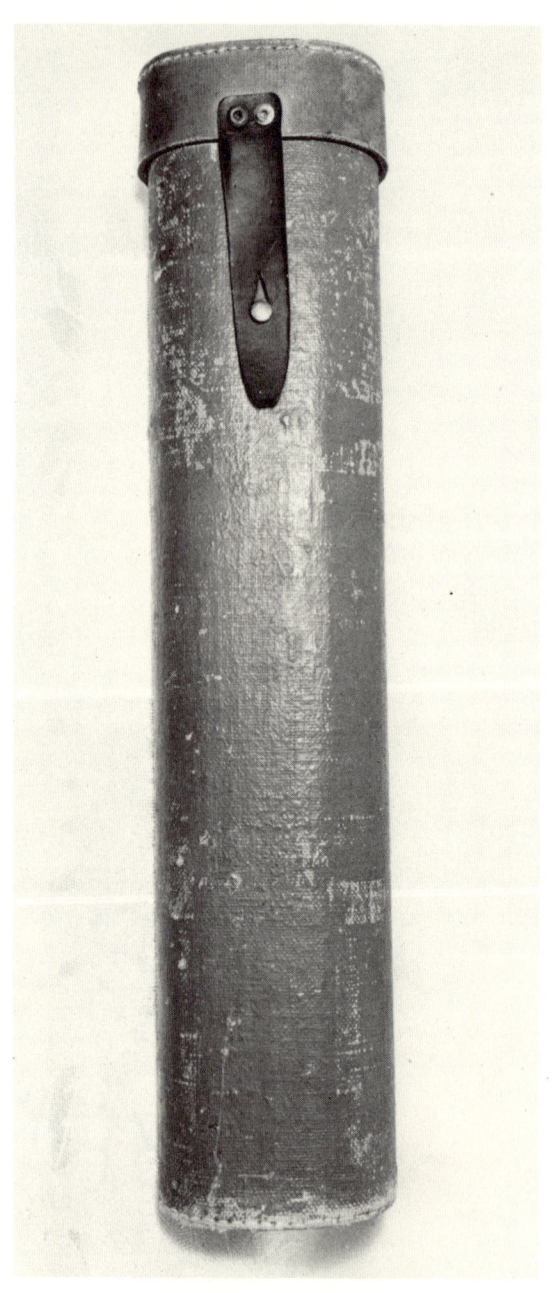

Telescopic sight case variant with fabric covering. Carrying cases were made from a variety of materials including formed metal, micarta, leather, and heavy cardboard.

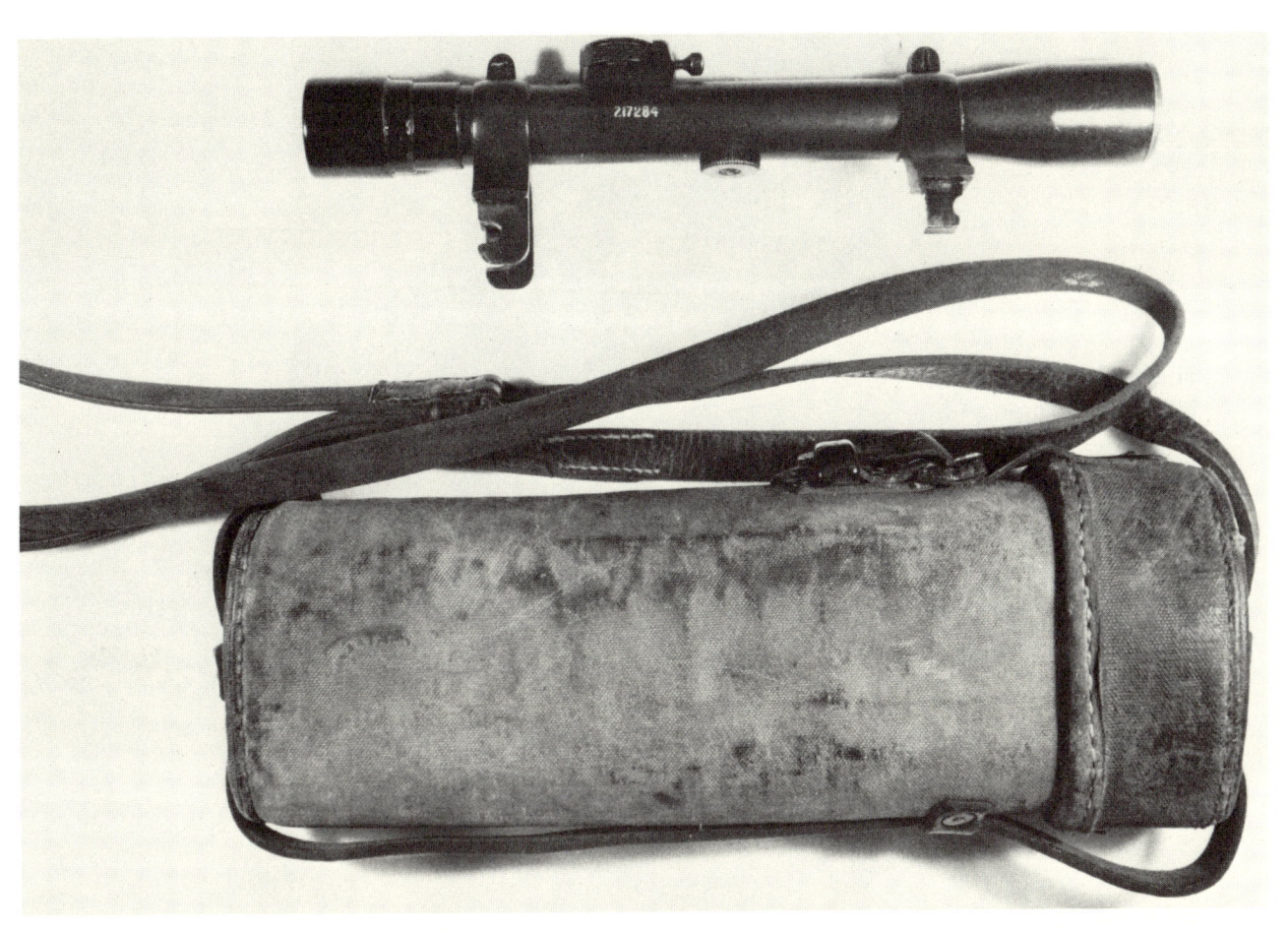

Model 1918 telescopic sight developed for British service use during World War I. Although a myriad of rifle scopes were also used by British snipers, this sight emerged as official issue late in the war (April, 1918). Note the claw-mount characteristics virtually identical to Imperial German sniping sights.

Authentic beyond all question, an original World War I German sharpshooting rifle in a severely pitted condition.

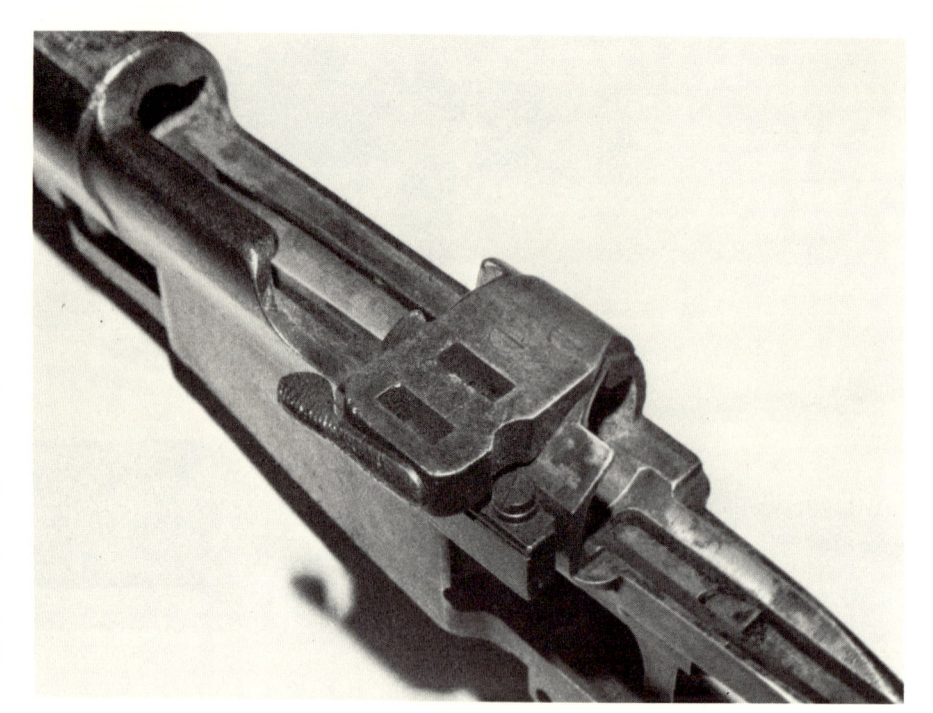

The remnants of an Imperial German sniping rifle (Spandau, 1916) reduced to relic condition by an axe-wielding German frau who smashed the weapon for firewood (stock) following World War II.

While not intended for sharpshooting purposes, but of interest: a World War I German Maxim machine gun (MG 08) with an optical sighting device cited as "an aid to accurate shooting out to 900 meters."

Issued in considerable numbers, the "low-power" MG 08 telescopic sights (approx. 2-power) provided an excellent field of view and ready target definition. Note the leather eye shield and objective lens cap.

Left view of MG 08 telescopic sight with textured green finish similar to that present on some World War I German binoculars and telescopic rifle sights.

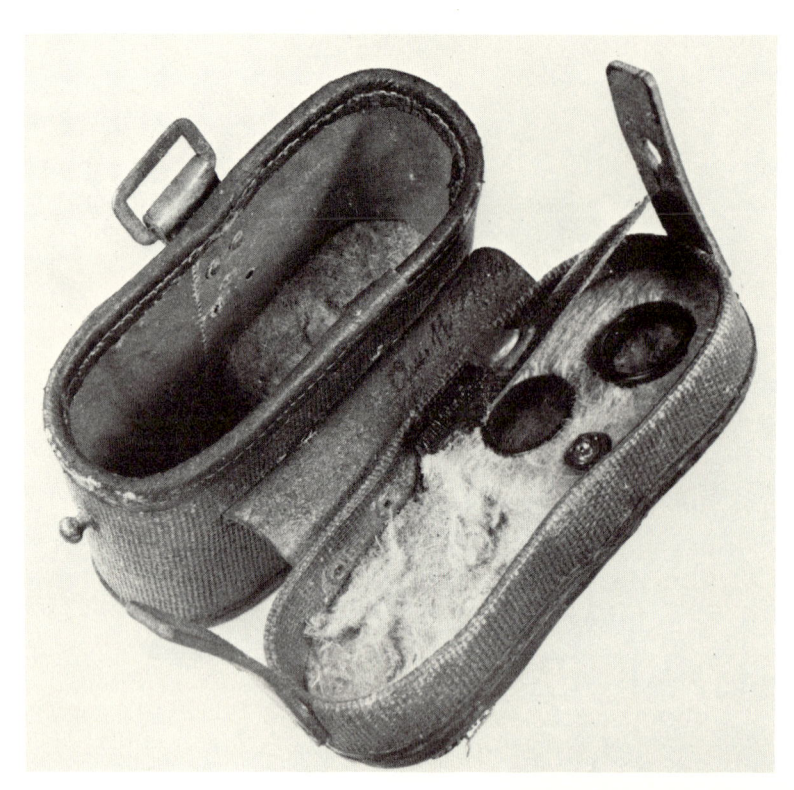

Carrying case for the MG 08 optical sight. Note the objective lens "filters" in place in the lid.

Early World War II photo showing a German marksman with a telescopic-sighted Karabiner 98b (K98b) typical of those originally fielded for use by the Reichsheer (German Army) following World War I. Note the early pattern scope carrying case slung from the belt.

CHAPTER III

Rifles for Sniping—Arming the Wehrmacht

The military forces of the Reich employed as their standard sniping arms the Karabiner 98k (K98k) series and the Gewehr-Karabiner 43 (G-K43) series rifles. These weapons respectively were the instruments of German sniping influence throughout the Second World War.

Of the two rifle series, the 98k was by far the choice of true sniping specialists, even after the introduction early in 1943 of the Selbstladegewehr 43 (self-loading Rifle 43), which was an evolution of German semiautomatic design that had commenced either just before or early in the war. There is nothing to indicate that either series weapons as issued to snipers for field service was ever manufactured solely for the purpose of sniping, that is, hand-built to exacting specifications using other than regular components. The known exceptions were variations of the 98k sharpshooting rifle that had an adjustable trigger.

The official military point of view toward sniping weapons of special configuration has always been one of particular dislike, due to the obvious difficulty in field supply and in repair of nonstandard parts. But the rifles chosen from weapons production and earmarked for sniping did receive careful preparation within the capabilities of unit armorers, after being tested for required accuracy. German Ordnance specifications stipulated that prior to acceptance for sniping purposes, a rifle must be tested and deemed capable of the accuracy requirements set forth by the Heereswaffenamt (Army Ordnance Office). This included sighting with the conventional open sight as well as the telescopic sight.

According to established practice:
Without consideration of the aiming error by the marksman the entire performance of the telescopic sight rifle rests upon the sum of the individual performance of weapon + telescopic sight + ammunition.

The rifle without telescopic sight:
Prerequisite for the selection of a rifle as a telescopic sight rifle is an unobjectionable hit pattern. The hit pattern is fired according to the effect-firing conditions.

The first effect-firing occurs at a range of 100 meters. Three shots are fired. A K98 or K43 rifle suffices in effect-firing if of the three fired shots the mean point of impact lies in a rectangle of 80 x 140 mm and all three shots are located in a circle of 120 mm diameter. (Cutting into the circle from outside is permitted). Toleration of wide shots is not allowed.

If this condition is not fulfilled, then the *second effect-firing* is repeated with 5 shots. The condition is fulfilled if with a K98k or K43 rifle all 3 of the 5 fired shots

Karabiner 98k (K98k) early Waffen-SS short side rail sniping system.

German marksman sighting a 98k short side rail sniping variant during winter action.

or the mean point of impact (according to Army Manual 242) lie in a rectangle of 80 x 140 mm and all 5 shots lie in a circle of 120 mm diameter. (Cutting into the rectangle or, respectively, into the circle from outside is permitted.)

The acceptance agency may not tolerate a weapon whose effect-firing pattern does not indeed completely fulfill the effect-firing conditions but which shows that it satisfies the above-named demands. In particular in the evaluation of hit patterns, a wide shot *incontestably* recognizable as such may be disregarded for the evaluation of the hit pattern if this otherwise satisfies all authorized demands as regards location of the point of impact and dispersion.

But even so, the acceptance agency is obliged to give a weapon for repair if by chance its effect-firing pattern satisfies the formal effect-firing conditions but does not let the location of the point of impact be clearly recognized or leads to the conclusion of a defect in the weapon.

Remark:

A relaxation of the effect-firing conditions as to the demands placed on accuracy has not ensued. For reasons of ammunition conservation the first effect-firing is carried out with 3 shots currently instead of with 5 shots as before.

From these rifles the best are selected for further use as telescopic sight rifles, that is only those which have an especially small dispersion and a good location of point of impact.

If the rifle has been selected as suitable for a telescopic sight rifle after passing the effect-firing conditions, then in conjunction therewith the trigger motion is once more tested and, if necessary, is reworked to a trigger pull of 1,5 to 2,5 kg.

The barrel position which has already been tested with each rifle is again specially tested with the telescopic sight rifle; it is necessary that the barrel does not stick in the stock and upper band and that it can swing freely upon discharge of the shot. This is decisive especially with the K98k.

Both tests are carried out responsibly by the Army acceptance points.

The rifle with telescopic sight:

Effect-firing ensues at a range of 100 meters.

Effect-firing is done with 5 shots.

Before examination whether a telescopic sight rifle satisfies the effect-firing conditions, the marksman corrects the setting of the telescopic sight with the help of single shots, whose position is determined through an additional telescope with strong magnification, continuing until point of aim and point of impact coincide at 100 meters range.

During the examination whether the rifle satisfies the following effect-firing condition, the location of the single shots may *not* be observed with the above-mentioned telescope with strong magnification.

A rifle K98 with telescopic sight or K43 with telescopic sight suffices in effect-firing if at least three shots of the 5 delivered shots lie in a circle of 70 mm diameter and all 5 shots lie in a circle of 105 mm diameter with a common center. The common center can travel within a circle of 120 mm diameter whose center lies in the middle of the lower ring. (Cutting into the circle from outside is permitted.)

The demands for hit exactness of a telescopic sight rifle must be suitably strict for its tasks. It is therefore permissible under no circumstances to bring a telescopic sight rifle to an accidental fulfillment of the effect-firing conditions through chance shooting. If in the meantime no correction has ensued, the effect-firing of a telescopic sight rifle with doubtful results may be repeated only once and in exceptional cases twice. If the effect-firing does not satisfy all conditions, the rifle may not be used as a telescopic sight rifle.

A stray shot *incontestably* recognizable as such may be disregarded for the evaluation of a hit pattern if the four other shots lie in a circle of 70 mm diameter (cutting in permitted) and the 5th shot is so far distant that it in no way can be claimed as normal dispersion.

It does not appear that any specific German arms manufacturer had an edge on the supply of sniping weapons other than in their

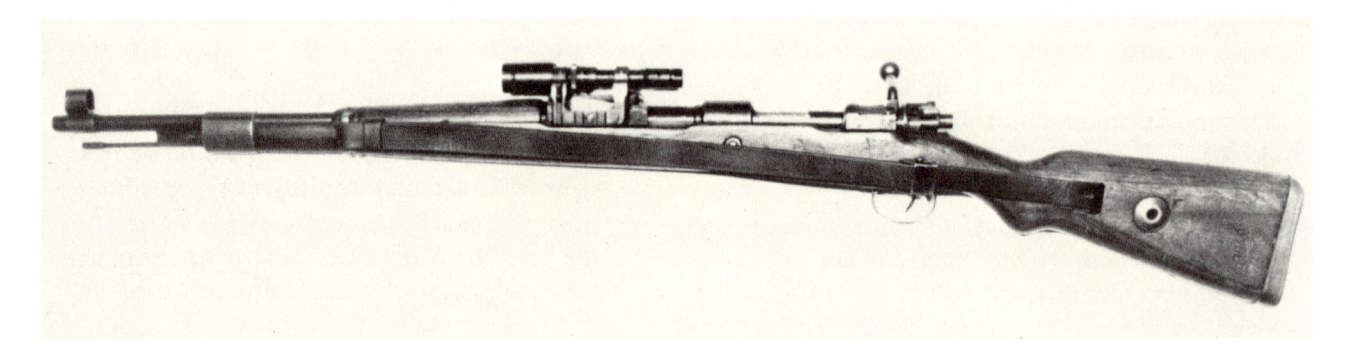

Karabiner 98k (K98k) with Zielfernrohr 41 telescopic sight.

The business end of a 98k-ZF 41 rifle. Issued to all German combat forces, this equipment saw extensive use during World War II.

capability to produce arms. Examples of the 98k in sniper configuration were based on weapons made by Steyr, Sauer, Gustloff et al, with many fabricated at the Mauser facilities.

The vast majority of 98k sniping arms fielded during this era were transformed into sharpshooting rifles from new production obtained directly from the various small arms manufacturers. A substantial but undetermined quantity of 98k rifles were also selected from existing stock at various army ordnance depots for conversion to sniping arms on an as needed basis with no regard for their origin or date of manufacture.

In addition to these practices, in many cases immediate lower echelon needs for sniping equipment were filled at the field level by unit armorers who, employing their own resources, appropriated whatever commercial telescopic sights and mountings were available in their area of operation. Despite the numerous military scope and weapon combinations developed by German Ordnance, it is of interest to note that 98k rifles were fitted with commercial components and used as sniper weapons throughout the war.

The question of whether the as manufactured 98k sniping arms attributed to the various manufacturers were assembled as such by the respective facilities and subsequently accepted by the military or made into sharpshooting arms as a separate military function has not been satisfactorily answered.

While hardly conclusive evidence, according to Steyr-Daimler-Puch, Werke Steyr (bnz) concerning their World War II involvement with the 98k claw-mount variant: "Especially good shooting weapons were selected for sharpshooting purposes and provided with a telescopic sight through the military service. Therefore the weapons were not declared as sniping rifles from the factory." In some cases, however, telescopic-sighted variations such as the 98k with the 1.5-power ZF41 type sights were selected and transformed into sharpshooting weapons at the Mauser (ar and byf) and Berlin-Lubecker Maschinenfabriken (duv) factories under military contract and close supervision. But on an overall basis, specific details relating the extent of this practice have remained obscure.

In addition to an endless variety of 4-power and 6-power commercial scopes originally intended for hunting purposes, the Zielfernrohrkarabiner 98k mounted the 1.5-power ZF 41 and ZF 41/1 sights, the 4-power commercial types produced under contract for military use, and to a lesser extent during the last days of the war, the Gw ZF 4-fach and ZFK 43/1.

There is little to be said about the shooting capability of a select 98k rifle, an accurate, durable, and trouble-free infantry arm even in special sniper trim. The venerable 98k served as both the standard of comparison and the principal German sniping weapon from 1939 until the Wehrmacht capitulated in 1945.

Despite planned obsolescence, high level efforts to reduce production and eventually replace the 98k with the Selbstladegewehr 43 during the latter months of the war, the K98k continued to be manufactured in quantity and converted for sharpshooting purposes even as the war ended.

It is of further interest to note frequent German Ordnance reference to the goal of bringing the performance standard of the G43 (Selbstladegewehr 43) and later the K43 (Karabiner 43) sniping system to the level of the telescopic-sighted 98k. As events transpired, however, this did not come to pass because these semiautomatic sniping systems remained plagued with operational difficulties throughout the war.

At one point during the First World War, the Germans faced an acute shortage of one-piece wood blanks suitable for rifle stocks. As a remedial measure, laminated stocks and handguards were eventually developed and used with most German small arms fielded during World War II. Bonded under pressure with phenolic resin, the layers of approximately 1.5mm thick beechwood produced an exceptionally strong stock, ideal for military use. This method of stock manufacture was advantageous to a sniping weapon since it was less susceptible to variations caused by moisture. Further experimentation utilizing plastic and plastic impregnated fabric stocks similar to what has become known as fiber glass was also conducted by German Ordnance.

Handguards fashioned from a tough plastic-like material, as noted on various Gewehr 41 rifles and G-K43 series semiautomatic rifles which bear the trademark "Durofol," are attri-

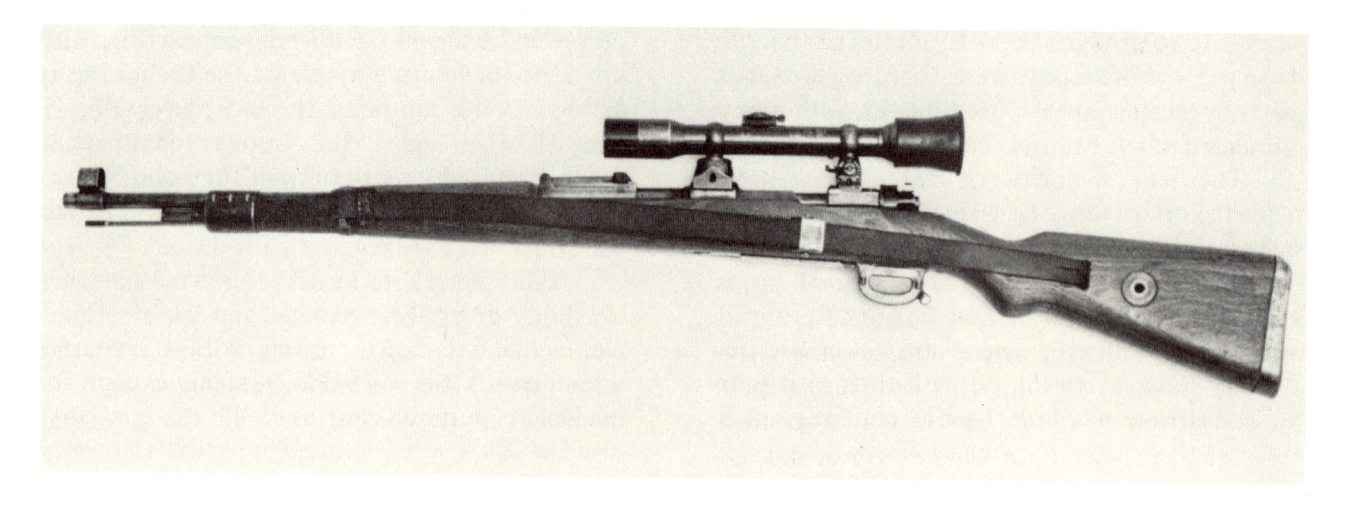

Karabiner 98k (K98k) turret mount sniping system.

Karabiner 98k turret mount sharpshooting rifle mounting a 4-power telescopic sight typical of those procured by the German military for sniper use.

buted to the German firm Durofol K.G., O. Brangs and Company, a manufacturer of molded and pressure-formed plastics. No complete stocks of this material have been reported on either the 98k or G-K43 series sharpshooting weapons, however. Although stocks of this type have survived the war, it is doubtful that manufacture ever progressed beyond the experimental stage.

Many of the World War I long-barreled Scharfschutzen-Gewehr 98 with their characteristic turned-down bolt handle and original 3-power and 4-power hunting scopes, were pressed into service once again in World War II. The postwar Zielfernrohrkarabiner 98b mounting Zeiss and various other commercial scopes, as utilized between the wars by the Reichsheer, also saw combat use by army and Waffen-SS marksmen during the early stages of the war when any weapon with telescopic sights was regarded as a veritable treasure.

In January 1943 design efforts intended to improve on the ailing semiautomatic Gewehr 41 system culminated with the introduction of the Selbstladegewehr 43, which incorporated features from both the G41 and the Soviet Tokarev rifles. Brought to fruition by the Walther firm in Zella-Mehlis, when accepted for general field use, the G43 was officially touted as the optimum infantry weapon.

German Ordnance had a great deal of difficulty affixing a succession of scope mounts to the Gewehr 41 and the K98k. They hit upon a particularly innovative solution by including a telescope mounting base as an integral segment of the G43 receiver. As time progressed, this feature and a fully adjustable trigger mechanism proved the only salient features for sharpshooting purposes.

When the German military decided in 1942 that the myriad of optical rifle sights then in use by their forces required standardization, the Army High Command (OKH) assigned the Army Ordnance Office (Heereswaffenamt) and subordinate bureaus the task of developing a common telescopic sight for use on all service rifles. Directly responsible for weapon design, procurement, and acceptance, the Ordnance Office determined that the new sight, if possible, should be fabricated from "sheet metal" to facilitate its mass production, with calibration the only difference between models as applied to various rifles. With major assistance from private industry, design efforts culminated in late 1943 with the Rifle Telescopic Sight 4-power (Gw ZF 4-fach) reaching production status.

Even though provision for telescopic-sight use was a part of the original Gewehr 43 (G43) design concept, and the ineffectual 1.5-power sights had seen early testing with transitional examples of the G41-G43 system, except for evaluative purposes and limited field trial with the Gw ZF 4 sight, it is believed that extremely few telescopic-sighted Gewehr 43 weapons were produced between February and December 1943.

In mid-1943 Hitler again requested that a program be initiated to develop a telescopic sight for the Selbstladegewehr 43. Although the exact date of origin remains vague, the ZF 4 design had been established long before the Fuhrer's demand added impetus to the overall program. This point is substantiated by ordnance reference prior to Hitler's directive and by confirmed existence of ZF 4 equipped Gewehr 43 rifles concurrent with the directive.

By way of partial explanation, in many cases ordnance developments were kept under wraps until they had proved successful, or they were quietly dropped rather than chance the wrath of Hitler or the whim of the High Command.

In any case, developmental work and initial production have been credited to the Voigtlander firm (ddx), who also bore the brunt of the subsequent difficulties encountered with the ZF 4 sighting system. In deference to Voigtlander, problems with the ZF 4 system were attributed to incessant high level demands for increased production rather than to flaws in the basic design.

Various contemporary sources cite the manufacture of 12,000 Gw ZF 4 sights by the close of 1943. But according to Jn 2 (German Infantry Branch, Inspectorate 2) documentation dated May 1943, production of the Rifle Telescopic-Sight 4-power was to ensue in October 1943 beginning with 500 units, followed by a quantity of 1,000 in November and December, with an increase of 1,000 sights per month until a satisfactory manufacturing process could be

Karabiner 98k (K98K) long side rail sniping system.

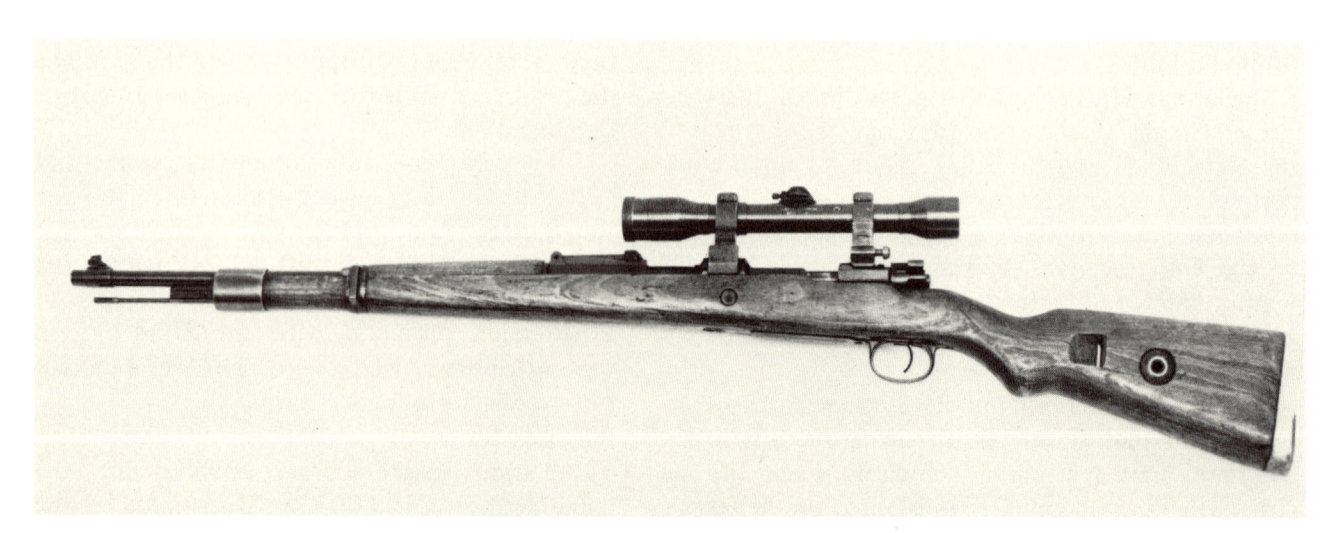

Karabiner 98k (K98k) claw-mount sniping system.

assured. It was also stated that: "A further increase of production cannot yet be introduced, since first of all only 50,000 units have been ordered and the total requirement was given by AHA/Jn 2 as 150,000 units." Supporting reference further mentions that only 9,000 Gw ZF 4 sights had been delivered by March 1944.

The Fuhrer Adolf Hitler had at one time envisioned the wholesale manufacture and distribution of telescopic-sighted versions of the Gewehr 43; but both the rifles and the ZF 4 sights were plagued with production problems that limited their actual field use in a sharpshooting capacity well into 1944.

To those within the German Ordnance system possessing the knowledge to recognize the requirements necessary for fielding an efficient sniping system, the most acceptable feature of the Selbstladegewehr 43 rested with the manufacturer's capability of having sharpshooting rifles ready for field use in considerably less time than was required for the preparation of an equal number of 98k sniping rifles.

Nevertheless, a verbal struggle between those duty bound to maintain the status quo, retention of the K98k as the principal Wehrmacht sniping arm, and the leading proponents of a semiautomatic sniping system continued through the end of the war.

While the reliability and capability of a select telescopic-sighted 98k was a formidable point of contention, opponents of the 98k were quick to point out that if acceptable accuracy could be obtained, the self-loading rifle offered the following advantages:

A. In case of target movement or errors on the part of the sniper, a second shot could be fired immediately.
B. Engaging multiple targets.
C. There was no discernible movement with a self-loader. Consequently, the possibility of detection was reduced.
D. Semiautomatic fire capability enabled the sniper to defend himself or engage the enemy as circumstances warranted.

From January 1944 to April 1945, two areas created problems that affected both the ZF 4 telescopic sight and the G-K43 series weapon system, the manufactured quality of the scope and the accuracy performance of the rifle. The combination of these two newly developed components proved an almost insurmountable task for the German Ordnance system.

Official ordnance tests conducted during February 1944 at the Walther facility with ten G43 rifles selected at random from regular production and mounting ZF 4 sights revealed "excessive shot dispersion" at only 100 meters. In addition to less than optimum accuracy with the G43 rifles employing both standard and telescopic sights, every one of the Voigtlander ZF 4 sights was found deficient for one or more of the following reasons:

target prong-disproportionate to cross-beams.
cross-beams not level.
target prong inclined to right or left.
blurred glass.

Particularly distressing was the knowledge that the test items, the rifles and the sights, were considered regular production.

Under the auspices of the Infantry Branch (Jn 2), evaluations with the Rifle 43 and Rifle Telescopic-Sight 4-power were made at Zella-Mehlis 25 March 1944. As directed in this case, highly trained SS marksmen were employed for shooting exercises with the test rifles.

As with the prior tests, select ammunition was used, but the same negative results were obtained with unacceptable shot dispersions of up to 15.5cm at 100 meters and 41cm at 300 meters.

For comparative purposes, rifle number 023 (Walther manufacture) fitted with an experimental barrel (740mm length) was fired under the same conditions at 100 and at 300 meters. The results were somewhat better at 100 meters (8cm vertical, 6.5cm lateral) but disappointing at 300 meters (28cm and 29.5cm).

There was no circumventing the conclusion that the Selbstladegewehr 43 sniping system in a production state was indeed far from combat ready. From that point forward, "It must be demanded that a head target be hit at 300 meters, that is, that the dispersion may not exceed 7 cm for 100 m."

In accordance with this demand:
Pruf 2 and Pruf 8 are requested to establish values for weapon and telescopic sight production through corresponding selection

of acceptable parts through an improvement of manufacturing and administration in order to clearly achieve the goal of 7 cm dispersion at 100 m. . . .

From this point forward, only the best telescopic sights will be chosen and assembled onto properly firing G43 rifles.

Events leading to difficulties with the semiautomatic sniping system can be traced to January 1944 when the Reich Ministry for Equipment and War Production, without any regard for an unproved system, demanded a substantial increase in the output of ZF 4 sights for the new infantry rifle (25,000 units monthly). The Main Committee for Precision Instruments and Optics moved to establish the new requirements as directed:

> The Main Committee for Precision Instruments and Optics has immediately planned for this new demand. The circle of delivery firms had to be expanded correspondingly and all preparations have been made for a prompt start of production.

In view of the disappointing preliminary results with the Selbstladegewehr 43 sniping system, Jn 2 responded adamantly.

> An intended monthly output of 4-power rifle scopes is again planned which according to the Infantry Branch can be achieved only at the cost of the quality of execution. The Infantry Branch has already taken a position against this and stresses once again that the application of snipers is to be vindicated only with a faultless telescopic sight and correspondingly good weapon. The telescopic sights now already coming from mass production do not meet this requirement. The quality will sink further if several firms are joined in a production increase. For the initial equipment a quantity of 150,000 4-power rifle scopes is considered to be sufficient. The imposition of further series before trials of the 4-power rifle scope at the front must be declined.

To emphasize the demand for increased production, the AHA/Jn 2 (General Army Office, Infantry Branch, Inspectorate 2) was advised:

> The planning of 25,000 4-power rifle scopes in a month rests upon a direct order of the Fuhrer. According to a decision of the Chief of Armed Forces Command/Army Staff (II) the granting of contracts according to this Fuhrer demand has to ensue at once. In order to be able to bestow the contracts to the production firms, a quick delivery of an order of instruction for *300,000 4-power rifles scopes* is requested immediately.

The battle lines regarding the suitability of the Selbstladegewehr 43 sniping system had been drawn, and as typified by the preceding excerpts taken from official correspondence in the early months of 1944, this controversy continued until the end of hostilities in the spring of 1945.

There were some high level factions more interested in supporting pragmatic efforts to field satisfactory sniping arms than in placating Adolf Hitler. This is evidenced by the following statement from the Chief of Army Equipment and Commander of the Replacement Army (Ch H Rust u. BdE) dated 5 May 1944.

> The Infantry Branch is called upon to grant an order of instruction for 300,000 4-power rifle scopes. That means that with the already ordered 150,000 4-power rifle scopes, 450,000 scopes of a completely new pattern, about which no front line experiences have been submitted, are to be placed into production. A burdening of the already strained optical industry to this extent with the prerequisites for careful production will not be defined according to this decision.
>
> A decision about the total quantity of 4-power rifle scopes to be produced is requested with reference to the following points:
> 1. Up to now no troop experience reports are on hand about the 4-power rifle scope. The first 500 usable scopes went on 1 May to the sniper training companies and will be removed to the front with trained snipers at the beginning of June.
> 2. Trial results up to now show that only scopes carefully produced and accepted

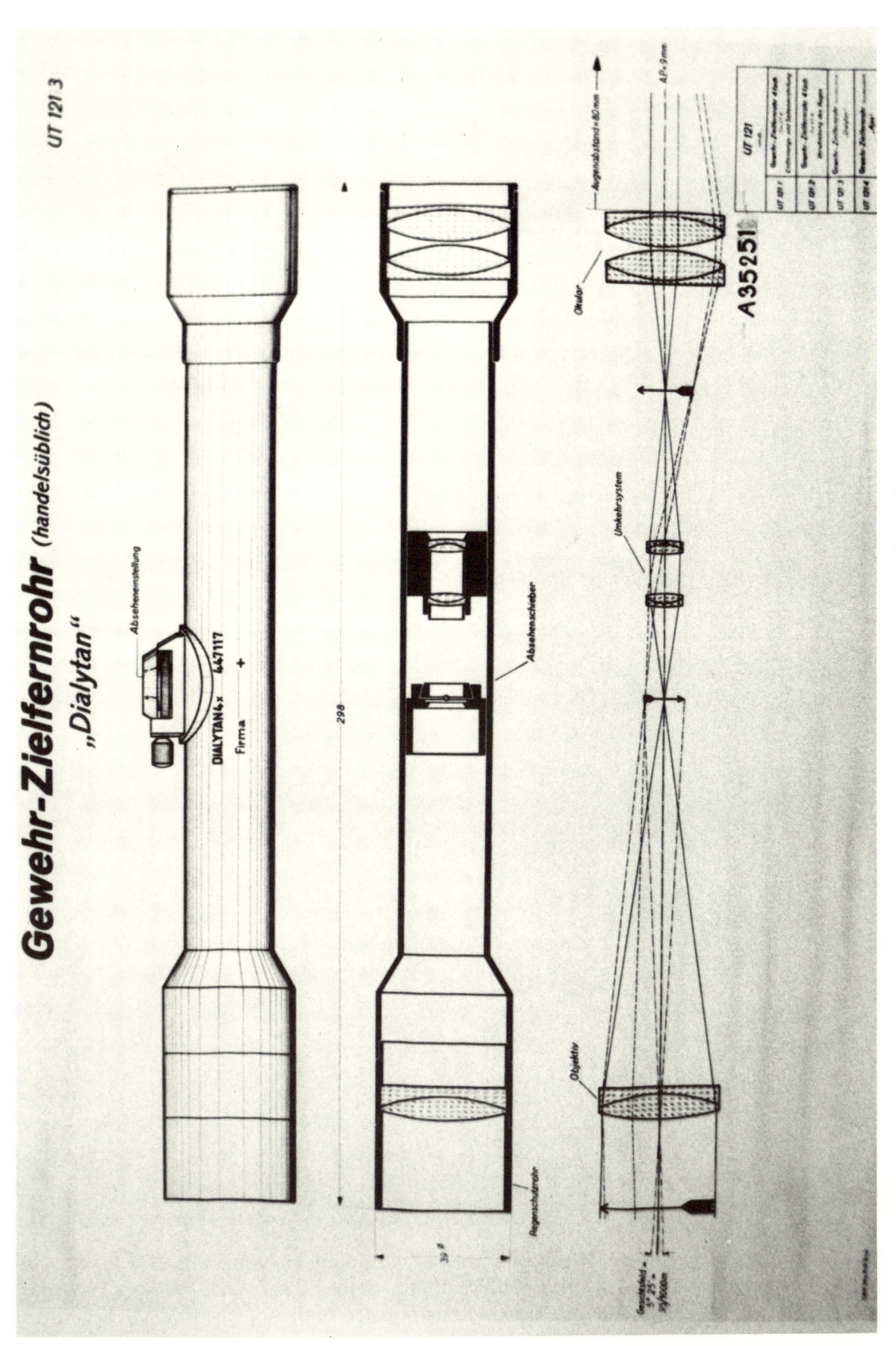

German Ordnance identification sheet showing a 4-power rifle scope typical of those procured under military contract from various commercial firms for fitting to 98k sniping rifles. Most sights of this type were manufactured under direct license from the Hensoldt firm. Although dimensions and characteristics varied through the course of production, the scopes were approximately 298mm in length with a 39mm objective, ocular and 26mm center tube diameter.

Developed for both regular and sharpshooter use, a Gewehr 43 (G43) with Gw ZF 4-fach telescopic sight is shown in sniper trim.

Semiautomatic Gewehr 43 with Gw ZF 4 telescope assembly during field evaluations held in October 1943. Brought to fruition by the Voigtlander firm, difficulties with the "sheet metal" design ZF 4 sights prompted ordnance recommendations that production be terminated until Adolph Hitler advised otherwise.

under the strictest examination fulfill to some extent the requirements to be placed on a good weapon. The scoped rifle must enable the single-shot engagement of a head target at least at a range of 300 meters with certainty.

If 25,000 scoped rifles shall be produced monthly the required quality will never be reached.

The Infantry Branch has the firm belief that the success with a smaller but more *carefully produced* quantity of 10,000 4-power rifle scopes monthly will be far greater than with the intended higher planning.

The marked increase in German sniper activity found the Karabiner 43 (K43), designated as such in April 1944, hurriedly involved as a combat sniping weapon on a limited scale. In standard form the semiautomatic rifle proved infinitely less troublesome than its telescopic-sighted counterpart; however, effective field application for sharpshooting purposes necessitated careful selection of both the rifle and the sight.

From the beginning of the war the German sniper had access to the high-grade 7.92mm ammunition vital to his mission. This was issued in three basic types: standard ball, armor-piercing, and tracer. Chance lots of quality ammunition were coveted since uniform ammunition to ensure consistent results was of paramount importance to the sniper.

As it was stated in July 1944:

The weapon performance is largely dependent upon the quality of the ammunition. Besides the unique weapon performance of the K43 rifle with telescopic-sight 4-power and the K98k with telescopic-sight for commercial use (Note: WWII German Ordnance reference to conventional 4-power hunting-type telescopic-sights both commercial and military contract), the normal SmE ammunition allows no precision shot over 300-400 meters as must be demanded of sharpshooters at these and further ranges. Until now it was possible to equip the sharpshooter with select sS ammunition which at the present time is only available in small bulk or not at all. The General of Infantry therefore demands the production of suitable sharpshooter ammunition at a volume of 20 million cartridges monthly.

By 1945 ammunition intended for sniper use was made available.

The ammunition for sniper use is the *Rifle Effect-Firing Ammunition sS*. This ammunition is *specially produced*, which must be expressly emphasized, since it is not possible through mere selection of ammunition from continuous production to find an ammunition which satisfies the demands to be made of it. The rifle effect-firing ammunition sS is accepted with regard to dispersion and impact point location (the normally produced rifle ammunition only with regard to dispersion)."

Lesser known but highly significant was the limited use of explosive observation ammunition against opposing personnel, B-Patronen.

On the strength of increasing employment of explosive projectiles by the Russian infantry the Fuhrer has released the use of B-cartridges by snipers for the region of the eastern front. The Reichsfuhrer SS has ordered expedited use. . . . The taking of B-cartridges to other fronts is forbidden. The snipers are to be instructed that they have to give up the B-cartridges upon employment to other fronts. 25 February 1945

Although the 98k possessed the ability to function reasonably with ammunition whose quality began to decline late in the war, this was not the case with the K43 where marginal ammunition aggravated the susceptibility to jam. For sniper use this problem was avoided by hand loading each round.

It is quite understandable that the true German sniper was suspicious of any weapon having more mechanical motion than he deemed necessary to fulfill his assigned mission. Consequently, among those German specialists having actual combat sniping experience, given the option, the 98k remained their first choice.

Even though gradual improvements to both the ZF 4 telescopic sight and the K43 weapon system alleviated some of the problems, the

Experimental plastic stock and handguard developed for the G-K43 system. Intended to supplement and eventually replace their wooden counterparts, stocks of this type did not reach production status. The rifle, a K43 variant (ac 45) has the scope mounting base (rail) machined from the receiver. Except for a limited number of special "high comb" stocks issued for sharpshooter use, woodwork for the semi-automatic rifles remained the same for both regular and sniping applications.

Original Heereswaffenamt photo (17 November 1942) is one of the earliest known illustrations of the Selbstladegewehr 43 system. Shown in transitional configuration with a Gewehr 41 receiver, an experimental "bell-jar" flash suppressor is attached to the threaded muzzle. Note bolt handle on the right side of the assembly, selector-lever above the trigger guard at the stock line (full and semiautomatic fire), and 25-round light machine gun magazine (MG 13-Dreyse) with 10-round modified LMG magazine separate. The original G41 integral 1.5-power scope mounting rails are barely visible.

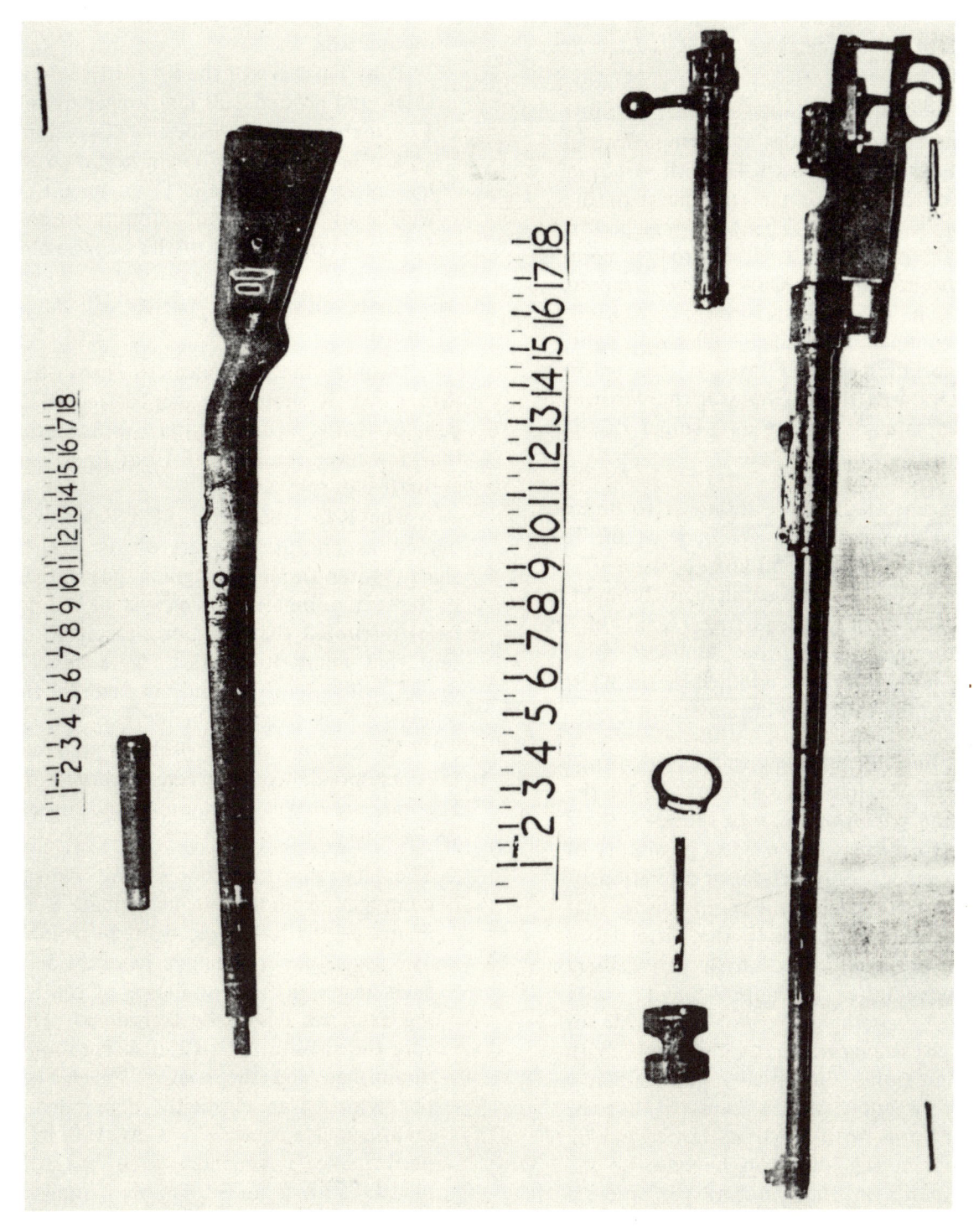

A photo from U.S. Army ETO Technical Intelligence Report III, dated 17 January 1945 depicts a "plastic composition" 98k rifle stock. According to the report the dimensions of this stock were identical to the regular wooden version. As it was then stated, "This weapon had been subjected to considerable abuse and moisture with all exposed metal encrusted with rust and deep pits. All surfaces covered by the stock and hand guard were perfectly preserved. Although the stock had received much abuse, there were few dents and scratches, such as a wooden stock would show under similar treatment."

semiautomatic sniping system was still plagued by operational deficiencies during the last days of 1944.

> Further simplification in production must be striven for. Besides this simplification the question of the K43 with 4-power telescopic sight for the equipment of the sniper is burning. The shooting performance up to now is designated as completely insufficient. Only a few weapons approach the performance of the K98k with commercially available telescopic sight. The speediest improvement of the entire realization of the K43 and of the 4-power telescopic sight for use as a sniper rifle is urgently demanded.

Concentrated efforts intended to improve the ZF 4 design and the accuracy of the K43 saw the development and introduction of the 4-power ZFK 43/1 telescopic sight brought to fruition by the Carl Zeiss firm (Jena). In line with attempts to increase accuracy, an exhaustive series of tests conducted by the Walther firm prompted the following statement.

> According to everything that has happened from the breech side back to improve the shooting performance with the K43, there exists from the weapon side a possibility of improvement only yet in an alteration of the barrels. Under the prerequisite that from the barrel blank to the barrel fabrication is accomplished satisfactorily, there are possibilities of improvement only through *lengthening of the barrels* or through *strengthening* of the same. With both measures, individually or together, a higher weapon weight must always be allowed for. With the trials carried out up to now in this direction noteworthy improvements in the shooting performance have not resulted with barrel lengths of 620-680 mm (opposite 550 mm with the normal K43) which were determined as the most suitable barrel lengths according to the absorption of vibration, as well as with somewhat strengthened barrels.
>
> The trials with thicker barrels yielded no improvement of shooting performance at short range (100 meters), yet a substantial increase is to be expected from 200 meters on upwards.
>
> Thus a pattern of the K43 with a 32 mm thick and around 100 mm longer barrel at 175 meters range yielded a dispersion of 10 cm in width and 7 cm in height. At 300 meters 15 cm width and 17 cm height. The weight of the weapon amounts to about 5,3 kg (overall length of the weapon about 1220 cm).
>
> Trials with thicker barrels will be continued.

Despite the presence of heavy-barreled K43 rifles, it should be emphasized that weapons of this type did not reach issue status, and that few ever reached the front lines even for evaluative purposes.

The K43 rifles with ZF 4 sights were included as an integral part of the Wehrmacht sharpshooter training program, but deficiencies in the semiautomatic sniping system had in some cases disrupted that training process to the extent that many K43 rifles were either withdrawn from use or set aside in favor of the 98k whenever they were obtainable.

An example of the problem was recorded in correspondence between Armored Troops Section, Jn 6/WuG (Training) and Jn 2 dated 28 February 1945.

> The sharpshooter schools have increased complaining about the inadequate performance of the telescopic sighted carbine 43. The level of performance has subsided to such an extent that an easing of the training exercises has to be considered. Therefore the significance of the sharpshooter is diminished and the goal of "each shot a hit" which was explicitly demanded; remains unattainable. . . . At 100 m the performance is generally good, but at 200 m it falls off in an intolerable manner. . . . One of the main error sources is seen in the mechanical adjusting of the telescopic sight. A change in the range and lateral adjustment does not shift in the expected amounts with the majority of the telescopic sights. Besides this, the telescopic sight is sensitive to jolting after being adjusted and thus highly intolerable. . . . The exercises carried out by the Sharpshooter

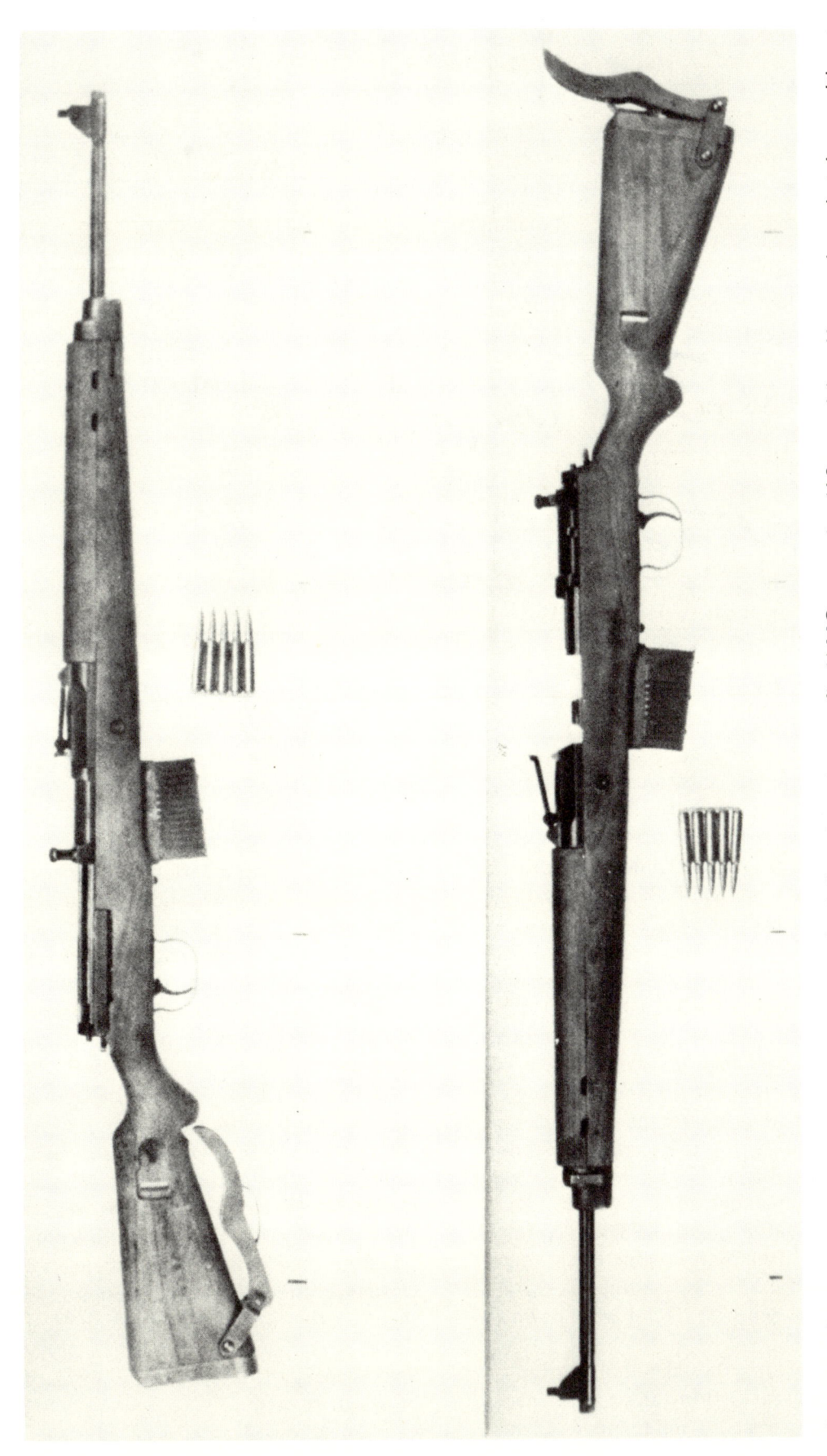

Gewehr 43 in standard trim with selector-lever for full and semi-automatic fire. Note the "recoil absorbing" butt attachment, modified LMG magazine (10 rounds) and integral receiver base without the center recess.

800 meter test firing:
12 shots

No.	Weapon	Serial number	Telescopic sight	Prior firing performance	Dispersion Height-Width in	cm	Remarks
1	K43	1658	Voigtlander	good	121	60	12 shots
2	K43	1322	Voigtlander	average	82	60	12 shots
3	K98b	3385	Kahles, 4X	good	66	66	12 shots
4	K98k	37325	Ajack 4X90	good	94	79	12 shots
5	K98k	9232	Dialytan 4X81	average	198	70	12 shots
6	Telescopic sight rifle	51937	Ajack 4X90	good	83	96	12 shots

1000-1200 meter test firing:
12 shots

No.	Weapon	Serial number	Telescopic sight	Height-Width 1000 meters		Height-Width 1200 meters	
1	K43	1658	Voigtlander	90	109	187	134
2	K43	1322	Voigtlander	124	111	174	119
3	K98b	3385	Kahles, 4X	65	103	—	—
4	K98k	37325	Ajack 4X90	85	131	89	120
5	K98k	9232	Dialytan 4X81	167	85	270	68
6	Telescopic sight rifle	51937	Ajack 4X90	—	—	—	—

School of the Panzertruppen can not be accomplished with the telescopic sight carbine 43.

Whereas remedial measures were implemented in an effort to improve the performance of both the ZF 4 sights and the K43 rifles, incessant demands for increased production during the final days of the Third Reich neutralized whatever progress had been made. When flaws in the manufacturing process had been located and duly rectified, hurried production created problems that were previously unknown.

Of all the surviving World War II German Ordnance documents relevant to sniping weaponry noted thus far, one of the most noteworthy reports was made to the Commander-in-Chief, AHA/Jn 2 by the Infantry School under date of 21 March 1945. The school was in charge of comparative testing of the "K43 telescopic sight carbine and K98k with telescopic sight. The firings were conducted at ranges of 800, 1000 and 1200 meters using select ammunition on days when the wind was not considered a factor.

Significant from the standpoint of the exceptionally long-range firing along with specific mention of the weapons and telescopic sights employed, the results are presented in part as originally recorded.

By the end of 1944 it was concluded that only "specially constructed rifles firing special ammunition" could satisfy the demand for consistent long-range hits out to 1000 meters. Judging from the mention of such equipment in pertinent German Ordnance documents dating from early 1945, the development of a special "hand-built" semiautomatic rifle appears to have been a definite consideration. With conditions as they were at that point, however, it is believed that the extent of these endeavors was limited at best.

The least known aspect of German sniping was the use of noise suppressors, or as they are more generally known, silencers. Several variations of these special devices were reportedly used with 98k sniping rifles, particularly during the latter stages of the war. At one point in their scramble for operational silencers, German Ordnance directly copied a design in use against them by the Soviet army on their Mosin-Nagant rifles. As expected, an effective device for the 98k appears to have been the primary purpose of design efforts.

Once the advantages of rifle silencers had

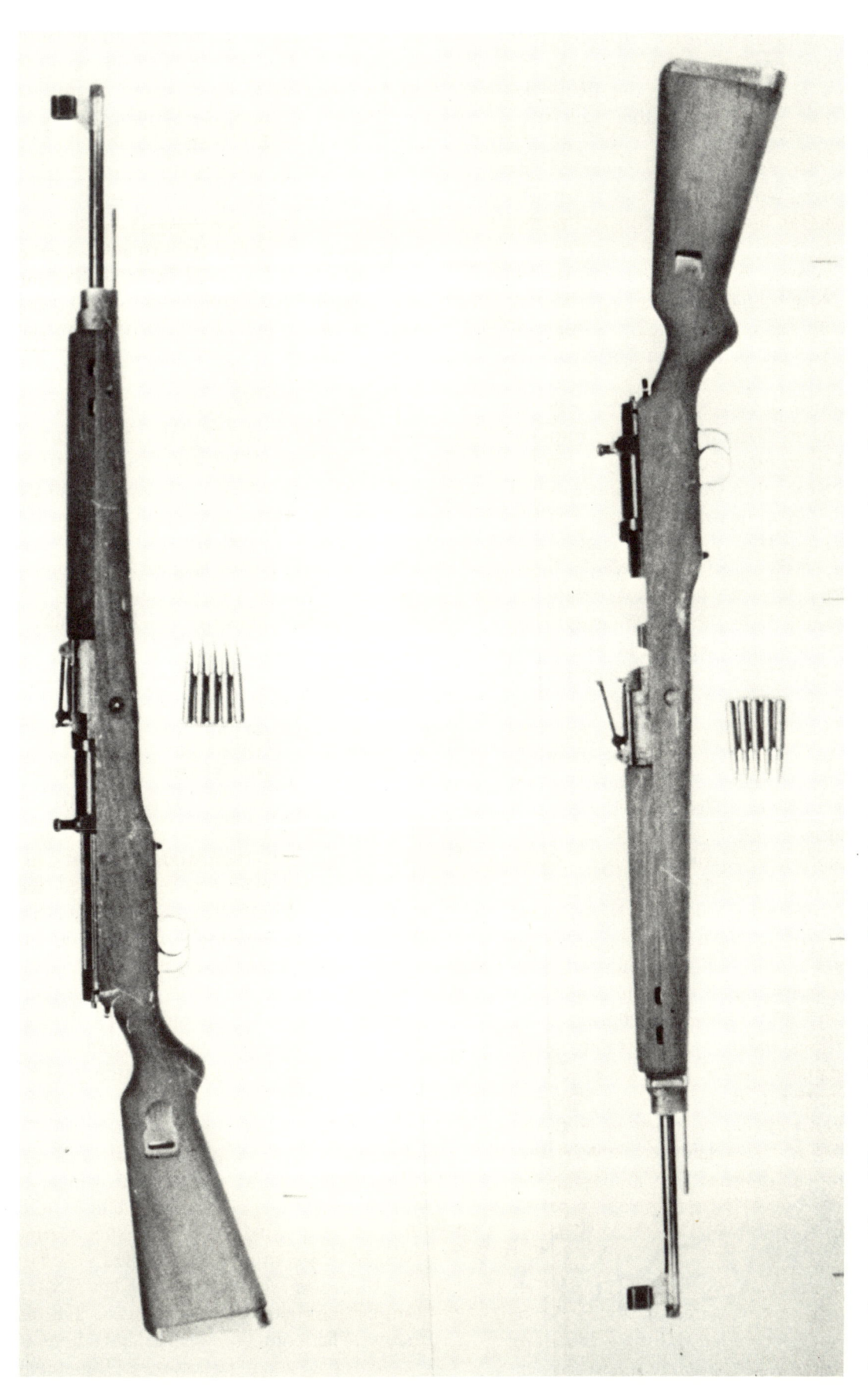

An early transitional Gewehr 43 rifle with a G41 receiver assembly, a combination of both systems. The receiver has the ZF 40 telescope mounting rails on either side of the rear sight assembly. This semiautomatic rifle has a 550mm barrel.

been fully demonstrated, Hitler requested that silencer development be conducted with all possible haste. While attending the field trial of an innovative 98k silencer, the work of a noncommissioned officer, Hitler was so impressed that he spontaneously awarded its designer, Walter Wolff, 10,000 Reichsmarks in appreciation for his contribution to the war effort of the Reich.

Records cite numerous design efforts for efficient small arms noise suppressors dating back to World War I when the advantages of such devices became evident on both sides of the trenches.

Even though the Allied use of rifle silencers has been documented, and Imperial German marksmen reportedly employed similar equipment, the extent of German silencer use on the western front remains obscure. Experts abroad maintain that an unspecified number of silencer equipped Gewehr 98 rifles were utilized against the Russians with impressive results. It is further contended that the German application of telescopic-sighted rifles, and to a much lesser degree small arms silencers, had such a profound effect on the Russian military that they were major factors in the subsequent development of an efficient sniping system and fully operational silencers by the Soviet army well in advance of World War II and the German invasion of Russia.

The evolution of German silencer efforts during World War II included the successful development of the 7.92mm Nah-patrone (close-range) cartridge for use in silencer equipped shoulder weapons. This ammunition possessed the necessary low velocity characteristics compatible with current German devices. The velocity of the standard 7.92mm Patrone-sS heavy-pointed bullet was 2,575 fps, while the Nah-patrone cartridge for silencer use was reported to have a relatively low velocity of 900 fps. Although use of the Nah-patrone cartridge limited the sniper's range, since his position was not given away by the weapon's report or the crack of a supersonic projectile, the advantage was obvious.

Later in the war when a great deal of sniping activity was relegated to defensive or static conditions (especially during the long withdrawal from Russia), German snipers reportedly urgently requested all operational versions of the Schalldampfer (ordnance designation for silencer). German design permitted the sniper to attach easily the silencing device to his sharpshooting rifle, then chamber the Nah-patrone round to be ready for action.

Advantageous or not, with a capability for engaging targets at 600 meters, subsonic ammunition made most Wehrmacht specialists extremely wary because of the range limitations. They considered close-quarter shooting somewhat less than prudent. In reality, however, no amount of National Socialist fervor could supplant the desire to escape detection.

Despite the serious consideration and fielding of silencers for the 98k and early Selbstladegewehr 43 system, the use of these devices did not reach extensive proportions.

In addition to the Arado Aircraft firm, Walther, Mauser, and Gustloff et al, were reportedly involved with silencer development or manufacture throughout the war.

In the final days of the Third Reich, even with the end obvious to most of the German military, the K43 controversy remained a subject of inter-bureau debate as evidenced by the following correspondence.

Subject: Production Volume for Telescopic Sight Rifles   2 April 1945

To AHA/Staff Ib

Jn 2 represents the standpoint that many more successes are to be achieved with 15000 accurately shooting telescopic sight carbines monthly than with 25000 telescopic sight carbines of mediocre quality. Six accurately shooting snipers within a company sector have more kills than ten trigger-happy types with scatterguns. The sniper's targets disappear if the bullet is on target. They are only hit if the first shot connects. If this is only possible on the basis of the given shooting performance, then not every head which has disappeared is a certain kill. The crippling terror and infallibility of the German sniper is then quickly diminished.

According to the K St N Div 45, 156 sniper carbines are provided to each division. The combat engineer battalion and field replacement battalion is not allowed

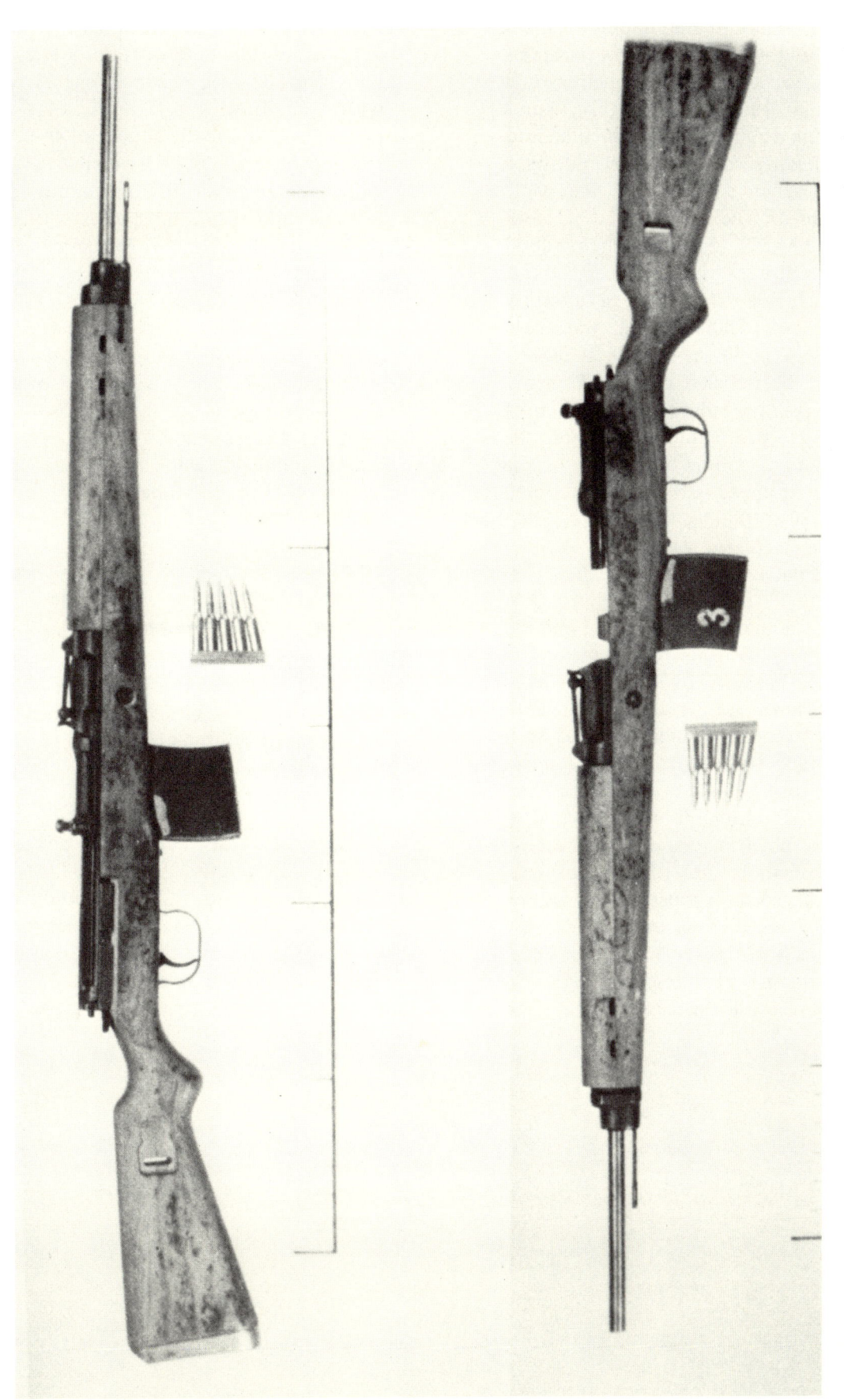

German Ordnance experimental semiautomatic Gewehr 43 modified to fire 7.62 x 54R (Russian) service ammunition with removable Tokarev magazine. Note the absence of a front sight assembly. The barrel length was cited as 550mm.

for in that, because the equipping of these two units with telescopic sight weapons brings little success and cannot be regarded as a pressing need. If one roughly figures 275 divisions of all kinds, which come into line for fielding snipers, there follows a total requirement for 43000 telescopic sight carbines. The requirement is estimated to be 25% covered. Thus 32000 telescopic sight carbines remain lacking, which can be covered with a monthly production of 15000 good telescopic sight carbines in three months, if 30% is figured in as a deficiency through enemy action etc. With a monthly production of 25000 units the equipping would be reached in two months. The value of the sniper thus equipped is thereby established at scarcely 50%. Besides, these snipers will succumb as increased casualties, whereby the apparent advantage is more than cancelled.

At the moment there are about 30 sniper training companies with a maximum performance capacity of 200 pupils each monthly. Limits are set here in terms of manpower to a further increase, because it is now already difficult to release suitable marksmen from the units in this number. Equipping of unsuitable marksmen, delivery to untrained marksmen or direct resupply of mediocre telescopic sight weapons to engaged units must lead to failure. The sniper concept then quickly goes the same way as it did in 1918 and after 1928, where shooting with telescopic sights declined because the successes with open sights were demonstrably at least as good.

Covering the deficiencies in telescopic sight carbines as a result of enemy action will, after drawing off of the 6000 weapons required for the supply of the sniper training companies, be possible at once with the remaining 9000 weapons. There remains here even a satisfactory buffer for production deficiencies, increasing of the equipment assets and equipping of Volkssturm and Werewolves.

For the above reasons it is necessary to avoid production planning for telescopic sight carbines which will never result in accurate weapons. A monthly output of 10000 K43 scoped K43 or K43 scoped K43/1 and 5000 K98k with scope (o) or scoped K43 and scoped K43/1 is considered correct.

By order
Freytag

Note: Telescopic-sight reference explanation:
K43 scoped K43 . . . . ZF 4
K43 scoped K43/1   ZFK 43/1
K98k with scope (o) . . . . 4-power commercial
   type military contract sights
K98k scoped K43 . . . . ZF 4
K98k scoped K43/1   ZFK 43/1

Regardless of continued demands for 25,000 units monthly, the principal ZF 4 manufacturer, Voigtlander (ddx), had only managed to attain a production level of approximately 12,000 sights per month when the war ended, and the Karabiner 43 in any form, never reached the front lines in anticipated quantities.

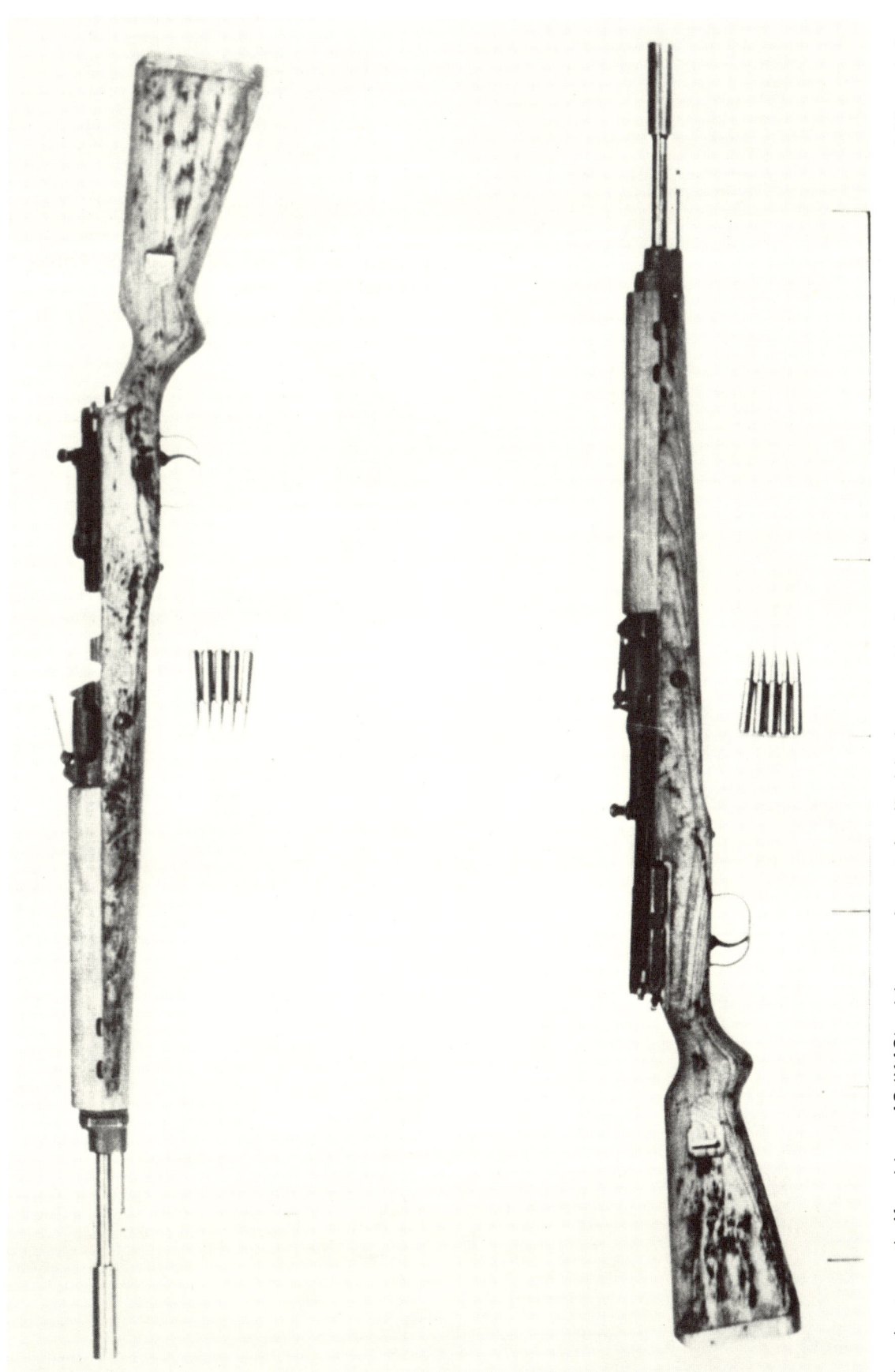

Semiautomatic Karabiner 43 (K43) with experimental cylindrical flash hider. According to German Ordnance documents: "Through experiments with the most varied conical and cylindrical forms and spiral-formed gas sluices, it was determined that the most effective flash suppressor for the sniper rifle K43 is a conical attachment, whereby the cone angle amounts to 14° and the length amounts to 55 m." Such devices were not known to have reached production status however. The barrel length was cited as 550mm.

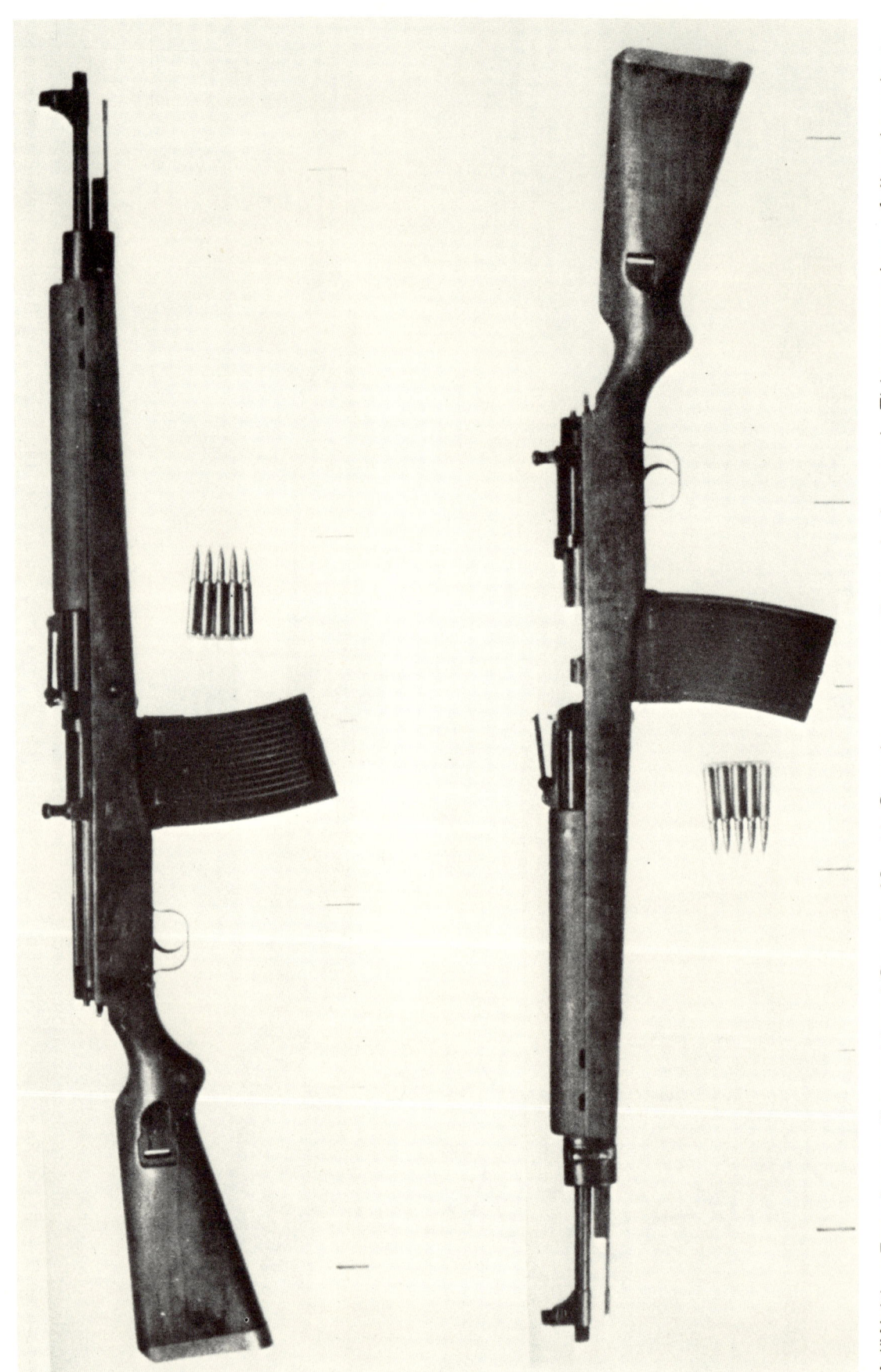

A "Walther Experimental" transitional Gewehr 43 with Gewehr 41 receiver assembly. Although early prototype Walther G43 rifles were referenced as the "G43 (W)," it is not known if they were actually marked as such. This weapon has a full and semiautomatic fire capability.

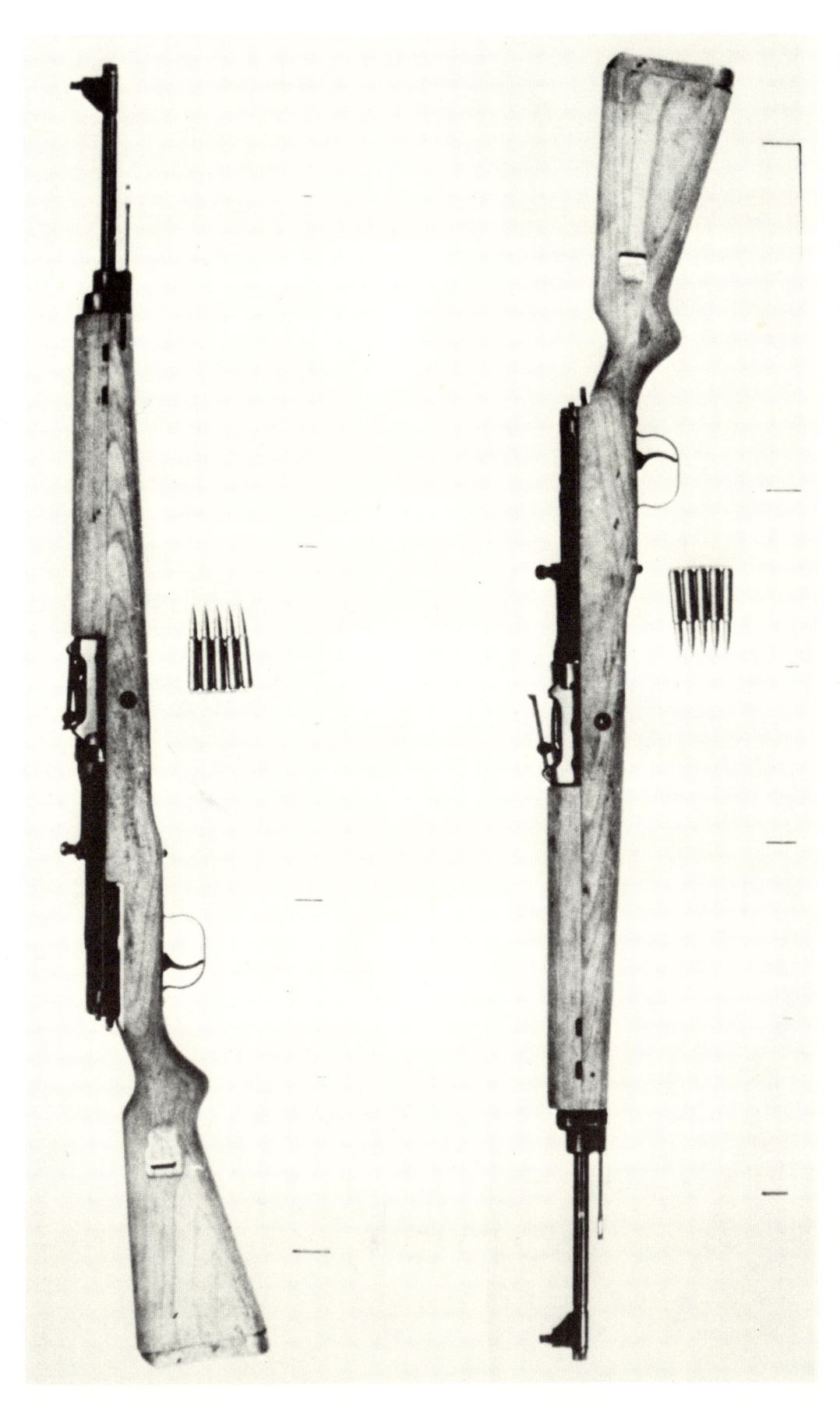

An unusual example among the unique: a semiautomatic Gewehr 43 with a "stamped receiver." According to reports, Berlin-Lubecker Maschinenfabriken (duv) and Mauser fabricated an unknown quantity of these weapons for evaluation purposes. The barrel length is 550mm.

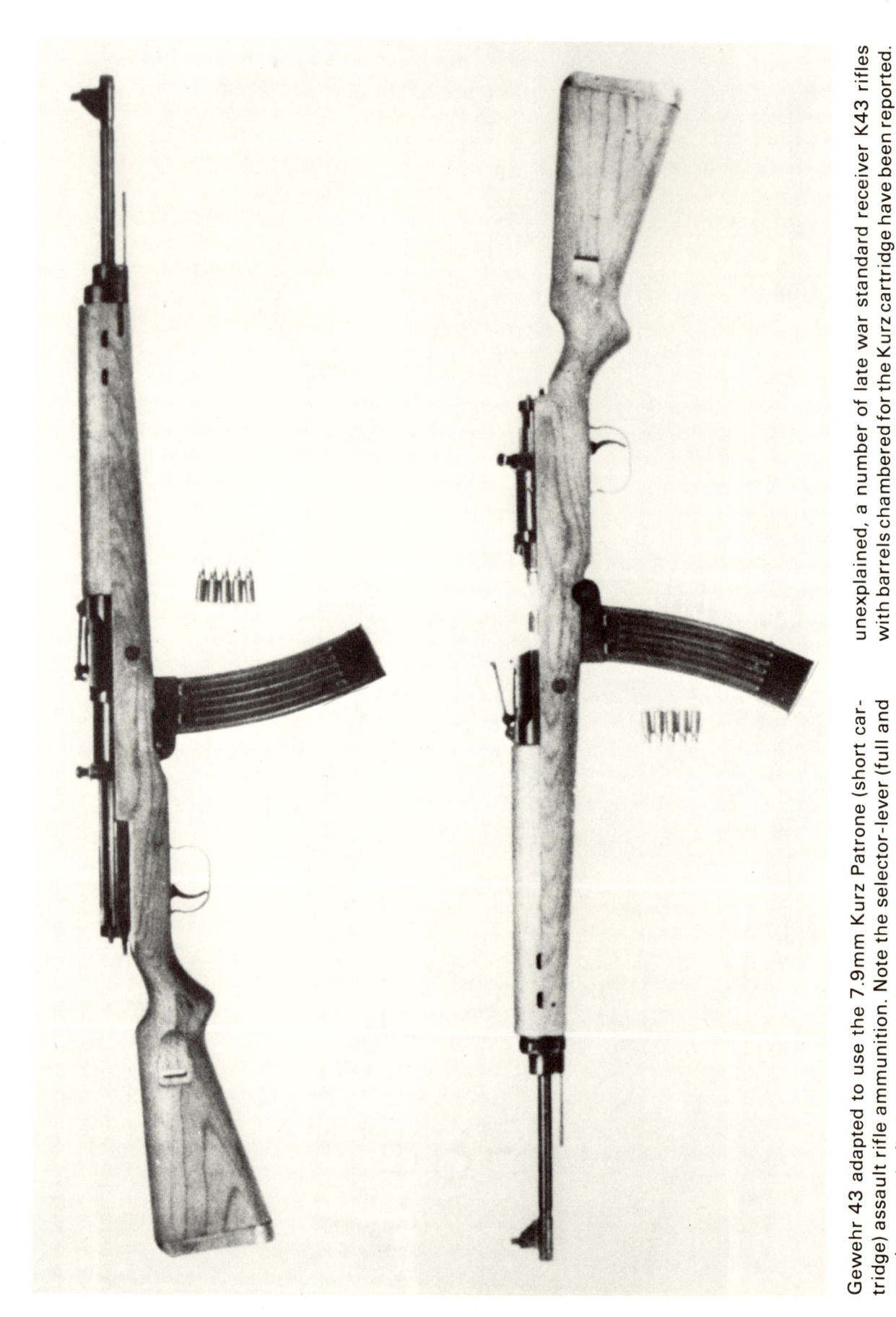

Gewehr 43 adapted to use the 7.9mm Kurz Patrone (short cartridge) assault rifle ammunition. Note the selector-lever (full and semiautomatic fire) and removable assault fire magazine. While unexplained, a number of late war standard receiver K43 rifles with barrels chambered for the Kurz cartridge have been reported.

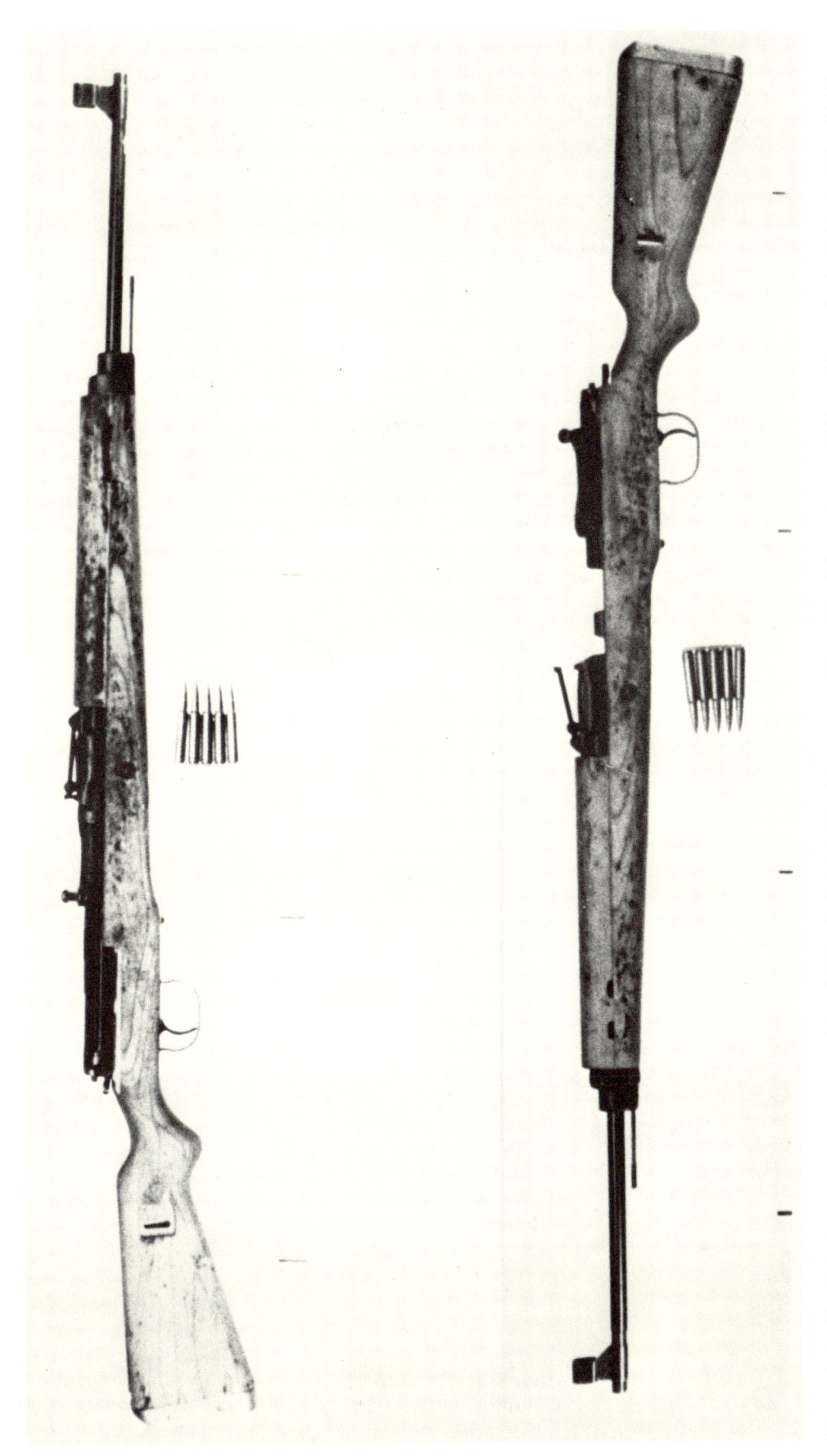

Extensive efforts to improve accuracy of the Selbstladegewehr 43 system included evaluations of both longer (lengths to 740mm) and heavier (diameters to 32mm) barrels. In this case the G43 was fitted with a barrel 622mm in length.

Schalldämpfer für Gewehr 91/30.

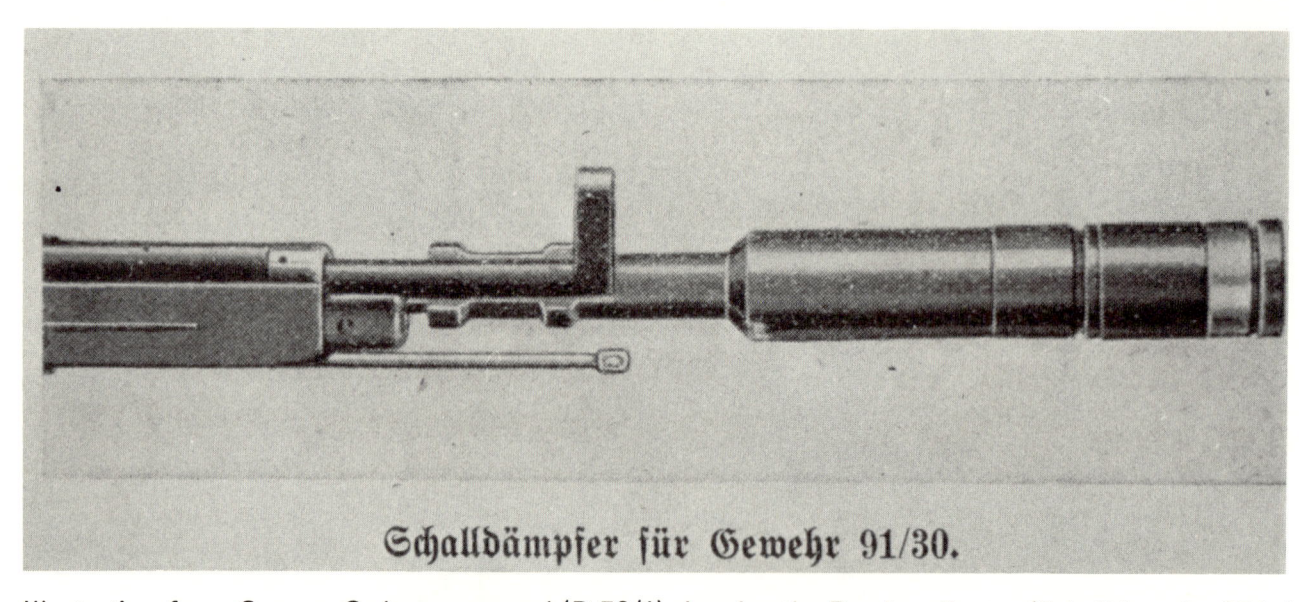

Illustration from German Ordnance manual (D 50/1) showing the Russian silencer (Schalldampfer 254 r) used with the Mosin-Nagant Model 91/30 bolt-action rifle. In addition to the use of captured silenced rifles by German patrols engaged in search and destroy missions against Soviet partisans early in the war, Russian silencers served as the design base for evolving German devices.

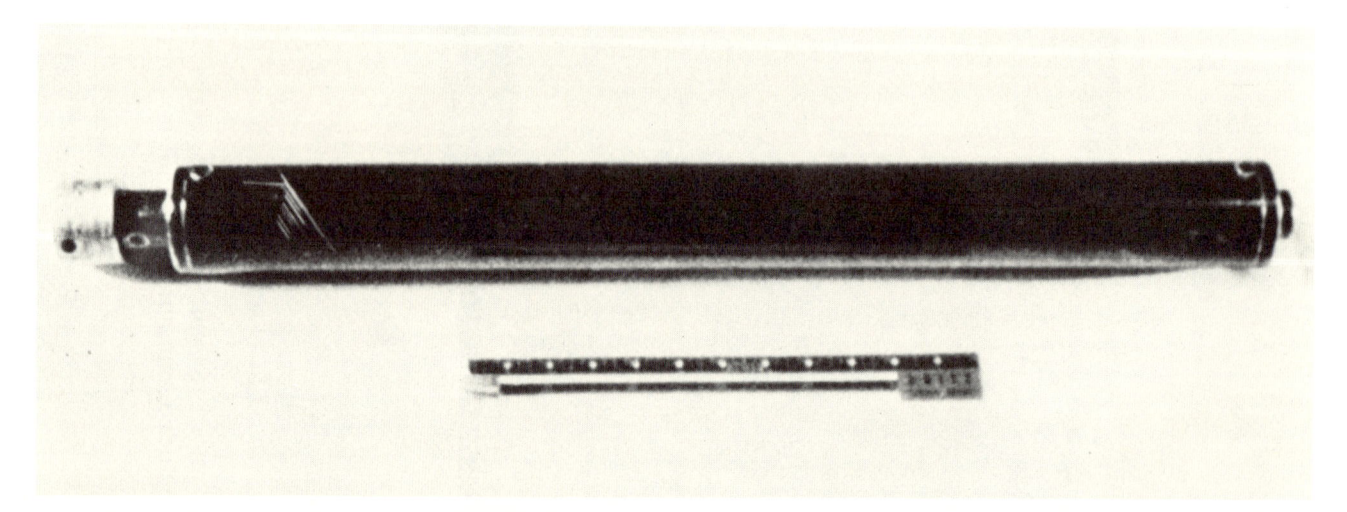

An extremely rare Gewehr 43 silencer that threads onto the muzzle bears the designation, "Schalldampfer 7X WaA 449 zs" silencer, serial number, Waffenamt, and production mark (unknown). Both flash hiders and silencers were a serious consideration with the original Selbstladegewehr 43 system.

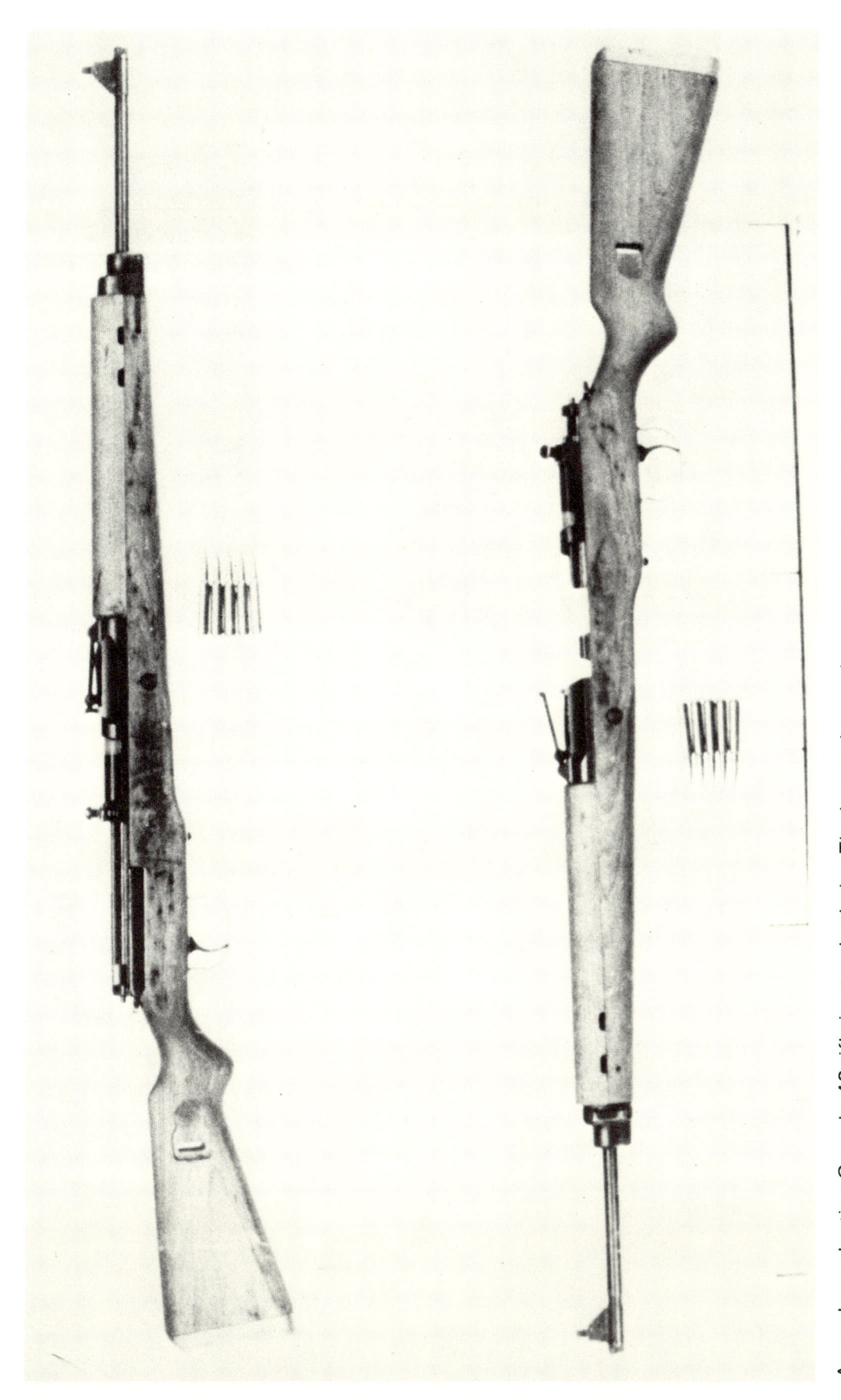

An early production Gewehr 43 rifle in standard trim. The integral scope mounting base does not have the machined recess found on the vast majority of the G-K43 series rifles.

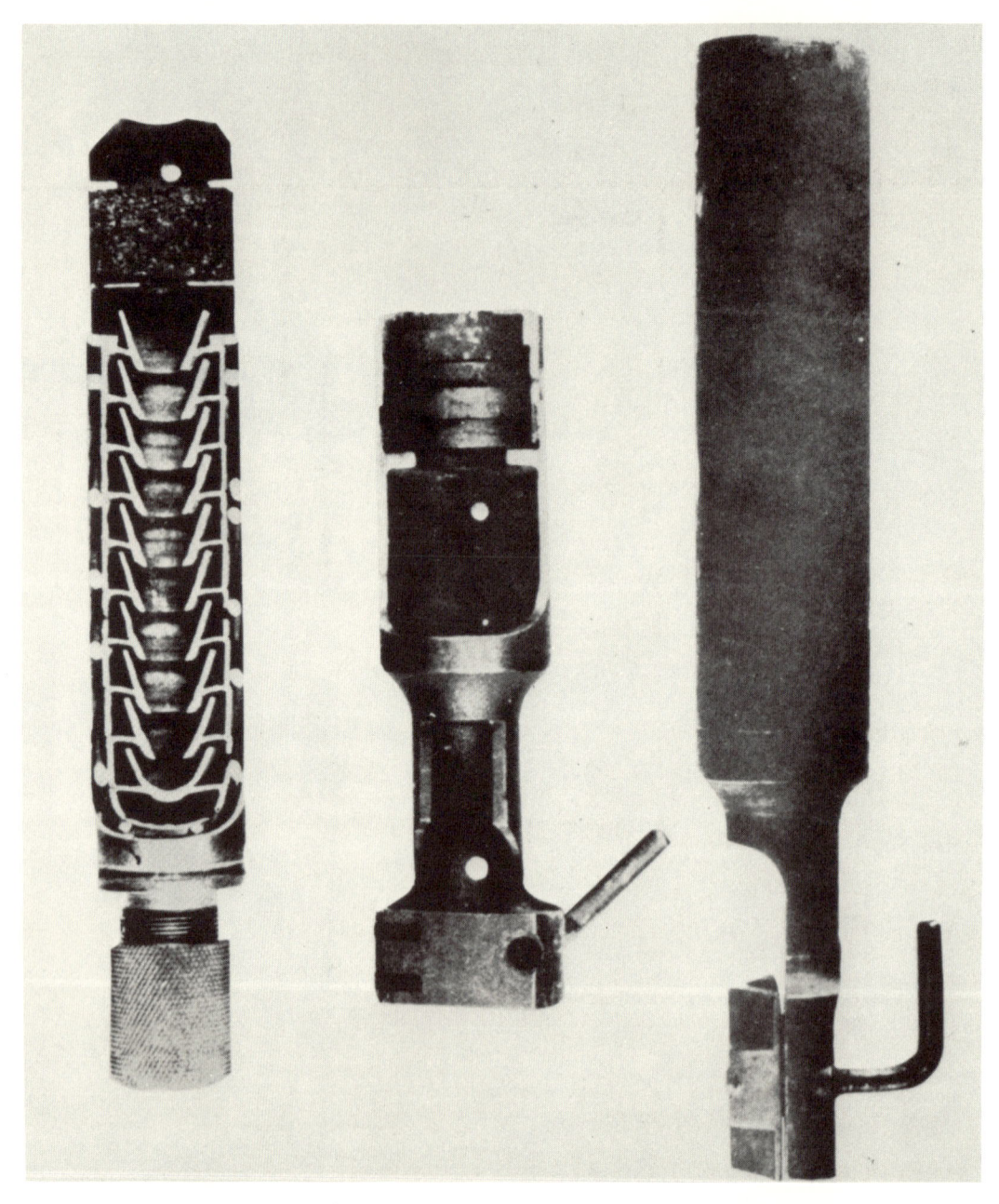

Variations of Schalldampfer (silencer) intended for the K98k rifle. The cup type grenade launchers, which have similar silhouettes and attaching mechanisms, have often been confused as silencers. Despite considerable design efforts, silencers for the 98k and Gewehr 43 system (G-K43) saw restricted field use, particularly for sniping purposes.

Although late war German sniper use of explosive ammunition against Soviet personnel is rarely mentioned, this copy of an original telegram states in part: "The Fuehrer has released B-cartridges for use only in the East."

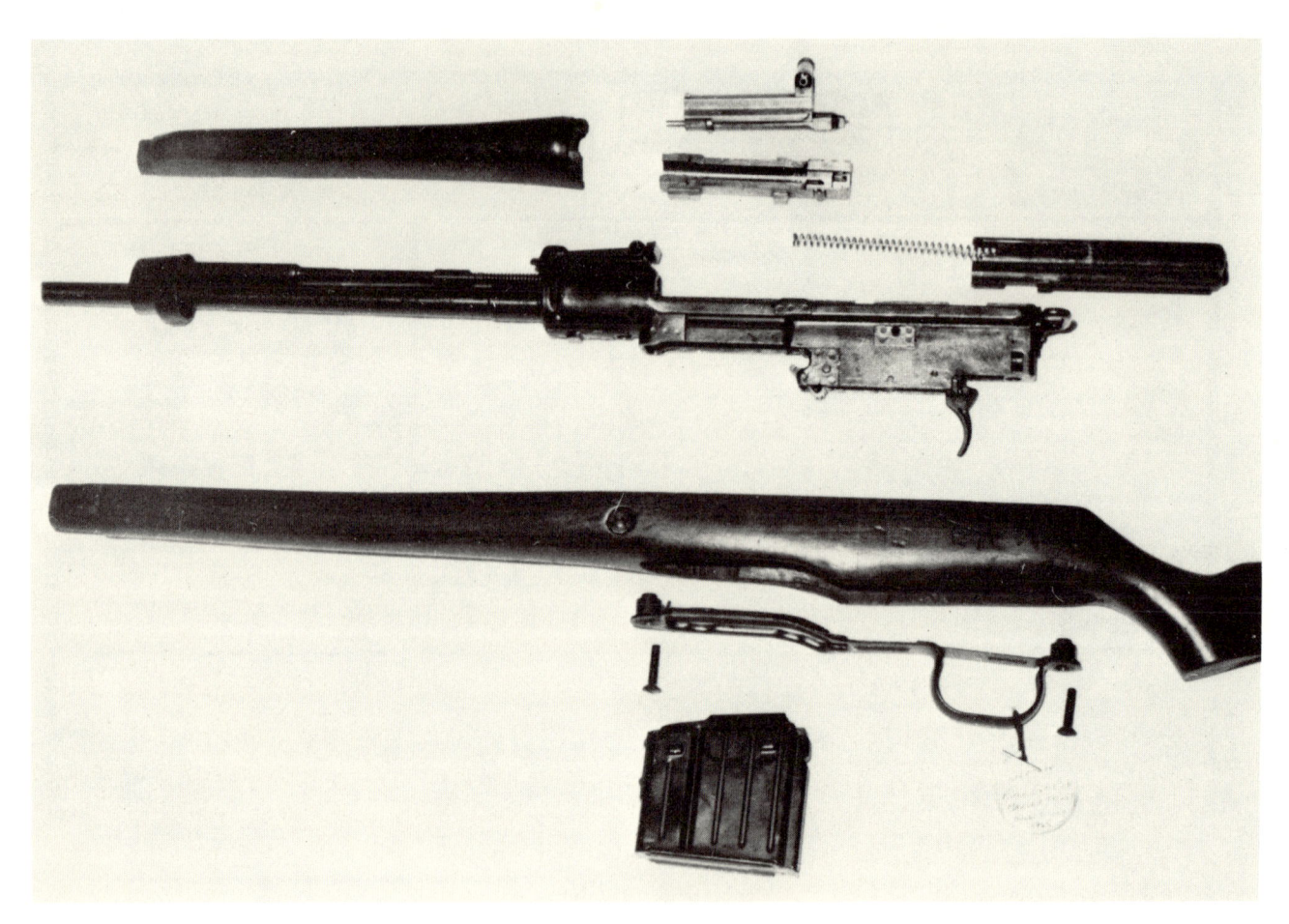

An experimental Mauser semiautomatic rifle design dating from about 1944 intended as a replacement for the K43 system. There are no markings on this unique tool-room model.

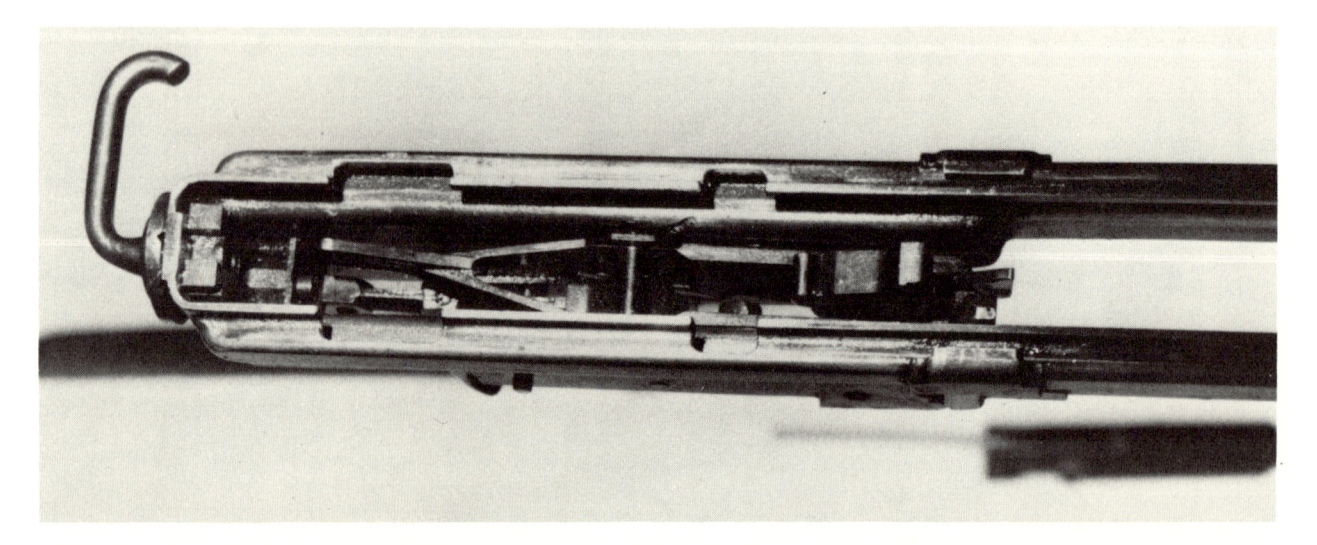

Close view of the experimental Mauser semiautomatic rifle showing the receiver fashioned from pressed steel (stamped). This design was never adopted or placed in production.

RIFLES FOR SNIPING 111

Left view, experimental Mauser semiautomatic rifle. Referenced as the "Gerat 03" this variation was credited to Altenberger, the noted Mauser designer.

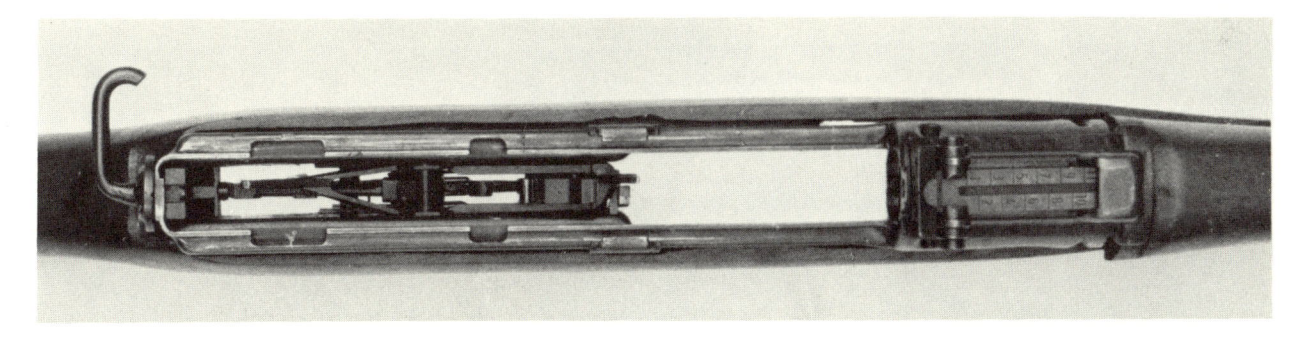

Top view, Mauser experimental rifle. The formed rod extending from the rear of the receiver rotated to lock or unlock the metal action housing (cover) and also served as the safety lever.

Right view, experimental Mauser semiautomatic rifle. No provision was made for mounting telescopic sights on this weapon.

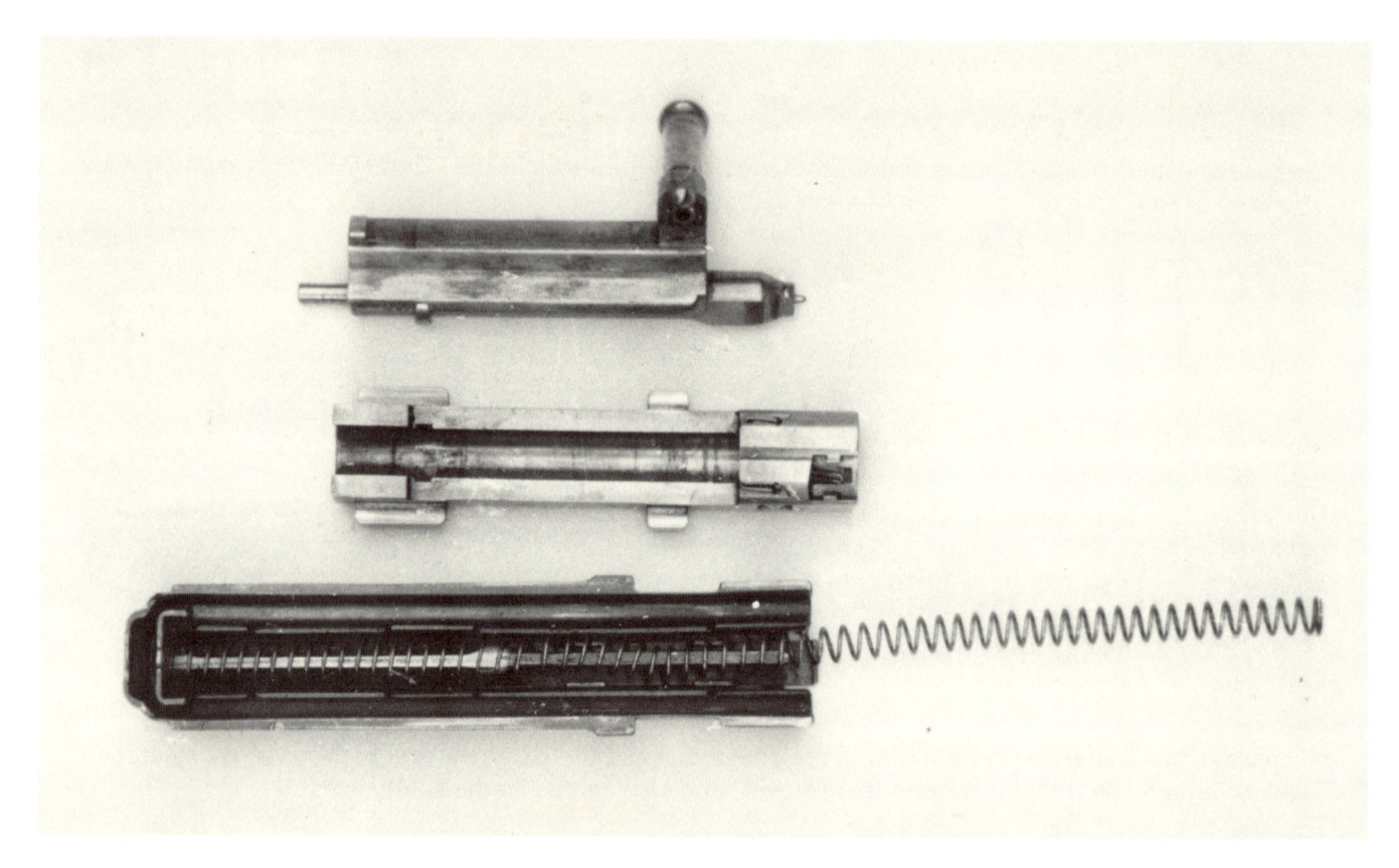

Mauser experimental rifle action housing, recoil spring, spring guide, and bolt assembly. A gas-operated delay blow-back semiautomatic system with the gas port located 241.30 mm from the muzzle. When fired, the operating rod moved the top half of the two-piece bolt to the rear permitting the "roller-type locks" to retract into the bolt unlocking the breech.

CHAPTER IV

Combat Experiences— The Eastern Front

The following presentation first appeared in the official Austrian military publication entitled *TRUPPENDEINST* (Troop Service) in the year 1967 as a part of a series of articles written by a recognized European authority on military sniping, Austrian Army Officer Captain Hans Widhofner.

Among persons questioned were the two most proficient German snipers of World War II with the comments of another accomplished sniper added to obtain a well-rounded picture concerning the use of sharpshooters in the German army. It was fortunate that the soldiers questioned were not only ready to give information, but were also able to illustrate and explain the point of respective question due to their former excellent training and enormous military knowledge.

It is interesting to note that in most cases actual field practice closely parallelled the training philosophy expounded at the official level during the latter months of the war, emphasizing the point that snipers were indeed a serious consideration of the German High Command.

Without question, the significance of this document places it near the top of the list of pertinent contemporary works relevant to German military history.

Captain Widhofner questioned the three seasoned snipers individually. They are designated in the order "A," "B," and "C." All were members of the Third Mountain Division of the late German army.

A: Matthias Hetzenauer of Tyrol fought at the eastern front from 1943 to the end of the war, and with 345 certified hits was the most successful German sniper.

B: Sepp Allerberger of Salzburg fought at the eastern front from December 1942 to the end of the war, and with 257 certified hits was the second best German sniper.

C: Helmut Wirnsberger of Styria fought at the eastern front from September 1942 to the end of the war and scored 64 certified hits. After being wounded, he served for some time as an instructor in a sniper training course.

Questions Asked Of The Snipers

1. Weapons used?
    A: K98 with six-power telescopic sights, G43 with four-power telescopic sights.
    B: Captured Russian sniper rifle with telescopic sight; cannot remember power, K98 with six-power telescopic sights.
    C: K98 with 1.5-power telescopic sights, K98 with four-power telescopic sights, G43 with four-power telescopic sights.
2. Telescopic sights used?
    A: Four-power telescopic sight was sufficient up to a range of approximately

The business end of a German sharpshooter rifle, a view seldom seen or survived. Note the use of mosquito netting to alter the helmet form.

A view of a typical World War II German reticle pattern (vertical pointed post with horizontal side bars) as seen through the eye of a sniper. Although prevalent in military scopes dating from this era, this pattern was not developed solely for sharpshooter use.

400 meters. Six-power telescopic sight was good up to 1,000 meters.
B: Used a captured Russian rifle with telescopic sight for two years; yielded good results. Six-power telescopic sight mounted on K98 was good.
C: 1.5-power telescopic sight was not sufficient; four-power telescopic sight was sufficient and proved good.

3. What is your opinion on a further increase of magnification?
A, B: Six-power sufficient. No need for stronger scope. No experience with greater magnification.
C: Four-power is sufficient in both cases.

4. At what range could you hit the following targets without fail?
Head   A: up to approx. 400 meters
                 B: up to approx. 400 meters
                 C: up to approx. 400 meters
Breast   A: up to approx. 600 meters
                 B: up to approx. 400 meters
                 C: up to approx. 400 meters
Standing man   A: 700-800 meters
                 B: up to approx. 600 meters
                 C: up to approx. 600 meters

5. Do the ranges indicated by you apply only to you, i.e. the best snipers, or also to the majority of snipers?
A, B: Only to the best snipers.
C: To me personally as well as to the majority of snipers. A few outstanding snipers could obtain hits at longer ranges.

6. What was the range of the farthest target you ever fired at, what kind of target, size?
A: About 1,100 meters, standing soldier. Positive hitting not possible, but necessary under certain circumstances in order to show the enemy that he is not safe even at that distance, or superior wanted to satisfy himself regarding capability.
B: 400 to 700 meters.
C: About 600 meters, rarely more. I usually waited until target approached closer for better chance of hitting. Also, confirmation of successful hit was easier. Used G43 only up to about 500 meters because of poor ballistics.

7. How many "second shots" (additional shots) were necessary per ten hits?
A: Almost never.
B: One to two. Second shot is very dangerous when enemy snipers are in the area.
C: One to two at most.

8. If you had a choice, what weapon would you use and why?
a. Repeating rifle such as K98k:
A: K98. Of all weapons available at that time it had the highest accuracy for permanent use; besides, it did not jam easily.
B: K98
C: K98
b. Semi-automatic rifle such as G43:
A: No G43. It is suitable only up to about 400 meters; inferior precision.
B: No G43. Too heavy.
C: Yes, if it does not jam easily, and if its accuracy is not inferior to the K98.

9. Today, if you had the choice between a K98 and a semi-automatic rifle that does not easily jam and has the same accuracy as the K98, which weapon would you take and why?
A: K98. Snipers do not need a semi-automatic weapon if they are correctly used as snipers.
B: Semi-automatic loader, if its weight does not increase.
C: Semi-automatic loader. Faster firing possible when attacked by the enemy.

10. Were you incorporated into a troop unit? All three belonged to the sniper group of the battalion; "C" was the commander of this group. They were numbered up to 22 men; six of them usually stayed with the battalion, the rest were assigned to the companies. Observations and use of ammunition as well as successful hits had to be reported daily to the battalion staff. In the beginning the snipers were called up out of the battalions; as the war continued and the number of highly-skilled snipers decreased, they were often assigned and given their orders by the division.

In addition, a few marksmen in each company were equipped with telescopic sights. These men did not have special training but were able to hit accurately up to about 400 meters and carried out a great deal of the work to be done by "actual snipers." These specially-equipped riflemen served in the company as regular soldiers. This is why

German sharpshooting specialists engaged in Russia were to quickly realize the worth of their Red Army adversary. Russia, summer 1943.

they could not achieve such high scores as "snipers."

II. Strategy and targets?
  a. Attack:
  A, B, C: Always two snipers at a time; one shoots, the other spots. Usual general order: elimination of observers of the enemy's heavy weapons and commanders, or special order, when all important or worthwhile targets were eliminated; for example, anti-tank gun positions, machine gun positions, etc. Snipers closely followed the attacking units and whenever necessary eliminated enemies who operated heavy weapons and those who were dangerous to our advance.
  A added: In a few cases, I had to penetrate the enemy's main line of resistance at night before our own attack. When our own artillery opened fire, I had to shoot at enemy commanders and gunners because our own forces would have been too weak in number and ammunition without this support.
  b. Attack during night:
  A, B, C: As far as we can remember, no major attacks during night were conducted; snipers were not used at night; they were too valuable.
  c. Winter attacks:
  A: Clothed in winter camouflage I followed behind the front units. When the attack slowed down I had to help by shooting machine gunners, anti-tank gunners, etc.
  B, C: Good camouflage and protection against cold was necessary. No extensive ambushing possible.
  d. Defense:
  A,B,C: Usually on my own within company detachment; order: fire at any target or only "worthwhile targets." Great success during enemy attacks since commanders can often be recognized and shot at long range due to their special clothing and gear such as belts crossed on chest, white camouflage in winter, etc. As a consequence, enemy's attack was prevented in most cases. (Shot the respective leaders of enemy's attack eight times during one day!) As soon as enemy snipers appeared we fought them until they were eliminated; we also suffered great losses. As a rule, the sniper watched for worthwhile targets at the break of dawn and remained in position until dusk with few interruptions.

  We were often in position in front of our own lines in order to fight the enemy more successfully. When access to our position was known to the enemy, we were forced to remain without provisions or reinforcements at such advanced positions. During alarm or enemy attack, a good sniper did not shoot at just any target, but only at the most important ones such as commanders, gunners, etc.
  e. Defense during night:
  A, B, C: Snipers not used during night; not even assigned to guard duty or other duties. If necessary, he had to take position in front of our lines in order to fight the enemy more effectively during the day.
  f. Did you score successful hits by moonlight?
  A: I was often called to action when there was sufficient moonlight since reasonably accurate sniping is possible with a six-power telescopic sight, but not with standard sights.
  B, C: No.
  g. Delaying action:
  A, C: In most cases four to six snipers were ordered to rear guard and eliminate any enemy appearing; very good results. Used machine guns for rear guard only in emergencies since snipers delayed enemy's advance by one or two hits without easily revealing his own position.
  B: No actual use of snipers; actual sniping not possible in mobile warfare since everybody shoots at appearing enemy.
  h. Delaying action during night, in winter: No information.

12. In what warfare could the sniper be most successful?
  A: The best success for snipers did not reside in the number of hits but in the

As the fortunes of war often required, a German marksman is shown using a captured Russian Mosin-Nagant M91/30 rifle with 3.5-power P.U. scope. Eastern front, 1943.

Sighting his target, the sun to his back, the sniper makes good use of standard practice.

damage caused the enemy by shooting commanders or other important men. As to the merit of individual hits, the sniper's best results could be obtained in defense since the target could be best recognized with respect to merit by careful observation. Also with respect to numbers, best results could be obtained in defense since the enemy attacked several times during a day.

B: Defense. Other hits were not certified.

C: Best results during extended positional warfare and during enemy attacks; good results also during delaying action.

13. Percentage of successful hits at various ranges?
Up to 400 meters   A: 65%   C: 80%
Up to 600 meters   A: 30%   C: 20%
Up to 800 meters   Remainder
Additional information:

A: This is why about 65% of my successful hits were made below 400 meters since targets could be better evaluated with respect to merit at that distance. At longer ranges hitting was still possible but I could not distinguish if target was worthwhile.

B: Do not remember. Mass of hits was below the range of 600 meters.

C: Shot mainly within range of 400 meters due to great possibility of successful hit. Beyond this limit hits could not be confirmed.

14. Do these percentages and ranges apply to you personally, or are they valid for the majority of snipers?

A: This information is applicable to the majority of snipers as well as to the best snipers, for:
the majority of snipers could hit with absolute certainty only within a range of 400 meters due to their limited skills; the best snipers could hit with reasonable certainty at longer ranges; they in most cases, however, waited until the enemy was closer or approached the enemy in order to better choose the target with respect to its merit.

B: Information is applicable to all snipers known to me in person.

C: Information is applicable to myself as well as the majority of snipers.

15. On the average, how many shots were fired from one position?
   a. Attack:
      A, B, C: As many as possible.
   b. Defense from secure position:
      A, B, C: One to three at most.
   c. Enemy attack:
      A, B, C: Depending on worthwhile targets.
   d. Combat against enemy snipers:
      A, B, C: One to two at most.
   e. Delaying action:
      A, B, C: One to two were sufficient since sniper was not alone.
      B added: During own attack as well as enemy's attack, hits were not confirmed.

16. What else is especially important in addition to excellent marksmanship?

A: Besides the generally known qualities of a sniper it is especially important to be able to outwit the enemy. The better "tactician of detail" wins in combat against enemy snipers. The exemption from commitment to any other duties contributes essentially to the achievement of high scores.

B: Calmness, good judgement, courage.

C: Patience and perseverance, excellent sense of observation.

17. From what group of persons were snipers selected?

A: Only people born for individual fighting such as hunters, even poachers, forest rangers, etc., without taking into consideration their time of service.

B: Do not remember. I had scored 27 successful hits with Russian sniper rifle before I was ordered to participate in sniper training courses.

C: Only soldiers with experience at the front who were excellent riflemen: usually after second year of service; had to comply with various shooting requirements to be accepted in the sniper training courses.

18. In what sniper training courses did you participate?

A, B, C: Sniper courses in the training area Seetaleralpe.

C: I was later assigned to the same courses as instructor.

Taking advantage of a brief rest, a German sniper team cleans their equipment on the eastern front, summer 1944. Such teams achieved the greatest success with one man shooting, the other observing on an alternate basis.

A sniper fires through a trench loophole at the advancing enemy during defensive action in Poland. Note the textured helmet finish and wire for attaching twigs and foliage as camouflage.

19. Was it advisable to equip the sniper with binoculars? What magnification did the binoculars have?
    A: 6 x 30 enlargement was insufficient for longer distances. I later had 10 x 50 binoculars which were satisfactory.
    B: Binoculars were equally important as the rifle.
    C: Every sniper was equipped with binoculars. This was useful and necessary. An enlargement of 6 x 30 was sufficient up to a range of about 500 meters.

20. Would you prefer a periscope which allows observation under full cover?
    A: Was very useful as a supplement (Russian trench telescope).
    B: No.
    C: Were used when captured.

21. Were scissor stereotelescopes (positional warfare) used?
    A, C: Yes, when available. Were used mutually by sniper and artillery observer.
    B: No.

22. What type of camouflage was used?

	A:	B:	C:
Fake tree stump	—	—	—
Camouflage clothing	yes	yes	yes
Camouflage of face and hands	yes	yes	yes
Camouflage of weapon in winter	yes	yes	yes

(White cover, white wrapping, white paint)

   B added: For two years I used an umbrella which was painted to match the terrain. In the beginning I was always camouflaged; face and hands; later on less extensive.

23. Did you use technical means to mislead the enemy?
    A: Yes; stuffed dummies, etc.
    B: Yes; for example, dummy position with carbines which could be fired by means of a wire-pull.
    C: No.

24. Did you use protective shields in positional warfare?
    A, B, C: No.

25. What is your opinion on the use of tracer ammunition?
    A, B, C: If possible, they should not be used at all in combat since they easily reveal the position of the sniper. Tracer ammunition was mainly used for practice shooting as well as ranging at various distances. For this purpose every sniper carried with him a few tracer cartridges.

26. Did you use observation ammunition, i.e. cartridges that fired projectiles (bullets) which detonate upon impact?
    A, B, C: Yes; upon impact a small flame as well as a small puff of smoke could be seen which allowed good observation of impact. By this method we could force the enemy to leave wooden houses, etc., by setting fire to them.
    Observation cartridges were used up to a range of about 600 meters; their dispersion was somewhat larger than that of heavy pointed cartridges (heavy pointed bullet).

27. How did you overcome side wind?
    A. By own judgement and experience; when necessary, I used tracer ammunition to determine wind drift. I was well prepared for side wind shooting by my training at Seetaleralpe where we practiced often with strong wind.
    B: By own judgement. We did not shoot when side wind was too strong.
    C: No explanation since snipers do not shoot with strong wind.

28. Can you recall the rules pertaining to your behavior when shooting at moving targets?
    A, B, C: No; important are one's own judgement and experience as well as fast aiming and firing.

29. Do you have any experience with armor-piercing rifles?
    A: Yes; several times I have fought against a "machine-gunner with protective shield." I could hit small targets only up to 300 meters since dispersion was considerably larger than with the K98. Besides, it was very heavy and clumsy and was not suitable as a sniper weapon. I did not use it against unarmored targets.
    B, C: No.

30. What was the method by which your hits were certified?
    A, B, C: By observation and confirmation by an officer, noncommissioned officer or two soldiers. This is why the number of certified hits is smaller than the actual score.

An alternate view of a German sniper team member maintaining his sharpshooting issue. The typical turret-mount 98k has a 4-power military contract sight. Note the leather caps which protected the lenses when the sight was not in use.

In addition to extensive use of captured Soviet bolt-action sniper rifles as shown, German marksmen also employed scope-sighted semiautomatic Russian Tokarev rifles. Note the helmet camouflage and position of the sniper next to the branch, an accepted practice.

Nested in the foliage, a sniper uses his telescopic sight to observe enemy activity. German snipers were known to direct mortar and artillery fire.

Garbed in winter camouflage clothing so necessary at the eastern front, a German sniper fires from a trench, steadying the rifle by grasping the butt with his left hand.

In addition to this proficiency with rifle scopes, the best German sharpshooters were also quite deadly with standard sights.

CHAPTER V

Camouflage — A Special Skill

During the Great War in the early days of trench warfare, the German sharpshooter became a byword for skill and cunning. German sniping development and camouflage methods combined to create monumental consternation among the Allies. These Germans plied their deadly trade with telling effect from all manner of carefully prepared camouflaged positions: among ruined houses, through loopholes, even on top of bodies of the dead in a quest for the firing position offering the best advantage.

The Germans made good use of their painted canvas sniper robes, which enabled them to lie out between the lines virtually undetected. Countless numbers of the enemy, daring a chance look over the top, were rendered "hors de combat." Their marks were small but when they hit they usually killed their man with a single shot.

The reemergence of the German military machine during the thirties brought forth the lessons of a war not long past, and camouflage experience proved no exception. Developments through the progression of the second major conflict found the German soldier at the end of hostilities by far the better equipped with camouflage field dress than perhaps any other troops.

There was an endless variety of issue and improvised jackets, smocks, and coveralls, based on a countless number of camouflage patterns. Through the war each service branch adopted numerous outfits and accessories closely relating to their combat theater and season. Ultimate refinements of their use were usually dictated by field expediency. In the various theaters of war, the German soldier fought in virtually every conceivable terrain and climate from Russia to North Africa. This prompted his active pursuance of effective camouflage garb, no matter how irregular it may have been at the official level.

The opening of the eastern front in 1941 and the subsequent Soviet sniping activity gave impetus to the German High Command for revision of their views regarding personal camouflage. There evolved a multiplicity of camouflage outfits, including special suits for the German sniping specialist with features deemed advantageous for his activities, such as loops for attaching natural foliage. The best known sniper outfit was the hooded jacket having a brown and green geometric pattern with reversible lining of white for use in snow.

However practical a reversible winter lining appeared to be, this feature did create problems for the snipers. They found it quite difficult to conceal all traces of white when the regular camouflage pattern was in use. The jacket hood proved especially troublesome since no amount of care would obscure entirely the winter lining from possible enemy detection. As a rule, the sniper concerned himself with the season at hand and improvised accordingly.

Since the individualistic temperament required of a sniper left the ultimate choice and mode of camouflage to his discretion, to cite an overall preference for one or another specific

Elaborate sharpshooter camouflage such as this was usually limited to defensive positions. France, 1944.

type would be speculative.

Particular emphasis was paid to disguising the head, hands, and facial area, since even under conditions favoring the sniper, these areas were the most frequently exposed and discernible to the enemy. The German sniper used somewhat basic methods to alter the appearance of these areas; dirt, grease, or soot rubbed over the skin to darken and tone down reflections were used in conjunction with hoods, gloves, and veils fashioned from available or applicable material.

The German service helmet, because of its distinctive configuration, presented an extremely difficult silhouette to disguise. The average German infantryman recognized this fact and utilized a great variety of nets, cloth covers, and foliage bands fashioned from chicken wire and cut-up inner tubes. In addition to these measures, the sniper used veils that provided an effective method of distorting the helmet form. These veils were not restricted to use only by snipers since they permitted reasonable vision and minimal discomfort for all troops. Veils were not a specific issue item; camouflage cloth, gauze, and mosquito nets were generally employed for this purpose. The Waffen-SS, however, made frequent use of a peculiar type of string-fringe face veil, which combat photos indicate were in general use.

The sniper's weapon and scope were also given adequate camouflage protection. This was most frequently accomplished by using strips of white cloth wrapped the length of the weapon for winter activity and burlap or camouflage cloth strips for greener seasons. A method popular among German snipers of the First World War was also employed to some extent, to daub paint of various brown and green hues on the weapon (pattern painting). Affixing pine boughs, twigs, and straw were also used in altering the weapon's outline.

The sniper's hide or specific area of concealment called for supplementary measures in conjunction with the personal use of camouflage. Climatic conditions called for great proficiency to successfully prepare a position that would remain free from detection. The sniper selected camouflage material timely for the prevalent weather and proper for the terrain. Insufficient as well as excessive camouflage only served to attract the attention of the enemy. The mastery of fieldcraft was paramount to the effectiveness of the German specialist. These points were dramatically emphasized during extensive static sniping that evolved on the eastern front. Numerous accounts exist of bitterly contested duels between German and Soviet sharpshooters, at ranges limited to only a few meters in some cases, with the victor owing his survival to self-discipline.

The early Wehrmacht sharpshooter instruction manuals contained little more than basic reference to the application of individual camouflage. In contrast, the late war manuals offered a comprehensive guide to the most efficient use of all types of camouflage for sniping purposes. A progression of combat experiences was the instructor.

In addition to basic sniper issue, binoculars and compass, a shelter triangle (Zeltbahn), which was normally issued for use as a waterproof cape, was utilized to drag away the diggings (dirt) from a freshly prepared position and to serve as an improvised sling for firing positions in trees. In place of the standard leather belt, which produced a detectable sheen, the sniper was admonished to use only the tropical issue web belt (Africakoppel). To emphasize this, it was stated: "If there is no Africa belt, then a cord shall be used; the black belt is forbidden." The infantry knife 42 (Jnfanteriemesser 42), which served the sniper primarily as a close combat weapon, was also intended as a tool for cutting foliage and shrubbery; the folding entrenching shovel (Spaten) was provided for "position construction and to arrange the rifle mounting."

In a detailed training manual on proven techniques for preparing an undetectable sniping position in buildings, open terrain, and wooded areas during each season, the most effective application of what had evolved as regulation issue sniper dress (camouflage jacket, pants, and helmet cover) was also thoroughly explained. The sniper was instructed to wear camouflage clothing at all times (training and combat). The manual further states: "If no camouflage clothing is available, fatigue dress must be imprinted or sprayed with appropriate camouflage colors. Camouflage dress must always be worn."

The German sharpshooter was guided by

A marksman taking careful aim with his 98k-ZF 41 issue. Note the binoculars near the rifle and regulation issue camouflage helmet cover with loops for attaching foliage.

A lightweight hooded camouflage smock (pullover) typical of those worn by both snipers and regular ground forces.

a basic rule: "Camouflage 10 times, shoot once!"

Although the course of instruction on camouflage varied considerably, the following German document represents a typical approach to sniper camouflage practiced in late 1944.

Good camouflage is the best life insurance!
What must I demand from good camouflage?
It must make me completely invisible and not hinder me in the performance of my task.

The sniper must be a master of camouflage!
When you camouflage yourself, bear in mind that the simplest is always the best. What is the simplest?

The natural camouflage.
Example.
   a. You can disappear completely behind a bush and yet observe and destroy your enemy.
   b. You must exploit light and shadows and thereby remove yourself from enemy view.
   c. The use of hollows, slopes, ridges, high tufts of grass, even a mole-hill can make you invisible. Always avoid conspicuous points, for you would also expect the enemy to be behind them.

Artificial camouflage can help you further.
What does one understand by artificial camouflage?
The camouflage jacket, the shelter triangle, the camouflage hat, gloves, face mask, color material for wrapping the rifle, etc. However, you must often reach for camouflage of a natural kind.

Natural camouflage of an artificial kind.
You will find camouflage such as branches, grass, moss and plants of all kinds in your immediate surroundings.
Here above all: The simplest is the best!

Examples for movement.
a. *The camouflage fan*, which is produced from auxiliary means that are to be found in almost every terrain offers you the best protection against being detected in open terrain.

Production: A long branch fork of 40 cm, a short fork about 25 cm, and a cross staff which corresponds in length to your shoulder width.

These three pieces are bound together and covered with a camouflage net corresponding to the surrounding ground cover.

Working your way forward with the camouflage fan must be done slowly in centimeters. If you come into terrain with different ground cover, then the fan is to be immediately re-camouflaged.
   a. *The grass mat* camouflages you against observation from the air and also permits your working forward for a favorable shooting range.

Production: Three camouflage nets are bound together, then spread over four stakes and braided with corresponding ground cover. While braiding, what must appear as the surface is the same as what appears in nature; thus braid with the grass roots *downward*.

While braiding, you must copy nature exactly!
After the ground cover is braided, the camouflage mat is put on in the following way:
One third of the mat is put over the head, so that even the face is completely covered. Observation ensures through two small openings, which are easy to make by spreading the camouflage material with the finger. The camouflage must hold fast to the body. The upper third is held fast with two loops which reach through the arms, the middle third is tied at the waist, the lower third is fastened at the thigh. The boots and rifle are to be camouflaged in the same manner.

Example for the ambush position.
Three to four camouflage nets are bound together and staked up, small branches of fir, pine, leaves, birch, juniper, etc. are then braided into the nets. Fastening of this camouflage is accomplished in the same manner as used with the grass mat, except that this method only camouflages the sniper from the front.
With this camouflage you can place yourself into open terrain—naturally, corresponding to your surroundings—and thereby observe the enemy unhindered.
   a. Pine-man.
   b. Leaf camouflage.
   c. Birch-man

For scope and binoculars an improvised reflection guard can be produced from cardboard. This inhibits the lenses from reflecting in the sun. For camouflaging the entire binoculars or

A hollow road marker serves as a sniping position in this case. (From sniper training illustration, 1944.)

Although carefully prepared, any escape from this lofty "hide" would be difficult for the sniper.

rifle, a camouflage cover can be placed over both.

Camouflage of the position.
The shoveled-out earth is removed with the shelter triangle. The immediate surroundings must not be disturbed by the entrenching work. The foxhole is covered to correspond with the surrounding ground cover. The weapon is camouflaged out to the muzzle. Instead of the camouflage net you can also use the shelter triangle. It has the advantage that:
  a. In winter the position remains warm.
  b. The camouflage can be covered with topsoil which scarcely dries up in the summer.

In no case lay grass, moss, leaves, etc. thereupon, since wind and sunshine become betrayers for you.

If in spite of continuous observation you have not recognized the enemy, you must lure him forth by an expedient deception. Never repeat a deception, always attempt something new. A deception must be natural, it is most effective when movement is included. It must be of simple means and simple to build. Turn to a deception that is not easy to produce only if no success is attained through simple means.

Always have the deception carried out by the observer, while the sniper lies in firing position waiting upon the target.

Deceptions are employed chiefly in defense positions.

Examples:
  a. *Observer.* The head and body are stuffed moss, grasses, etc. A camouflage net is used for the head form, cardboard or a red-brown cup of birch-bark serves as a face. A stick with cross-piece serves as the backbone and shoulder. A thin staff moves arms and binoculars. The improvised glass is produced from wood, the lenses from clear discs.
  b. *Marksman digging in.* A marksman betrays himself by throwing earth from cover and by flashing of the spade. In his pause from work one sees tobacco smoke rising up, then the deception appears from cover.
  c. *Marksman shooting from a firing embrassure.* Head filled with grass, fastened upon sled; the marksman appears by pulling on a cord as if moving himself sideways, with the view toward the enemy. By anchoring the rifle and firing a shot with a cord the deception becomes more effective.
  d. *Fork dummy.* Head moveable to the front or to the side, has the same effect as under c.
  e. *Tree-climber.* A deception used for luring an especially well-camouflaged opponent. Three cables move the body, arms and legs corresponding to the climbing movement of a person.

Principle: The dummies used for deception purposes must be carefully produced and manipulated so authentically that the opponent will still be deceived at a close range.

The grossly exaggerated accounts of vast numbers of German snipers confronting the advancing Allies from virtually every tree in Europe are, no doubt, responsible for the myth existing to now, that any person firing any weapon from cover qualifies him as a sniper, a misconception perpetuated by the news media. While German training manuals did depict riflemen in trees, this was merely standard training for the German soldier. Tree slings were reportedly issued for this purpose, and it is more than likely that they were utilized primarily by observers and casual riflemen. Even though various sharpshooter training courses did elaborate on the best methods of climbing trees to secure an advantageous firing position, this was an effort to prepare the sniping specialists for any eventuality rather than an exercise resulting in self-destruction.

There were, no doubt, circumstances that may have necessitated the use of a tree "hide" by true snipers, but when one considers the extreme difficulty in obtaining a position required for accuracy while so precariously perched and the difficulty in withdrawing from such a position, it seems doubtful that an accomplished sniper would make extensive use of such positions.

The manner in which the German Scharfschutzen accomplished his mission was not simple, but rather the fruition of excellent fieldcraft training and practice. Of interest and tribute to the competence of these unique specialists is the fact that very few true German snipers were either taken or reported killed; the accomplishment of either was eagerly pursued by the Allied soldiers.

In addition to issue fabric camouflage helmet covers, an elastic or rubber band was often used for attaching vegetation as typified by this drawing.

As part of overall sniper training, the German specialists were instructed to exercise imagination when preparing their "hide." (From sniper training illustration, 1944.)

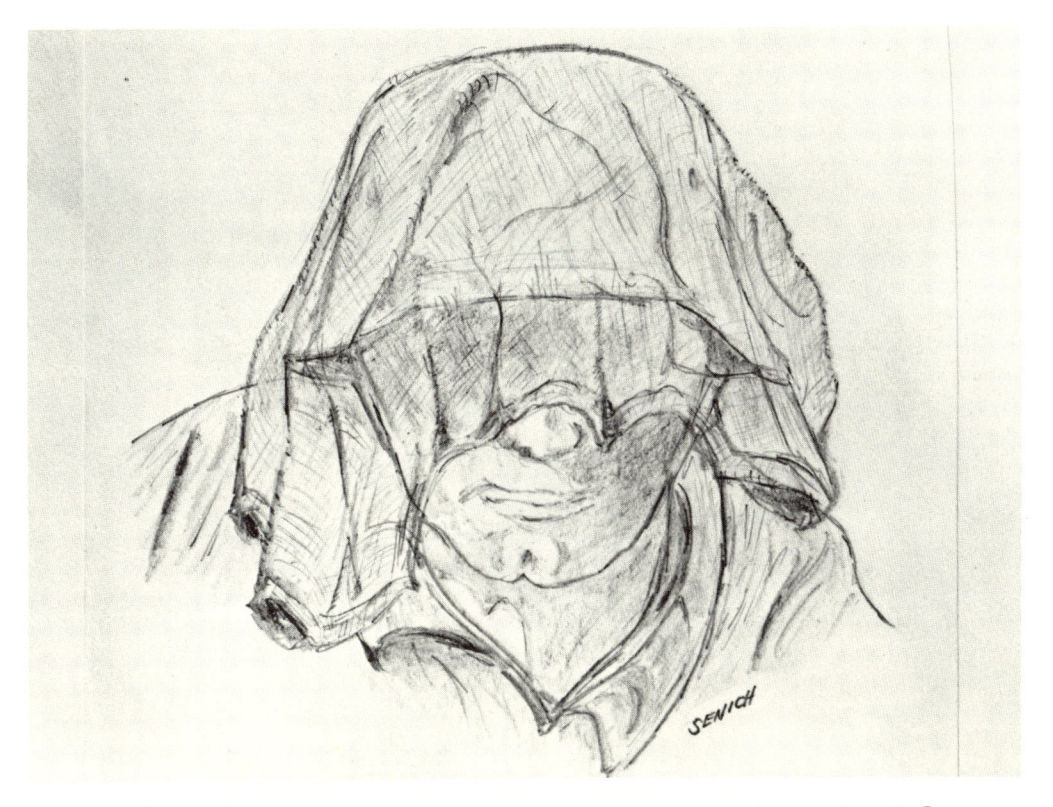

Light, efficient "sniperveils" such as this saw considerable use by all German field forces. The familiar outline of the helmet or cap was easily disguised by use of the veil originally intended as protection from mosquitos.

A prime example of winter camouflage. The sniper has wrapped his scope-sighted 98k with white fabric.

Close view of a camo garbed Waffen-SS rifleman. There was no mistaking the unique SS camo patterns.

German sniper training illustration, circa 1944.

Variant camouflage smock (pullover) without a hood. Although hooded jackets with special loops for attaching natural camouflage (vegetation) were fielded for sniper use, they were not limited to this application.

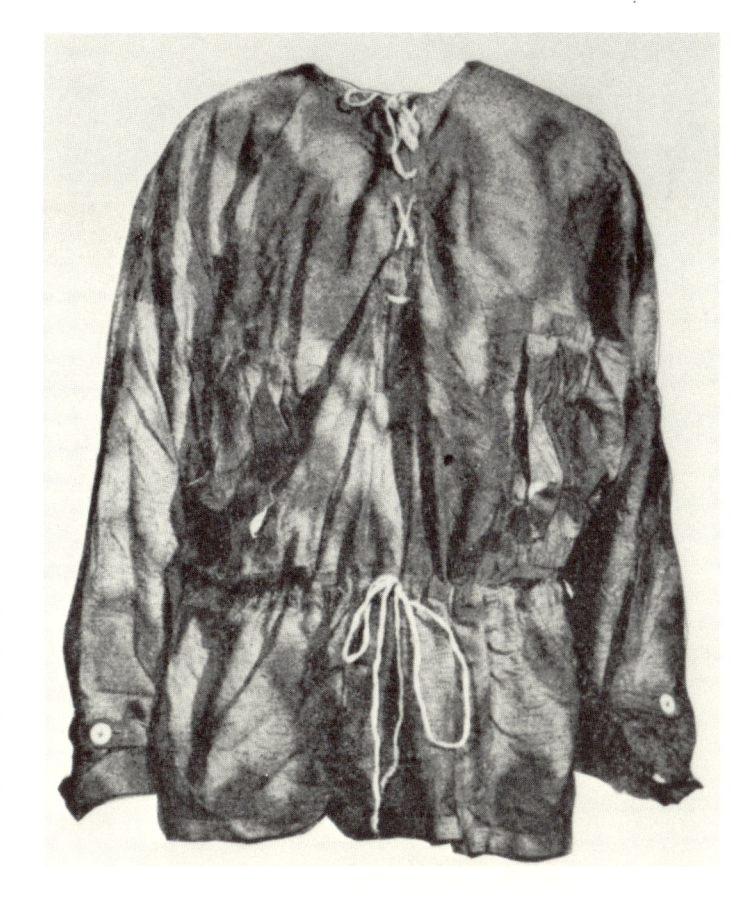

An improvised mask of gauze permitted the sniper or observer reasonable freedom from enemy detection during winter activity.

Waffen-SS combat personnel made effective use of a string-fringe face mask in addition to camouflage helmet covers and jackets.

Covered with pine boughs, a German sniping specialist is shown during training exercises in September 1944.

Waffen-SS sniper-spotter team track a likely target. Note the early 98k short side rail sniping rifle and distinctive camo garb, one of many patterns employed by SS combat personnel throughout the war.

One of two methods employed for climbing. A wire stretched between the sniper's shoes was used for scaling trees as shown.

CAMOUFLAGE — A SPECIAL SKILL 141

Spent cartridge cases driven into the tree trunk at appropriate intervals were also used by German snipers for climbing purposes.

The shelter triangle (Zeltbahn) employed as an improvised tree sling. Impractical perhaps, but a point of German sniper instruction nonetheless.

An interesting photo sequence; No. 1—Outfitted in "mottle pattern" coveralls, a captured Waffen-SS sniper blends with the wall he passes.

No. 2—British soldiers march the SS marksman off to internment.

No. 3—British soldiers shown searching the SS sniper following his capture. The rifleman providing cover holds a No. 4, Mark I(T) sniping rifle (British) with telescope removed. Note the SS "runes" on the German helmet.

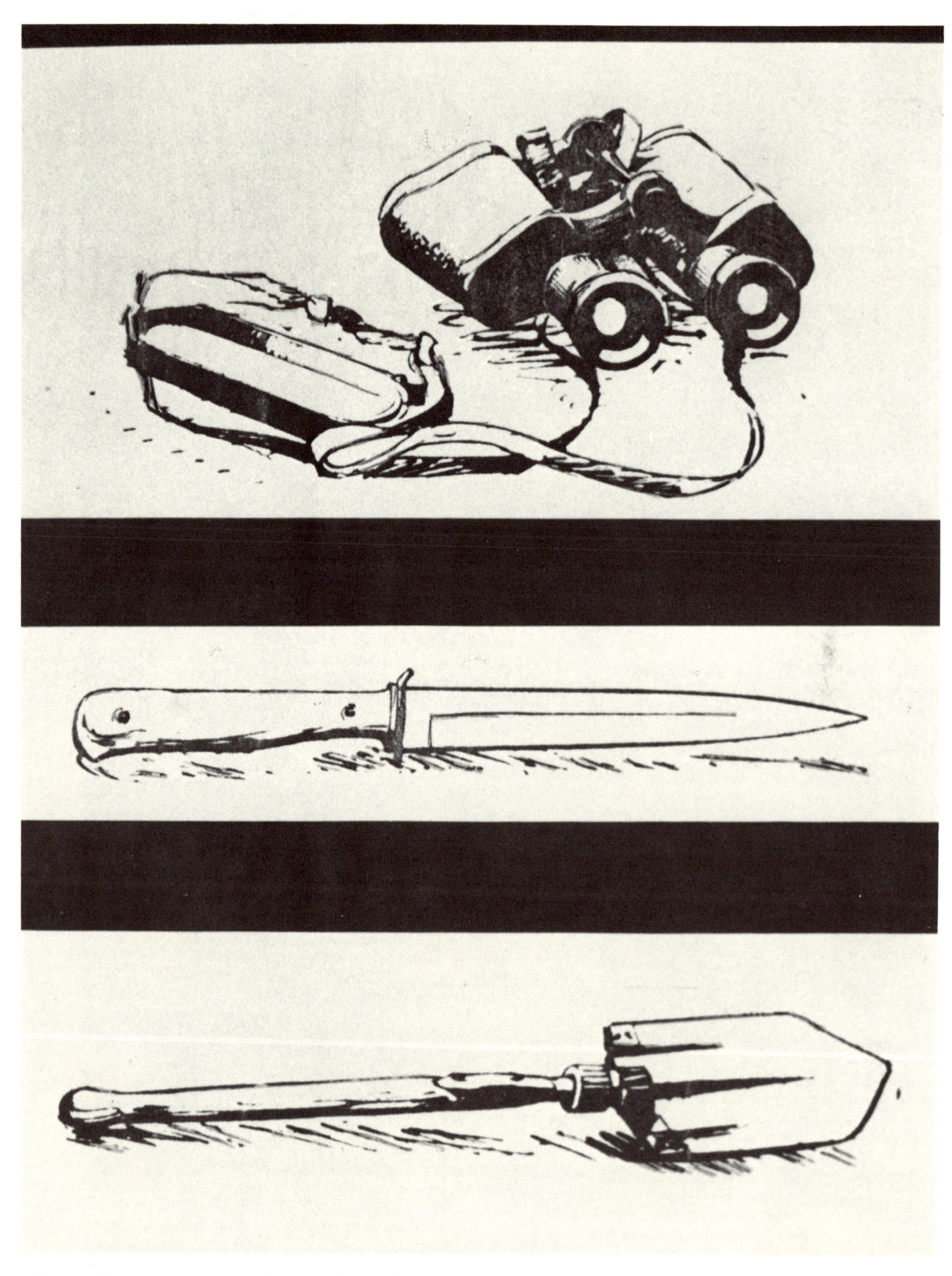

A part of regulation sniper issue: binoculars, infantry knife 42, and short-handled folding entrenching tool, as noted in a late war Panzergrenadiere sharpshooter training manual.

Despite the unique appearance, this variation of face mask was no doubt most uncomfortable.

While generally limited to long-standing, defensive positions, thorough preparations were frequently required to conceal a sharpshooter. (From sniper training illustration, 1944.)

CHAPTER VI

The Sharpshooter Award

As was customary in the armed forces of the Reich, special service badges were awarded those individual soldiers who distinguished themselves in action. The sniping specialists, having gained considerable recognition, proved no exception; on July 1943, the Army High Command, Chief of Army Equipment and Commander of the Replacement Army (Oberkommando des Heeres, Chef H Rust u. BdE.) issued an appropriate decree.

It is intended to introduce a special badge, to be worn on the uniform, for snipers.

Suggestions from the troops as to the form and manner of wearing are desired.

The badge shall not exceed 6 x 4 cm, yet deviations are permissible with otherwise usable suggestions.

It can consist of textiles (embroidered or sewn) or of metal (fastenable) and shall be awarded to all noncommissioned officers and men who have fulfilled the shooting requirements of the sniper class.

Suggestions for the Sniper Badge with simple colored sketches are to be submitted to the Deputy General Command A.A.K. Section IIa0 up until 15 Aug. 1943.

It was intended that the sniper badge replace the Iron Cross (Eiserne Kreuz), which until then was the only visible decoration awarded for sniping prowess, and also show the field troops that German forces had specialists counteracting the well-known Soviet marksmen.

Although a number of design proposals were submitted and rejected, an acceptable design was finally approved. As events transpired, the German military found themselves confronted with a dilemma of their own making: should the badge be a special class of shooting decoration or a standard combat badge. They were also concerned about the Soviet practice of decorating their sharpshooters; Soviet sharpshooting had been highly publicized in Germany as a repugnant form of bounty hunting.

Realization that the approved design drew entirely too much attention to the reason for its existence delayed establishment of the sniper award until the following year. Nevertheless, in revised form The Sniper Badge of the Army (Das Scharfschutzenabzeichen des Heeres) was finally brought to fruition by order of the Fuhrer on 20 August 1944.

The point was specifically made that only those men who were actually trained as snipers and who had fulfilled the field requirements could receive the award.

Orders of the Fuhrer
Order of the fuhrer regarding the establishment of a sharpshooters badge of distinction.
The Fuhrer   Fuhrer Headquarters,
                                  20 August 1944.
1.
In recognition of the great effort of the single rifleman with weapon as a sharp-

Führerbefehle

34. Befehl des Führers über die Einführung eines Scharfschützenabzeichens.

Der Führer.   F. H. Qu., den 20. 8. 1944.

1.

In Anerkennung des hohen Einsatzes des Einzelschützen mit Gewehr als Scharfschütze und zur Würdigung der hierbei erzielten Erfolge führe ich für das Heer und die ⚡⚡-Verfügungstruppe das Scharfschützenabzeichen ein.

Das Scharfschützenabzeichen wird in 3 Stufen verliehen.

2.

Die Durchführungsbestimmungen erläßt der Gen. d. Inf. b. Chef Gen. St. d. H.

Adolf Hitler

Durchführungsbestimmungen zum Führerbefehl vom 20. 8. 1944 über die Einführung eines Scharfschützenabzeichens.

Der Führer hat ein Scharfschützenabzeichen für das Heer und die ⚡⚡-Verfügungstruppe eingeführt. Hierdurch soll der hohe Einsatz des Schützen mit Gewehr und seine Erfolge im gezielten Einzelschuß gewürdigt und gleichzeitig ein Ansporn für eine Steigerung der bisher erzielten Leistungen gegeben werden. Dementsprechend ist das Scharfschützenabzeichen nach folgenden Grundsätzen zu verleihen:

1. Das Scharfschützenabzeichen wird durch den nächsten truppendienstlichen Vorgesetzten mit den Befugnissen mindestens eines Regimentskommandeurs auf schriftlichen Vorschlag des Einheitsführers an solche Soldaten verliehen, die als planmäßige Scharfschützen ausgebildet und eingesetzt sind. Dem Beliehenen ist eine Urkunde über die Verleihung auszustellen und die Verleihung in die Personalpapiere einzutragen (siehe Anlage 2).

2. Das Abzeichen (siehe Anlage 1) ist in 3 Stufen unterteilt und wird auf dem rechten Unterarm getragen. Sofern ein Soldat ein Funktionsdienstgradabzeichen besitzt oder neben dem Scharfschützenabzeichen verliehen bekommt, ist dieses unter dem Scharfschützenabzeichen zu tragen.

3. Es werden verliehen:

   Die 1. Stufe für mindestens 20 Feindabschüsse, die ab 1. 9. 1944 erzielt wurden (Abzeichen ohne besondere Umrandung);

   die 2. Stufe für mindestens 40 Feindabschüsse, die ab 1. 9. 1944 erzielt wurden (Abzeichen mit Silberkordel umrandet);

   die 3. Stufe für mindestens 60 Feindabschüsse, die ab 1. 9. 1944 erzielt wurden (Abzeichen mit goldgelber Kordel umrandet).

Im Nahkampf erzielte Abschüsse werden nicht angerechnet. Im übrigen muß der Feind bewegungsunfähig geschossen sein und darf nicht die Absicht gezeigt haben, überzulaufen oder sich gefangen zu geben.

4. Über jeden Abschußerfolg ist bei der Einheit eine Meldung und Bestätigung durch mindestens 1 Zeugen einzureichen. Die Einheiten legen auf Grund der Meldungen Scharfschützenlisten gemäß anliegendem Muster an (Anlage 3). Ein Auszug aus der Scharfschützenliste ist bei Versetzungen der neuen Einheit zusammen mit den sonstigen Papieren zu übergeben. Eine rückwirkende Anrechnung von Abschüssen erfolgt nicht, um unnötigen Schriftverkehr zu vermeiden. Es wird vielmehr vorgeschlagen, die bisherigen Leistungen durch die Truppe bei der Verleihung von Eisernen Kreuzen mit bewerten zu lassen.

O. K. H., 20. 8. 1944
Gen. d. Inf. b/Chef Gen. St. d. H.

Zusatzbestimmungen des O. K. L. (LP) zum Führerbefehl vom 20. 8. 1944 über die Einführung des Scharfschützenabzeichens.

1. Laut O. K. W., 29 b 28. 14/9215/44 WZA/WZ III c, vom 14. 12. 1944 hat der Führer entschieden, daß das Scharfschützenabzeichen den im Erdkampf eingesetzten Soldaten aller Wehrmachtteile verliehen werden kann, wenn die Voraussetzungen für die Verleihung erfüllt sind.

2. Im Bereich der Luftwaffe erfolgt die Verleihung des Scharfschützenabzeichens auf Vorschlag des Einheitsführers durch den nächsten truppendienstlichen Vorgesetzten mit den Befugnissen mindestens eines Geschwaderkommodore bzw. Regimentskommandeurs. In Zweifelsfällen ist Entscheidung bei O. K. L. (LP) zu beantragen.

3. Die vorstehend bekanntgegebenen Durchführungsbestimmungen des Gen. d. Inf. b. Chef Gen. St. d. H. gelten in vollem Umfange auch für die Luftwaffe.

4. Besitzzeugnisse gemäß Anlage 2 sind von den Verleihungsdienststellen selbst zu beschaffen.

5. Der Bedarf an Abzeichen ist auf dem Bekleidungsnachschubwege zu decken. Anforderungen sind an das örtlich zuständige Luftgaukommando (Verwaltung) zu richten.

O. K. L., 2. 1. 1945.
Az. 29 g 10/LP Ausz. u. Diszpl. (I A).

L. V. Bl. S. 28

The original sharpshooter award orders of Adolf Hitler, as taken from World War II German records.

shooter and in order to honor the successes obtained therewith I establish for the Army and SS-Reserve the sharpshooter badge of distinction.

The sharpshooter badge will be given in three degrees.

2.

The command is issued by the General of the Infantry, Chief of the General Staff of the Army.

Adolf Hitler

Command by order of the Fuhrer of 20 August 1944 regarding the establishment of a sharpshooter badge.

The Fuhrer has established a sharpshooter badge for the Army and the SS-Reserve. With it the great effort of the rifleman with weapon and his successes in the aimed single shot shall be honored and at the same time it shall be an incentive for an increase in the efforts obtained up to this point. Accordingly, the sharpshooter badge is to be granted according to the following guidelines:

1. The sharpshooter badge will be granted by the next (in troop order) superior within the troop with a minimum rank of Troop Commander after a written suggestion by the unit leader of those soldiers, who were trained and put in action as regular sharpshooters. The selected is to be given a document regarding the granting of the badge and this document is to be entered in the personal file. (see enclosure 2)

2. The badge (see enclosure 1) is divided into three sections and will be worn on the lower right arm. Inasmuch as the soldier has a service rank badge, or is granted one besides the sharpshooter badge, it is to be worn below the sharpshooter badge.

3. It will be granted:

The first degree for at least 20 enemies shot, accomplished since 1 Sept. 1944. (Badge with no special border)

The second degree for at least 40 enemies shot, accomplished since 1 Sept. 1944. (Badge border with silver cord)

The third degree for at least 60 enemies shot, accomplished since 1 Sept. 1944. (Badge bordered with golden-yellow cord)

Shots made in close combat will not be counted. Also the enemy must fall immovable and must not have shown the intention to give himself up or allow himself to be taken prisoner.

4. Notification handed in to the unit regarding each successful shot must be accompanied with certification by at least one witness. The units will set up sharpshooter lists according to such notification and confirming to the enclosed pattern (enclosure 3). An abstract from the sharpshooter list is to be given to the new unit along with all personal papers in case of transfer. A retroactive accounting of shots does not follow in order to avoid unnecessary paper work. It is suggested, however, that the efforts of the troops be considered toward the granting of Iron Crosses.

Army High Command, 20 August 1944
General of the Infantry, Chief of the General Staff of the Army.

It is interesting to note that when first presented in the original sharpshooter award order, the expression "SS-Reserve" (SS-Verfugungstruppe) was inadvertently used in place of "Waffen-SS," the official reference for this organization since 1940. This oversight was amended by decree of the Army High Command under date of 28 September 1944 and all mention of "SS-Reserve" was deleted and changed to "Waffen-SS."

The recipient of the sharpshooter award was considered the elite among German riflemen and to display this badge as authorized was undoubtedly done with great pride. Perhaps the stringent requirements set forth limited actual issuance to an infrequent occurrence because very few individuals having this badge were killed or captured by the Allies.

Captain C. Shore in his classic work, *With British Snipers to the Reich*, comments:

It is perhaps interesting to record that throughout the campaign in N.W. Europe from D-Day to May 8th, 1945 and afterwards in Germany from Westphalia to the

The Sharpshooter Badge (official design) in final form as it appeared in the General Army Notices dated 7 September 1944. As a result of Soviet troops executing any German prisoner (sniper) displaying the Sharpshooter Award, an official late war directive (1945) recommended removing the special badge when capture was emminent.

Original Sharpshooter Badge design as approved by the Supreme Commander of the Armed Forces in 1943. Eventually deemed as too distinctive, this design was never brought to fruition.

The Sharpshooter Badge with gold border for 60 kills. Actual size about 70mm x 53mm. Prior to its award, the Iron Cross had been given for sniping proficiency. It was intended that the badge replace the Iron Cross, which was thought to be awarded for too many achievements, and show the troops in the field that the German forces had specialists counteracting the well-known Soviet snipers.

Danish border I never met a German wearing the German sniper's badge, an eagles head (which we took to be symbolic of the observation side of the sniper's art); and despite the most exhaustive enquiries, and during my time at the Training Centre I was in a sound position for making such an enquiry, I never came across anyone who had met a Hun with the sniper badge.

The badge itself was comprised of an eagle's head of black and white in the center of a grey background having dark cinnebar green oak leaves. The outer ring was light green and the eye and beak of the eagle were ochre.

It is indeed a fortunate militaria student or collector who possesses an authentic German sniper's badge regardless of degree, and he most certainly would be in Valhalla to possess all three.

On 14 December 1944, eligibility for receipt of the sharpshooter award was expanded to include all ground combat units Heer, Waffen-SS, Luftwaffe, and Kriegsmarine.

Additions to the Regulations of the Air Force High Command (Air Force Personnel) to the order of the Fuhrer of 20 August 1944, concerning establishment of the Sharpshooter Badge.
1. According to Armed Forces High Command document 29b28 14-9215-44 WZA-Wz IIIc, of 14 December 1944 the Fuhrer has decided that the sharpshooter badge can be given to all branches of the armed forces engaged in ground combat when the prerequisites are fulfilled.
2. In the area of the Air Force granting of the sharpshooter badge will be made upon suggestion of the unit leader to next (in troop order) superior with the minimum rank of Squadron Commander or Regiment Commander. In cases of doubt a decision is to be applied for through the Air Force High Command.
3. The preceding published regulations by the General of the Infantry and Chief of the General Staff of the Army are also valid to the fullest extent for the Air Force.
4. Certificates according to enclosure 2 are to be procured by the granting service units.
5. The supply of badges is to be covered under clothing supply placement. Requisitions are to be addressed to the Air District Command appropriate for the area (Administrative).

Air Force High Command (OKL) 2 January 1945

On 3 January 1945, the Navy High Command (Oberkommando der Kriegsmarine or OKM) joined with the Luftwaffe and officially announced the establishment of the sharpshooter award in accordance with existing requirements. The obvious purpose of the award extension was for inclusion of Luftwaffe personnel who by late 1944 were, for the most part, serving as field infantry. While Luftwaffe Field Divisions had been organized specifically for ground combat activity early in the war, the extent of Luftwaffe sniper training or of actual combat efficiency remains obscure. The elite Fallschirmjager (paratrooper) units as originally formed included a small number of men in each platoon trained and equipped with sniping weapons, the purpose being to supplement the then limited long-range fire power.

Captain C. Shore cites an interesting anecdote relating to Luftwaffe snipers.

> I know of one occasion when a Scottish Battalion met some real German snipers, and, according to the Battalion sniping warrant officer it was the only time they had come across true German sniping shooting. It was outside a small village, Francofonte in Sicily. The Battalion bumped rather suddenly into a German Parachute Regiment which was holding high ground, and their snipers made some truly remarkable shooting at ranges of 600 yards. They stuck to their positions and kept up their brilliant shooting even when subjected to heavy shell-fire. They were of course, well dug-in in previously prepared positions. When they were eventually driven out by the advancing troops, the majority of these snipers got away by use of the most excellent fieldcraft.

Issue paper (certificate) for the Scharfschutzenabzeichen (sharpshooter award). Approximate full size, 127mm x 177mm.

The German sniper in classic form: the telescopic-sighted rifle in one hand, the binoculars in the other. Note the Iron Cross and scope devoid of all finish.

THE SHARPSHOOTER AWARD

Left behind and isolated in a large mine field northeast of Caen, Gefreiter Kurt Spengler brought down a "considerable number" of enemy troops with his telescopic-sighted rifle before being killed by repeated shelling. Reported later by captured British personnel, Spengler's deeds were duly noted in a World War II German army publication.

THE GERMAN SNIPER 1914–1945

Prepared for action, a Fallschirmjager (paratrooper) Unteroffizier (NCO) wears the distinctive early pattern jump-smock. The bandolier contains ammunition for the scope-equipped 98k rifle.

THE SHARPSHOOTER AWARD

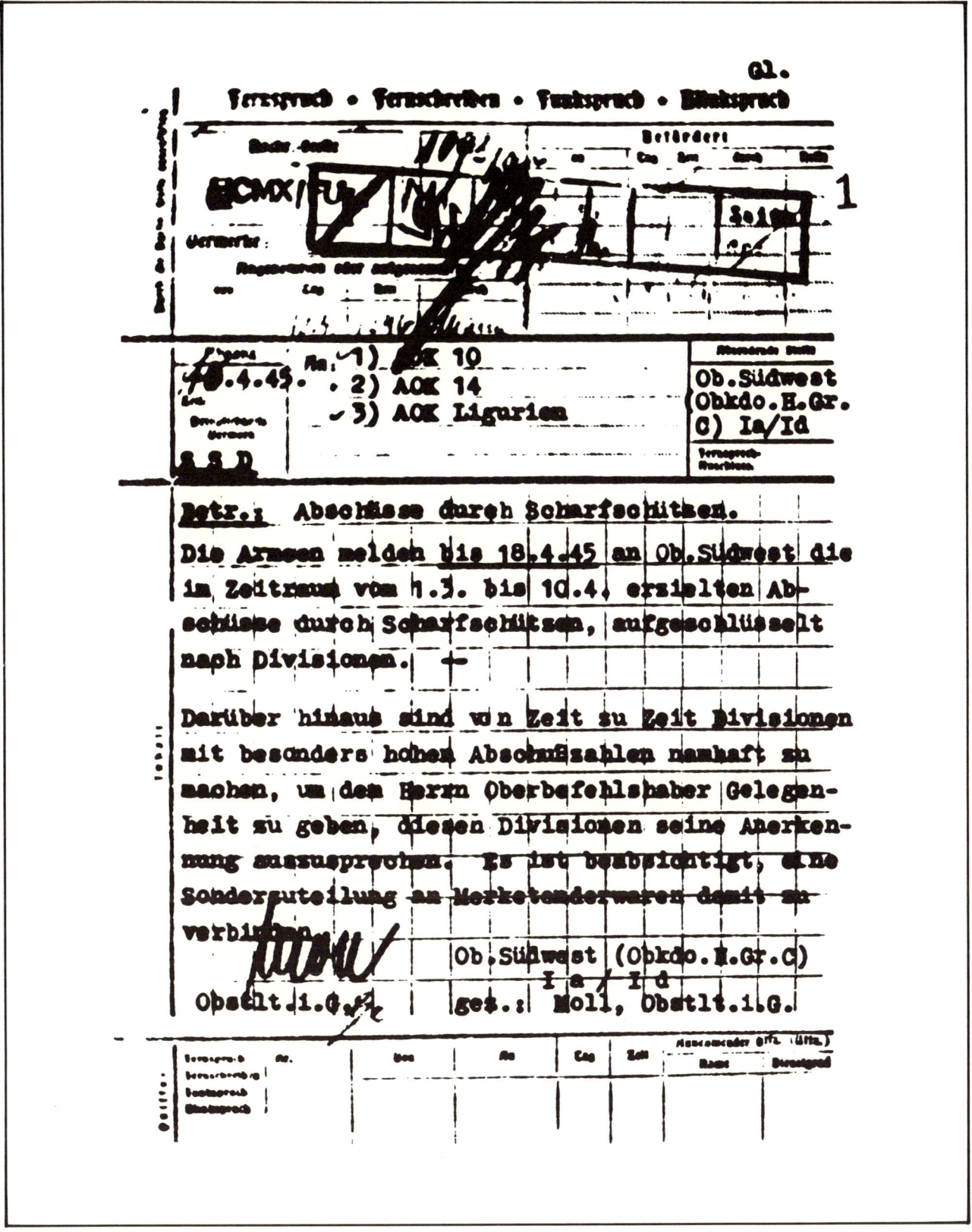

As directed by the High Command, Army Group C, a count of enemy casualties inflicted by the division sharpshooters between 1 March and 10 April 1945 were to be reported for the purpose of recognition and for the special distribution of canteen goods to those divisions with the highest count.

In response to the High Command directive (Army Group C), a total of 451 sharpshooter kills were reported by nine combat divisions. Note the 236 tally in one case.

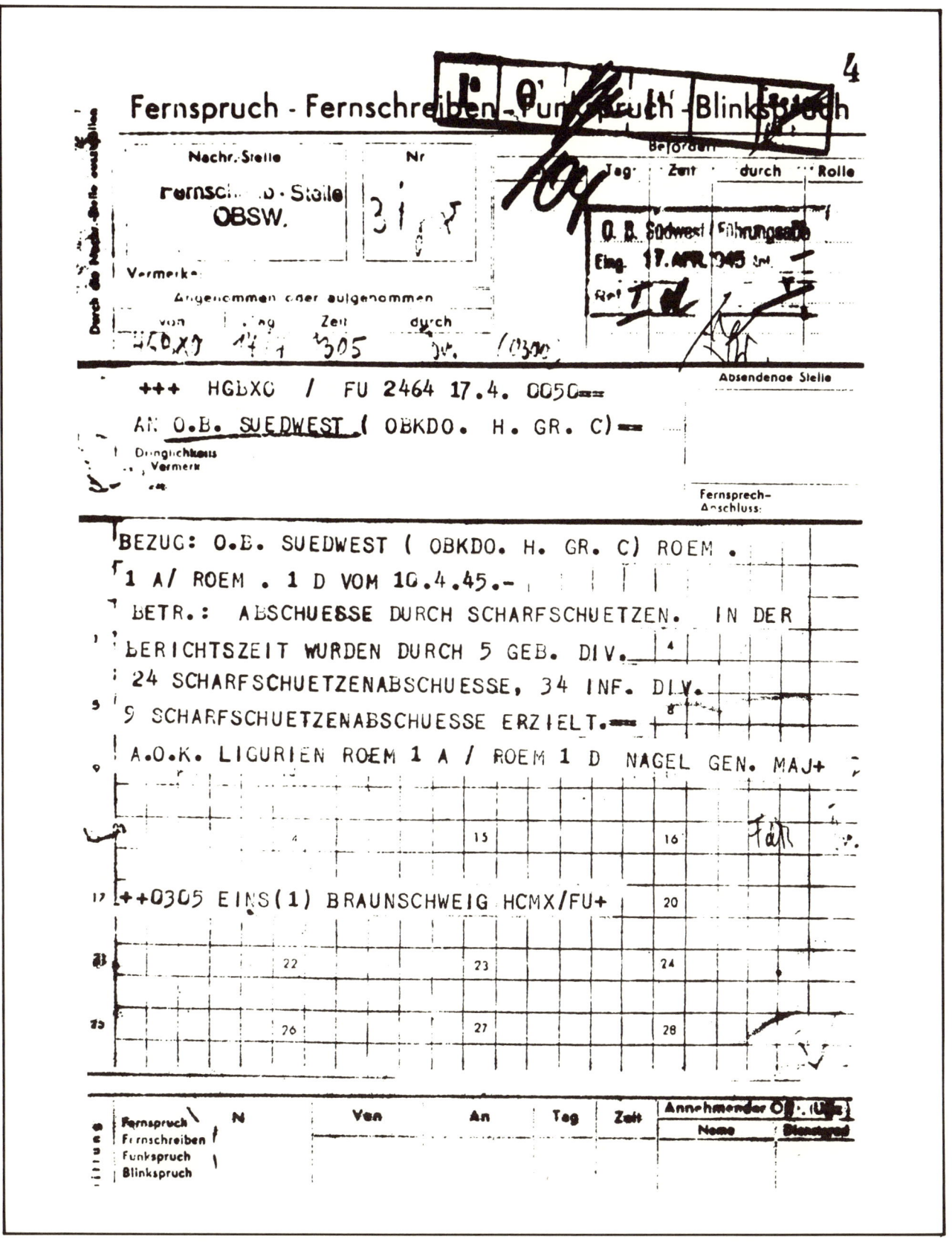

Further reply to the High Command directive (Army Group C) of 10 April 1945 included 5 Gebirgsdivision and 34 Infanteriedivision reporting 24 and 9 sharpshooter kills respectively.

CHAPTER VII

Karabiner 98k — Short Side Rail System

Adopted by the infant Third Reich for special applications during the mid-30s, the short side rail variation of the Karabiner 98k was originally intended for police use during civil strife (riot control).

As prepared for paramilitary use, standard 98k rifles were selected and tested for accuracy, after which a contoured "short" side rail having a dovetailed upper surface was carefully hand-fitted to the left side of the receiver.

The earliest rails were attached with three screws, and quite often the entire base would work loose from the shock of repeated recoil. The addition of three auxiliary locking screws and eventually two tapered pins was employed for added security. The U-shaped telescope mounting was a high grade machining with a single lever mechanism for locking.

The telescopic sights used on these early Reich sharpshooting weapons were usually 4-power Ajack, Zeiss, Hensoldt, or Kahles hunting scopes taken directly from commercial stock. Machined steel rings with two lock screws held the scope in its proper position on the mount.

Despite early deficiencies, the short side rail method of telescope mounting was refined and saw general service use between 1941 and 1945 in all combat theaters, with 98k rifles of various manufacture serving as the base weapon.

Even though the short side rail system is considered one of the more common World War II German sniping variations, those produced specifically for use by the early Waffen-SS combat units are viewed with considerable interest. Unique in their configuration, the SS short side rail variants were based on reworked Gewehr 98 receivers with barrels, stocks, and hardware of current 98k manufacture.

Although a quantity of reworked SS rifles was reportedly fabricated in 1939-40, in addition to the sharpshooting version, an unspecified number of standard shoulder arms of the same configuration (reworked Gew. 98 receivers) was also included in this total.

In spite of this involvement by various firms in this case, a number of original pieces having definite Mauser origins have been noted bearing a "S/243" (Mauser, Borsigwalde), "S42" or "42" (Mauser, Oberndorf) production mark stamped on the barrel (left side) or receiver (right side). Aside from the recognizable Mauser variants, however, the lack of identification (Waffenamts or production marks) on most SS rifles of this type makes it virtually impossible to ascertain the original firm responsible for their assembly.

As far as has been determined, commercial 4-power sights were utilized with SS marked Ajack or Hensoldt scopes. The receiver bases were stamped with the corresponding rifle number, but this practice did not appear to have been standard, since authentic examples without base numbering have been observed as well. Prior to its issue in original form, the mount was stamped and the telescope was engraved with the rifle serial number.

The SS use of reworked rifles as the basis for their early sniping issue rather than regular production K98k's amounted to the path of least resistance at that time. The concept and subsequent formation of full-fledged SS combat units was anything but well-received by the hierarchy in the regular German army. The SS were hard-pressed to draw sufficient materiel through normal supply channels; as a result, during early activity (1939-40) they were equipped with something less than the cream of German weaponry in many cases.

Therefore, as experts conclude, the early Waffen-SS use of reworked World War I rifles (both standard and sniping configuration) constituted a simple expediency.

Camo garbed Waffen-SS sniper-observer team armed with the short side rail 98k rifle.

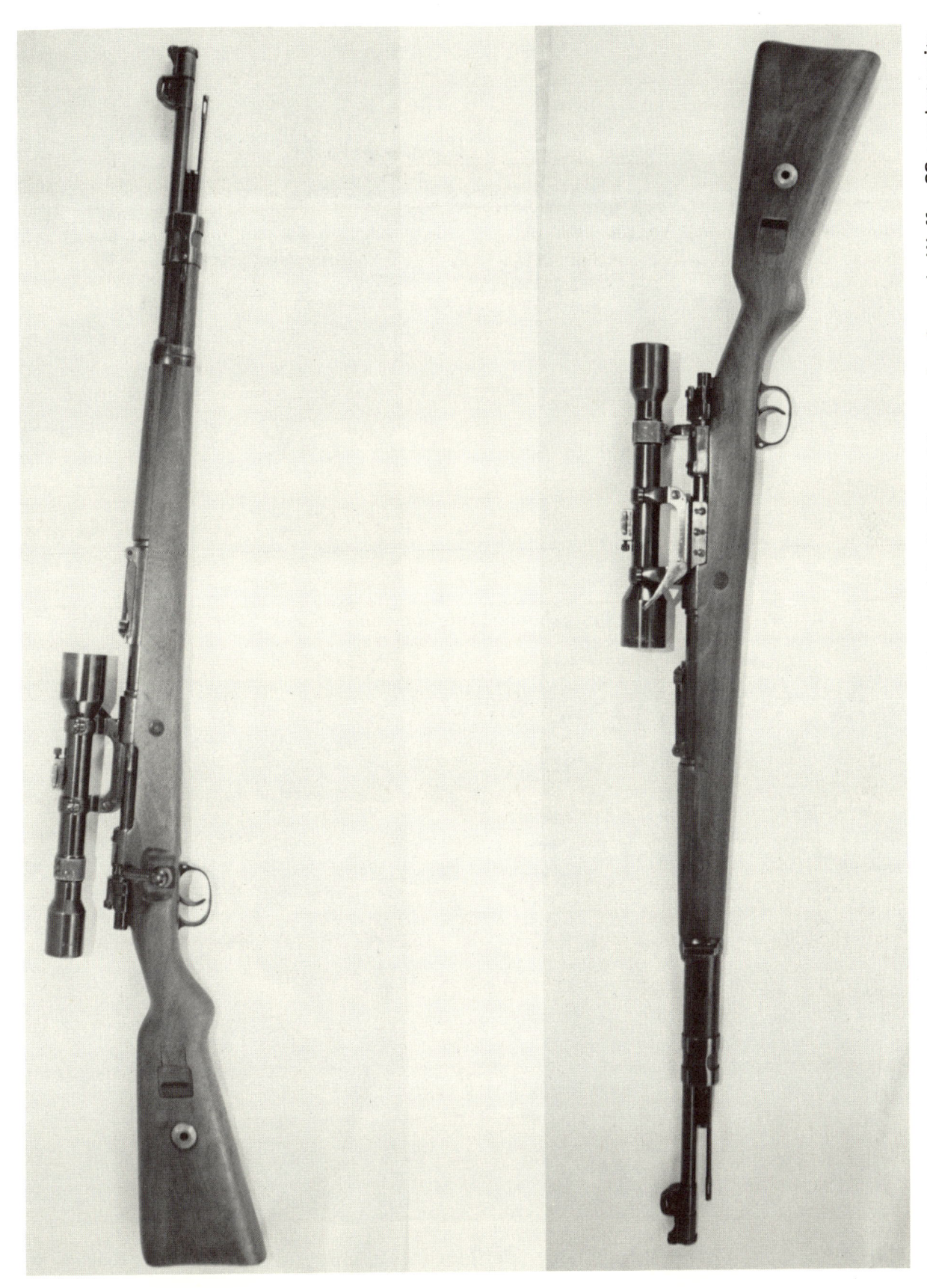

Short side rail sniping rifle (S/42) with 4-power Ajack telescopic sight as originally fielded for use by the early Waffen-SS combat units.

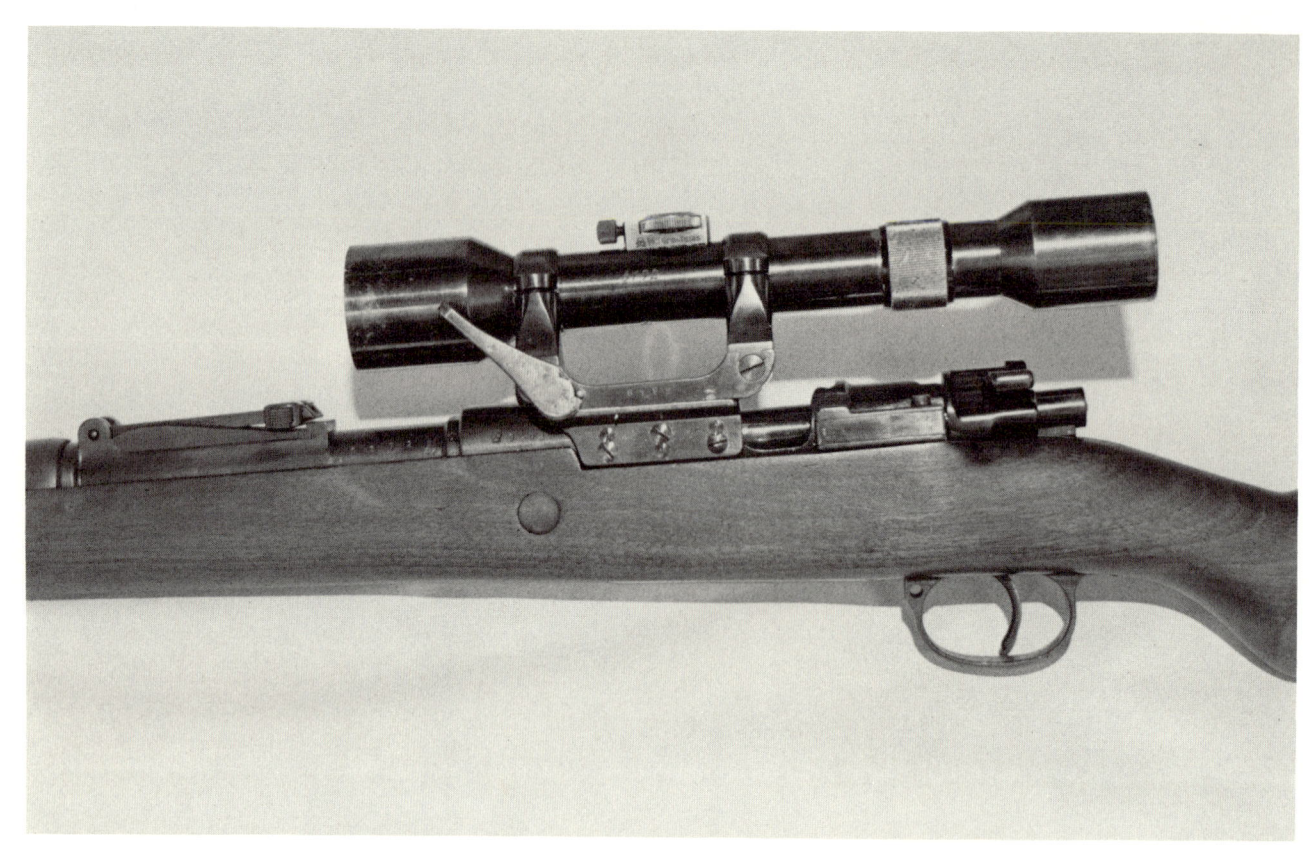

The "S/42" (Mauser, Oberndorf) SS sharpshooting rifle and finely finished SS marked Ajack sight. The knurled ring served to focus the telescope.

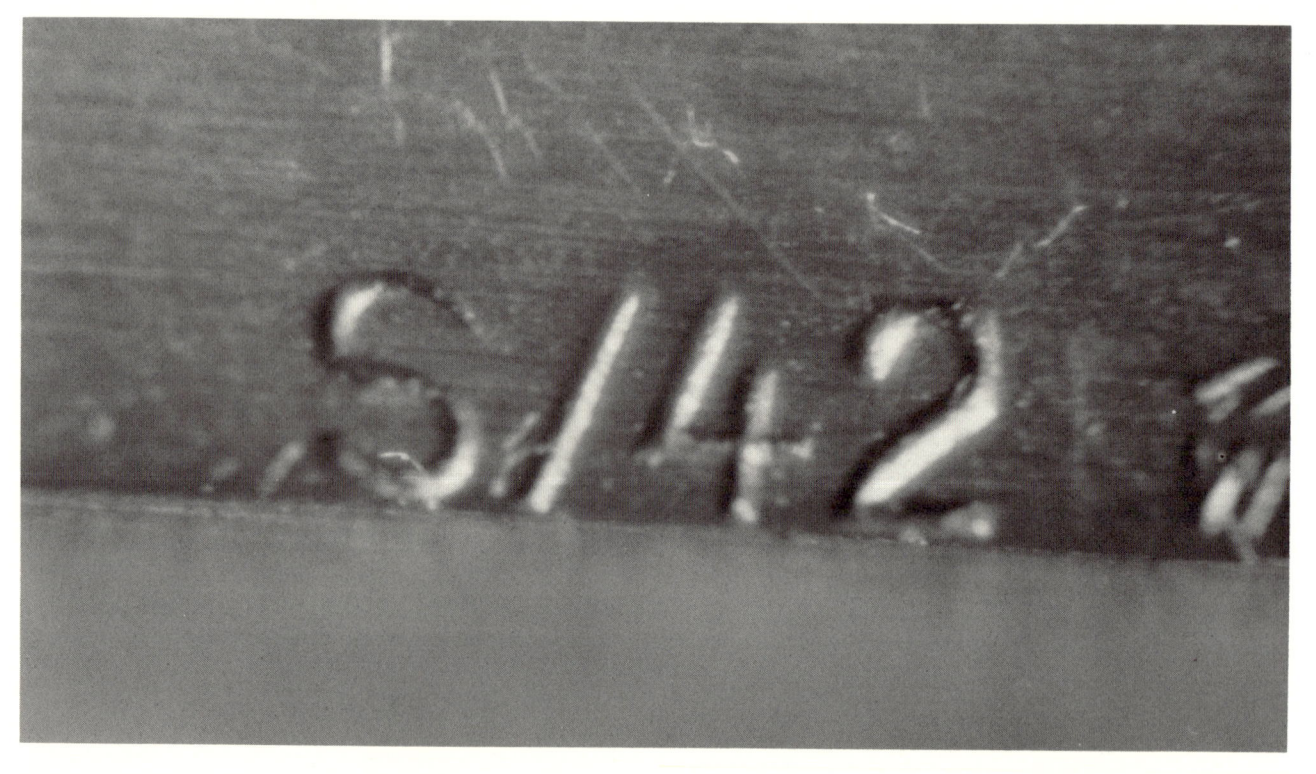

Manufacturer's code "S/42" (Mauser, Oberndorf) impressed on the right side of the receiver ring directly above the stock line along with "WaA 63" Waffenamt stamps.

KARABINER 98K — SHORT SIDE RAIL SYSTEM

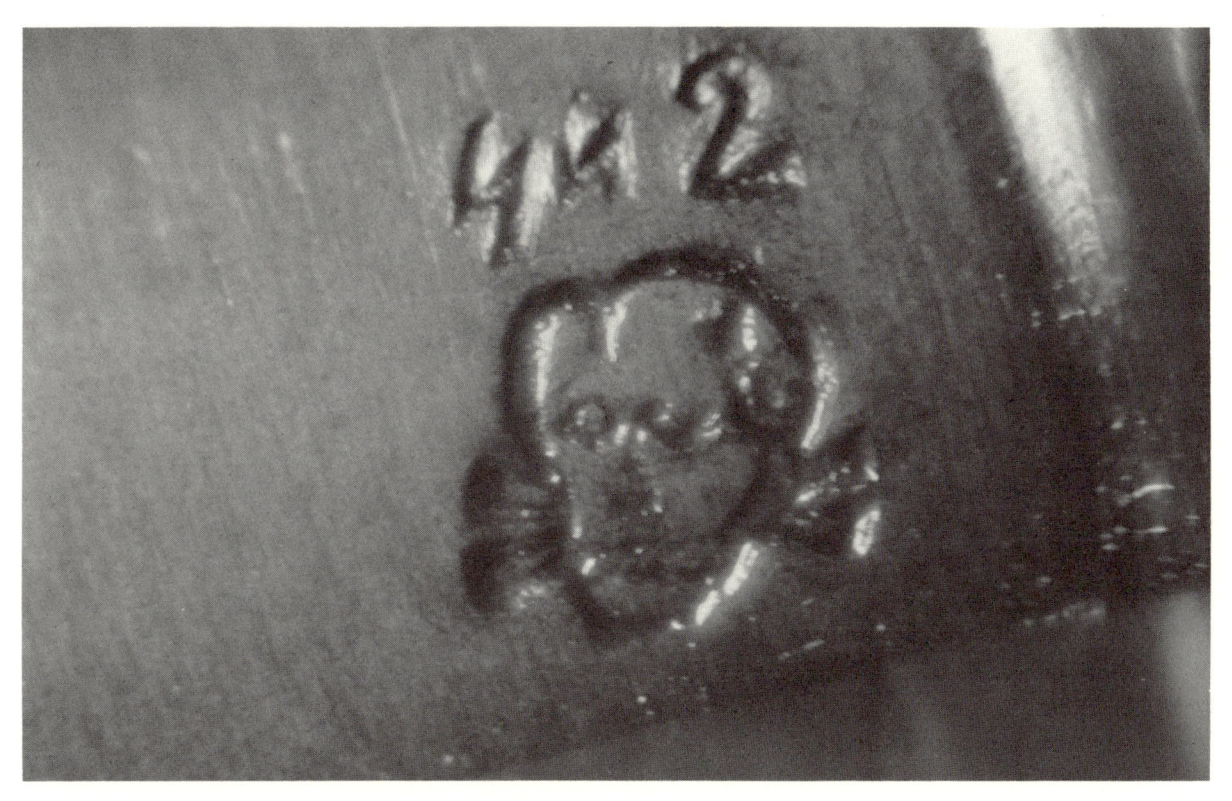

The Totenkopf (Death's Head) proof with SS "runes" and the numeral "2" as stamped on the left side of the barrels on SS Gew. 98 rework rifles. While unconfirmed, this marking is believed to represent the SS ordnance depot (2) responsible for their acceptance or distribution.

Close view of typical SS Totenkopf marking as they appear on the Ajack service glass (Dienstglas) procured for use with the short side rail rifle.

A close view of the Mauser origin (S/42, right side of the receiver ring) SS short side rail 98k rifle with receiver base removed. The top of the receiver ring was milled to remove original manufacturer's identification, a characteristic typical of early SS issue rework rifles. Note the "Gew. 98" legend and threaded base mounting holes.

KARABINER 98K — SHORT SIDE RAIL SYSTEM

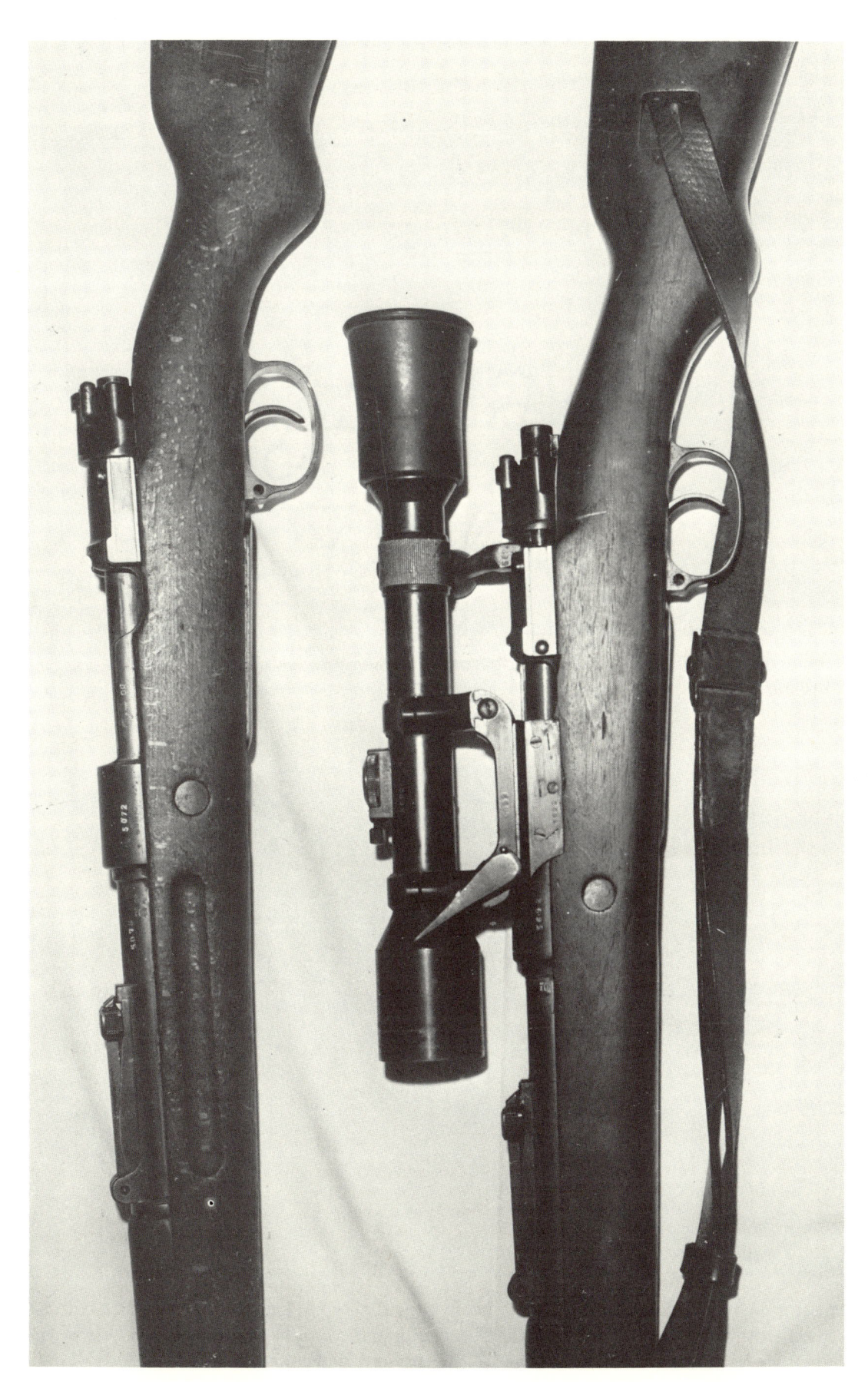

(Top) A standard 98K (rework) made for SS use. Note the "Gew. 98" legend and Death's Head markings on the barrel before the receiver ring. (Bottom) Excellent finish and workmanship are characteristic of the SS short side rail sniping variants. Matching serial numbers on the base, mount, scope (SS Dienstglas), and rifle make this a highly desirable collection piece. Note the bolt release latch from a Polish Radom, which indicates the practice of improvisation by German armorers.

Short side rail receiver base (Waffen-SS) showing the front stud over which the mount locking lever closes. Note the careful finishing of the screw heads.

Short side rail mount in half-locked position. The groove on the inside bottom of the lever is loosely fitted over the stud at the front edge of the base. A firm push downward on the lever will lock the scope in place. Each time the rifle is fired, the recoil increases the forward tightening action.

KARABINER 98K — SHORT SIDE RAIL SYSTEM

Under view of short side rail telescope mount. The raised ridge at the rear of the dovetail acts as a stop which prevents the mount from sliding too far forward on its base. Note the locking groove on the bottom of the lever. Large screw holds the front ring on the mount. This screw must be loosened to adjust windage; in order to move the rear of the scope the front ring must also be free to pivot.

Close view of SS Dienstglas (service glass) markings on a short side rail Ajack scope elevation adjustment base. The "1943" numbering corresponded to the rifle serial number. Early SS marked 4-power Hensoldt scopes have been noted as well.

The SS Totenkopf (Death's Head) stock marking located in this instance immediately behind the rear trigger guard screw.

KARABINER 98K — SHORT SIDE RAIL SYSTEM

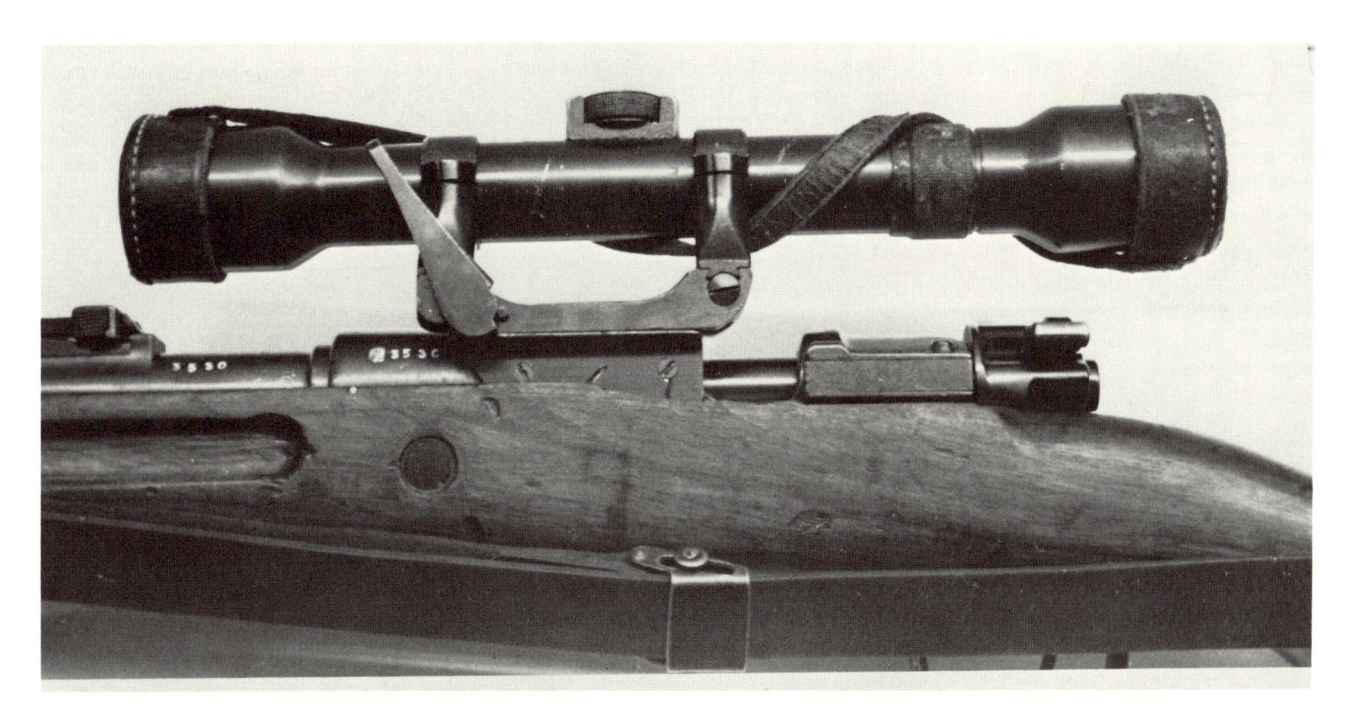

A typical, unidentified (no manufacturer or component assembly markings) early original Waffen-SS short side rail sniping rifle mounting a 4 x 81 Kaba commercial scope.

Close view of the SS short side rail variant. Note the SS Totenkopf, original Gew. 98 Imperial proof, rifle serial numbers, and barrel manufacturer's markings. In this case the receiver base was not numbered.

A World War II era 98k SS rework rifle in standard configuration captured by a U.S. Army Officer in Vietnam in 1967. Although two identical Totenkopf barrel markings have been noted on SS rifles of this type, those with only a single marking are most frequently encountered. The "five-pointed star" is virtually identical to that which appears on SVW P-38 pistols produced by the French and is believed to be a "pressure/stress" proof.

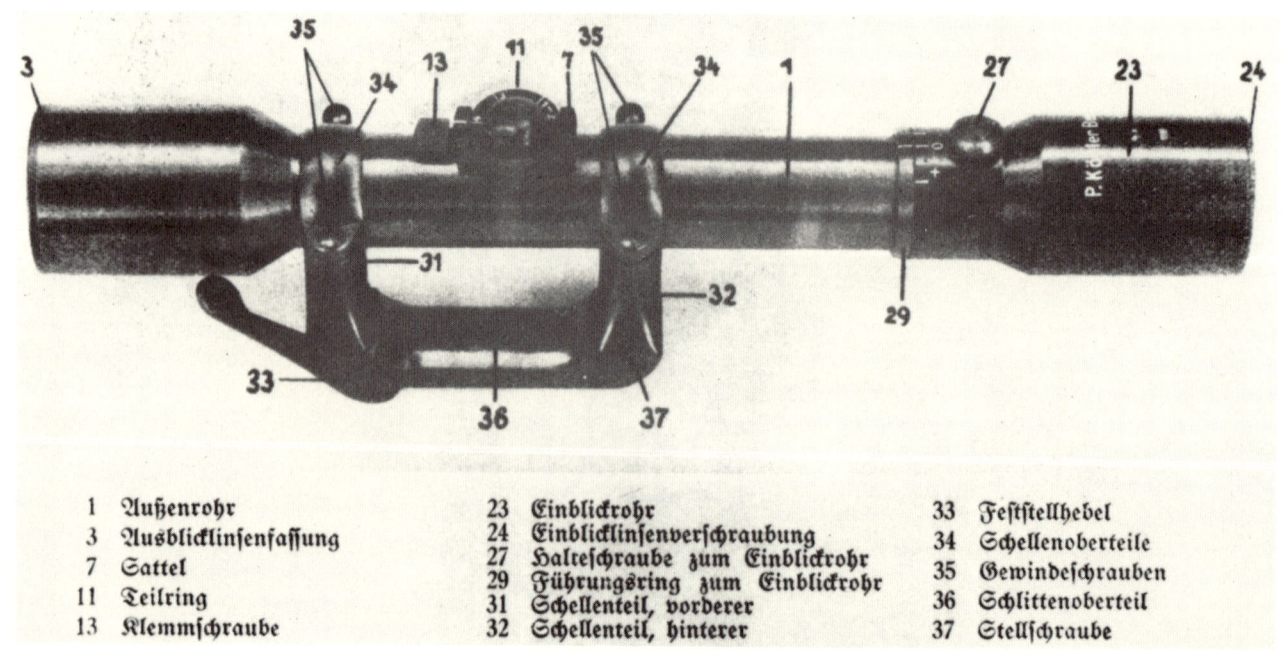

1 Außenrohr	23 Einblickrohr	33 Feststellhebel
3 Ausblicklinsenfassung	24 Einblicklinsenverschraubung	34 Schellenoberteile
7 Sattel	27 Halteschraube zum Einblickrohr	35 Gewindeschrauben
11 Teilring	29 Führungsring zum Einblickrohr	36 Schlittenoberteil
13 Klemmschraube	31 Schellenteil, vorderer	37 Stellschraube
	32 Schellenteil, hinterer	

An early illustration of a short side rail mount assembly with a P. Kohler commercial scope as taken from the 1941 publication *Waffen und Schiesstechnischer Leitfaden fur Die Ordungspolizei*. As then stated "The telescopic sight carries the factory number of the carbine to which it matches and with which it will be used. It may not be used on another carbine. The matching is done in every case at the Ordnance Department of the Technical Police Academy."

KARABINER 98K — SHORT SIDE RAIL SYSTEM

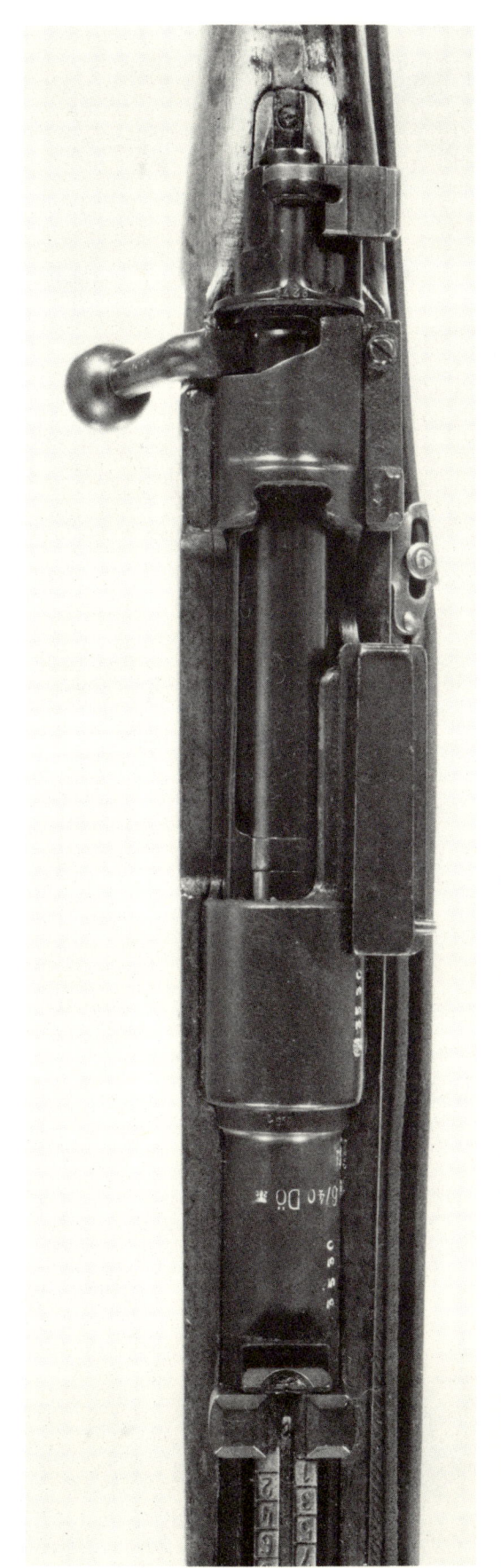

Top view of SS short side rail rifle illustrates positioning of the base on the reworked Gew. 98 receiver. Note the absence of any markings on top of the receiver ring (typical).

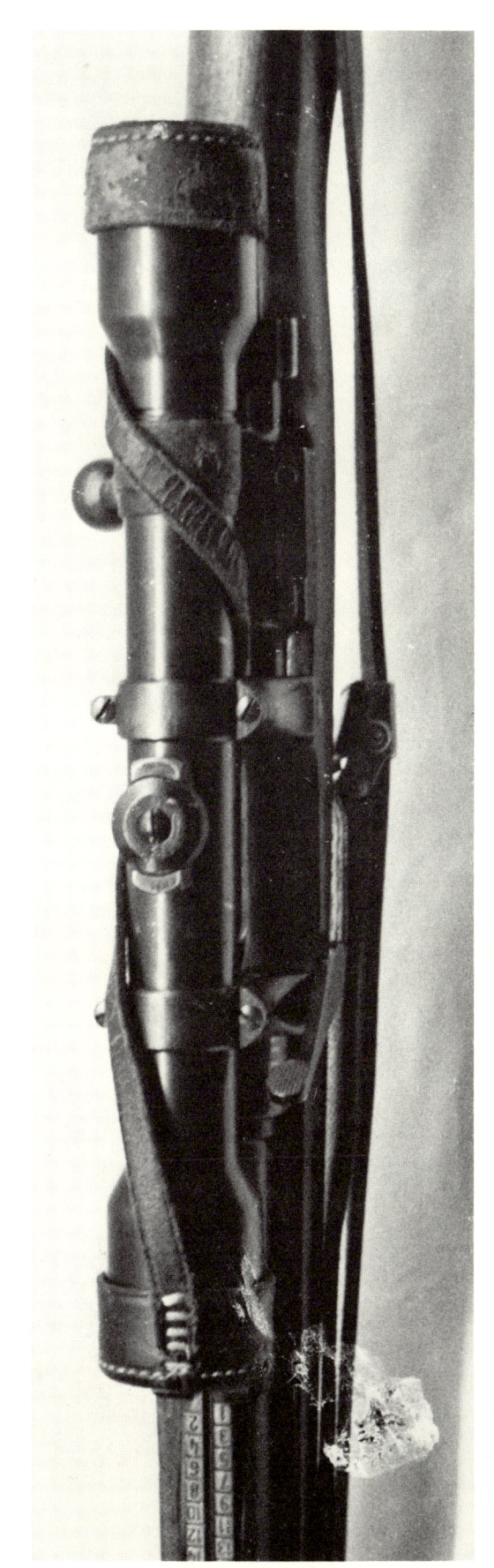

Top view of SS short side rail 98k with Kaba scope illustrates positioning of the sight in relation to the rifle centerline.

An excellent front view of the short side rail sniping system in combat use.

Typical side rail telescope mounting with "Weihrauch HWZ" logo. Believed to have been the prime contractor, Hermann Weihrauch Waffenfabrik, Zella-Mehlis, mounts have also been noted with the variant trademark HZW. All mounts do not bear these markings, however.

Right view of short side rail 98k. Note the skillful use of burlap sacking to disguise the head and helmet form. The sight is a commercial variant with a focus ring.

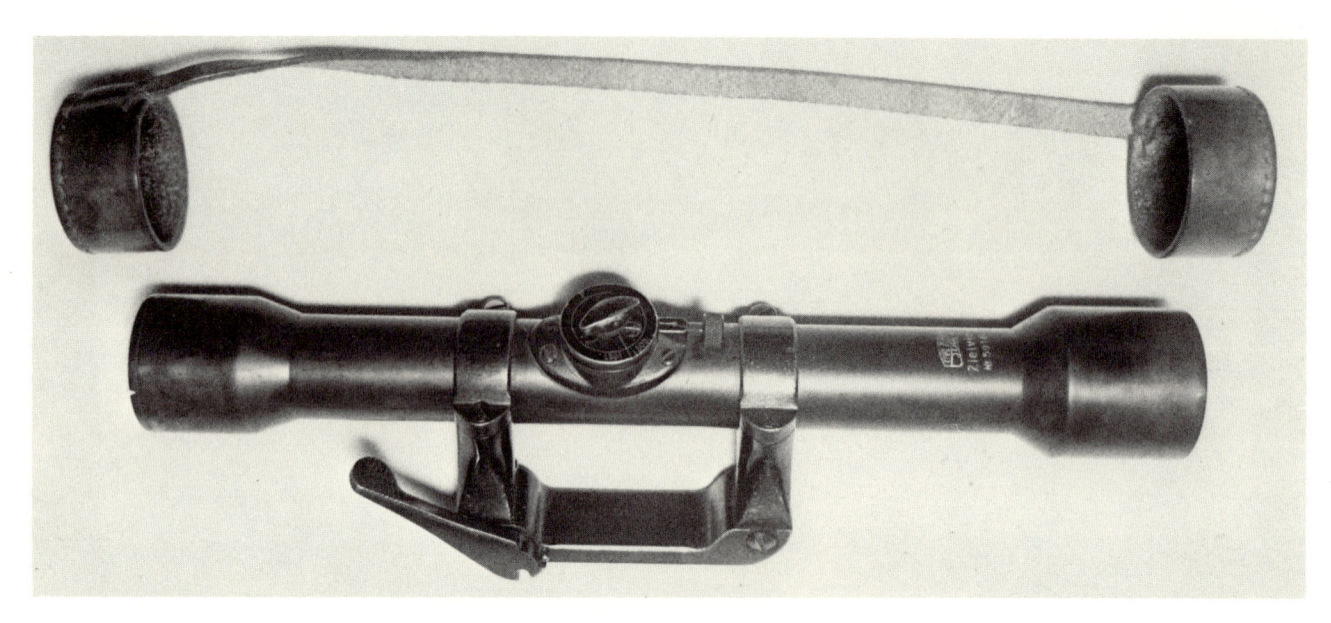

Variant side rail mount with a prewar 4-power Zeiss (Zielvier) telescopic sight. Short side rail mounts and bases are not always interchangeable due to slight manufacturing variations and the locking lever notch not engaging properly with the pin-protrusion on the base.

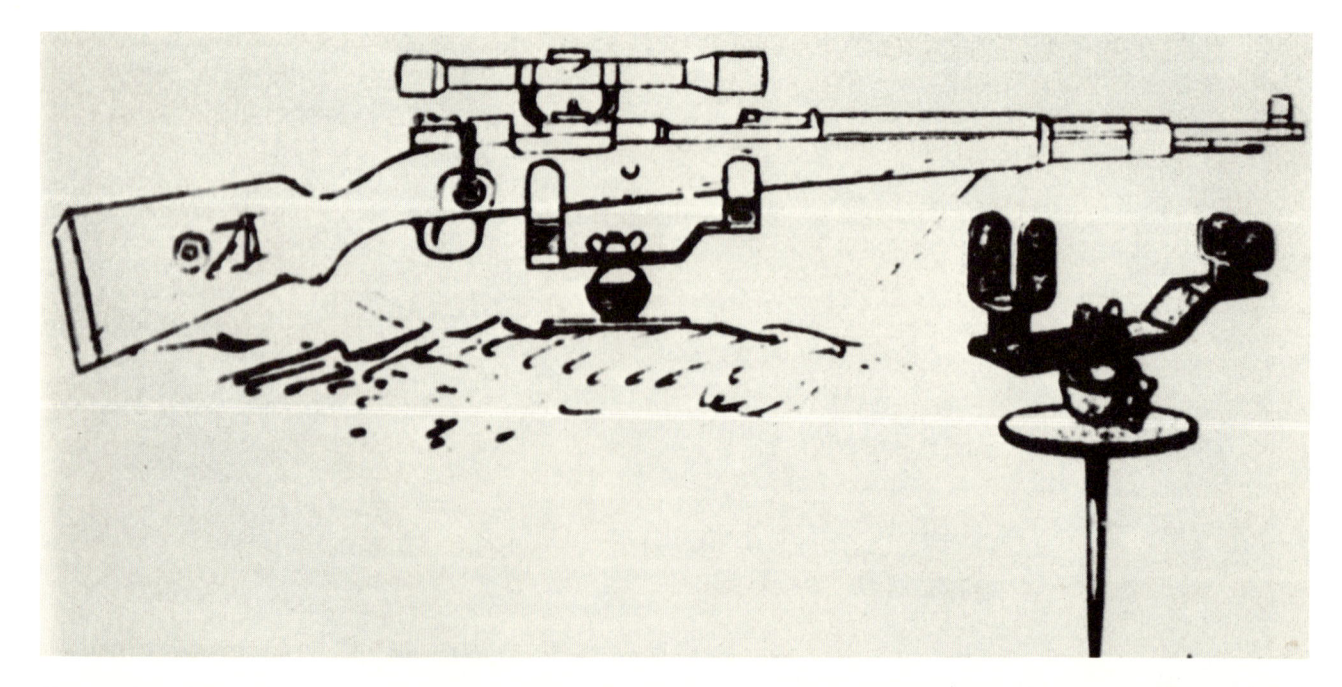

Illustration from World War II German sniper training manual H.Dv. 298/20 (October 1944) *Kampfschule der Panzertruppen Ausbildung und Einsatz des Scharfschutzen*, depicting a special apparatus designed for sighting-in the sniping rifles used for training purposes. In this case a short side rail type.

A World War II manufacture short side rail variant based on a J.P. Sauer (ce 43) 98k rifle. This is the first attempt at strengthening the short side rail base and mounting system. Two tapered pins were added to the base for strength. A special locking device was also added to the standard mount: a simple screw with a double-winged head which reaches through the mount engaging a recess in the top of the base. This rifle also reflects the first attempt to prevent the scope from moving forward in the ring assembly from repeated recoil. A steel bar was soldered and screwed to the side of the tube between the rings. This method proved time consuming and led to development of the circular recoil ring which appears on some military contract scopes. The sight is a 4-power Kahles.

Close view of the short side rail improvements. Note the two pins in the base, double-winged locking screw in mount, and recoil bar on the scope. The Sauer firm (ce) is believed to have been responsible for bringing the short side rail system to its final form.

Close view of improved short side rail receiver base showing circular recess.

Left view of improved short side rail receiver base. Note the location of the front pin in this case.

KARABINER 98K — SHORT SIDE RAIL SYSTEM

Receiver side of the improved short side rail scope mounting base showing pins which extended into the receiver wing for added security.

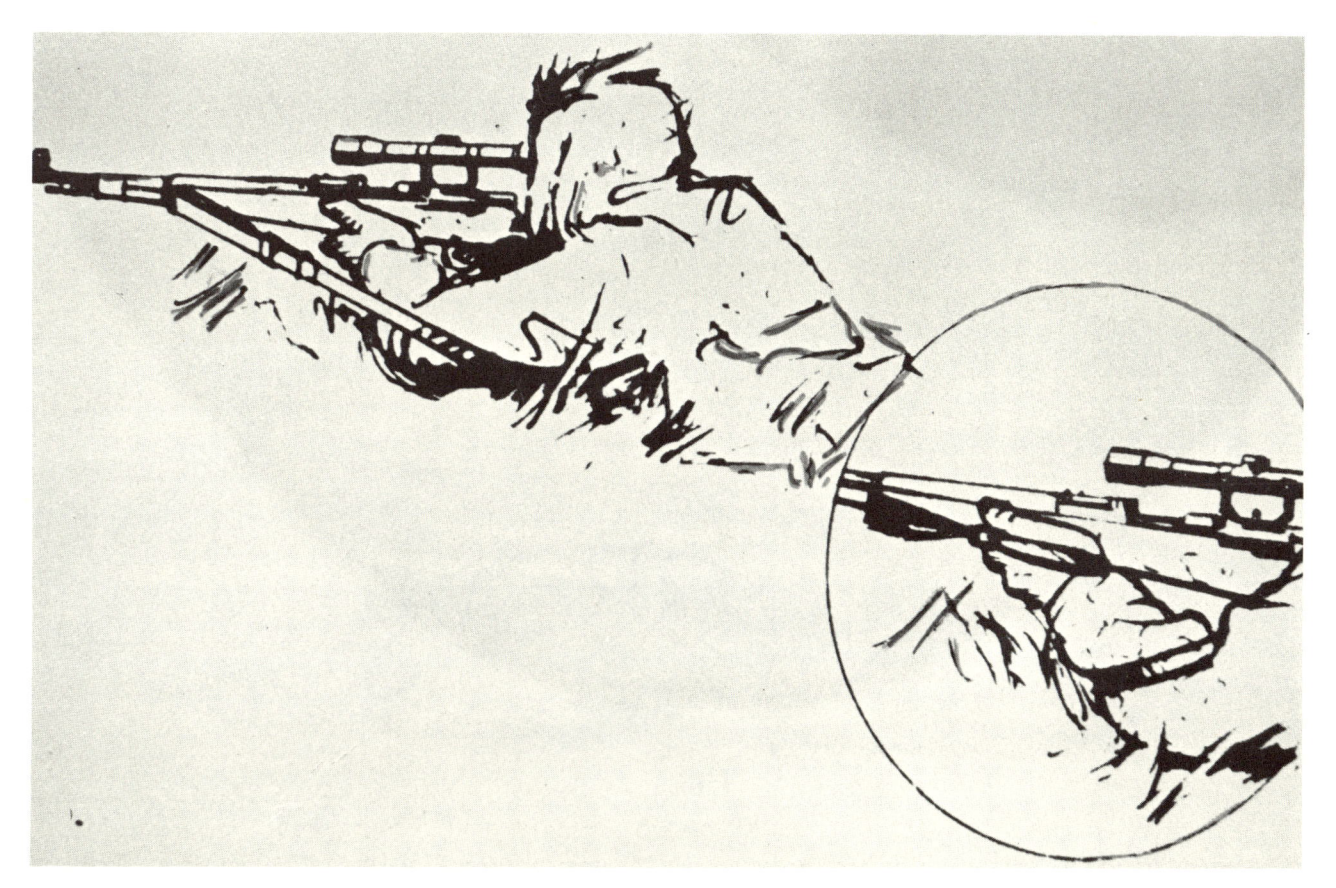

Extracted from 1944 German sniper training manual (H.Dv. 298/20) to illustrate the correct method of employing the rifle sling to its best advantage. The rifle is a short side rail model.

A short side rail telescope mount with a variant locking stud. Note this mount has only one grasping extension (single-wing) on the locking mechanism.

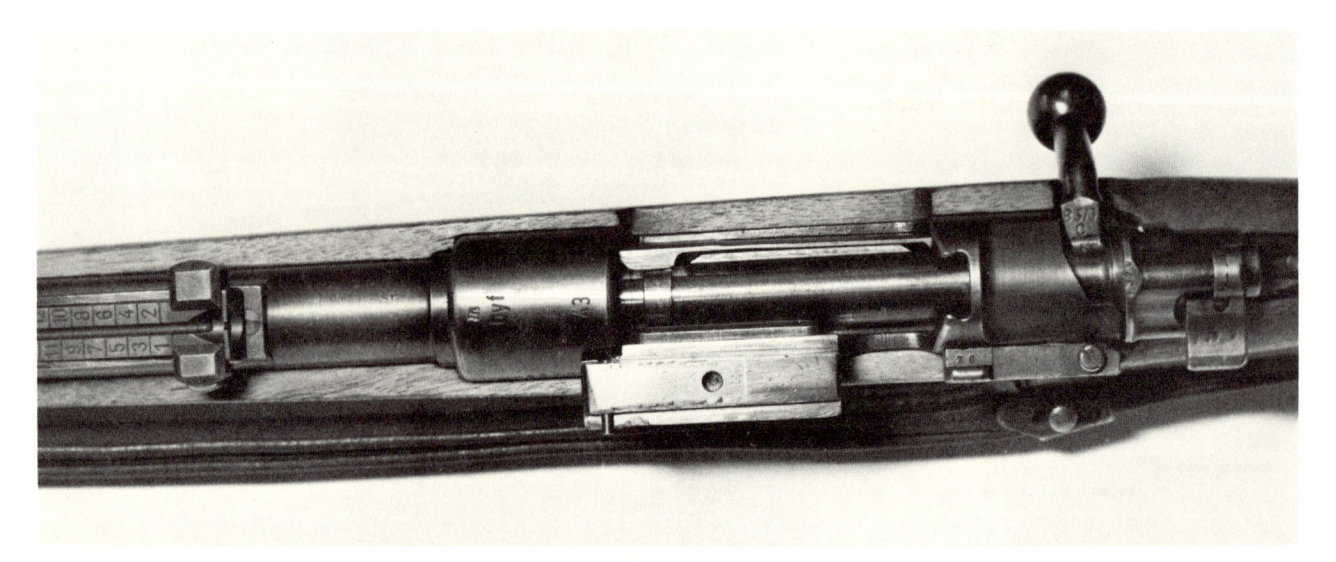

Top view of improved short side rail system with Mauser manufacture (byf 43) 98k rifle. Note the circular recess in the top of the base which engaged an auxiliary locking device (winged-screw) located in the center of the scope mount.

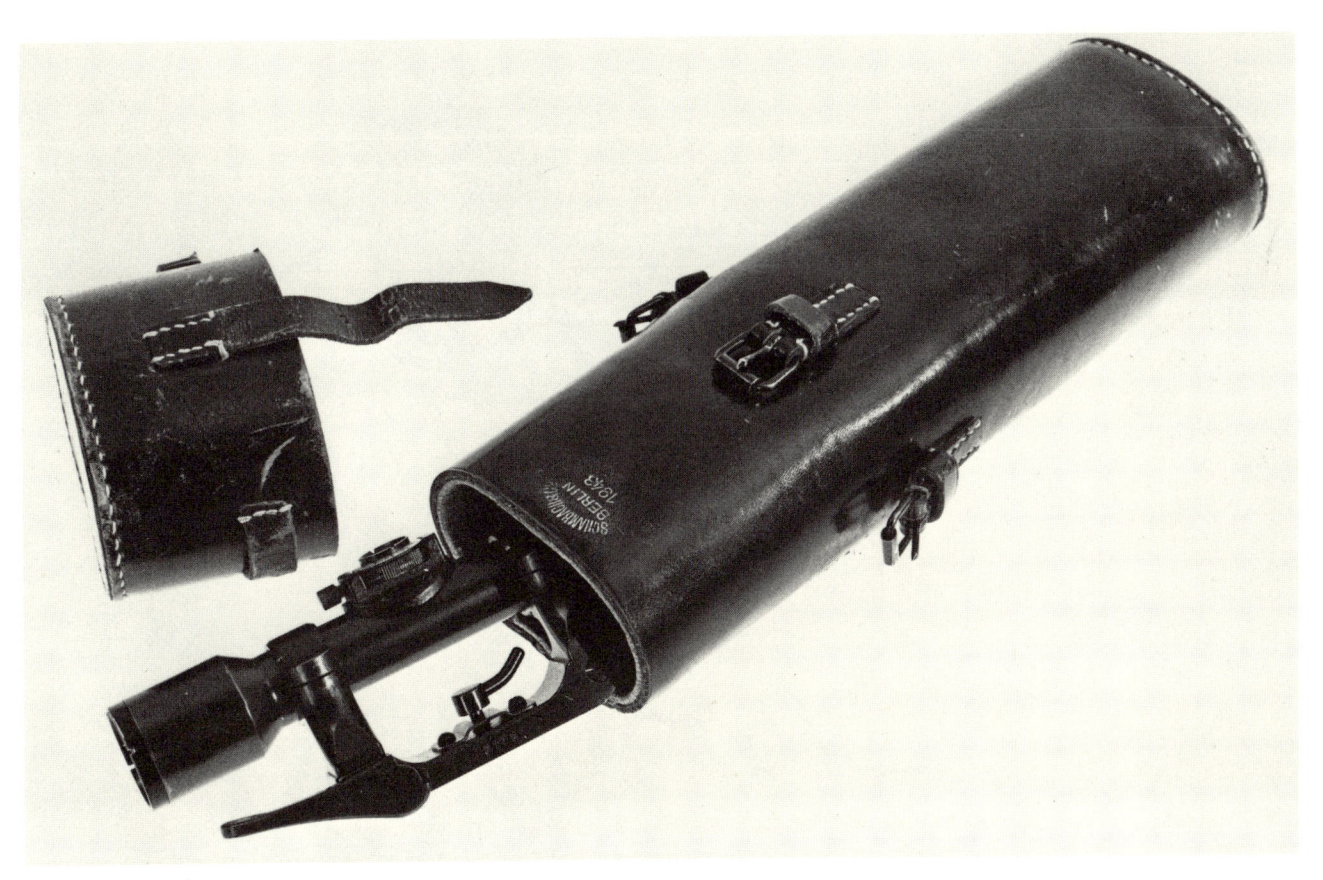

Leather scope carrying case (Schambach and Company, Berlin, 1943) as issued with some short side rail sniping rifles during the war. The 4-power Hensoldt scope is held by a single-wing rail mounting.

Short side rail telescope mount assembly with commercial markings as intended for the foreign market. The sale of telescopic rifle sights to German citizens was forbidden after 1941.

German sharpshooter training illustration (1944) showing one of many recommended firing positions.

Unique and extremely rare, an experimental J. P. Sauer (ce 43) 98k sniping variant designed to accept cartridge clips from the right side of the receiver at a 45-degree angle with the telescope in place. The Sauer firm's participation and developmental work with the side rail system was directly responsible for the evolution of the "long" side rail variation.

KARABINER 98K — SHORT SIDE RAIL SYSTEM

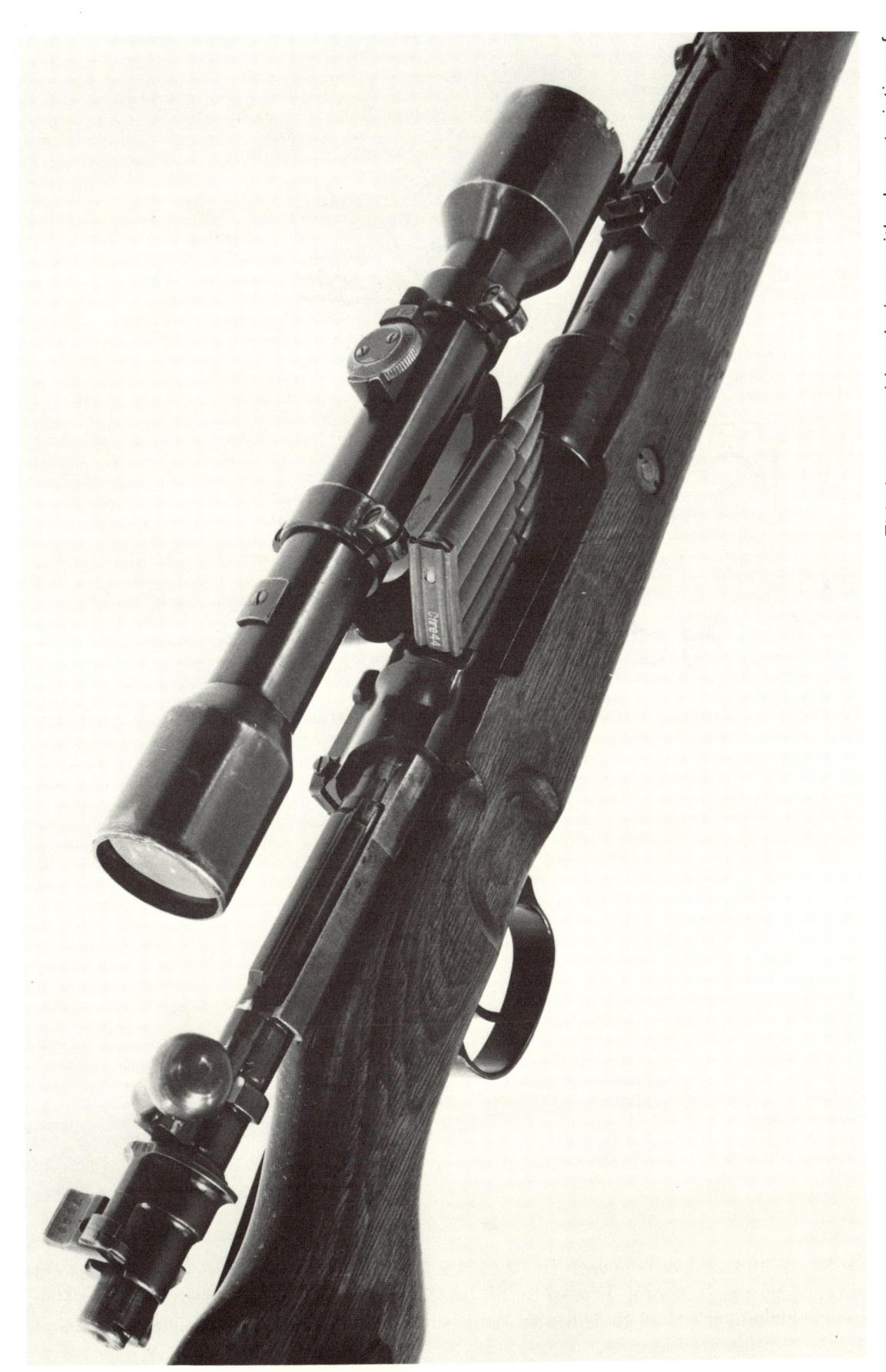

Close view of experimental J. P. Sauer 45-degree side-feed sniping rifle with loading clip inserted into the receiver. The mount bears the "HWZ" logo and holds a 4 x 90 military contract (blue +) Ajack scope. This is a transitional piece with characteristics of both the short and long side rail sniping systems.

A Gebirgsjager (German mountain troop) lance corporal sights his short side rail 98k from the remains of a Russian tank.

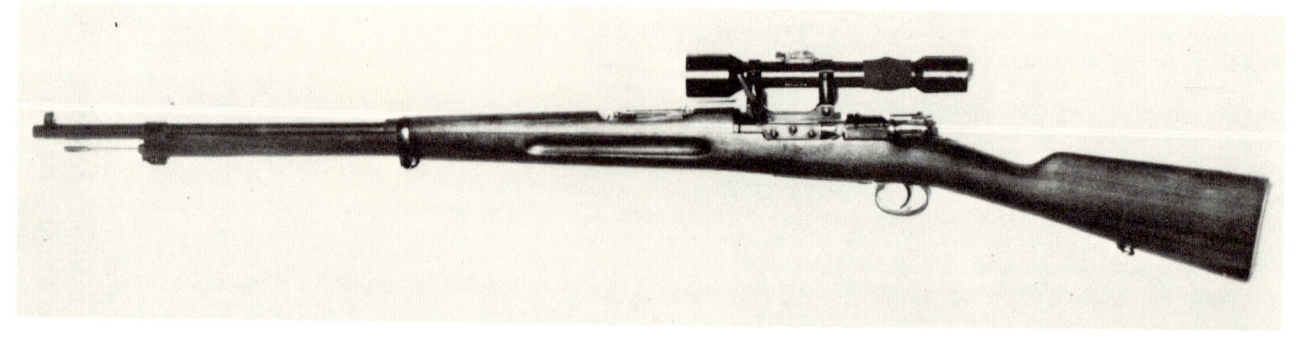

Swedish army Model 1941 sniper rifle with an Ajack 4 x 90 telescopic sight and side mount assembly as originally furnished by the Jackenroll firm in Berlin (scope and mounting components) before German military requirements brought a halt to deliveries. While similar to the German short side rail system used with the 98k, subtle differences do exist.

Close view of Swedish rail sniping rifle with AGA 3 x 65 telescopic sight (M 1942) designed to replace the German 4 x 90 Ajack system.

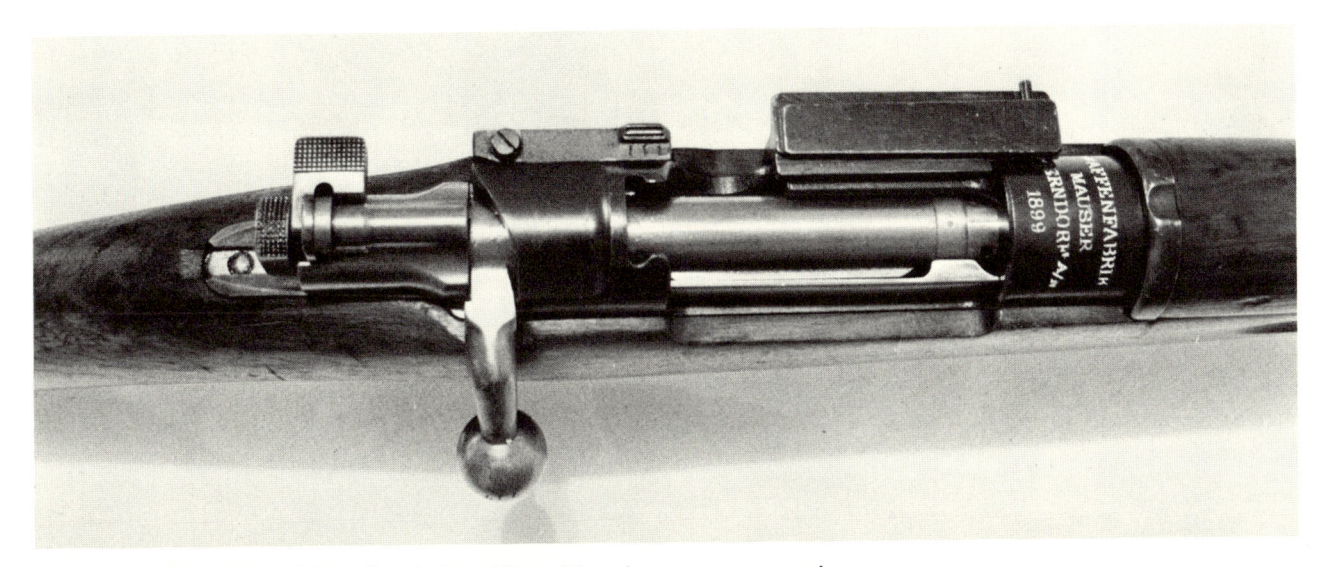

Top view of Swedish side rail sniping rifle with telescope removed.

CHAPTER VIII

Karabiner 98k — Zielfernrohr 41

Manufactured and fielded in great quantity from 1941 until the cessation of hostilities in 1945, the Karabiner 98k mit Zielfernrohr 41 (carbine 98k with telescopic sight 41) saw varying use by all branches of the Wehrmacht throughout the war.

From 1928 until the devastating effect of Soviet sniping forced a return to reality, high-ranking officials within the German military structure were convinced that the overall level of individual marksmanship had progressed to the point that further development of telescopic rifle sights was totally unwarranted. Despite this conviction, in what amounted to a major concession, the development of an optical device that enhanced the marksmanship capability of the average German rifleman was given official sanction.

A 1.5-power optical rifle sight design was brought to fruition before German military operations in Poland in the latter part of 1939. It was viewed with disdain by those recognizing the need for adequate sharpshooting equipment. As it was then stated, "While the Army Ordnance Office suggested four-power magnification, such a one with only 1.5 magnification was asked for by competent authorities and developed accordingly." It was later decided that from 1942 forward, six percent of all 98k rifles manufactured were to be equipped with base assemblies for mounting the 1.5-power sights.

The 98k-ZF 41 combination, while never seriously intended to serve as a pure sniping arm, was categorized as such and continued to be referenced for German "sharpshooter use" in various training manuals and pamphlets into 1943.

As noted in pertinent documents dealing with the ZF 41 sights, the 98k-ZF 41 assembly ceased to exist as a telescopic-sighted rifle within German Ordnance circles beyond 1 January 1944. Official decree, however, would not be handed down until the following May when efforts to eliminate confusing equipment reports relevant to combat losses (weapons) and supply resulted in an ordnance directive that was forwarded to all army groups under date of 1 May 1944. "The carbine with telescopic-sight 41 will no longer be reported or considered telescopic-sight rifles, but carbines instead."

Of parallel interest are additional ordnance papers from 31 May 1943 that specifically mention the delivery of 87,396 units up to that time and existing orders for 370,000 1.5-power sights. In anticipation of 4-power telescopic-sight manufacture, the following measure was to be implemented.

> The production firms will be informed that the final delivery date for ZF 41 and ZF 41/1 is 31 December 1943. It is considered to cancel possible quantities which are not delivered up to this point.

It should be emphasized that the lack of supporting documents beyond the aforementioned raises serious doubts as to whether this measure remained as cited or was subsequently amended. In any case, despite down-grading of the ZF 41 system, 98k rifles with provision for mounting the 1.5-power sights (base assemblies) continued to be manufactured through 1944 and into the early months of 1945.

The 98k-ZF 41 receiver base is a short dovetailed rail machined out of the left side of the rear sight band. There is a wedge-shaped groove machined in the face of the rail, running its length and ending in a rectangular locking recess. The scope mount has two beveled rollers in the center of its corresponding dovetail section. When the mount is slid onto the base, the rollers wedge into the groove with a tightening action. When the scope mount is pushed all the way forward, a spring-loaded locking latch engages with the recess at the front of the rail, securing the mount firmly to the base. The mount was designed to place the scope over the bore directly above the standard rear sight assembly.

Even though basic windage and elevation settings were established when the scope assembly was fitted to the rifle at the time of manufacture, the rifles were zeroed by ordnance personnel prior to their final issue. Rotating the graduated elevation adjustment drum encircling the scope body allowed the rifleman to compensate for range variations. Windage adjustments were made internally and were not readily effected in a combat environment.

Initially, the 1.5-power sights bore the "KF" marking with no model designation. Subsequent production was marked "ZF 40," "ZF 41," and "ZF 41/1." While evidence supports the opinion that the ZF 40 as originally marked was first intended for use with the Selbstladegewehr 41 (self-loading rifle 41 or G41), the specific difference between the ZF 40 and the 98k-ZF 41 as originally manufactured remains unknown.

As a point of interest, German Ordnance referenced use of the ZF 40 device with the Gewehr 41 as necessitated by the "relationship of the sight axis to the bore axis." Whether this was based on G41 telescope mounting practice or on an internal design characteristic that precluded the sight's use with the 98k rifle has not been satisfactorily explained.

Unfortunately neither German Ordnance documentation nor original ZF 40 sights predating the reworked examples have surfaced.

Equally perplexing but worthy of note are U.S. Army late war (1945) captured materiel reports which in one specific case make casual reference to a Gewehr 41(M) Mauser semi-automatic rifle "fitted with a ZF 41/1 telescope" as having been retrieved in combat. An early war limited production G41(M) in conventional form captured in 1945 would be of great interest by itself; an extremely rare variant with telescope mounting base and a ZF 41/1, in this case, was indeed quite significant.

The disassembly and careful dimensioning of a considerable number of 1.5-power sights from various manufacturers representing both early and late production, KF, ZF 40 (rework), ZF 41, and ZF 41/1 indicate production variables typical of all World War II German military hardware. The principal difference between the ZF 41 and ZF 41/1 devices consisted of a shift from the original complex manufacturing design to a simplified version (internal lense system), and as further delineated in supporting ordnance papers dating from early 1943, the newly modified production ZF 41/1 "was originally designated ZF 42, but later finally ZF 41/1."

The 98k 1.5-power scopes were designated ZF 41 and, when simplified internally, ZF 41/1. However, ZF 40 sights that were subsequently reworked, ostensibly for use with the 98k, have the original ZF 40 markings either lined-out or milled from the tube surface with the ZF 41 or, more commonly, the ZF 41/1 as the substitute designation.

The following is a listing of confirmed manufacturers that were involved with 1.5-power sight production through the course of the war.

Sights produced by "cxn," "dow," and "cag" are considered the most common, with the majority of the firms cited responsible for only limited quantities.

Although pre-1941 rifles with provision for mounting the early 1.5-power sights were reportedly based on 98k's produced by Mauser, and other manufacturers, weapons of this type classified as regular production were manufactured for the most part by Mauser,

KARABINER 98K — ZIELFERNROHR 41

The 98k rifle as originally fitted with 1.5-power ZF 41 telescope. Typical of those issued during the war, this weapon is coded "byf 44."

Right view of K98k "byf" code rifle with ZF 41 scope.

D 136

Karabiner 98k - Zf 41

(Zielfernrohrgewehr)

und

Zielfernrohr 41 (Zf 41)

Beschreibung, Handhabungs- und
Behandlungsanleitung

Vom 10.2.42

Front cover from the early World War II German Ordnance manual D 136 (1942) concerning the description and handling of the 98k rifle with the ZF 41 sight assembly.

production mark	manufacturer
cag	Swarovski, D., Glasfabrik und Tyrolit-Schleifmittel-Werke Wattens/Tirol
clb	Wohler, Dr. F.A., Optische Fabrik Kassel
cxn	Busch A.-G., Emil, Optische Jndustrie Rathenow
ddv	Oculus, Spezialfabrik ophthalmologischer-Justrumente Berlin
dow	Opticotechna G.m.b.H. Prerau (Czechoslovakia)
dym	Runge & Kaulfuss, Fabrik f. Feinmechanik und Optik Rathenow
eso	Optische Werke G. Rodenstock Munchen
fvs	Spindler & Hoyer, mechan, u. optische Werkstatte K.-G Gottingen.
fzg	Feinmechanik e. G.m.b.H. Kassel
gkp	Ruf & Co., Nachfolger der optischen Werke vorm, Carl Schutz & Co. Kassel
hap	Kohl S.-G., Max, Physikalische Apparate/ Laboratoriums Einrichtungen Chemnitz
jve	Optisches Werk Ernst Ludwig Weixdorf.
kay	Ford-Werke A.-G. Werk Berlin
kov	Etablissement Barbier Benard et Turene Paris
mow	Seidenweberei Berga C. W. Crous & Co. Berga/Elster

Oberndorf (byf) Mauser, Borsigwalde (ar) and Berlin-Lubecker Maschinenfabriken (duv) beginning in 1941. The greatest activity by "ar" and "duv" encompassed the period between 1941 and 1943. The Mauser Werke at Oberndorf is acknowledged to have been the prime contractor for weapons of this configuration until the end of the war.

A production figure of 86,000 telescopic-sighted 98k rifles attributed to the Mauser Oberndorf facility has been frequently cited in various contemporary offerings. However, they fail to mention that the *Combined Intelligence Objectives Sub-Committee Report* (CIOS Report XXXIII-IV), from which this figure was extracted, does not delineate what part of this total is comprised of the 98k-ZF 41 type rifles or the "turret mount" sniping variants also attributed to the Mauser "byf" code.

A little-known facet of German sharpshooter instruction involving ZF 41 sights entailed the applied use of 5.56mm rimfire caliber (.22 long rifle cartridge) bolt-action rifles (Kleinkaliber, or KK, or small caliber) fitted with 1.5-

THE GERMAN SNIPER 1914–1945

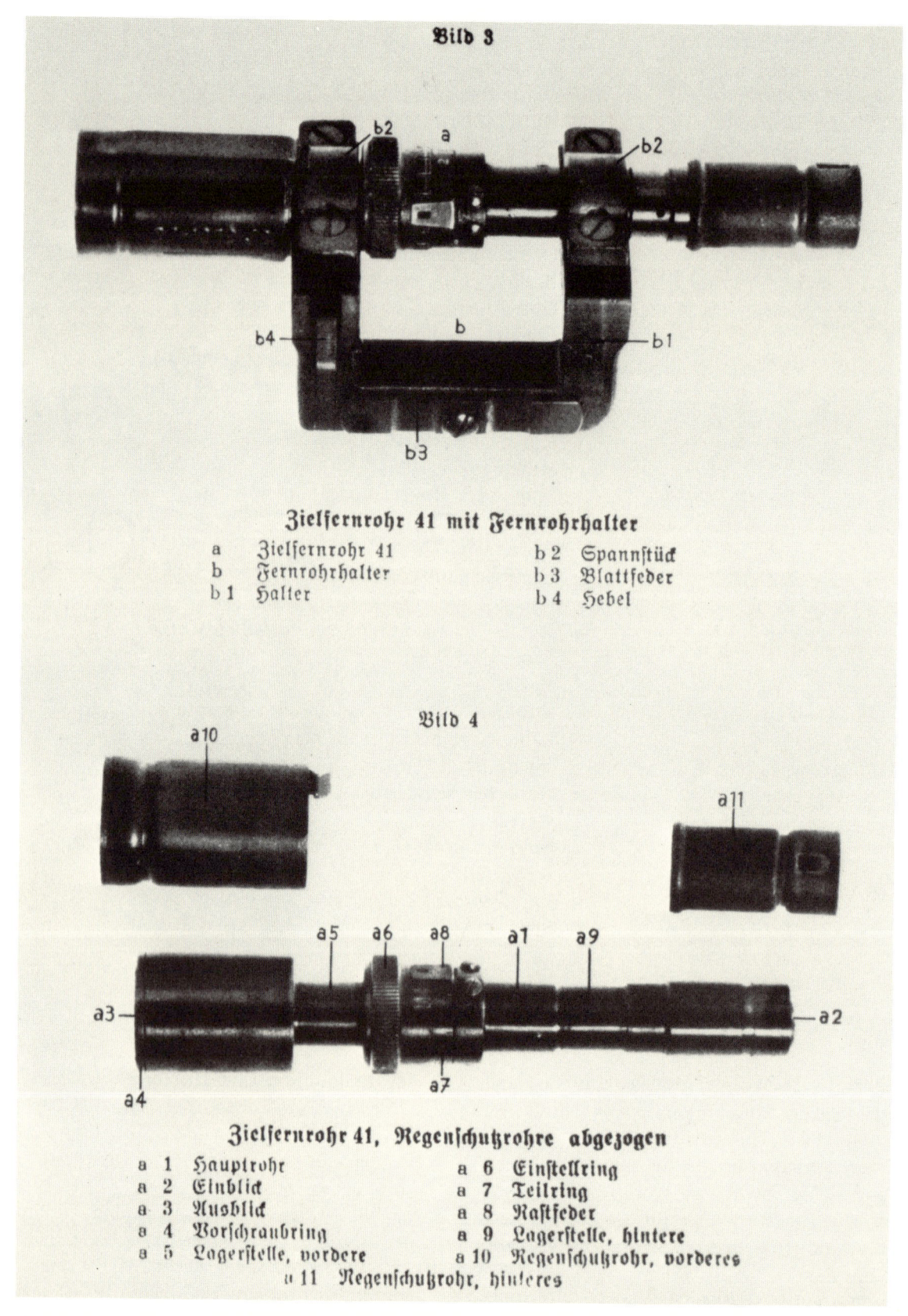

Bild 3

Zielfernrohr 41 mit Fernrohrhalter

a	Zielfernrohr 41	b 2	Spannstück
b	Fernrohrhalter	b 3	Blattfeder
b 1	Halter	b 4	Hebel

Bild 4

Zielfernrohr 41, Regenschutzrohre abgezogen

a 1	Hauptrohr	a 6	Einstellring
a 2	Einblick	a 7	Teilring
a 3	Ausblick	a 8	Rastfeder
a 4	Vorschraubring	a 9	Lagerstelle, hintere
a 5	Lagerstelle, vordere	a 10	Regenschutzrohr, vorderes
		a 11	Regenschutzrohr, hinteres

Original illustration from D 136 with the ordnance description of the ZF 41 sight and mount components.

power sights for training exercises up to 80 meters. As noted in a late war Panzer-Grenadier sniper training manual:

> The firing with the small caliber awakens and heightens the feelings of personal security. Confidence in the weapon is increased. This must also be preserved when transfer is made to the large caliber. . . . Shooting with the small caliber spares ammunition, but must not be carried too far, because the marksman gets too used to the lack of recoil and the lighter trigger pressure point. . . . The firing of the KK is carried out in open terrain without any special separate measures.

Of additional interest are original requisition documents from the main branch of the School of Guerrilla Warfare (Werewolves) located near Berlin. Under date of 1 January 1945, they request an adequate supply of "KK rifles with telescopic-sight 41" along with 20,000 rounds of KK ammunition for use in training sharpshooters in preparation for the final stand. A mythical organization as some contend? Not hardly!

German riflemen garbed in winter issue clothing during a lull in combat. Note what appears to be an early production 98k mounting a ZF 41 scope.

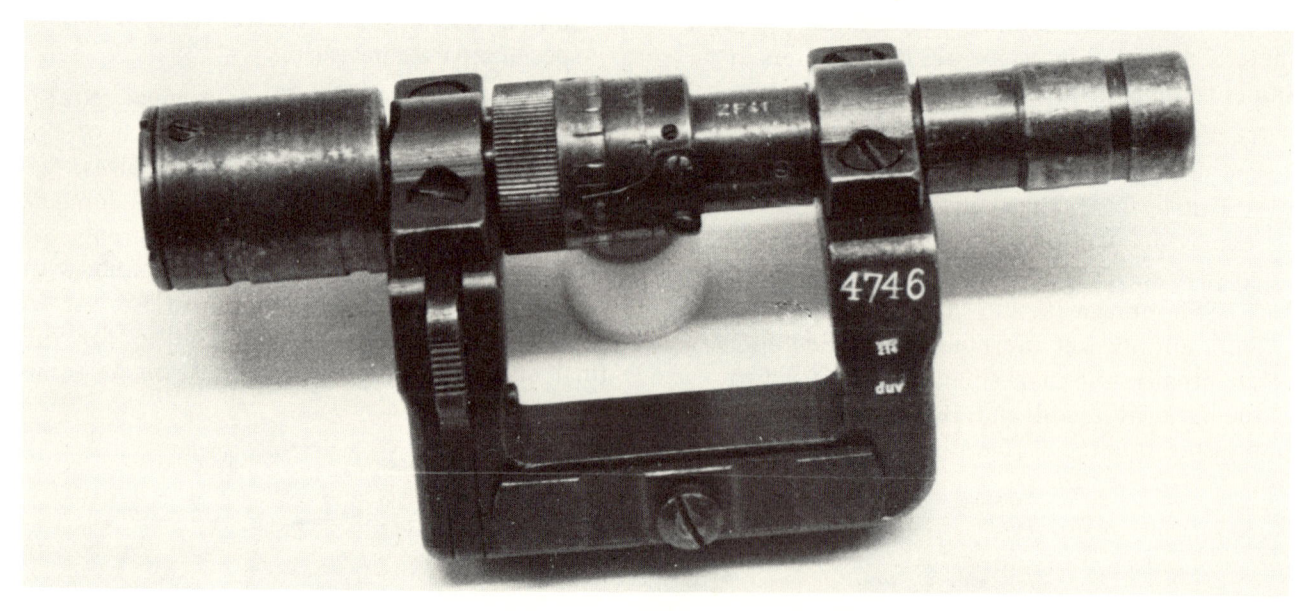

An early manufacture ZF 41 (cxn) sight and mount assembly. Note the 98k rifle number, "214" Waffenamt and "duv" stamp on the mount (typical). Both rain shields are missing.

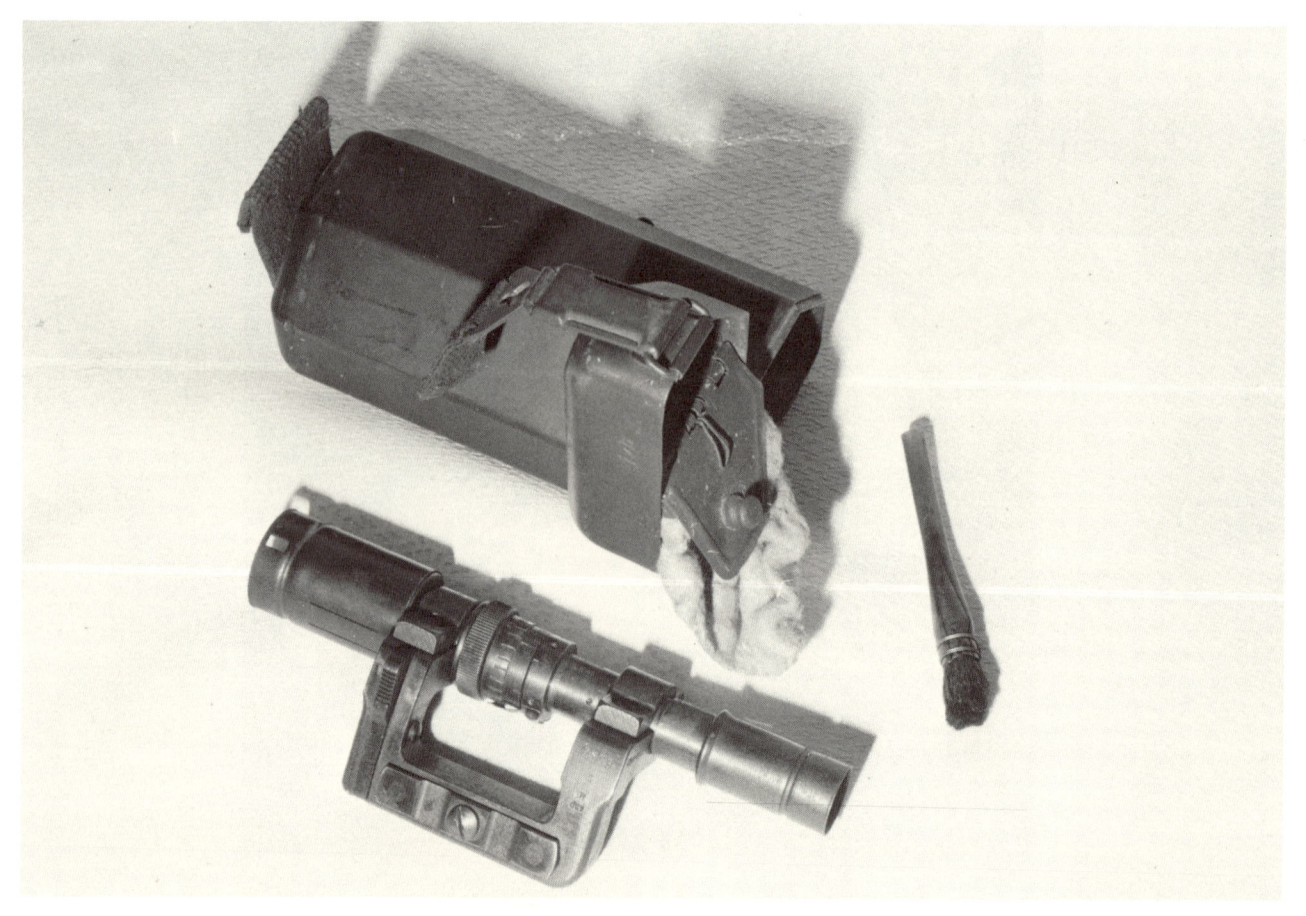

Zielfernrohr 41 (Telescopic Sight 41) with carrying case, original issue "Klarinol" cloth and lens brush.

An early ZF 41 sight base. This rail does not have the groove typical of later issue models. The "flat-side" base required considerable effort to mount and remove the scope assembly in subzero weather, and it is believed that German experience in the bitter Russian winters dictated the change in design. The 98k rifle is a "duv 41."

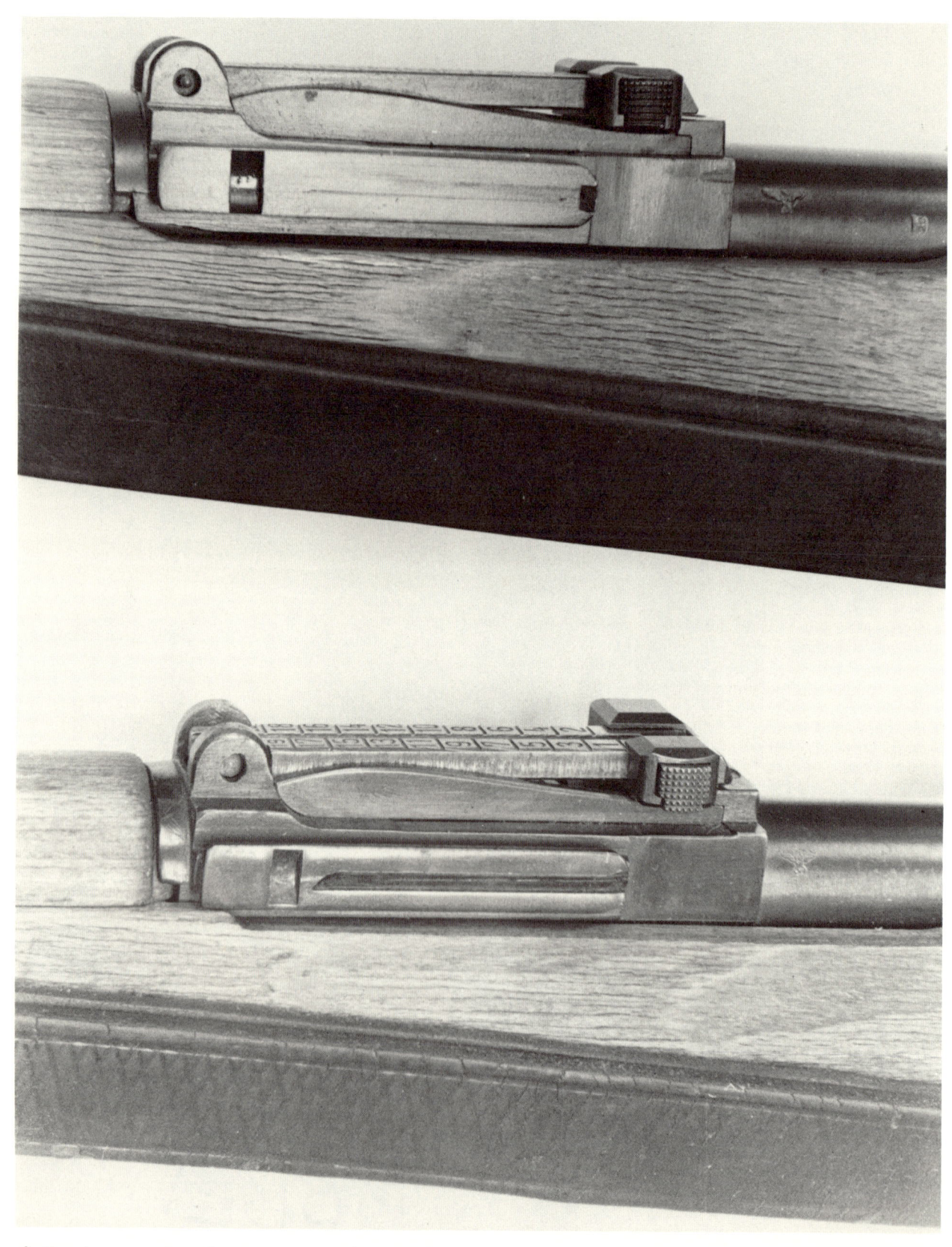

A direct comparison between an early manufacture "flat-side" 98k-ZF 41 scope mounting base (top) and the improved variant for use with bevel rollers.

KARABINER 98K — ZIELFERNROHR 41

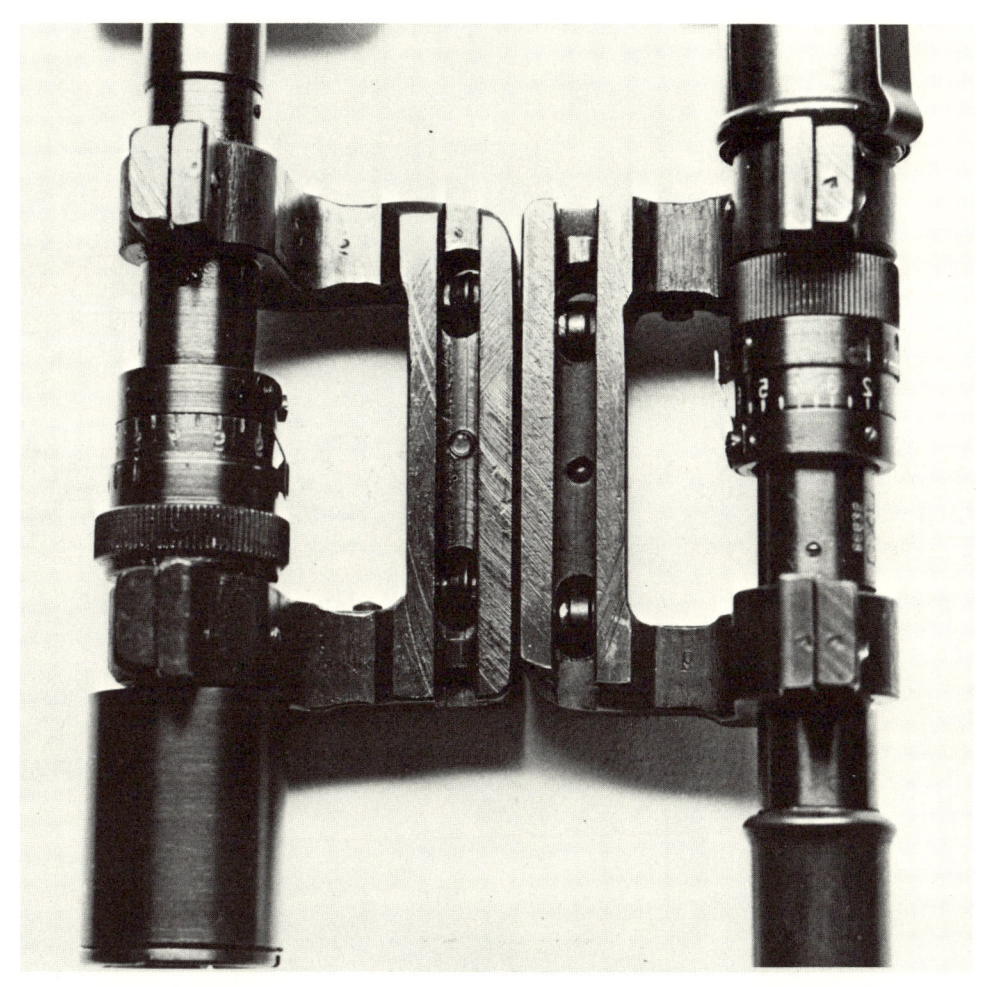

Comparison of the early (flat roller) and later model (bevel roller) system as used with the 1.5 power scope mounts. The simple change was effected without altering the existing mount configuration.

The two variations of ZF 41 series wedge rollers removed from the scope mount. Top is early and has flat-faced rollers. Bottom is the common type with bevel rollers.

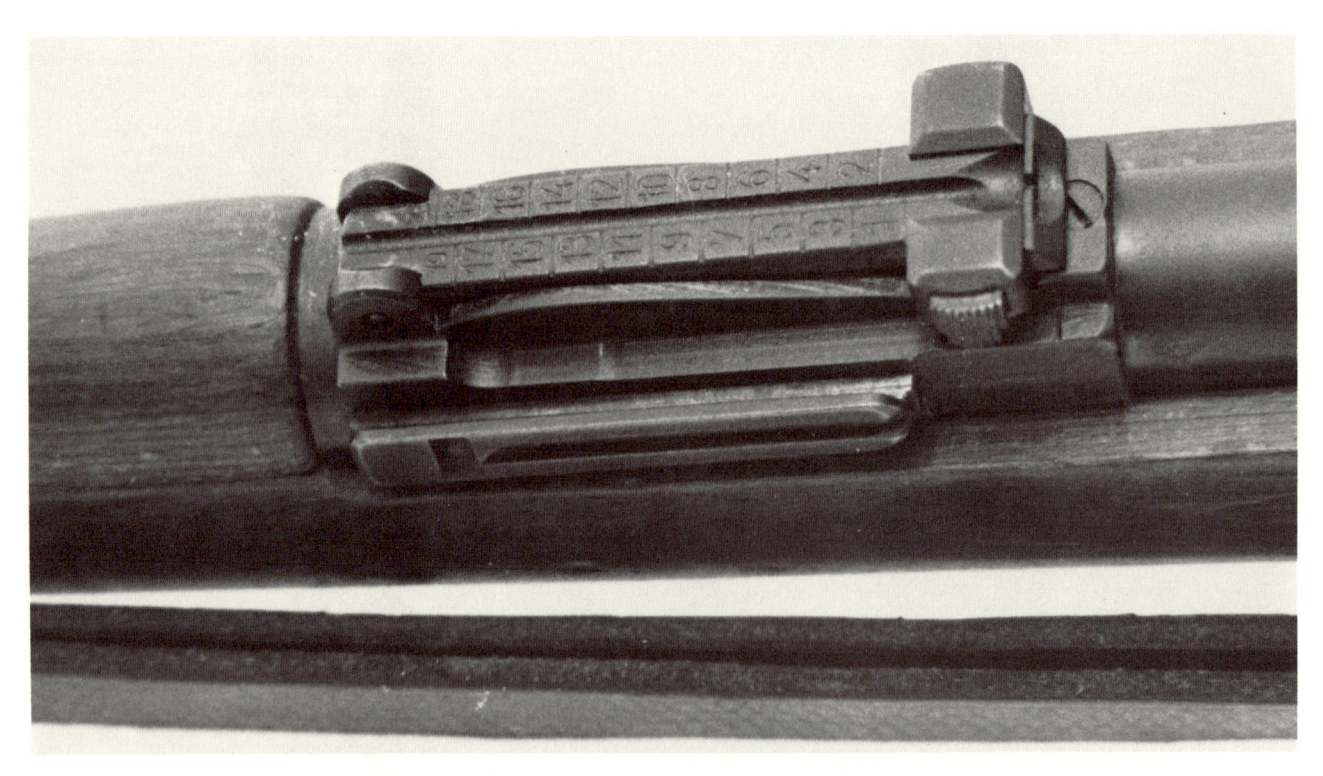

Detail of the improved 98k-ZF 41 scope mounting rail (base). Dovetail, bevel roller groove, and mount locking latch recess are visible.

Detailed view of ZF 41 type mount dovetail. The beveled rollers are visible in the back of the dovetail slot. Note mount locking latch just forward of the right roller.

KARABINER 98K — ZIELFERNROHR 41   197

Another view of the 98k-ZF 41 base showing the bevel roller groove and latch recess. The scope assembly slid on from the rear.

Typical ZF 41/1 (cag) scope and carrying case with objective and ocular rain shields removed. Some objective shields bear an ordnance stock number. Both sight and mount were normally given a commercial type blue-black finish for protection from the elements. In all cases, mounts were stamped with the "duv" code and "214" Waffenamt, while the legend "K98k-ZF 41" appeared on the back portion of the rear support facing the receiver.

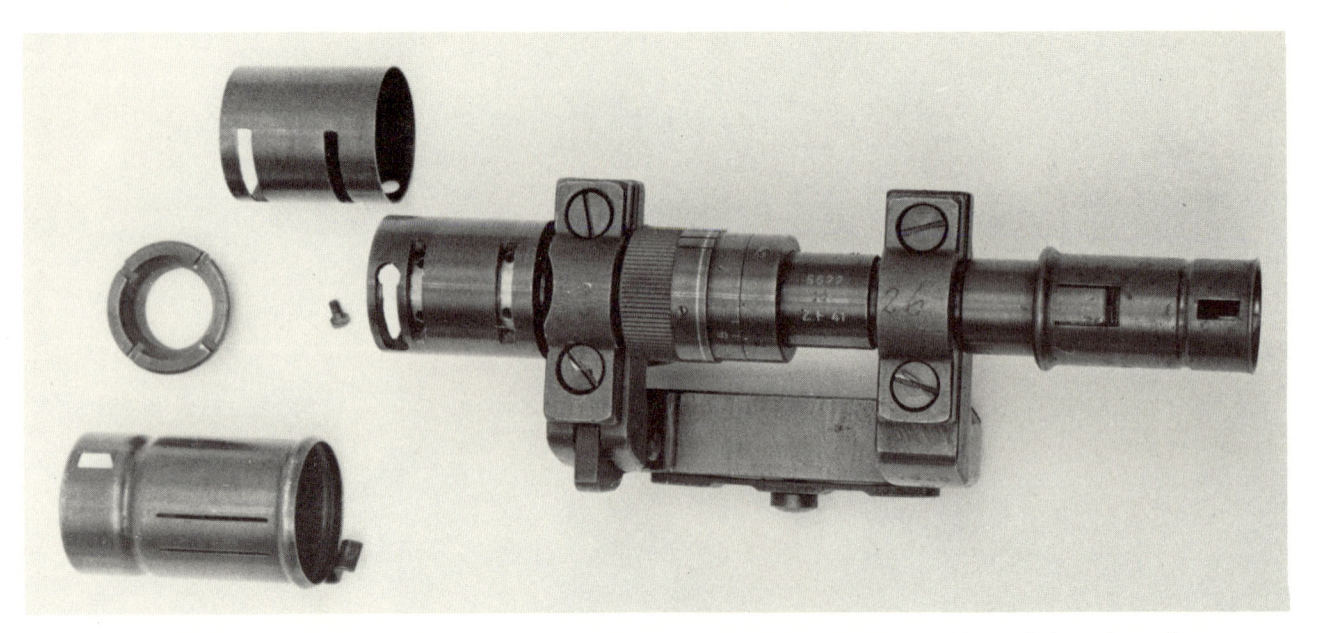

Windage adjustment for the 1.5-power sights. The rain shield, objective cell retaining ring, dust cover and screw have been removed and appear at the front of the ZF 41 sight. The eccentric sleeve (drum) can be seen in place under the two lateral openings. Pointed instruments are inserted into the small adjustment holes in the sleeve, rotating the assembly right or left to change windage as required. Note the "flush-knurl" elevation ring.

Close view of ZF 41 sight showing the protective dust cover on the objective end of the scope. Note the "raised-knurl" elevation ring.

The ZF 41 sight with protective dust cover removed and "adjustment pins" inserted into the eccentric sleeve (drum). Scopes were "zeroed" to their mating rifle under controlled conditions prior to issue for combat use.

Youthful late war German army conscripts. One is armed with a 98k-ZF 41 rifle.

An example of the extremely long eye-relief necessary with ZF 41 type rifle scopes. Focus adjustment was not provided with the 1.5-power sights.

An unusual example of the 1.5-power rifle scope with no model designation. Believed to be a transitional variant, the device, as manufactured by "cxn," bears a "KF" directly above the scope serial number. According to the early war (1942) German manual, *Maintenance of Weapons, Equipment and Ammunition in Winter*, D158:

> "Only a fraction of the optical equipment now in the hands of units is lubricated with non-freezing oil and vacuum grease . . . Instruments which have not yet been applied with a cold resistant grease and those which bear the mark KF can be expected to adjust only with great exertion due to solidification of the grease."

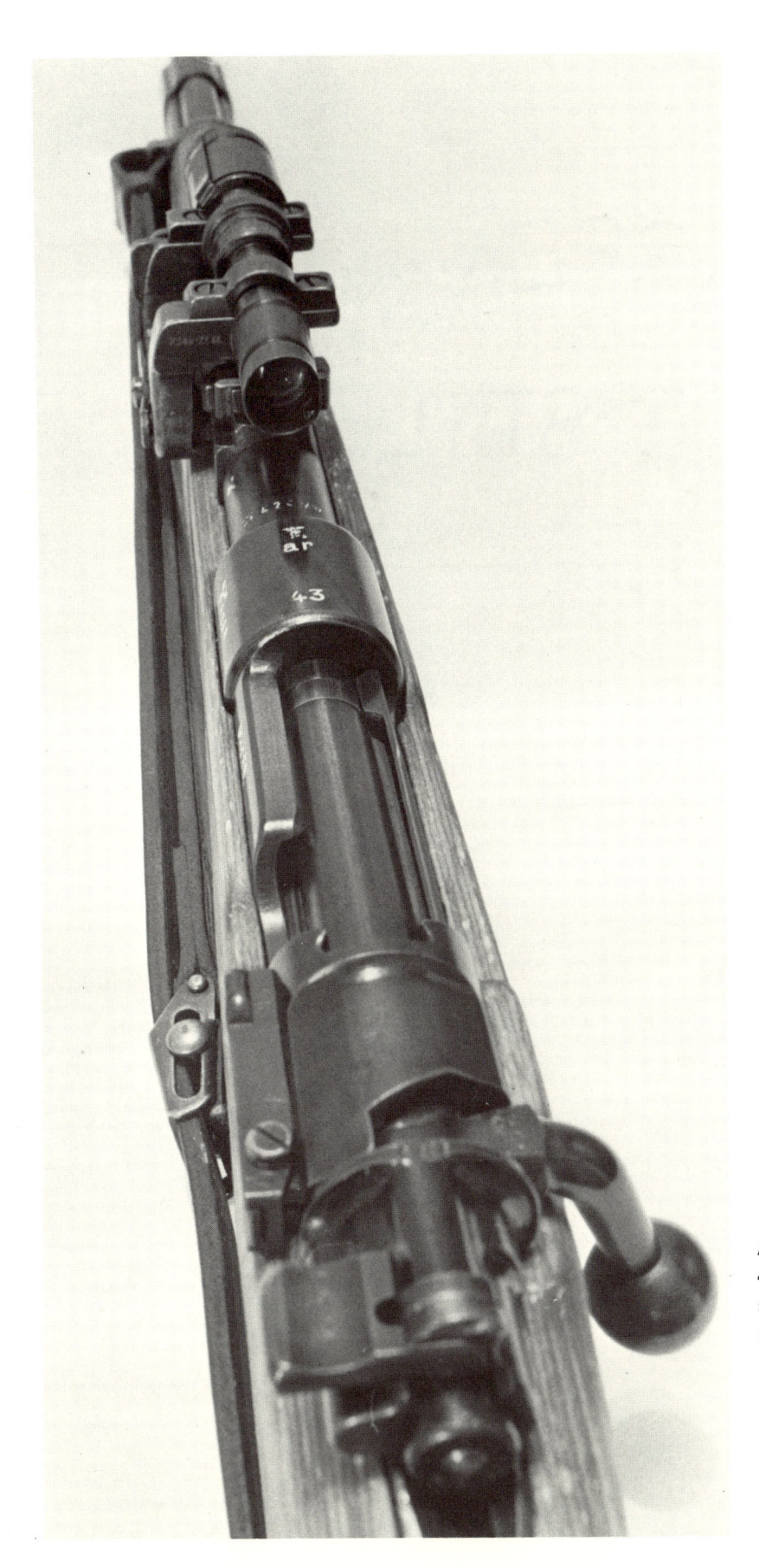

An unusual view of a 98k rifle (ar 43) mounting the ZF 41/1 rifle scope. The exceptionally long eye-relief is evident.

Opposite side of "cxn" sight showing the scope power (magnification) and field of view. Except for the manner of marking, these sights are comparable to the ZF 41 models in overall appearance. Of further interest are some of the early production ZF 41 scopes which also bear the "KF" marking in addition to the "ZF 41" designation.

A "cag" code ZF 41/1 reworked sight with the new designation applied directly over the milled portion of the tube. Late production sights (revised lense design) were stamped "ZF 41/1" on the scope tube as manufactured. Regardless of their designation or point in time, all sights of this type were 1.5-power.

KARABINER 98K — ZIELFERNROHR 41

A "cxn" code ZF 41 sight as reworked from a ZF 40 model. In this case the original markings were milled from the tube and restamped "41." The vast majority of reworked scopes are stamped "ZF 41/1," however. Elevation range drums were graduated from 100 to 800 meters (1-8). When the drum is rotated, the reticle moves up and down in the line of sight.

Typical ZF 41/1 (dow) ZF 40 rework sight. Note the original designation has been lined-out and re-stamped following modification. Even though a few early production scopes have been noted with a single vertical post reticle, most 1.5-power sights have the typical vertical pointed post with horizontal side bar pattern applied directly to the surface of the lens.

An early flat roller, "cxn" code 1.5-power sight. In addition to a "ZF 41" designation, this sight is stamped "KF" as well.

The ZF 41 series scopes, no matter how optically deficient for effective sniping, saw considerable use throughout the war.

Two variations of ZF 41 scope cases. The right one is an early type, "Continental Green" color with leather belt loop and bears commercial markings. Left case is the most commonly encountered: "Tropical Tan" color, web belt loop, and by "jvb." Original cases have also been noted painted "Luftwaffe Blue." Two thousand early production 98k-ZF 41 rifles were procured by the Luftwaffe Weapons Office before the end of 1941.

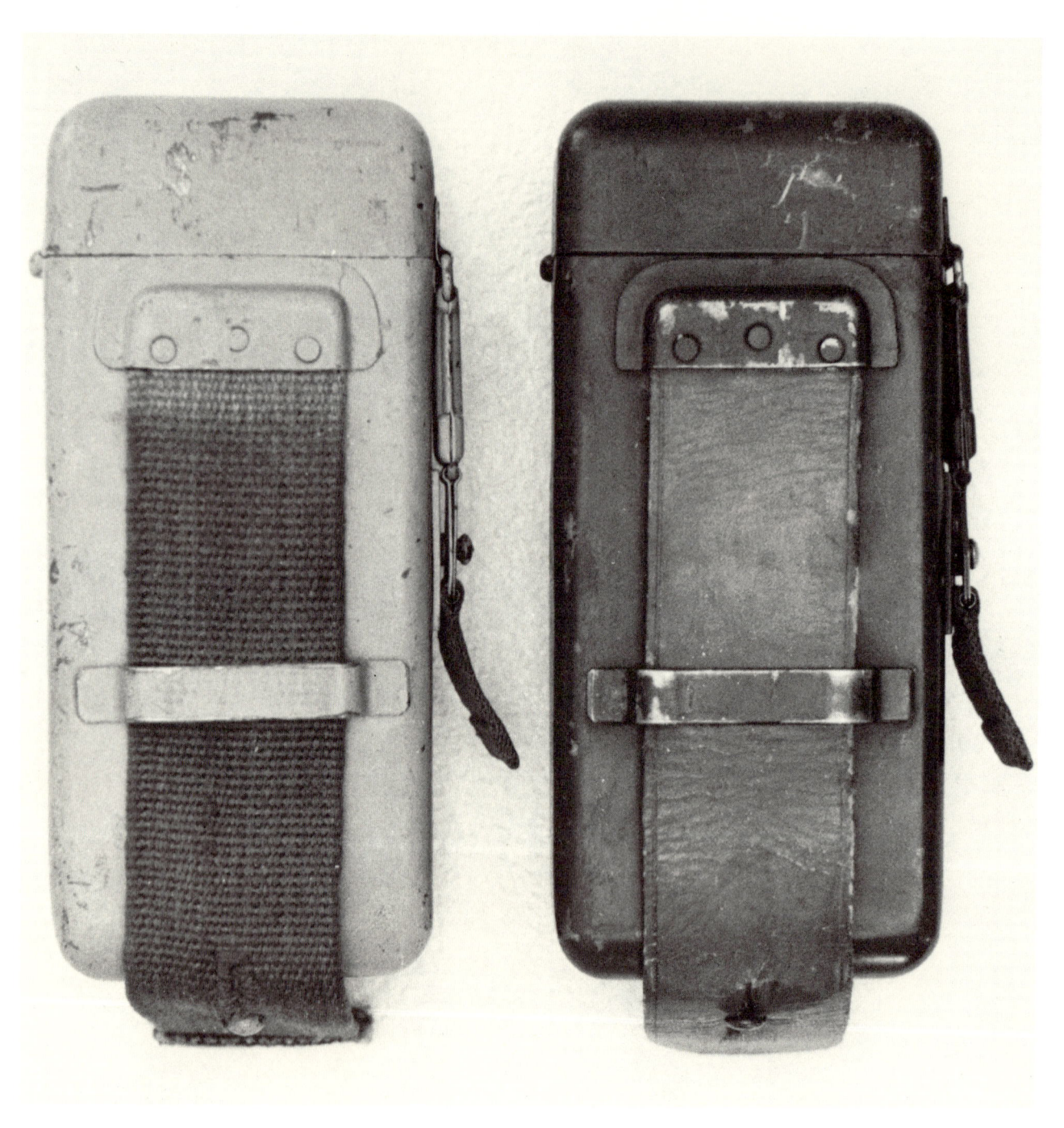

Back view of the two ZF 41 scope cases. Note the early leather belt loop on the right.

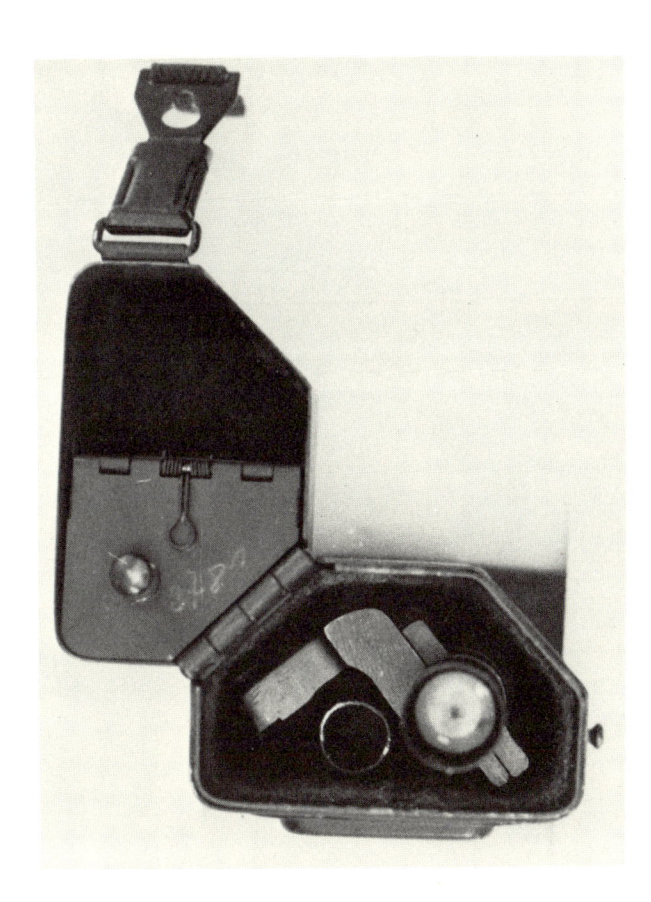

Internal view of a formed metal ZF 41 scope carrying case. The scope is in place, and the early cylindrical lens brush container is plainly visible. The brushes were generally lost and are seldom encountered. The spring-loaded lid in the top of the case opens into a small compartment containing a specially treated cloth known as the Klarinoltuch (Klarinol cloth) intended to prevent condensation from forming on the lenses.

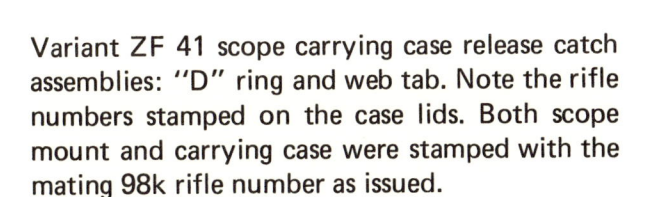

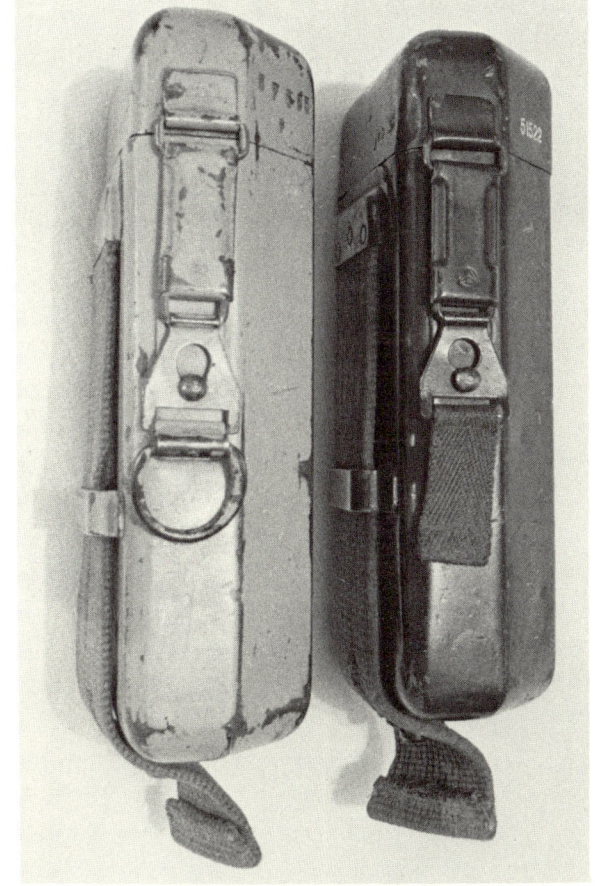

Variant ZF 41 scope carrying case release catch assemblies: "D" ring and web tab. Note the rifle numbers stamped on the case lids. Both scope mount and carrying case were stamped with the mating 98k rifle number as issued.

An example of nonstandard field application of "splinter-pattern" camo paint applied to a ZF 41 scope case.

KARABINER 98K — ZIELFERNROHR 41

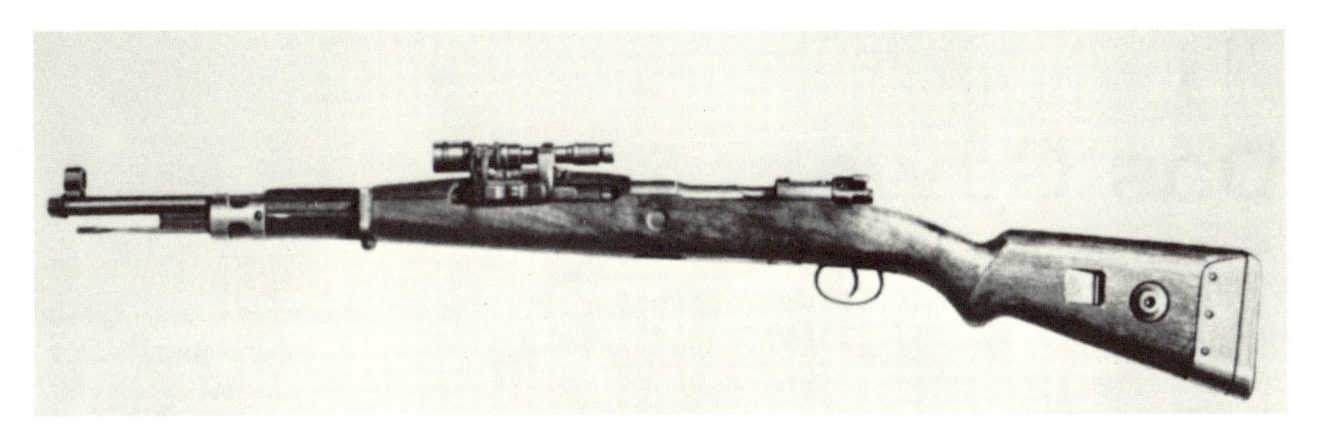

Unique and extremely rare, a Gewehr 33/40 fitted with the ZF 41 sight as originally illustrated in a World War II German weapons manual. Intended ostensibly for use by mountain units, scope-sighted versions of this rifle are virtually unknown. The weapon depicted is believed to be a prototype piece.

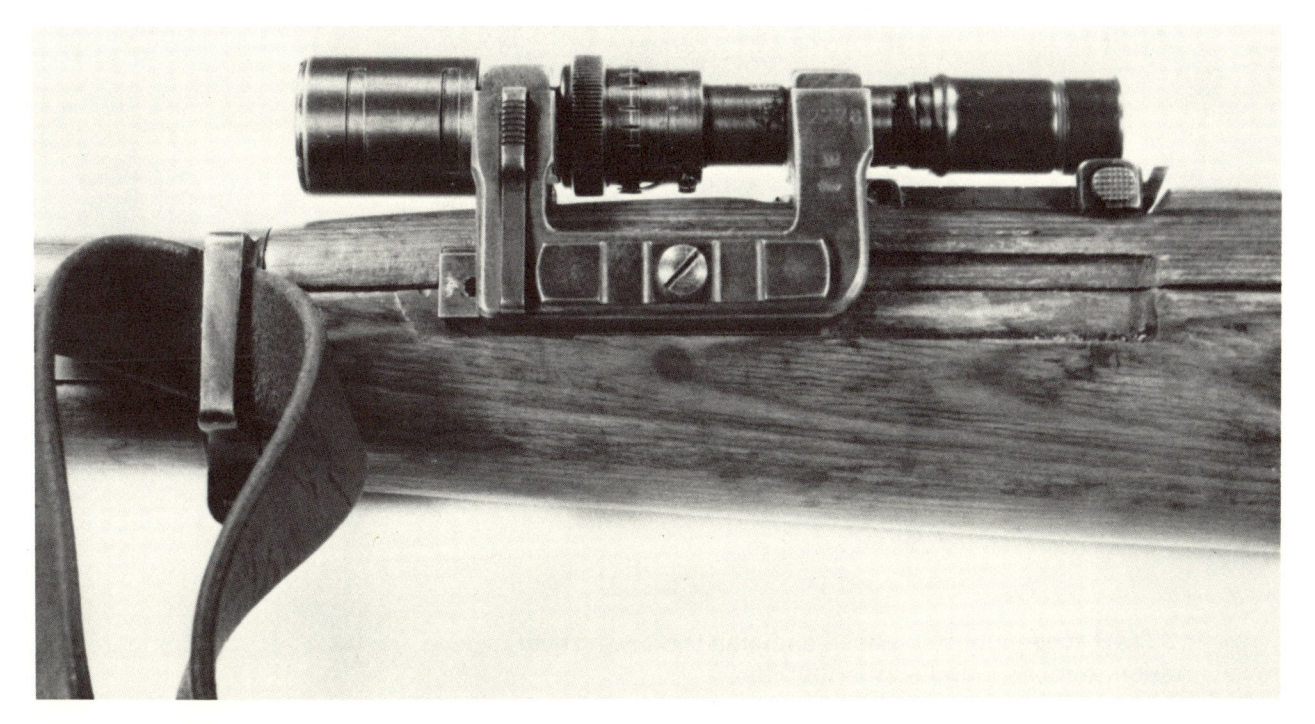

An uncommon Gewehr 33/40 (G. 33/40) with original base assembly for mounting the 1.5-power telescopic sight.

Top view of Waffenwerke Brunn manufacture (dot 1942) G. 33/40 with telescopic sight removed. Compare the base location to that shown in the ordnance manual illustration.

Gewehr 33/40 scope mounting base showing the bevel roller groove and latch recess. Compare this base variation to those found on the 98k rifle.

KARABINER 98K — ZIELFERNROHR 41

Telescopic sight Gewehr 33/40 with handguard and stock removed. The one-piece machined base assembly was soldered to the barrel. A small Eagle and Swastika (Hoheitszeichen) stamp is present at the front of the base.

Top view of Gewehr 33/40 scope mounting base assembly. The rifle serial number (7550) was engraved on top of the barrel band (base assembly) and is barely discernible. A Waffenamt marking (WaA 280) is also present on the right side of the band.

Gewehr 33/40 (dot, 1942) with 1.5-power telescopic sight.

An example of the KKW or Kleinkaliber Wehrsportgewehr (small caliber military and sporting rifle), .22 caliber (5.6mm) bolt-action rifle as originally fitted with a ZF 41/1 scope. In this case, the weapon was made by Gustloff and represents one of the type pressed into service for late war sharpshooter training.

Close view of Gustloff Werke 5.6mm rifle showing the scope mounting rail (base) with dimensional characteristics almost identical to those found on the 98k.

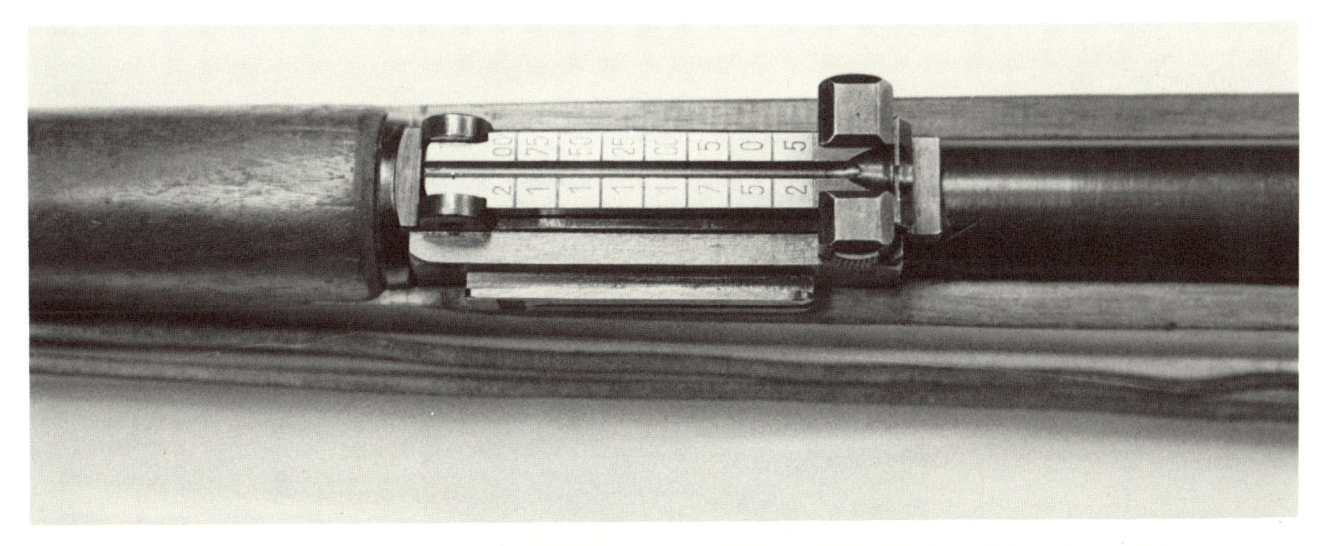

Top view of Gustloff KKW rear sight and scope mounting base assembly. Although considered an essential part of the sharpshooter course of instruction, small caliber rifles of this type were not intended for combat use.

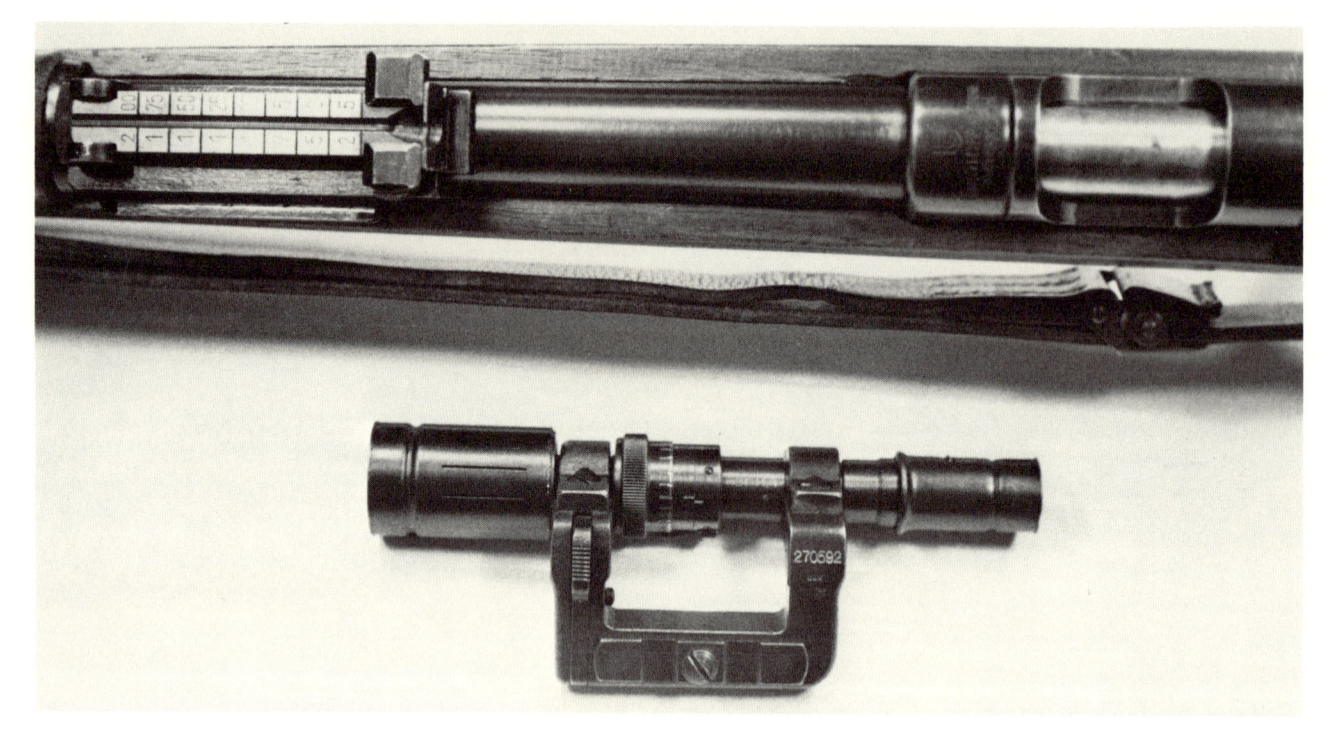

The KKW ZF 41/1 telescopic sight with corresponding rifle serial number (270592) stamped on the mount (typical).

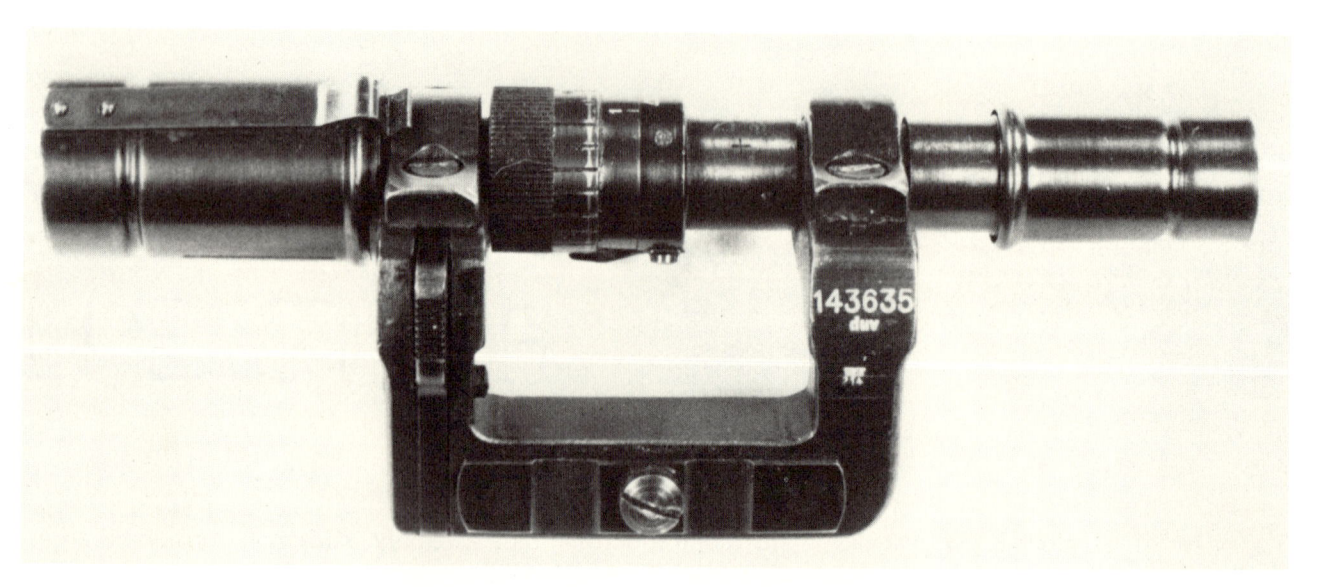

A "ddv" code ZF 41 sight originally paired with a Gustloff KKW rifle (143635).

CHAPTER IX

Karabiner 98k — Turret Mount System

Considered to be the most efficient design of all German sharpshooting weapons, the "turret" mount rifles were in general service from their introduction in 1939 through war's end.

In limited use during the Polenfeldzug (Polish campaign), the original issue turret mount Zielfernrohrkarabiner 98k (telescopic sight carbine 98k) was selected from various weapons at the Army Supply Offices at Spandau, Konigsberg, and Hannover for fitting with bases, mounts, and the 4-power Telescopic Sight 39 (Zeiss, Zielvier), the first type of rifle scope utilized with this variation.

Following installation of the telescope assembly and test firing, the rifle serial number was stamped on the left side of the front sight mount "to ensure that the telescope be only used with the weapon it was adapted to."

Drawing on their experiences from World War I, the German military recognized that precision telescopic-sighted rifles were more prone to damage from transit (shipping) than from combat attrition. Following delivery to its final distribution center, each rifle was test fired by competent ordnance personnel to verify combat fitness prior to issue. In the event the telescope or rifle incurred damage beyond the limits of normal field support, the weapon was returned to its point of origin for repairs or refitting of the sight to another rifle.

Although Mauser (byf) evolved as the principal manufacturer for weapons of this configuration, the J.P. Sauer firm (ce) was involved in limited production as well.

Two basic variations existed: the "low" and the "high" turret mount rifles, the actual difference being in the recess depth of the front base cone. The "high" cone recess is 6.35mm greater than that of the "low" type.

Overwhelming evidence indicates that both initial and very early production pieces (1939-40) were "low" turret models, but both variations were fielded during the war. The specific reason for the use of two telescope mountings in this case has not been revealed.

A wide variety of commercial and military contract rifle scopes were utilized with both types of mounts, but due to their point of original assembly and resultant variances, all sight assemblies were not fully interchangeable.

The front and rear bases of the turret mount system are affixed to the receiver with soft-solder and screws. The forward base, machined in the form of a hollow cone (German reference: ("pot-like turning"), has an undercut groove that provides a bearing surface for a tension spring on the underside of the front scope mount. The rear base dovetail has the configuration of an elongated sideways letter H.

The rear scope mount is fitted with a stop

Das Zielfernrohr 39
(Zielvier)
für den Karabiner 98k

Beschreibung,
Handhabungs- und Behandlungsanleitung

Vom 22. 1. 40

Berlin 1940
Gedruckt in der Reichsdruckerei

Front cover of the original German Ordnance pamphlet (D 134, January 1940) detailing the first 98k turret mount sniping rifles.

screw on the right and a wing clamp on the left that engages the side of the rear receiver base. Windage adjustments are made by loosening the lock screw on the side desired and tightening the screw on the opposite side of the mount. This action forces the rear scope ring to move to the loosened side thereby changing windage. Elevation adjustments are made with a graduated range drum located on top of the telescope.

A unique "tunnel" machined through the rear mount and front base permitted the sniper use of his conventional rifle sights with the scope in place, a simple but effective provision typical of German design.

Through the course of the war turret mount 98k rifles were subjected to rigorous combat duty by all branches of the Wehrmacht: Heer, Waffen-SS, Luftwaffe, and Kriegsmarine. The rifles were used by trained Navy marksmen to fulfill their regular duties on submarines and surface vessels, to provide support for landing parties, and on occasion, to detonate sea mines endangering passage of their craft.

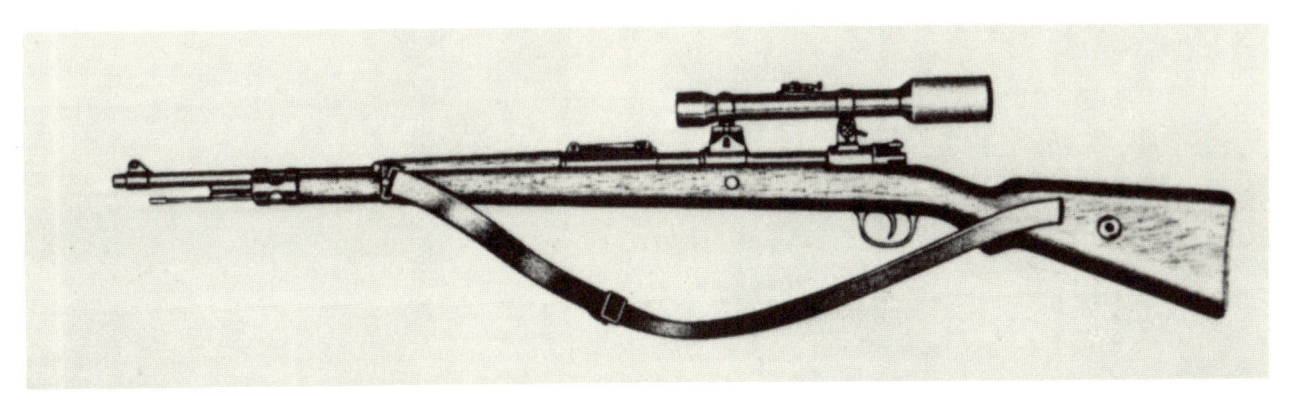

"Carbine 98k with Telescopic Sight 39 (Zielvier)" as shown in pamphlet D 134.

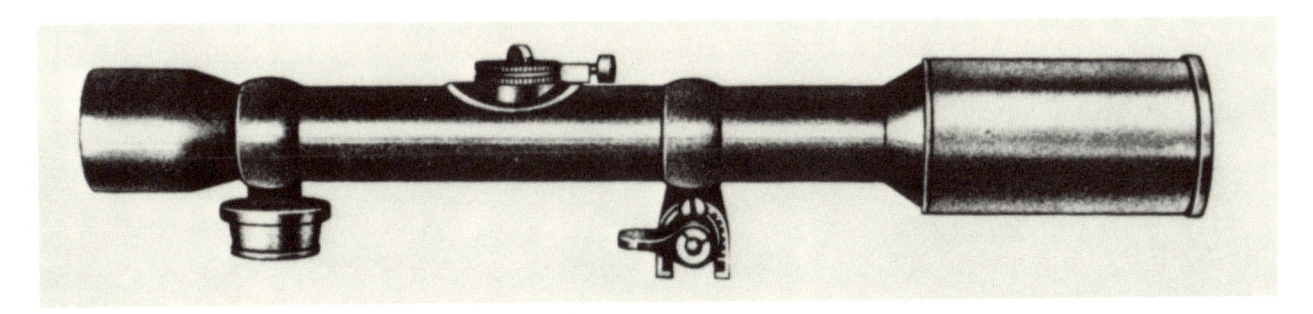

Original illustration from D 134 showing the early 4-power Zeiss "low" turret mount sight with an "eye protector of rubber."

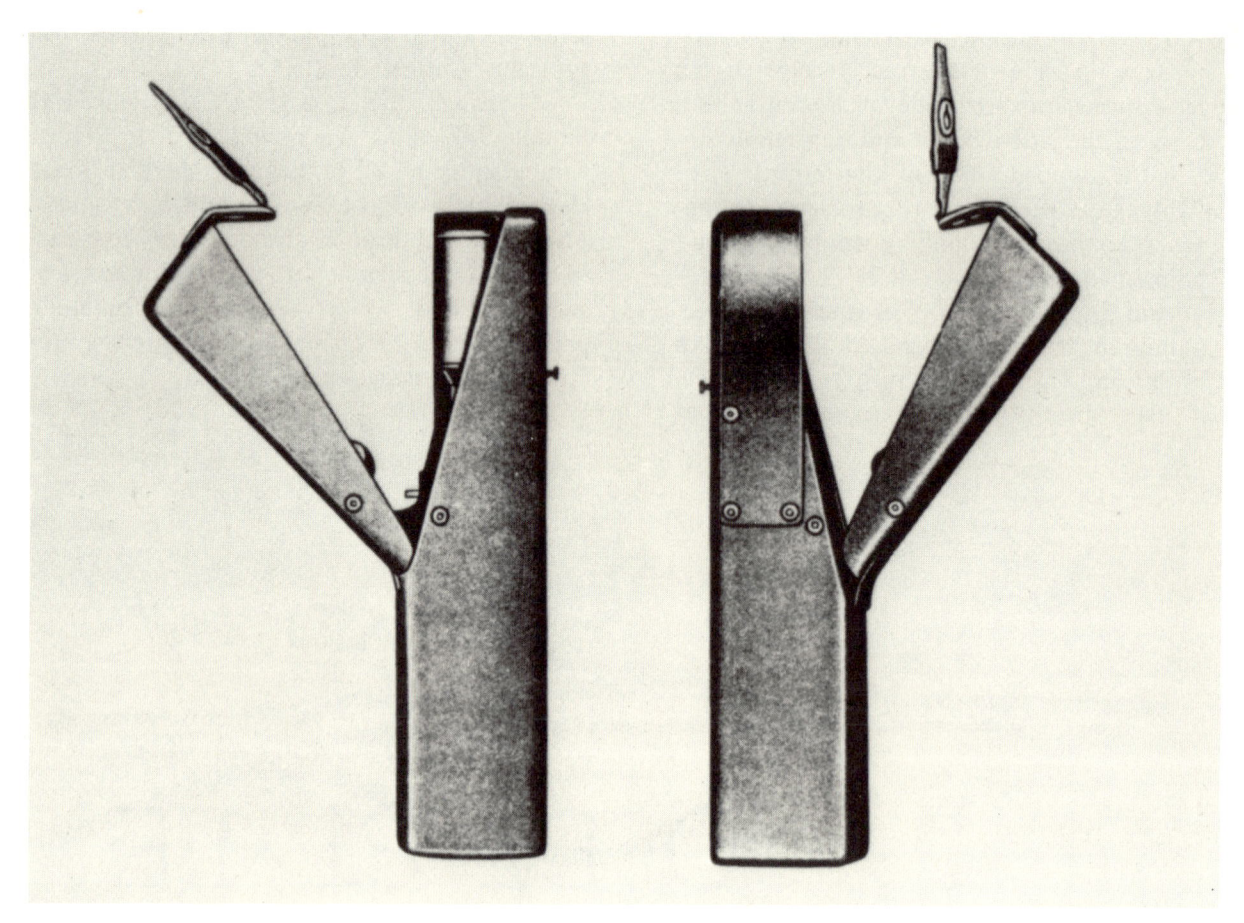

The first issue turret mount scope carrying case as noted in German Ordnance pamphlet D 134.

Extremely rare original issue Zeiss (Zielvier) 4-power telescopic sight fitted with "low" turret mounts as first fielded for German sharpshooter use in 1939. Note the 98k rifle number stamped on the protruding circular edge of the front mount, a practice (number location) that is characteristic of but not limited to early turret issue.

Zeiss "low" turret rear mount assembly and unit-issue telescope markings which were "Verboten" (prohibited) for security reasons after German forces rolled into Russia in 1941.

Top view of early "low" turret scope elevation drum (100-800 meters) and focus adjustment vane. Prewar commercial Zeiss-Zielvier sights were available with universal focus (no adjustment) and as illustrated. Note the barely discernible small screw head on the inner side of the rear mount wing-clamp which served to retain a "leaf-type" detent spring rather than the coil spring utilized on later turret systems.

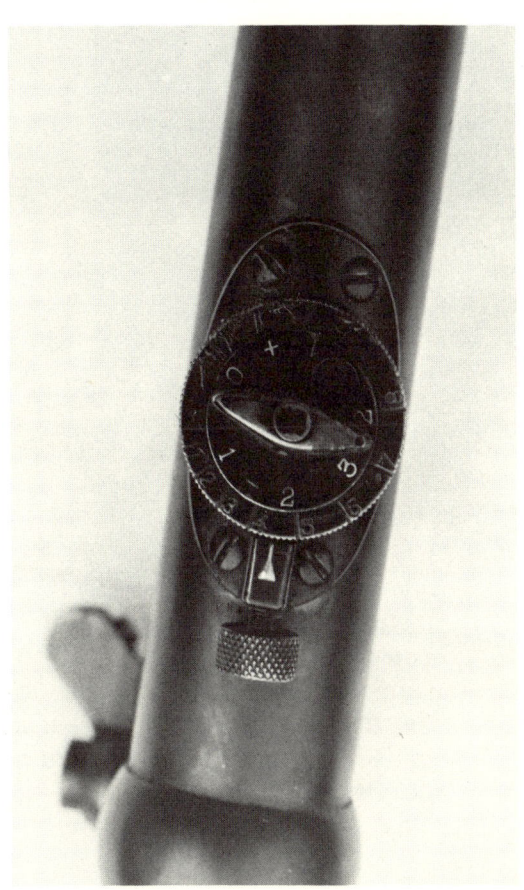

Top view of rear section of "low" turret Zeiss sight with manufacturer's legend, model designation, and scope serial number (factory).

Original military issue carrying case intended for the turret mount "Telescopic Sight 39 (Zielvier)." Formed from aluminum and covered with leather, the unusual case was only made in limited numbers. A small brush was included for cleaning the scope lenses.

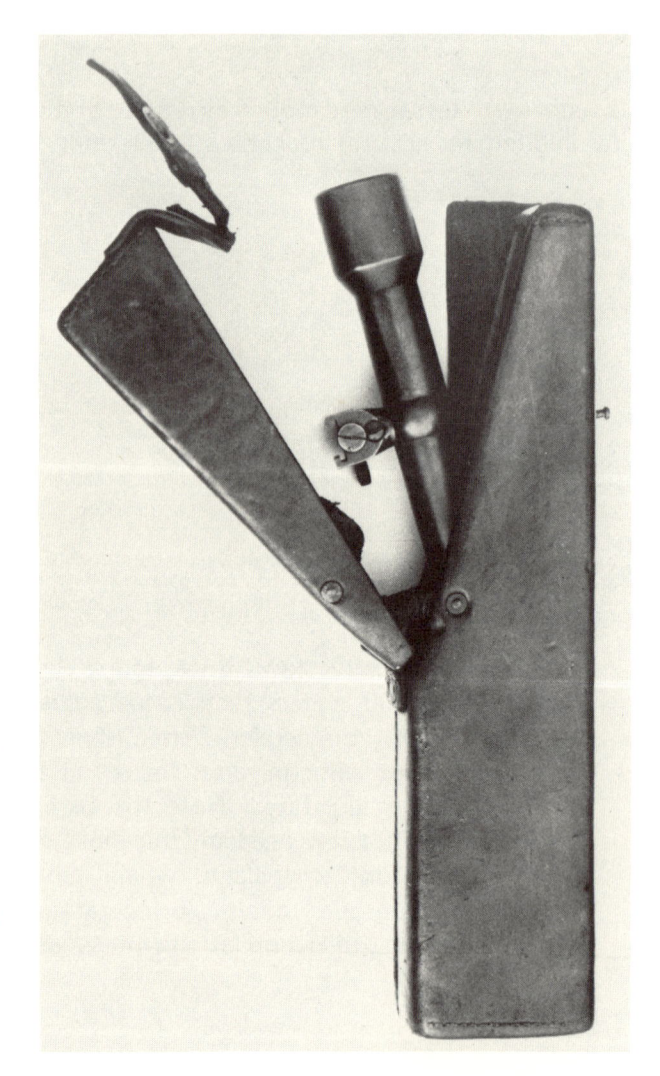

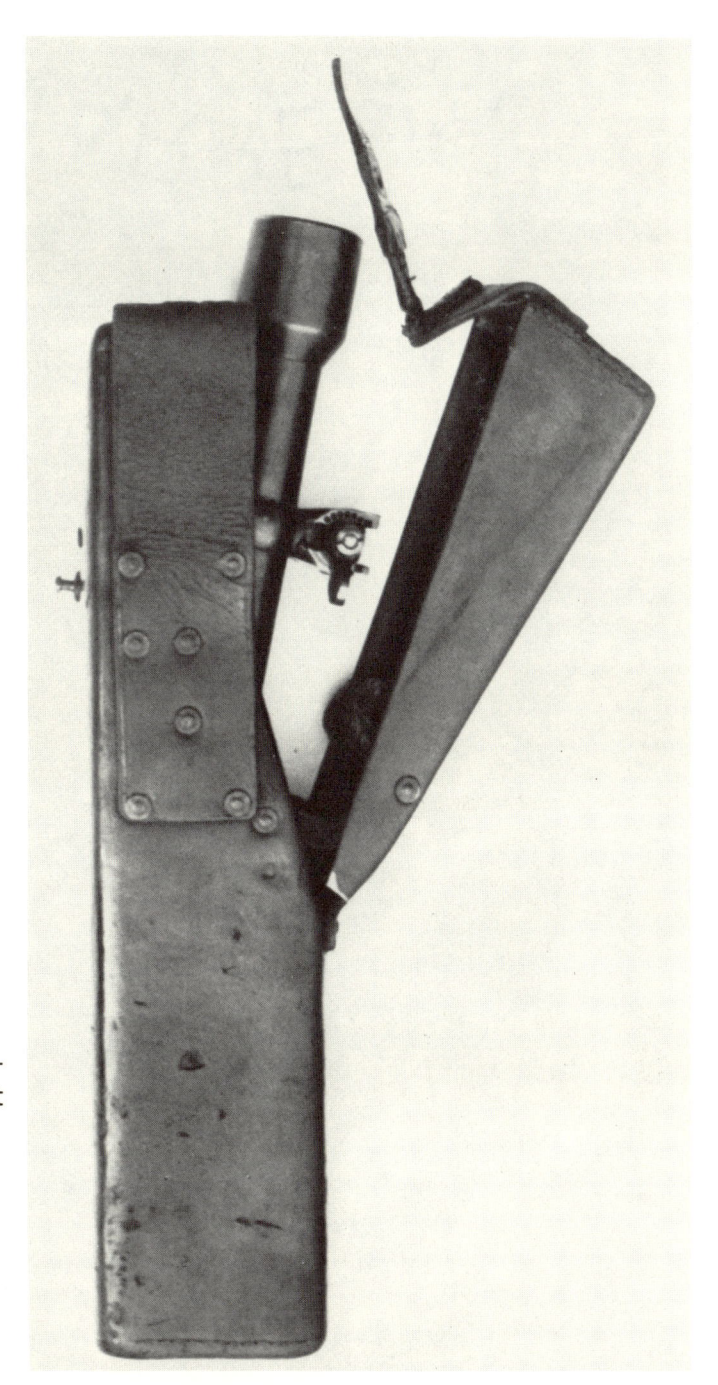

Rear view of carrying case for Zeiss turret mount scope. Note the leather belt loop fastened to the back of the case.

Top view of another early sharpshooter issue 4-power Zeiss turret mount scope with leather lens caps. Note the focus adjustment vane.

German paratroopers advance cautiously during the battle for Crete, 1941. Note the early issue turret mount sniping rifle at the head of the column.

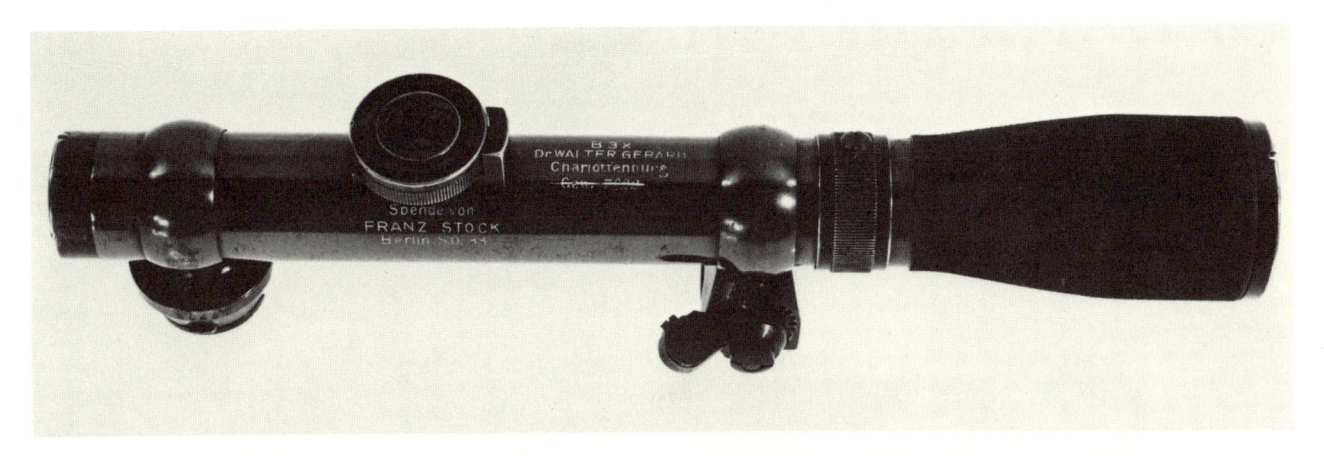

A venerable World War I era Gewehr 98, 3-power Gerard scope as fitted with early model turret mounts for use by the Reich. As an expendiency, many World War I scopes were pressed into service with early Wehrmacht sniping equipment. In this case, the Dr. Walter Gerard scope was originally donated to the Imperial German war effort by Franz Stock, Berlin. Note the lined-out World War I Gew. 98 rifle number; rear mount "leaf-type" detent spring screw; and 98k rifle number on the protruding circular edge of the front mount. Of immense historical significance, the World War I-World War II Gerard scope was mated with a "42" code (Mauser) 1940 date 98k rifle.

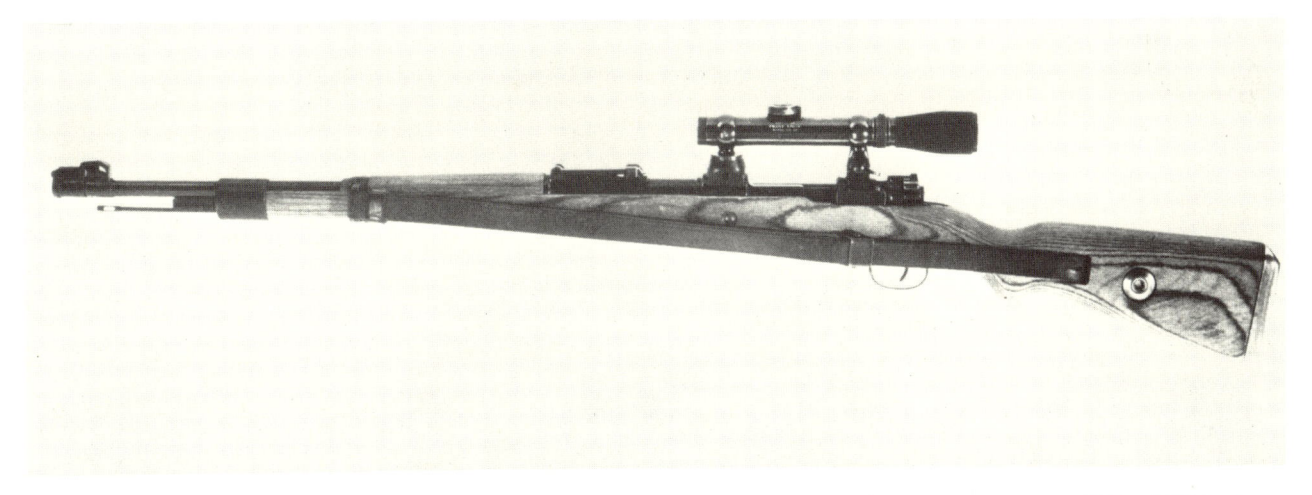

Mauser manufacture 98k rifle (42 code, 1940 date) with early "low" turret system and World War I era 3-power Gerard scope.

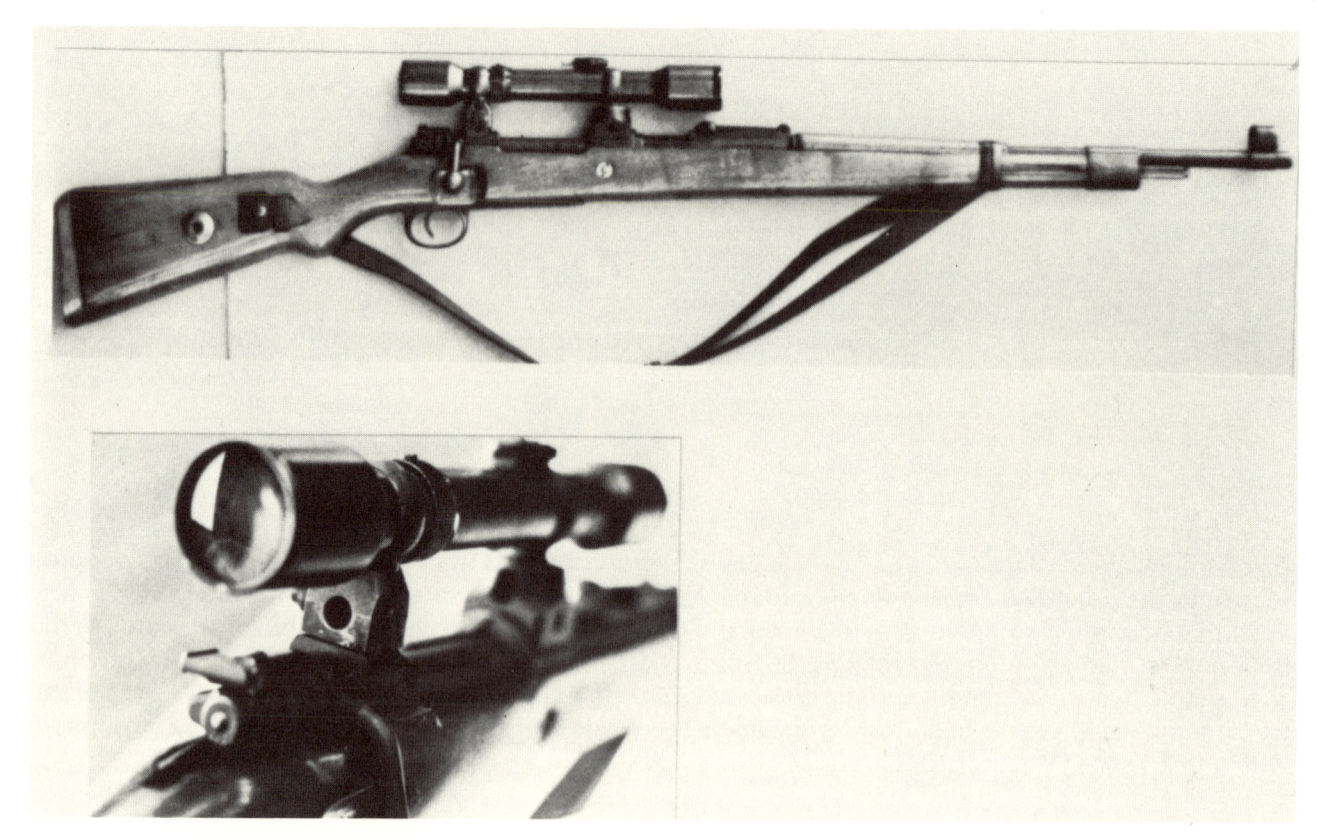

An early Mauser 98k turret mount sniping rifle equipped with a 4-power commercial hunting scope. Note the focus adjustment ring. (Lower left) Oblique view of "tunnel" through mount and base to permit use of the standard sights with scope in place.

Unique "split-ring" turret mount assembly of unknown origin. The sight, a 6-power prewar commercial variant with steel tube, aluminum objective, and ocular housings, bears no manufacturer's markings. Note the focus ring behind the rear mount.

KARABINER 98K — TURRET MOUNT SYSTEM

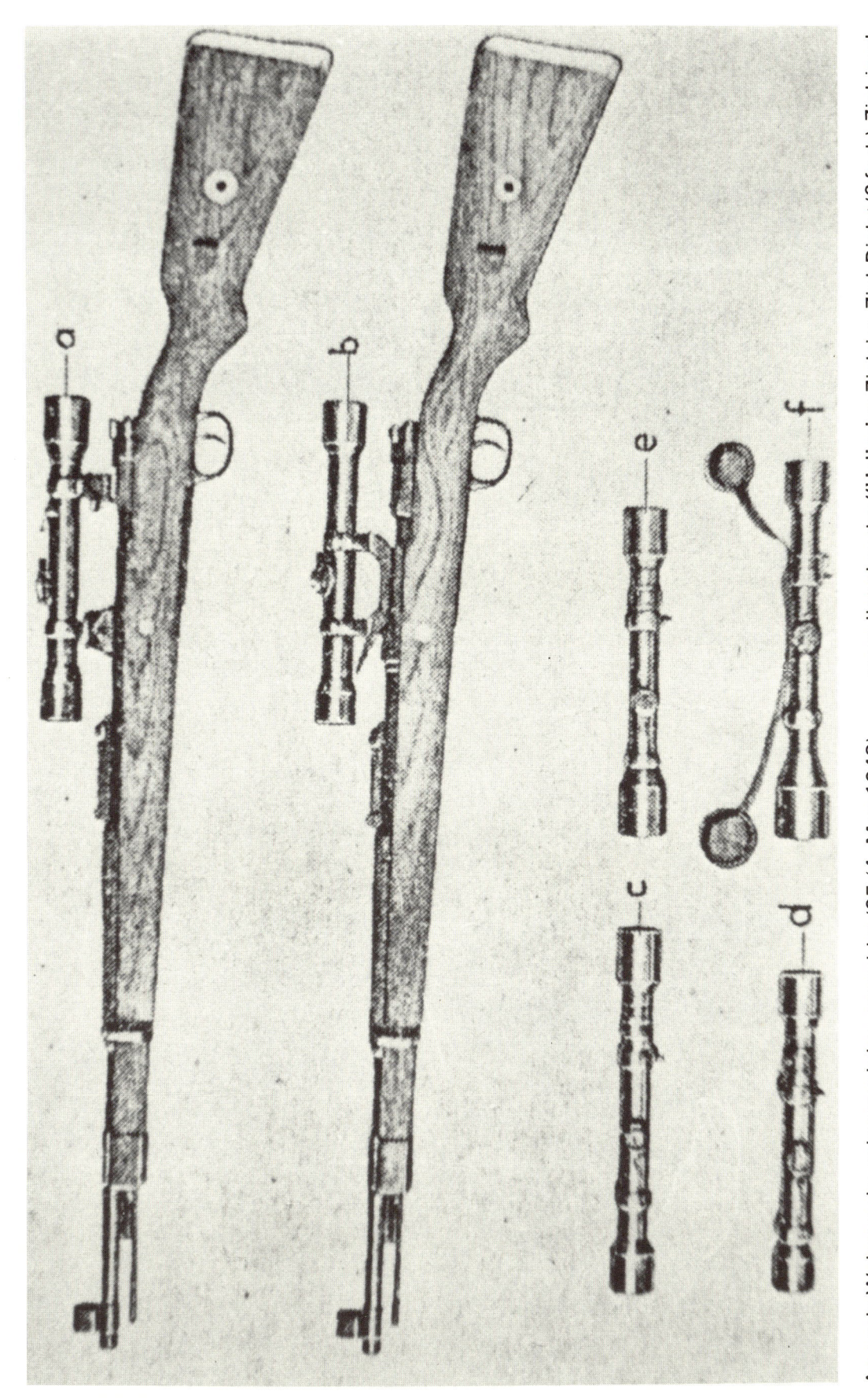

An early Wehrmacht sniper training pamphlet (25/4, May 1943) illustration with turret and short side rail mount 98k rifles. As originally captioned, the commercial scopes used with these rifles were listed as the "Heliavier, Zielvier, Ziel-Dialyt (6 fach), Ziel-jagd, AJAK, and Zielsechs (6 fach)."

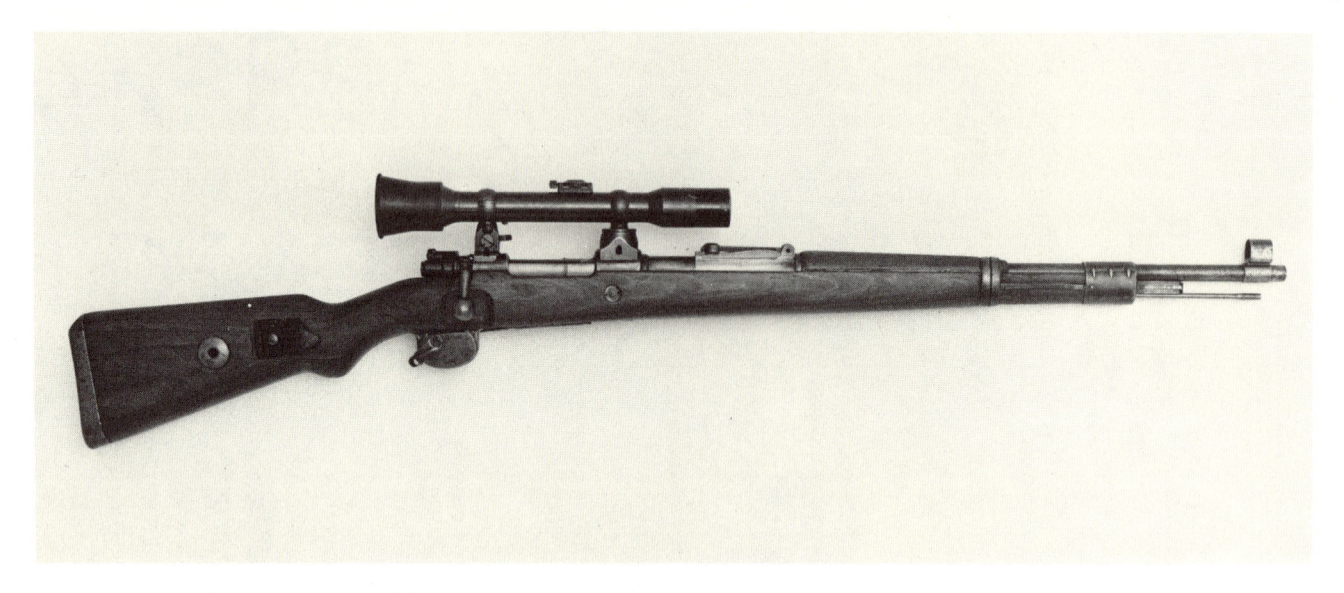

Mauser "high" turret mount 98k sniping rifle with 4-power Kahles scope (cad) and winter trigger.

Left view of Mauser "high" turret mount sniping rifle.

Close view of Mauser "high" turret variant with winter trigger assembly that allowed the rifleman to fire the weapon while wearing winter issue gloves or "arctic mittens." The scope is in position but unlocked.

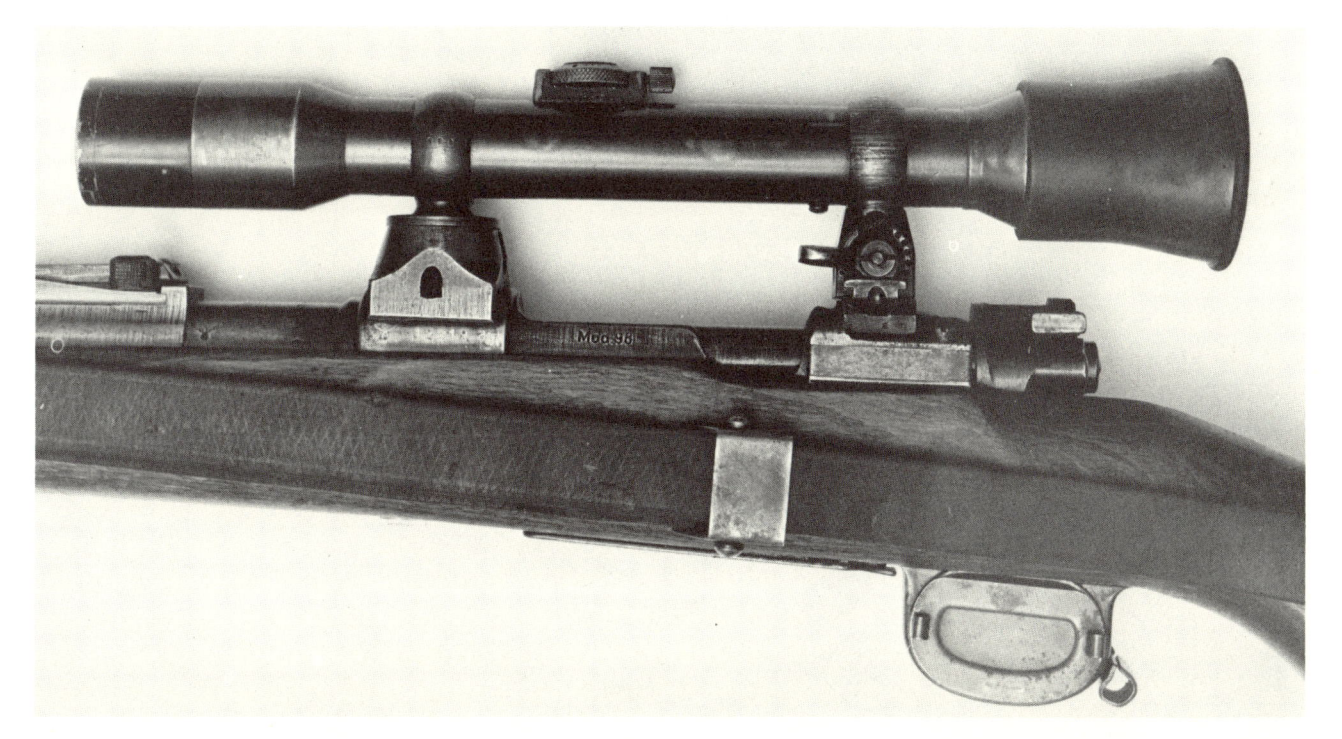

Left view of Mauser "high" turret rifle. Note the barely discernible small star (asterisk) and "135" Waffenamt on the rear base. While all Mauser origin rear bases appear to have had the "135" Waffenamt, not all bear the small star. Of further interest is the "milled" safety lever and "rolled" back edge of the rubber eye-guard. The rear mount wing clamp is in the unlocked (forward) position.

Outfitted in full winter camouflage, a German sniper has the winter trigger device on his turret mount 98k. This permitted the weapon to be fired with a gloved hand.

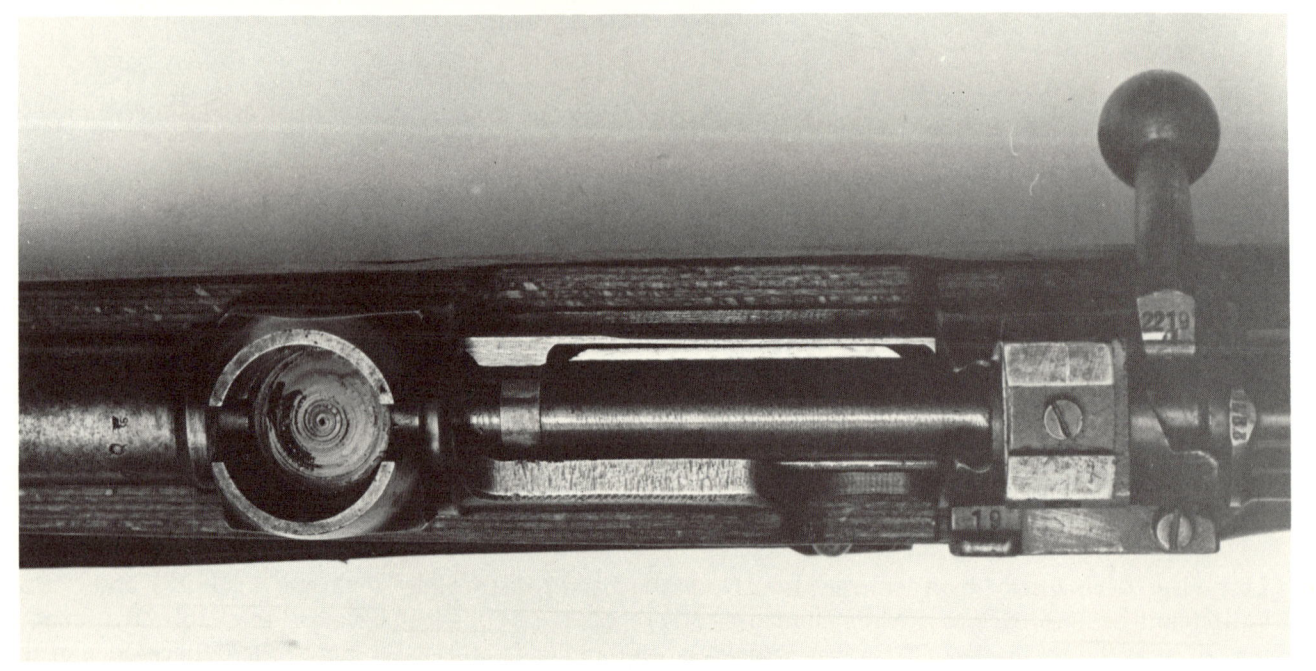

Typical Mauser "high" turret mount receiver bases.

KARABINER 98K — TURRET MOUNT SYSTEM

Underview of the "cad" (Kahles) "high" turret mount scope. Although rubber "eye protectors" were a part of the early issue, few have survived.

Mauser "high" turret front scope ring and mount assembly. Both front and rear rings were soldered directly to the telescope tube and as such presented no problem from the recoil action of the rifle. Note the rifle serial number on the front ring and tension spring in the slot.

Mauser turret system rear scope ring and mount. Note the extremely rough machine marks on the ring and small "rolled-pin" stop beneath the wing clamp. The five recesses function as progressive stops for the wing clamp when tightened.

A camo garbed army marksman demonstrates his 98k turret mount sniping rifle (from *Atlantic Wall*, 1944.)

Mauser "low" turret mount 98k sniping rifle with 4-power "bmj" code telescopic sight, leather lens caps, and "low" scope safety.

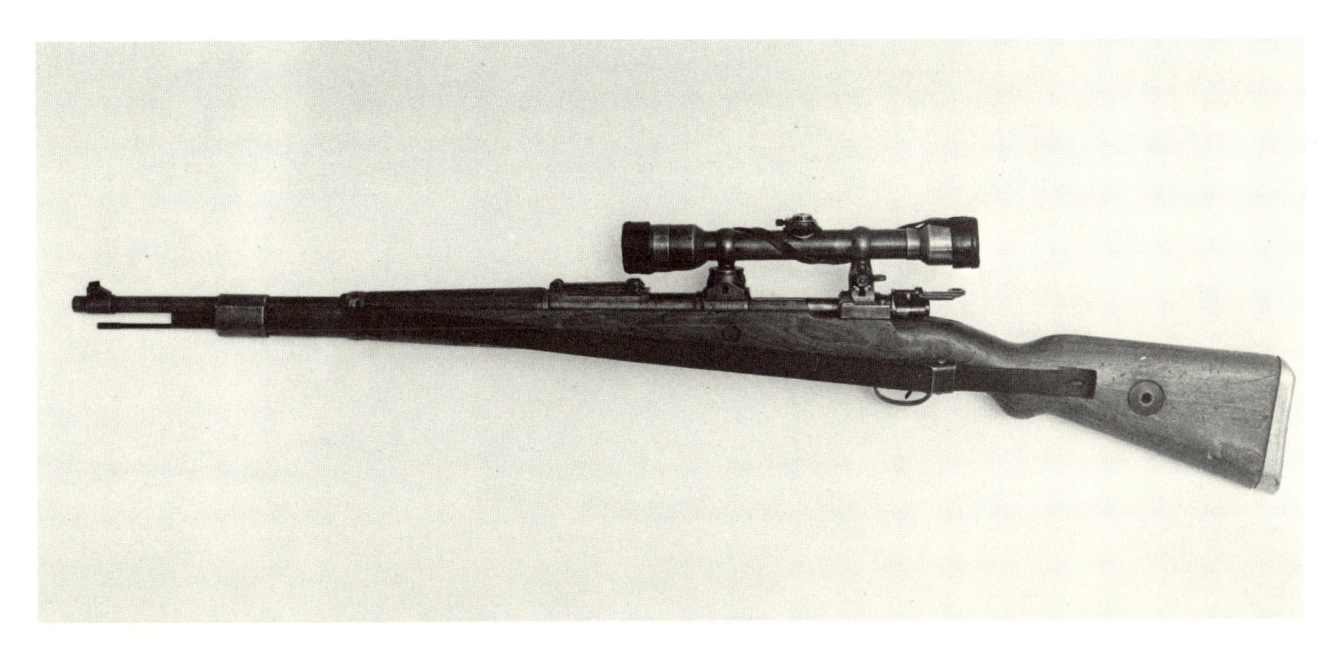

Left view of Mauser "low" turret mount sniping rifle.

Close view of Mauser "low" turret mount variant. Note the "135" Waffenamt on the rear base and the absence of an asterisk (star). The "low" scope safety was originally intended for the "bcd" code long side rail sniping rifle and most were applied by postwar collectors for their exotic appearance.

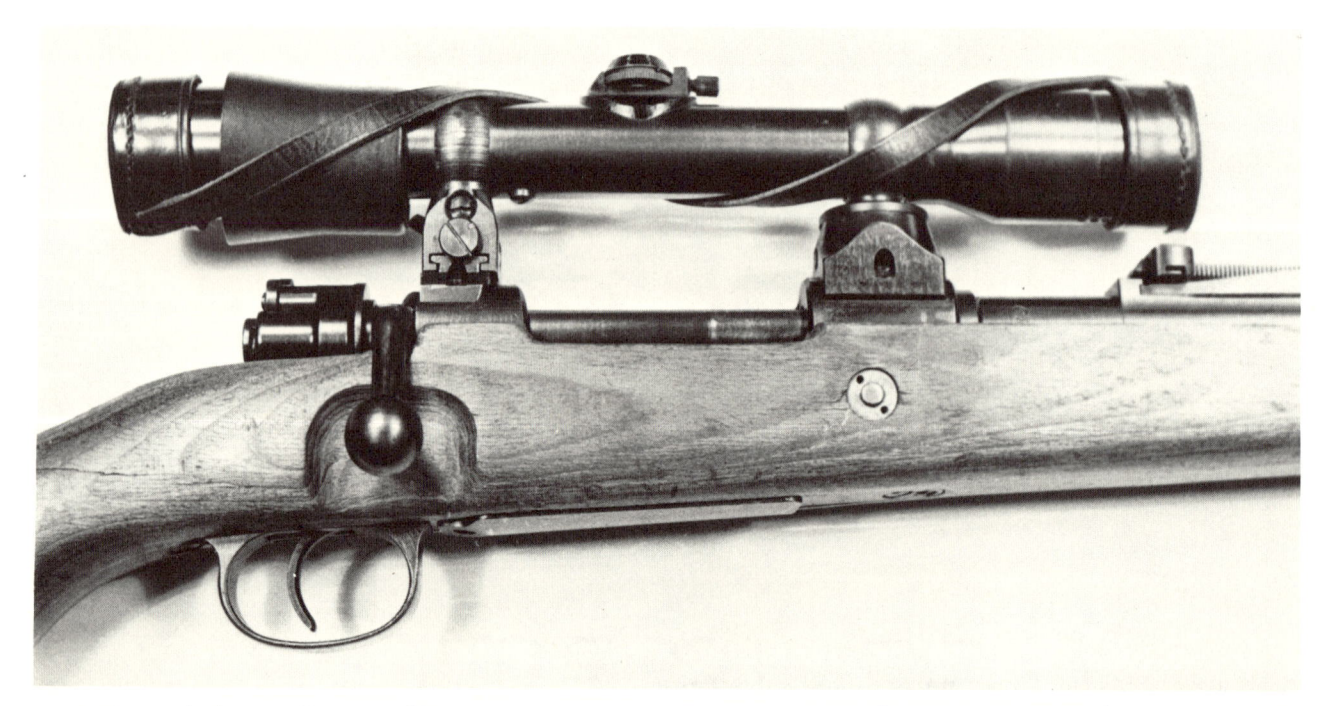

J. P. Sauer (ce) "low" turret 98k sniping rifle with 4-power "bmj" (Hensoldt, Wetzlar) telescopic sight, leather lens caps, and rubber eye-guard. To lessen the chance of loss, the rubber eye-guard was pushed forward when the lens caps were placed on the sight.

KARABINER 98K — TURRET MOUNT SYSTEM

Close left view of J. P. Sauer "low" turret front mount and base assembly. A Waffenamt (37), small Eagle and Swastika, and rifle serial number are stamped on the receiver ring directly below the base.

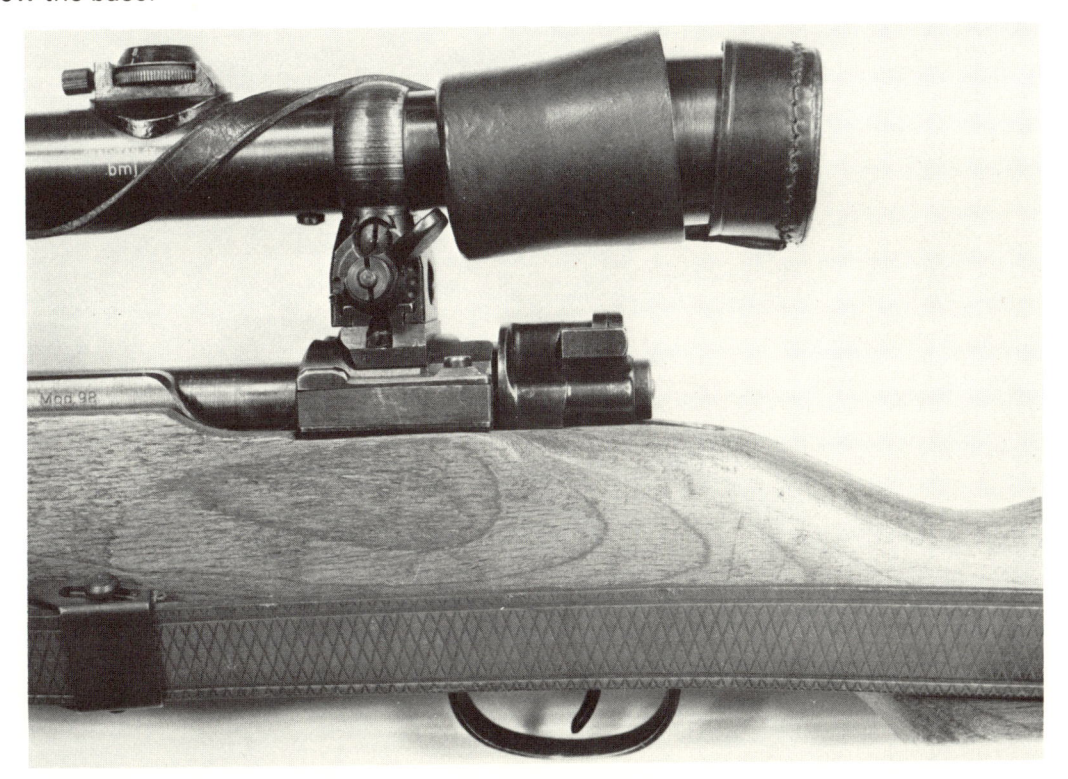

Close left view of J. P. Sauer "low" turret rear mount and base assembly. A "milled" safety lever was used to provide additional clearance with the sight, a feature noted on many original World War II German sniping rifles.

An excellent view of a turret mount sniping rifle in use. Russia, summer 1944.

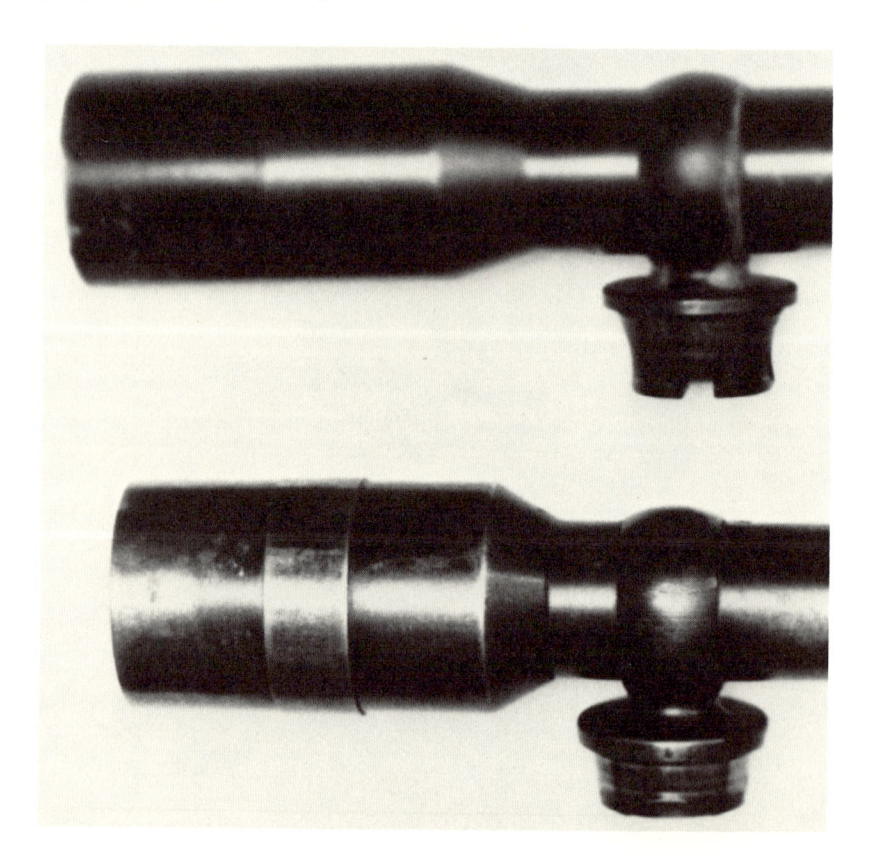

Comparison of "high" and "low" turret front ring and mount assemblies, showing difference in height of the section which fits into the front cone. Top is the "high" and bottom is the "low" mount.

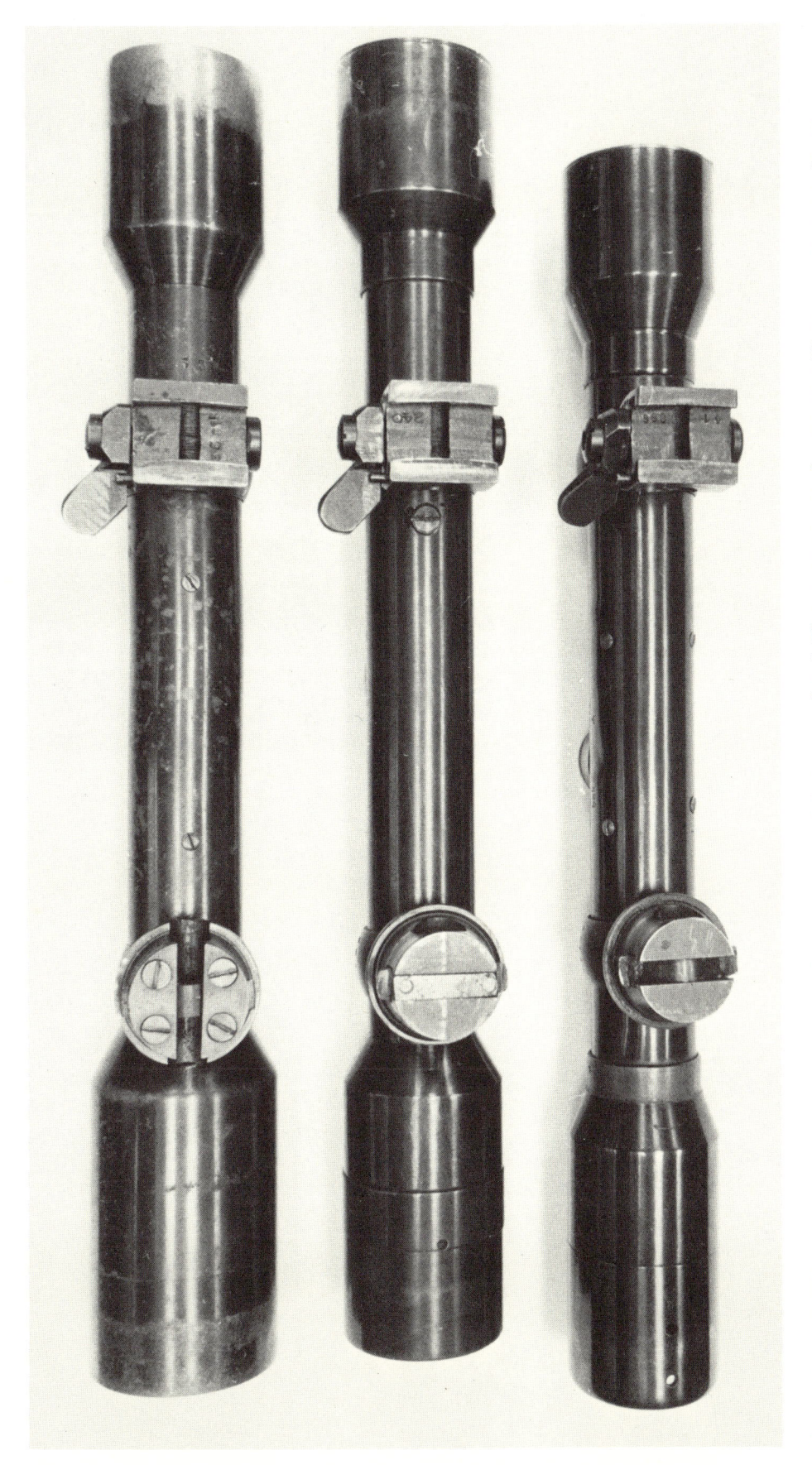

An underview of three turret system scope and mount assemblies. (top). The scopes are 4-power military contract variants. Note the four screws characteristic on "low" turret front mounts

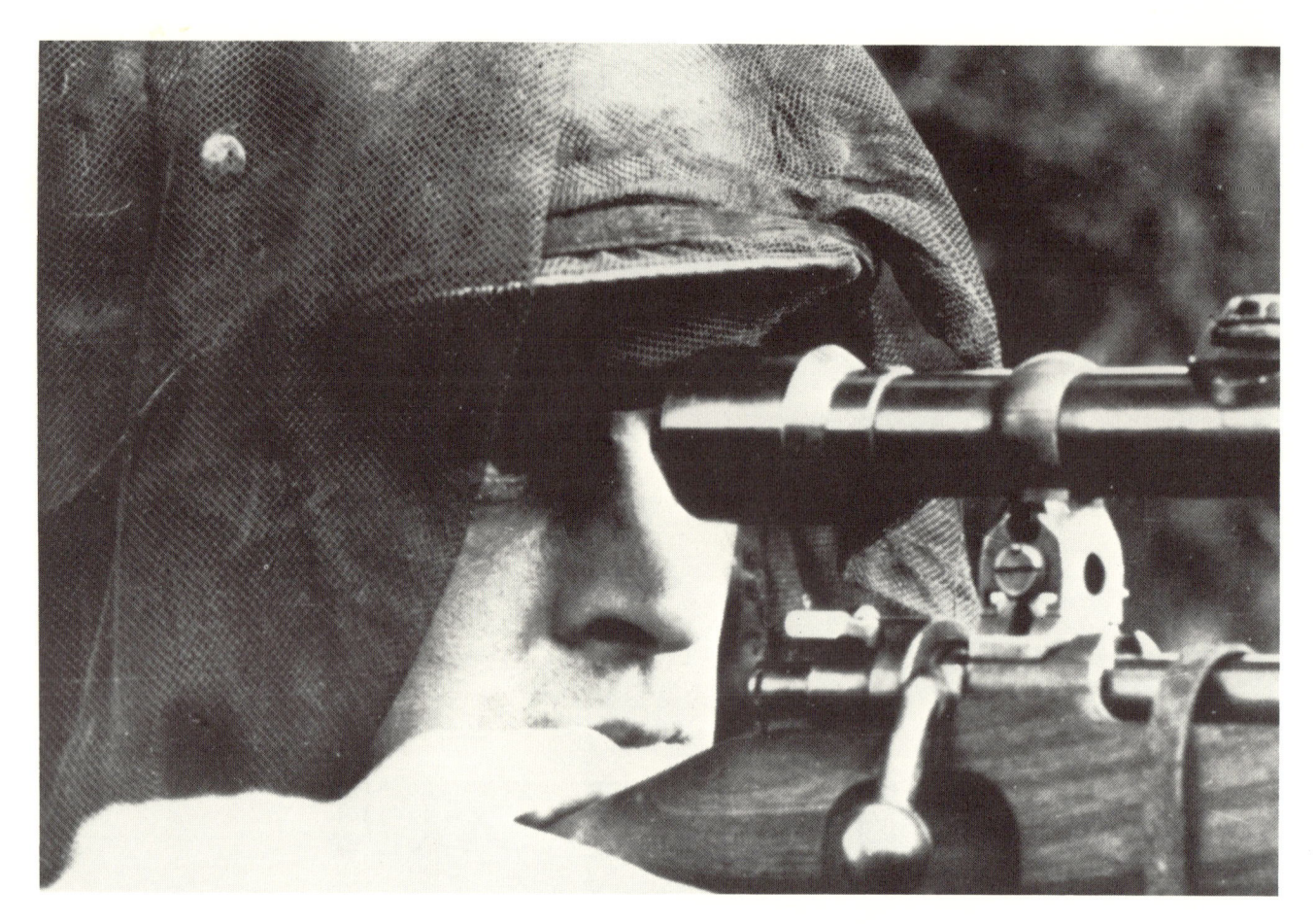

A study in intensity; the sniper sights his turret mounted scope. Russia, 1944.

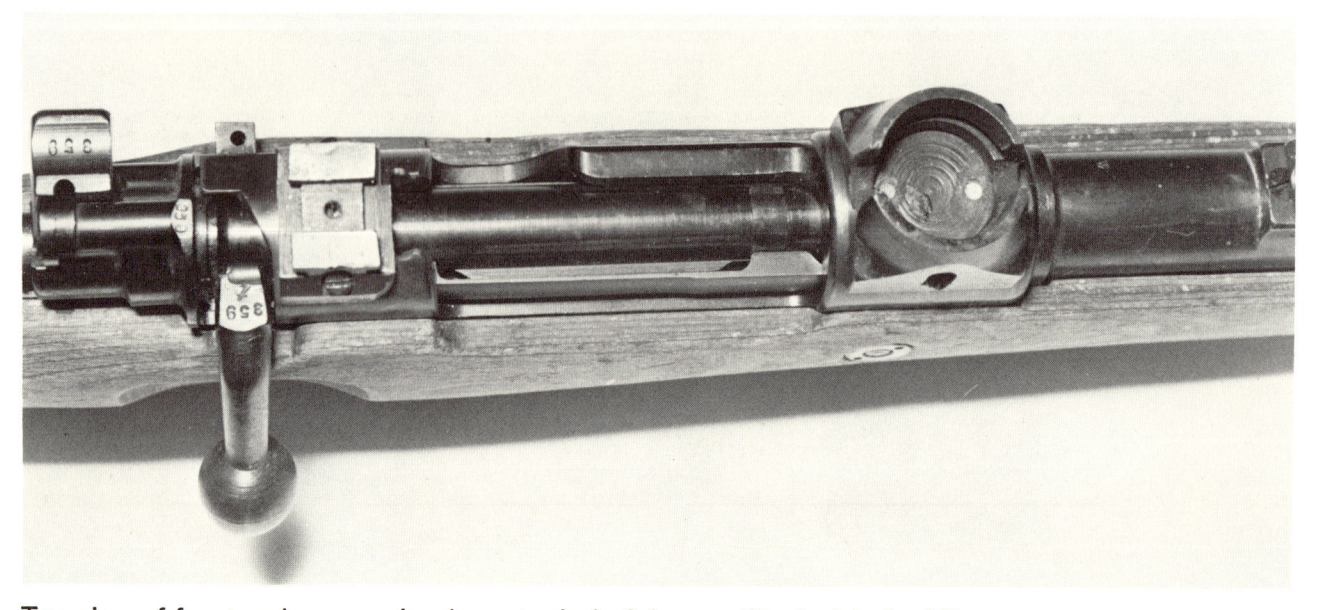

Top view of front and rear receiver bases typical of those utilized with the 98k turret mount system.

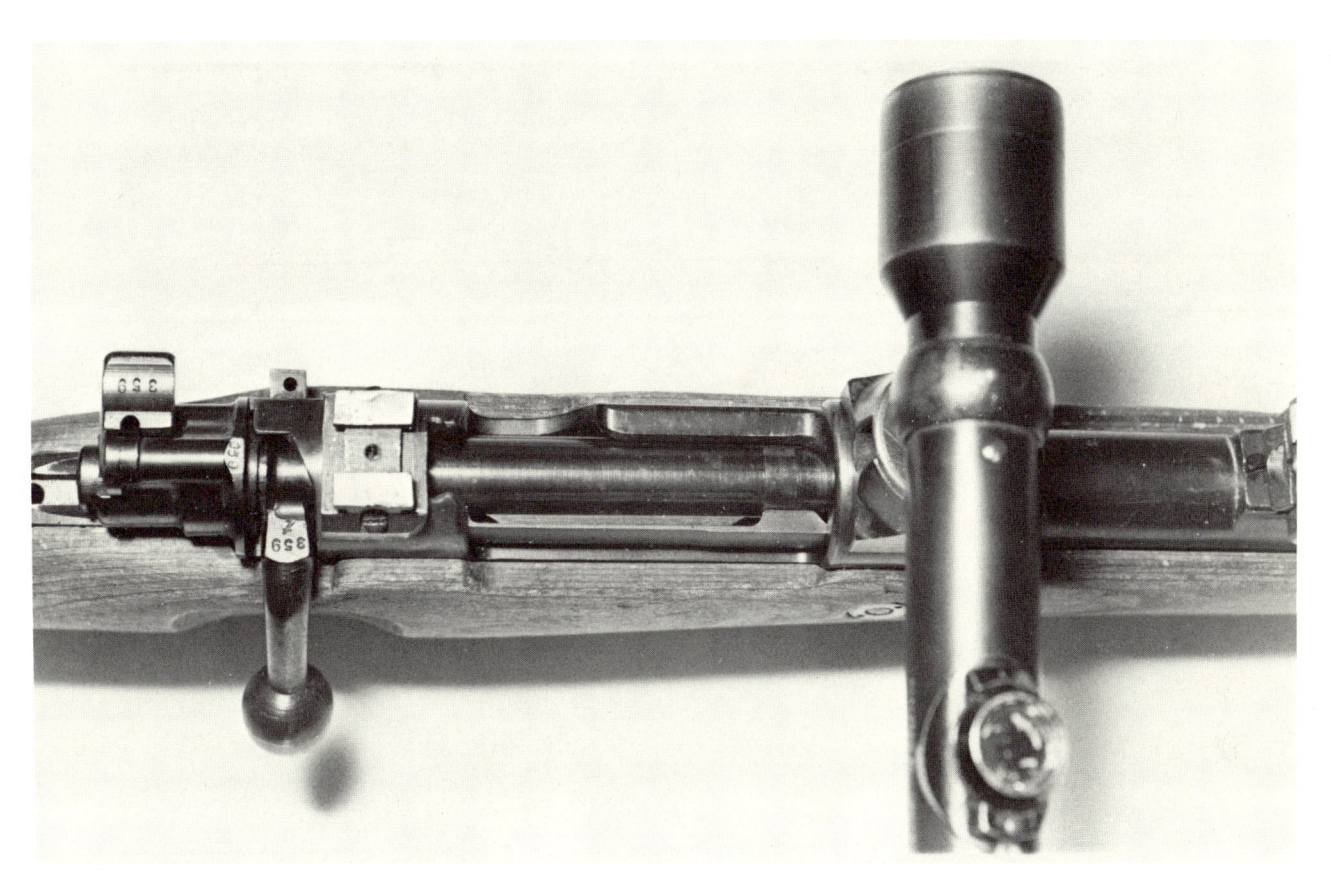

Top view of turret system with scope inserted into the front base cone at 90 degrees to the receiver when mounting the sight.

Top view of turret system with scope rotated toward receiver bridge to engage the rear base assembly.

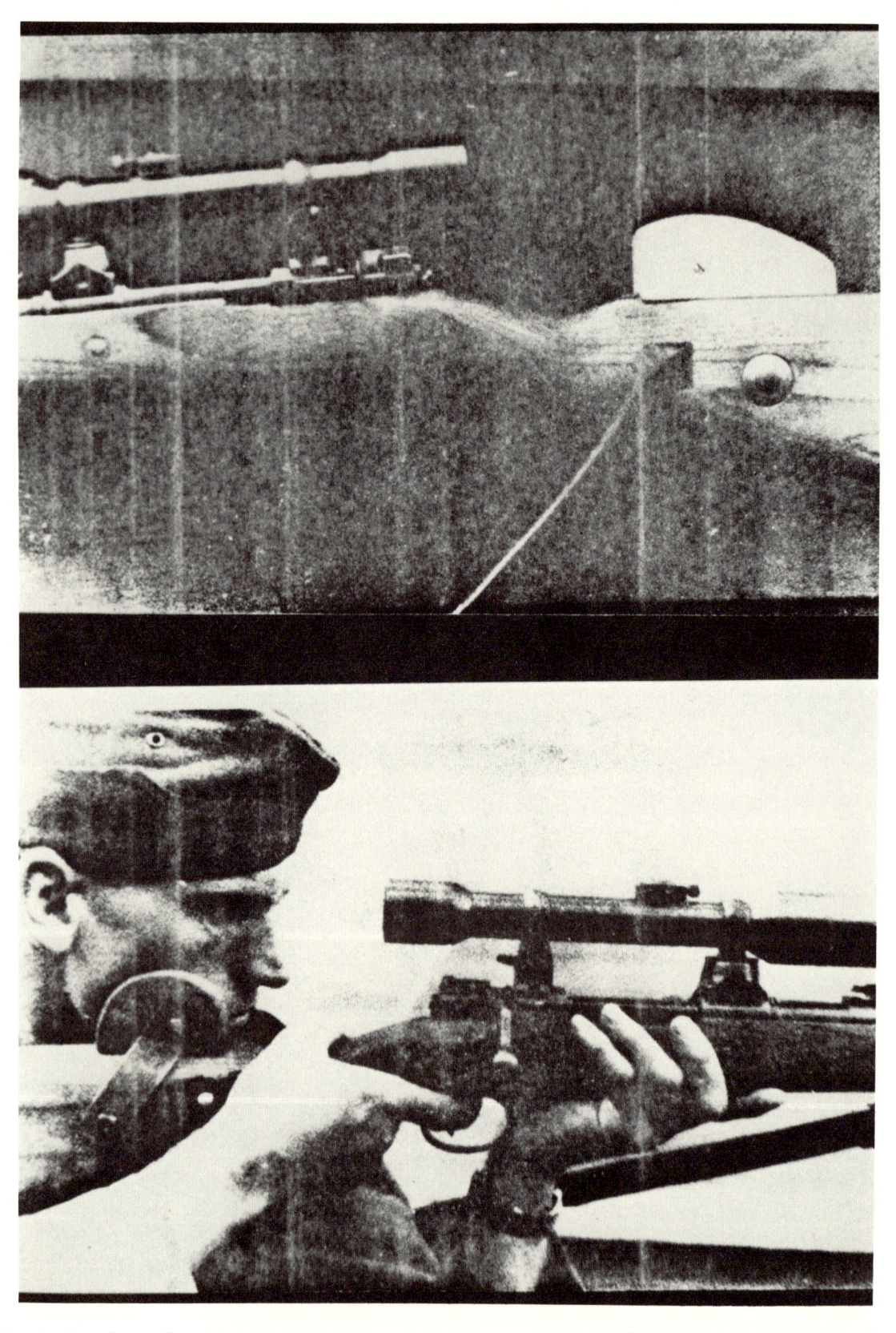

Illustration from German sharpshooter training manual H.Dv. 298/20 (October 1944) depicts a turret mount 98k with improvised cheek rest intended for training or for combat use as the individual marksman deemed necessary.

KARABINER 98K — TURRET MOUNT SYSTEM

Formed metal case for carrying the turret mount scope assembly when removed from the rifle. Note the "D" rings for attaching the shoulder strap.

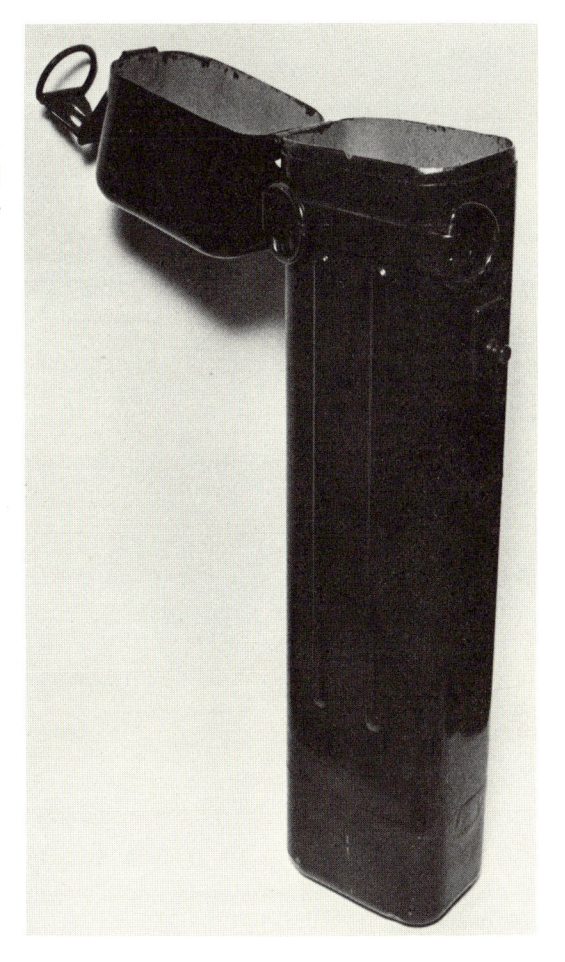

Turret mount 4-power military contract scope and variant metal carrying case with "D" rings and web belt loop. Note the rifle number stamped on the case lid.

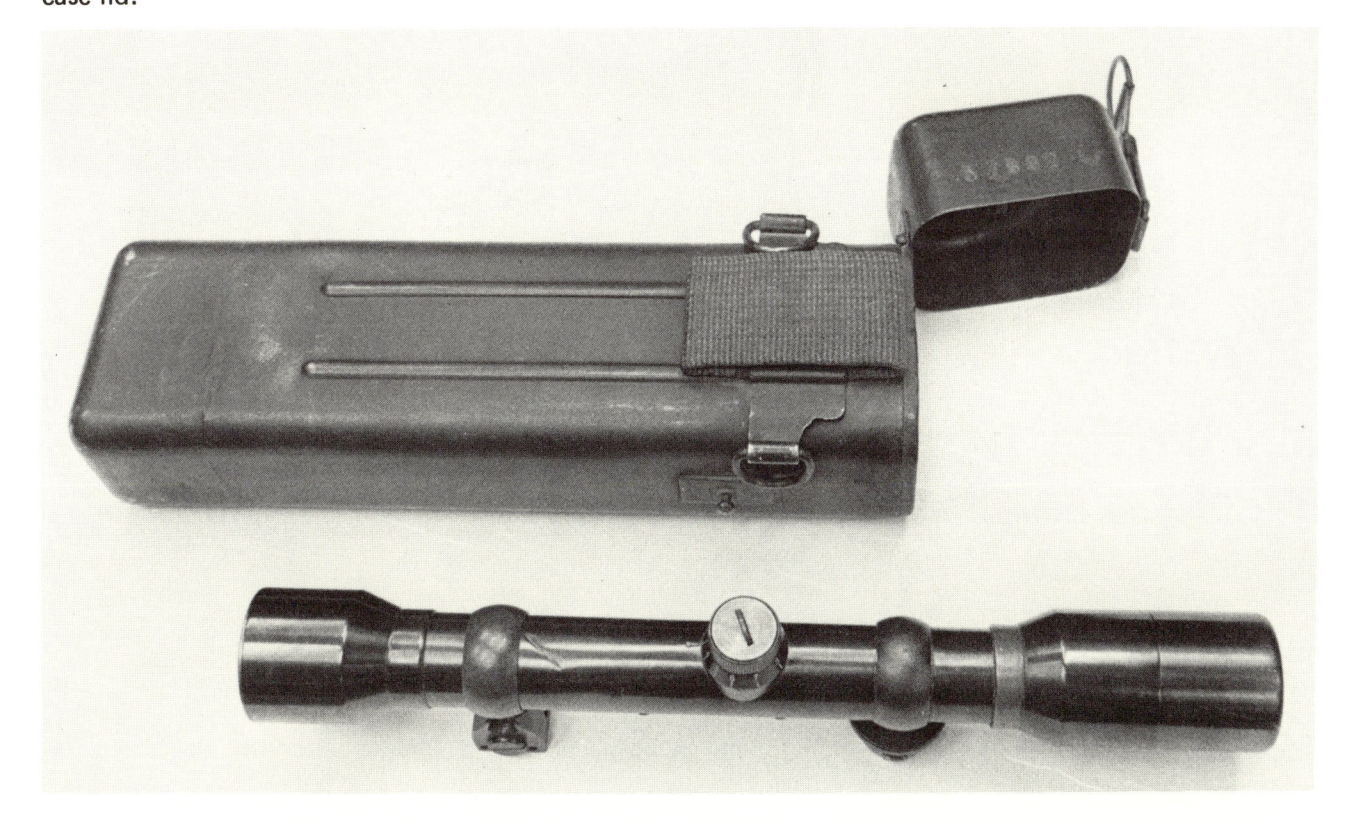

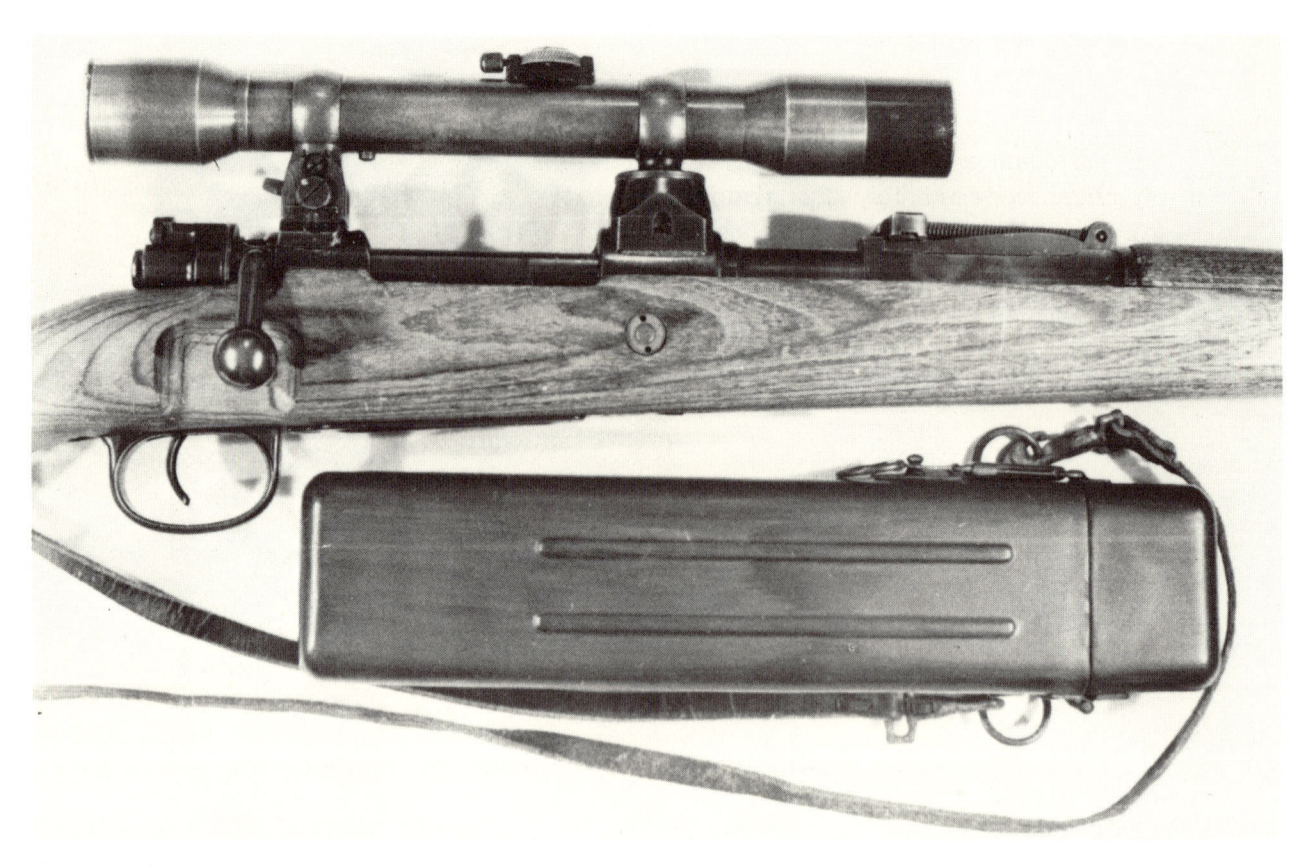

Typical 98k turret mount sniping rifle with 4-power military contract scope and carrying case. Note the original shoulder strap, an item rarely encountered.

Mauser turret mount rifle with Eagle and Swastika proof and "135" Waffenamt directly below the base on the receiver ring. The rifle serial number is not present in this case.

Typical Mauser turret mount receiver markings: Eagle and Swastika proof, serial number, and "135" Waffenamt.

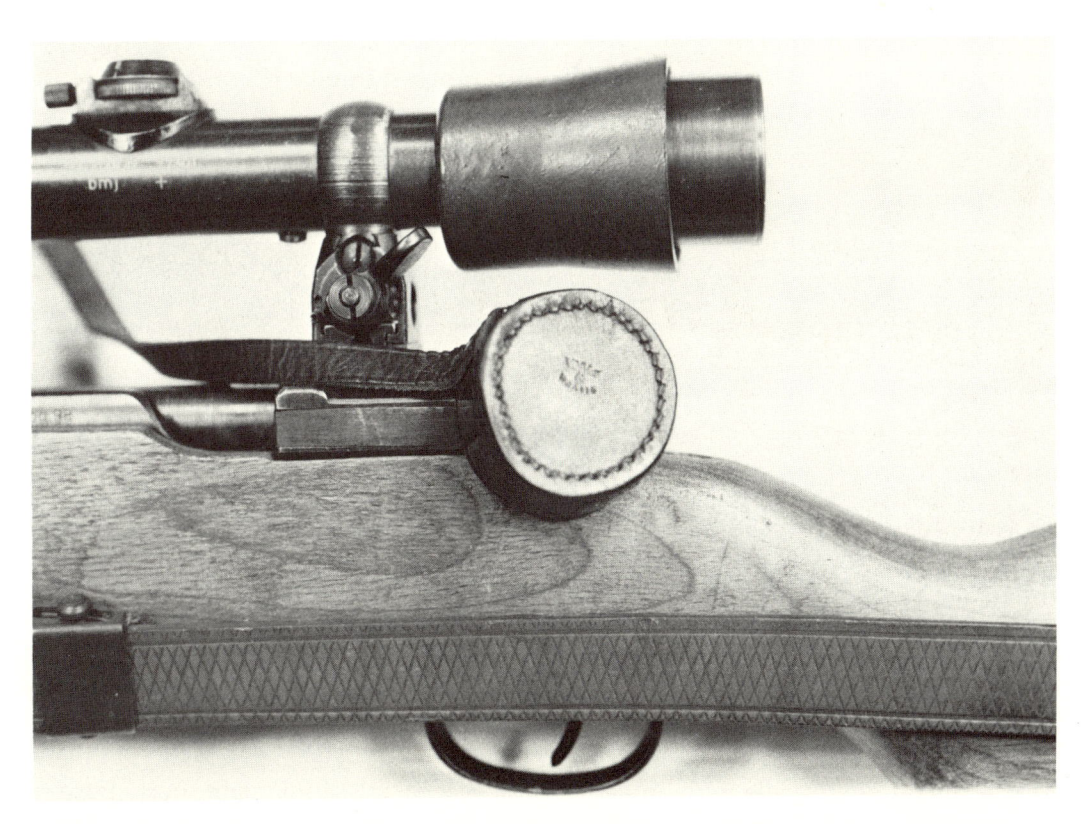

Close view of leather lens caps showing the Waffenamt (WaA 414) impressed on the ocular end. Note the "straight" back edge of the rubber eye-guard.

German sharpshooter, circa 1944, sighting the 98k turret mount sniping rifle.

Typical Mauser turret mount rifle with an early Ajack 4 x 90 telescopic sight. Early issue scopes rivaled the finest commercial type in quality and blue finish, but as the war progressed extremely rough examples were produced with a grey phosphate type finish.

A comparison of two Kahles (cad) turret mount scopes. The example at left is of late manufacture with two drain holes and arctic sun shield. The early variation at right has neither the drain holes nor provision for a sun shield. Note the large circular openings (tunnel) in the front and rear mounts which permitted use of the standard sights with the scope mounted on the rifle.

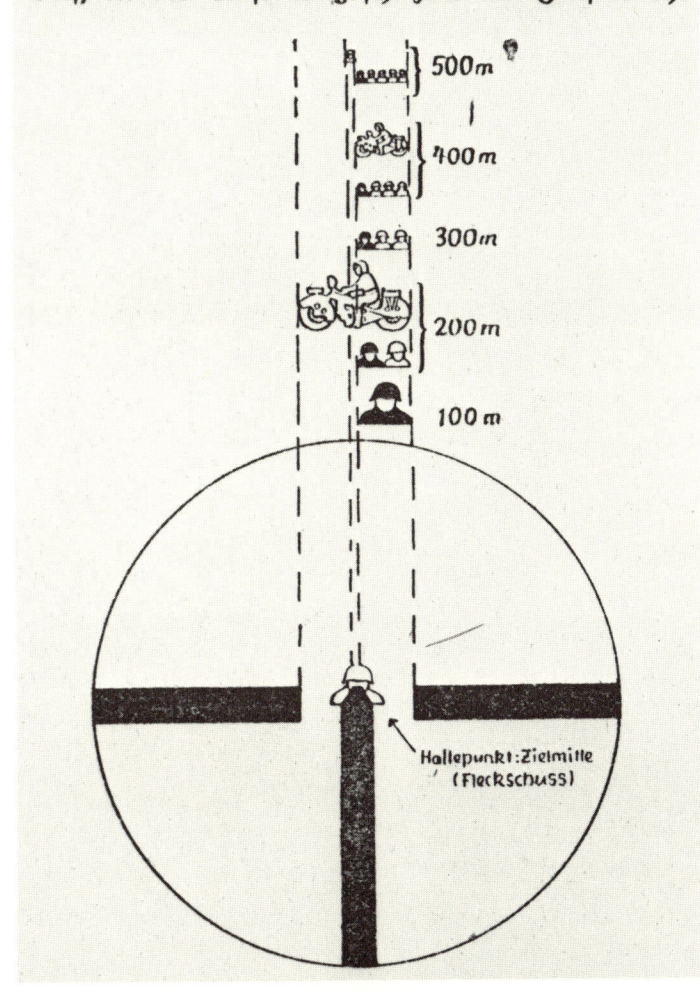

Beispiele des Entfernungsschätzens mit Zielfernrohr

The typical pointed post and horizontal side bar reticle pattern found in most German telescopic sights dating from World War II was also used for estimating target ranges. As noted in an early sniper training manual (1943) and literally translated, "Illustration to estimate distance using a telescope sight" . . . "Target Center (Shot to the center of the bull's eye)."

Wehrmacht snipers demonstrating their preparedness while awaiting the impending invasion (from *Atlantic Wall,* 1944).

CHAPTER X

Karabiner 98k — Long Side Rail System

The war was going badly for the Germans, with their forces clamoring for small arms in ever increasing quantities. From the experiences drawn from the Russian use of snipers, the High Command was prompted to seek additional weapons for sniping purposes. Early in 1944 the Gustloff Werke (bcd) began manufacture of the "long" side rail variation based on the revised 98k rifle.

The principal difference between the long side rail rifle and the standard 98k was the special enlarged receiver having a machined flat to accommodate a telescope mounting base of similar configuration. The base was securely held in position with three screws and two tapered pins in addition to three lock screws, which were employed as an added measure of security. Despite the prevalent use of two tapered base mounting pins through the entire span of production (early to late manufacture), variant long side rail receivers having only three internally tapped screw holes with no provision for the pins (holes) have also been noted and as yet are unexplained.

The scope mount has a long locking lever located in its center, acting as an extension of a threaded screw passing through the mount. Alternately moving the lever forward and backward locks or unlocks the mount on the receiver base. The pressure exerted against the base is positive and permits no movement. A secondary lock mechanism consists of a self-actuated spring-loaded latch. As the scope mount is slid on to the receiver base (rail) dovetail, the latch rides up until it engages with a small elongated recess in the forward section of the base. To remove the mount, the rear latch section is depressed, the long locking lever is moved to the rear, and the mount slid from the dovetail.

Early issue long side rail mounts utilized split rings of machined steel. As a production expediency, however, this type of ring was replaced with simple spring bands for holding the telescopic sight in place.

To counteract the tendency of the scope to move forward when the rifle was fired (recoil action), a recoil ring was attached to the scope tube immediately behind the front ring or band.

Windage adjustments (lateral movement) were made in the manner typical of German telescopic sights with the exception that instead of the usual small screwdriver, a special tool was required to actuate this mechanism.

A small but significant provision that exemplified the efficient overall design of the long side rail system was the inclusion of a sheet metal rail cover that protected the vulnerable edges of the receiver base dovetail when the sight assembly was removed from the rifle.

Aside from the early prototype pieces

Waffen-SS sniper team during late war winter action. The 98k long 4-power Hensoldt scope. Note the SS "runes" on the rifleman's side rail sniping rifle has the special safety, spring-band mount, and helmet.

The Gustloff (bcd) long side rail sniping system with 4-power "dow" telescopic sight.

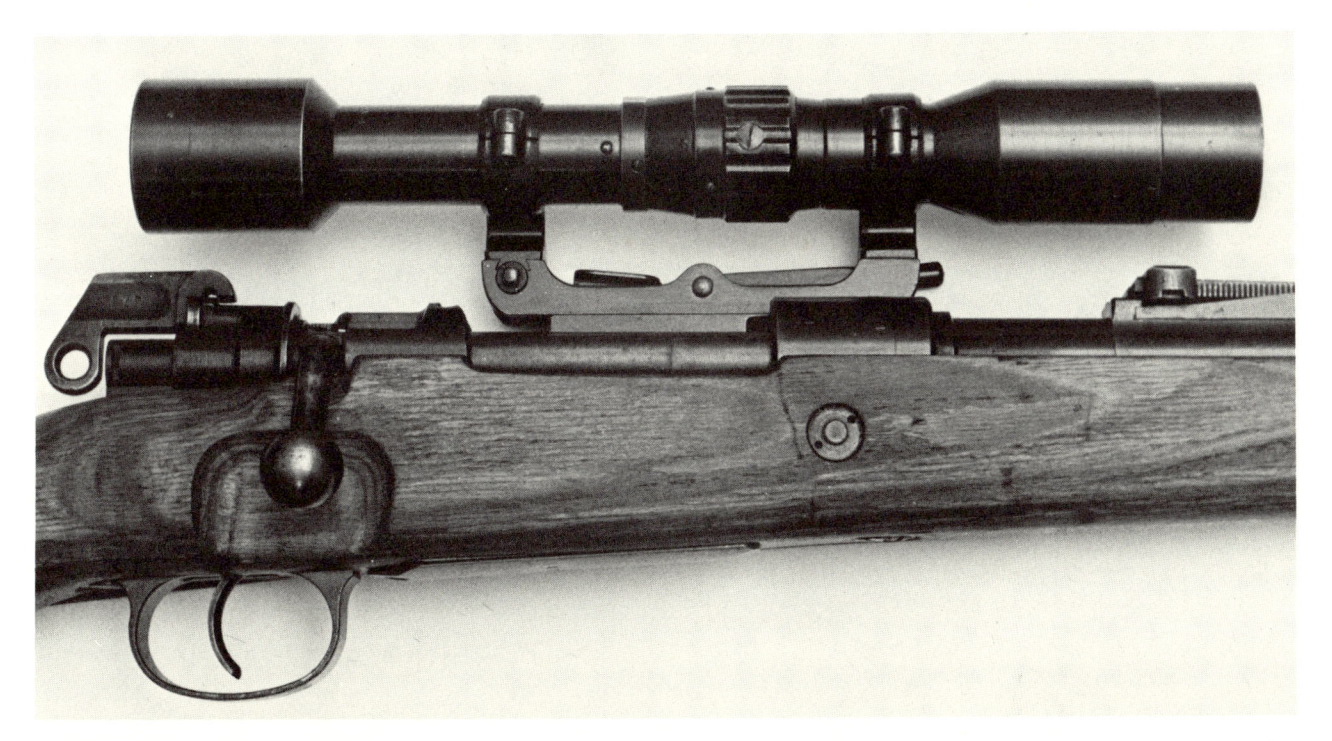

Close view of "dow" scope in place on the "bcd" rifle. Elevation adjustments were made by rotating the ring at the center of the telescope tube. Note the early manufacture machined steel scope rings.

(1943) directly attributed to the J.P. Sauer firm, sniping weapons of this configuration (special enlarged receivers) saw limited production and issue based on late war 98k rifles manufactured by Sauer (ce) and Mauser (byf). In addition, the Gustloff variants were produced in rather large quantities from 1944 to 1945.

While hardly intended for sniping purposes but still of interest are surviving examples of the Volkssturm Gewehr 98 (People's assault rifle 98), which was based on special enlarged Gustloff 98k receivers in some cases. Intended for use by the German people in a last ditch stand against the invading armies, variant VG 98 rifles utilizing the unique "bcd" receiver fitted with a fixed open rear sight on the receiver ring, machine gun barrels, welded front sights, and crudely finished stocks are particularly noteworthy.

Close view of the special enlarged 98k receiver showing the machined flat. The three large holes are tapped for the base holding screws, the other two for the tapered pins.

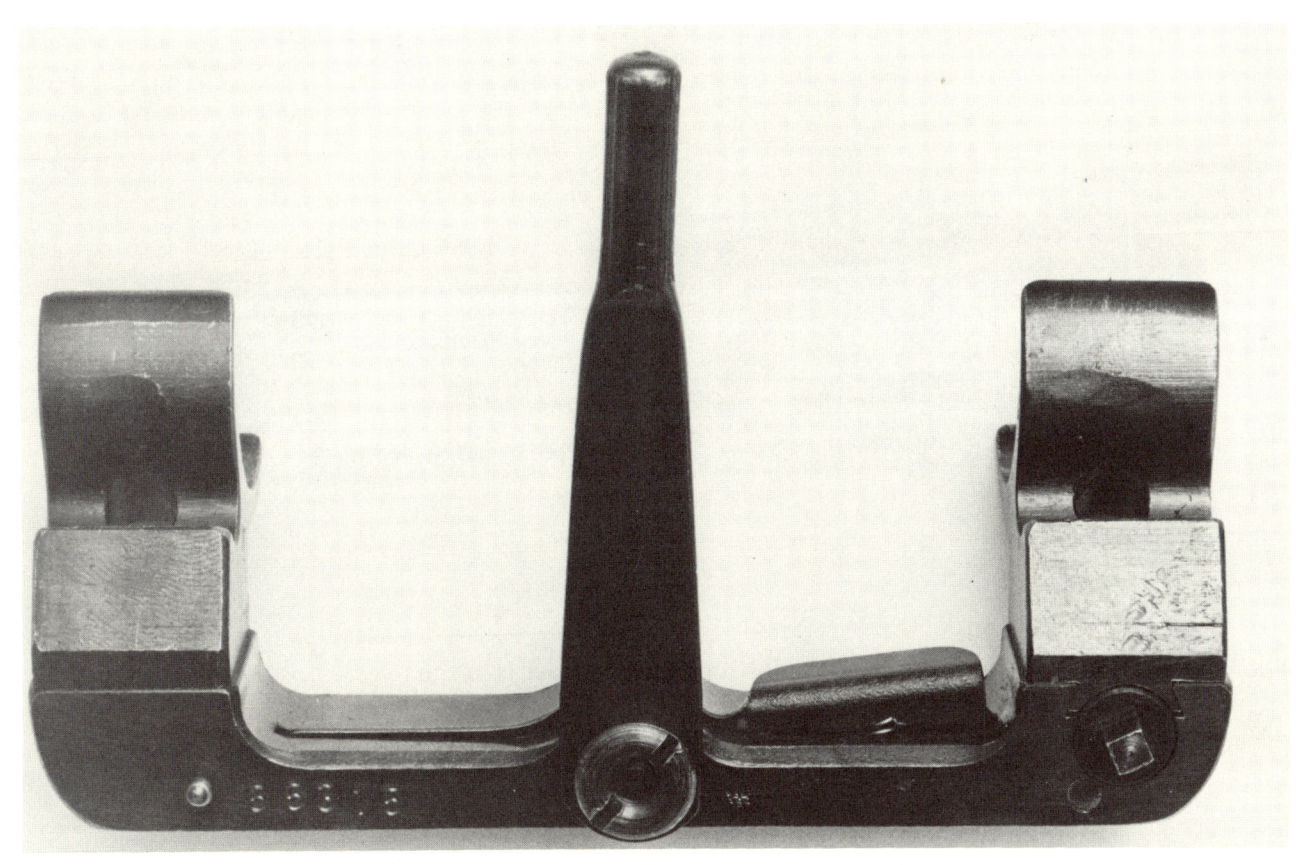

A long side rail telescope mount shown with the locking lever in the vertical position. Note the spring-loaded latch. The square stud at the mount's rear was provided for windage adjustments. As originally issued, the mount was stamped with the corresponding rifle serial number. This mount bears the "359" Waffenamt and serial No. 55315.

An underview of a long side rail mount showing the spring-loaded latch. Note the small projection that engages with the elongated hole in the receiver base.

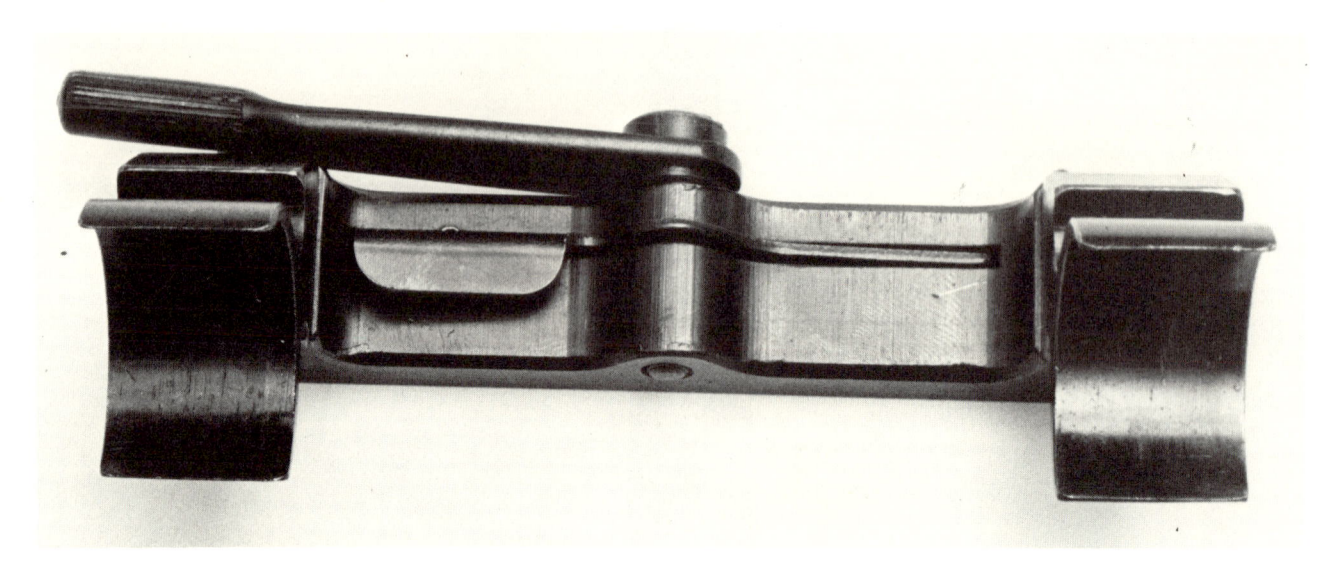

Top view of a long side rail scope mount. The long locking lever is in the release position (to rear).

The long side rail base as fitted to the receiver. The smallest hole, located at the top front, accepts the hook projection of the spring-loaded latch. As a rule, the receiver bases were not numbered. Note the "359" Waffenamt which appears on all "bcd" scope mounting bases.

Close view of reverse (receiver) side of a long side rail base. The smallest of the holes are for the locking screws. The stud at the base front is a plunger which, in conjunction with the notches at both ends, retains a sheet metal cover. This cover protected the vulnerable edges of the rail dovetail when the scope assembly was not on the weapon. The slightest damage to the rail surface and edges would prevent the mount from sliding on to the base.

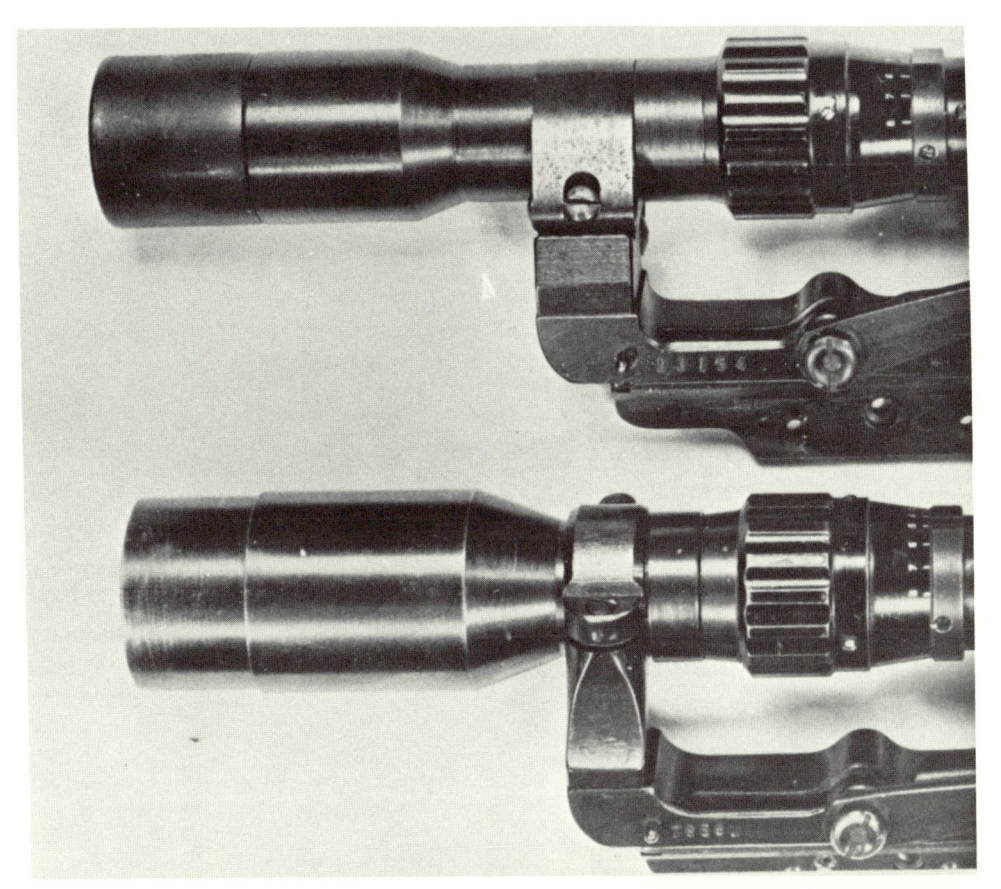

Two variations of the "dow" code (Opticotechna) 4-power center elevation adjustment scope with long side rail mounting. The top scope has a special aluminum objective tube extension (sun shade-rain shield) and is held by standard spring bands. The bottom example is a common "dow" variation with early steel rings. There are at least four types of this telescopic sight, with the variations being in length, size of objective lens, elevation adjuster, objective tube extensions, and fixed focus (universal) or focus adjustment.

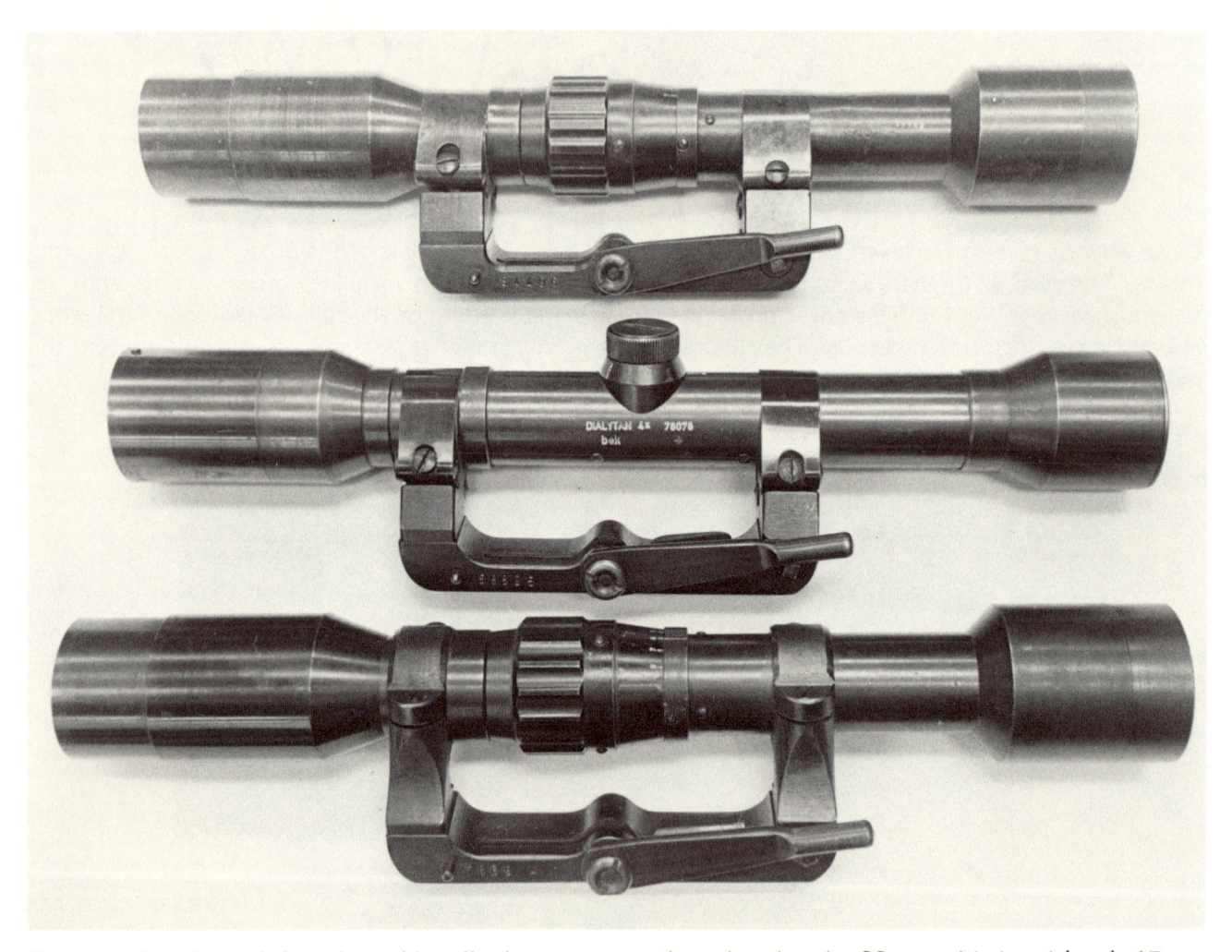

Comparative view of three long side rail telescope mountings showing the 22mm wide band (top), 17mm narrow band, and early manufacture machined steel ring type mount. Note the telescope "recoil rings" which overcame the tendency for the scope to work forward in the mount from the recoil action of the rifle.

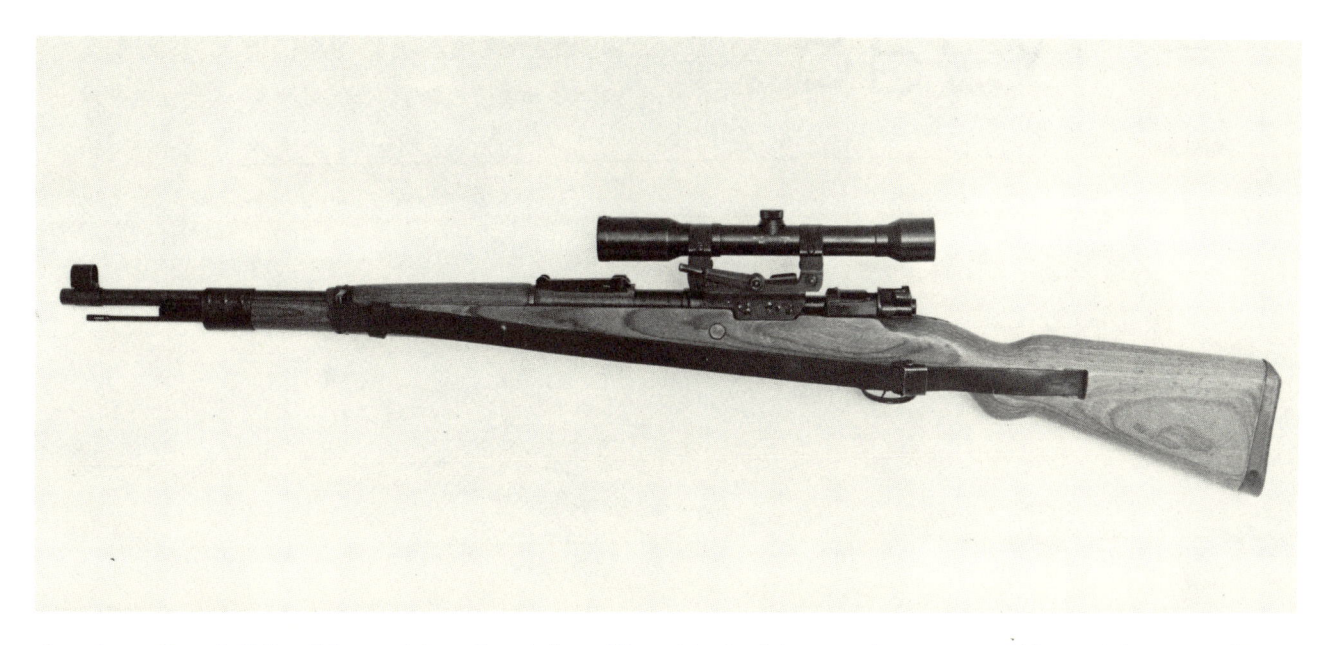

A unique Gustloff (bcd) long side rail sniping rifle with double set trigger assembly and 4-power "bek" code (Hensoldt, Herborn) military contract scope with "cap" type arctic sun shield.

Gustloff 98k long side rail sniping rifle with special double-set trigger assembly.

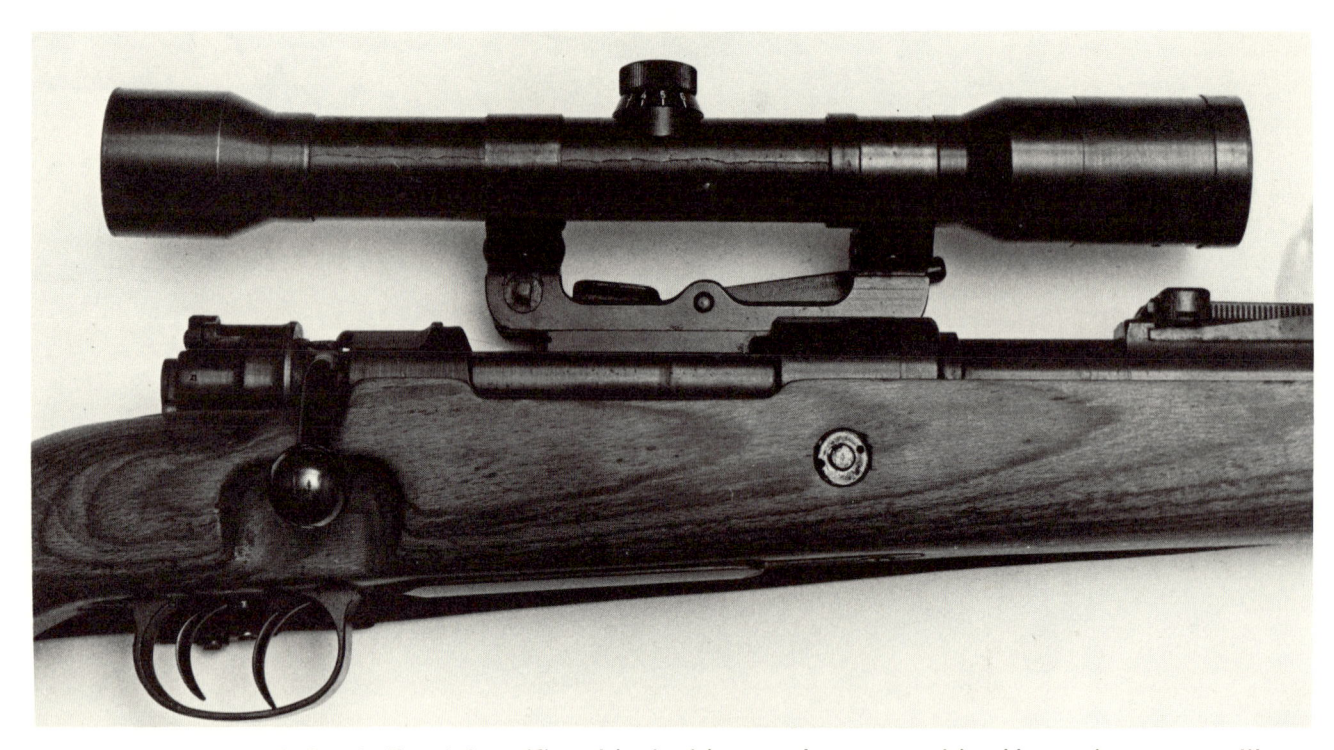

Close right view of Gustloff sniping rifle with double-set trigger assembly. Unusual on any military rifle, this feature permitted the "trigger-pull" to be adjusted for absolute minimum pressure. Such trigger mechanisms are usually found only on precision target rifles.

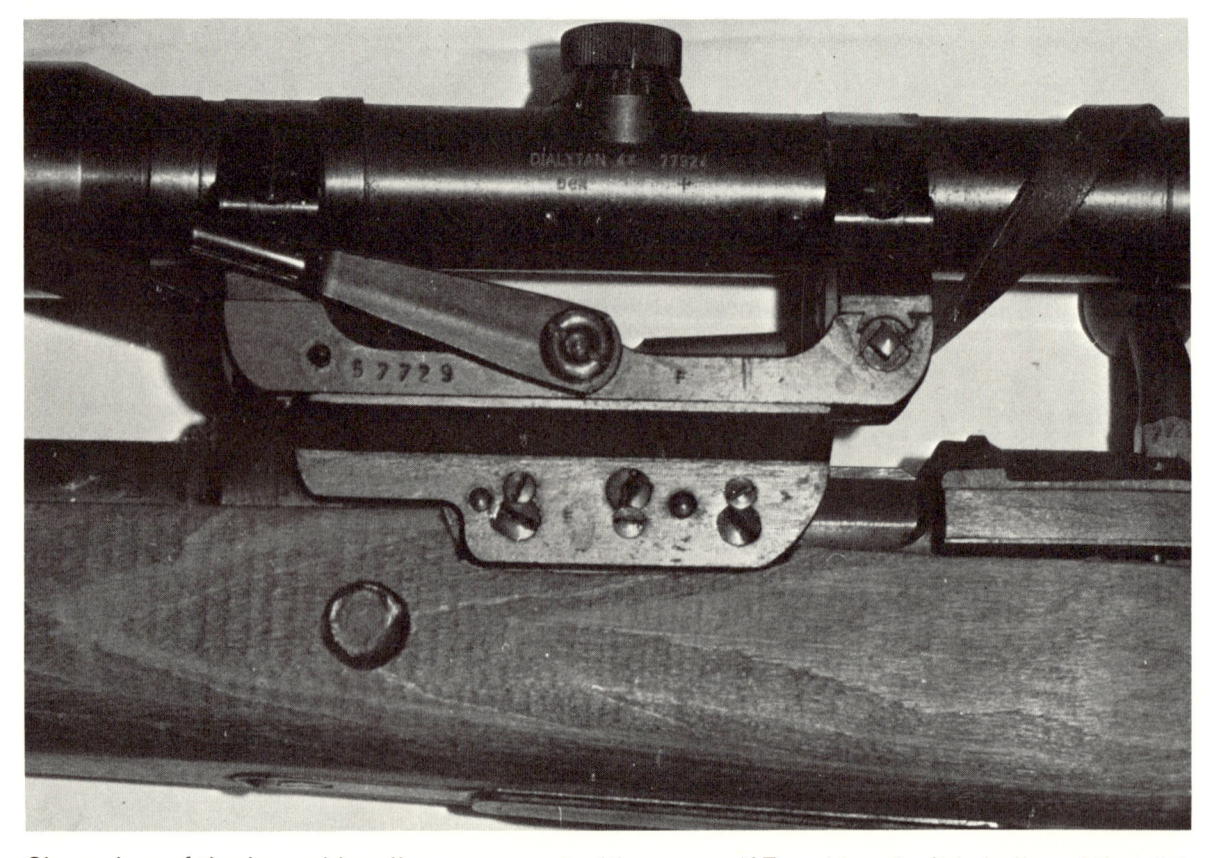

Close view of the long side rail scope mount with narrow (17mm) bands. It is believed that this type predates the common wide band variation.

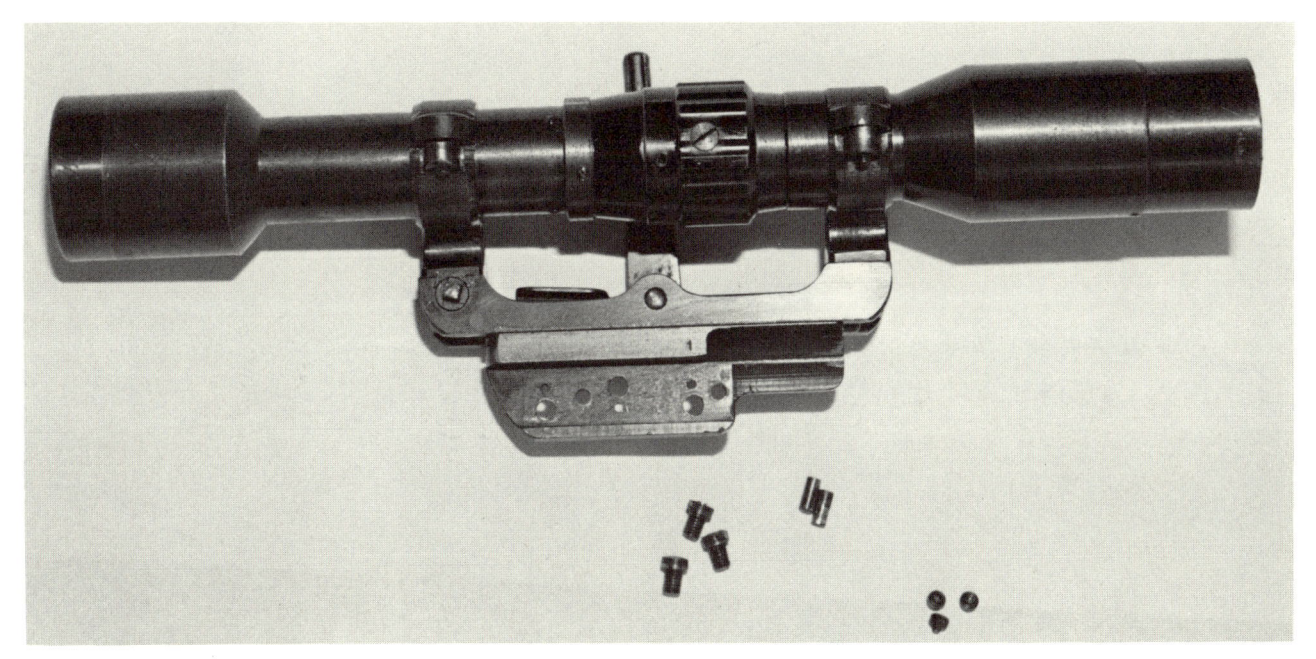

Receiver side of a "bcd" long side rail base, mount and "dow" telescopic sight. Note the three base mounting screws, lock screws, and tapered pins.

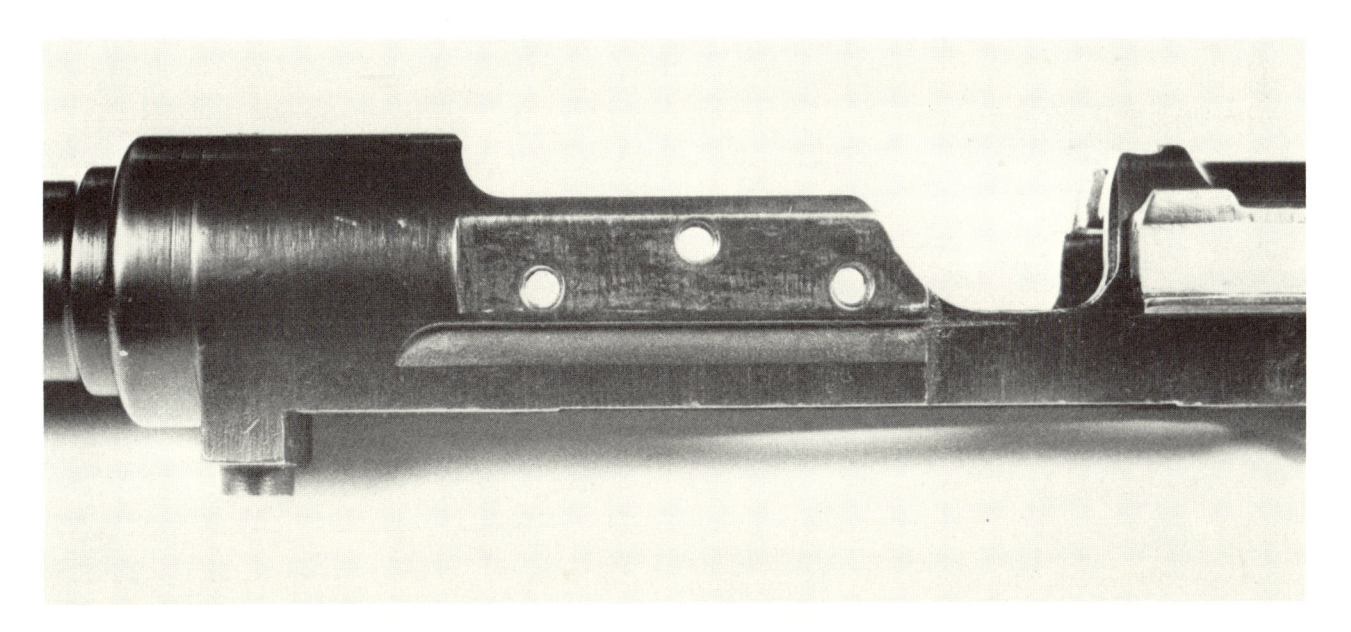

Variant "bcd" 98k special enlarged sniping rifle receiver having only three tapped base mounting screw holes. Compare this receiver with that of a standard 98k rifle.

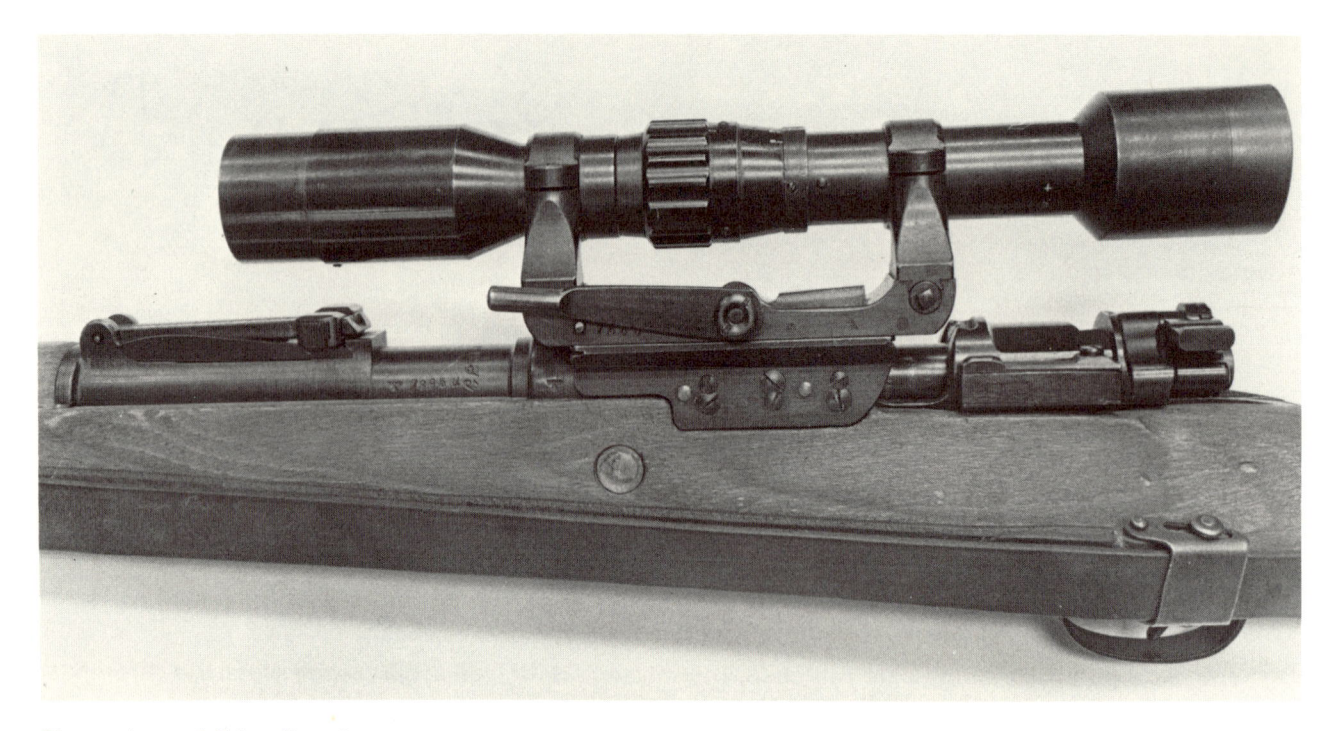

Close view of "dow" code scope with machined "full-split-ring" mount. Note the "milled" safety lever found on early issue "bcd" sniping variants. Intended to provide additional clearance with the sight, these were used in advance of the special "low" scope safety.

Although a considerable number of "three-hole" "bcd" receivers have been noted, corresponding receiver bases (no provision for tapered pins) are virtually unknown. Note the manner in which the stock was altered to clear the receiver base.

KARABINER 98K — LONG SIDE RAIL SYSTEM

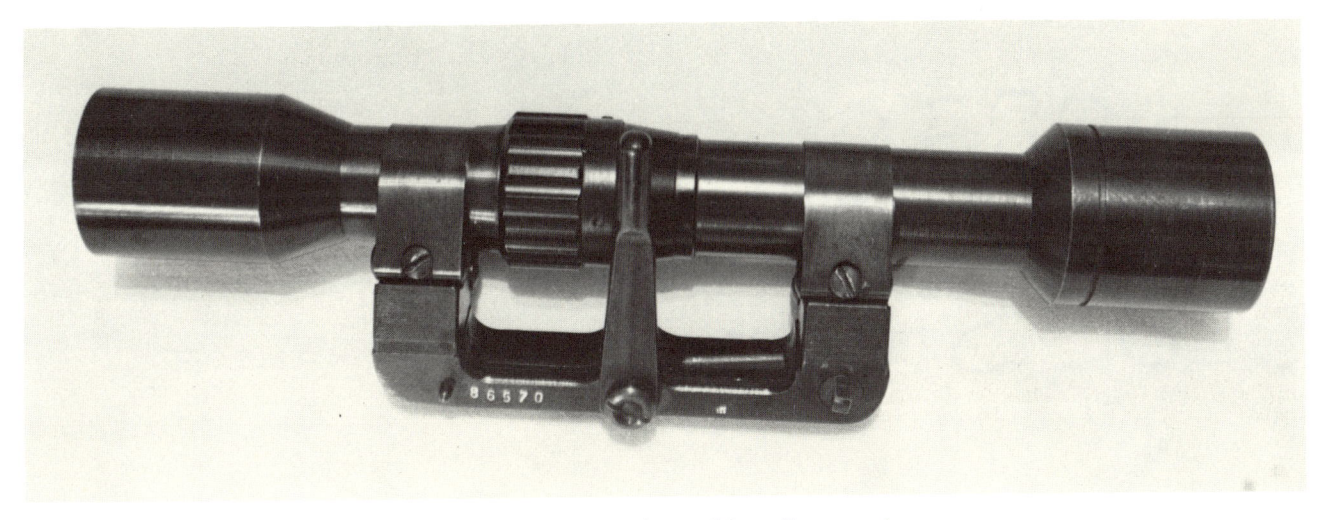

A typical "dow" telescope with 22mm wide band long side rail mounting.

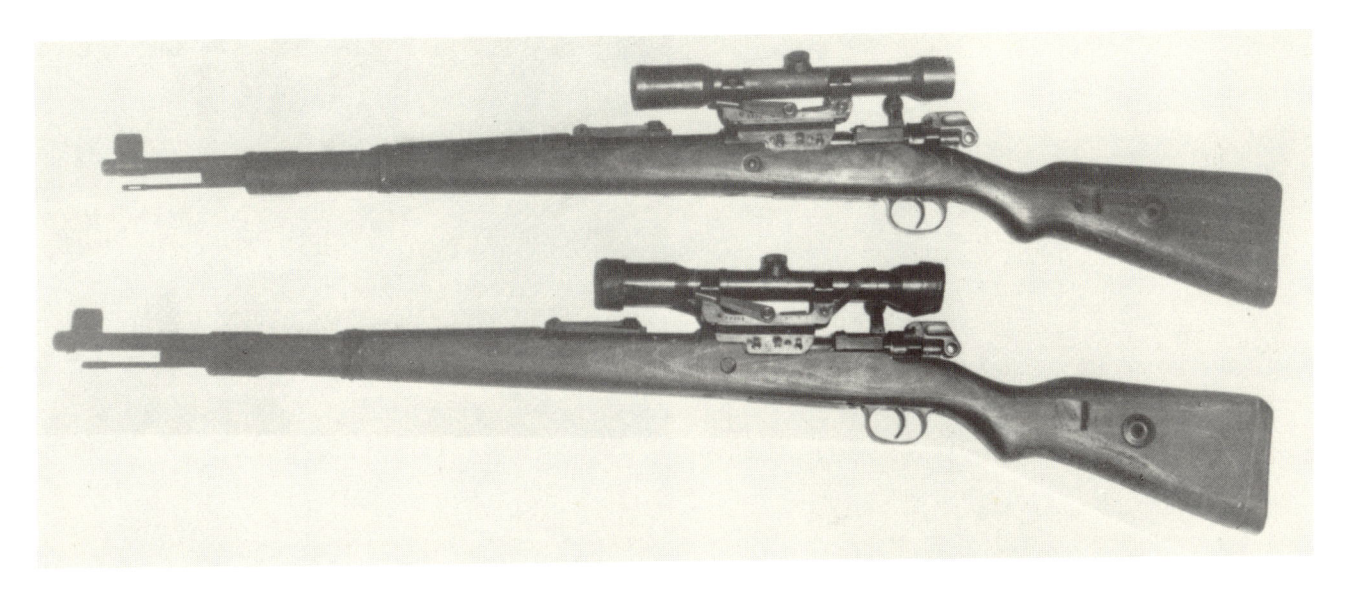

Two examples of finely finished "bcd" rifles. The top rifle is dated "45" and has a late grey phosphate finish "bek" scope with arctic sun shield. The bottom rifle mounts a blued "bek" scope with no shield provision and has 17mm bands as opposed to the wider 22mm bands above.

A Gustloff sniping rifle (bcd 4) with Opticotechna (dow) telescopic sight (factory no. 19707). The 4-power scope has the typical pointed post and horizontal side bar reticle pattern. Note the extremely rare "composition" lens caps originally issued with this sight.

Close view of unusual "dow" production mark on sight no. 19707. Note the "double-V" stamping in place of the usual "W." The blue + marking indicates reliable function in sub-zero weather.

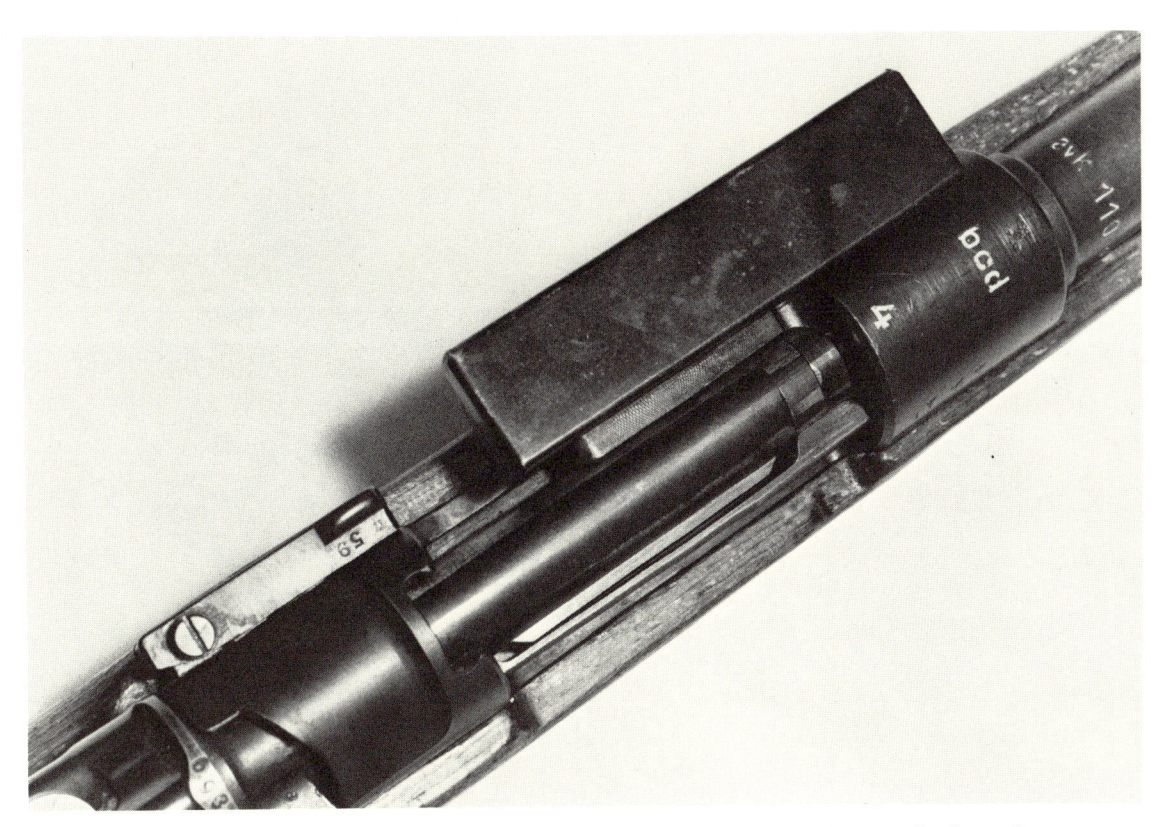

Sheet metal rail (base) cover designed to protect the edges of the dove-tail when the scope assembly was removed from the rifle.

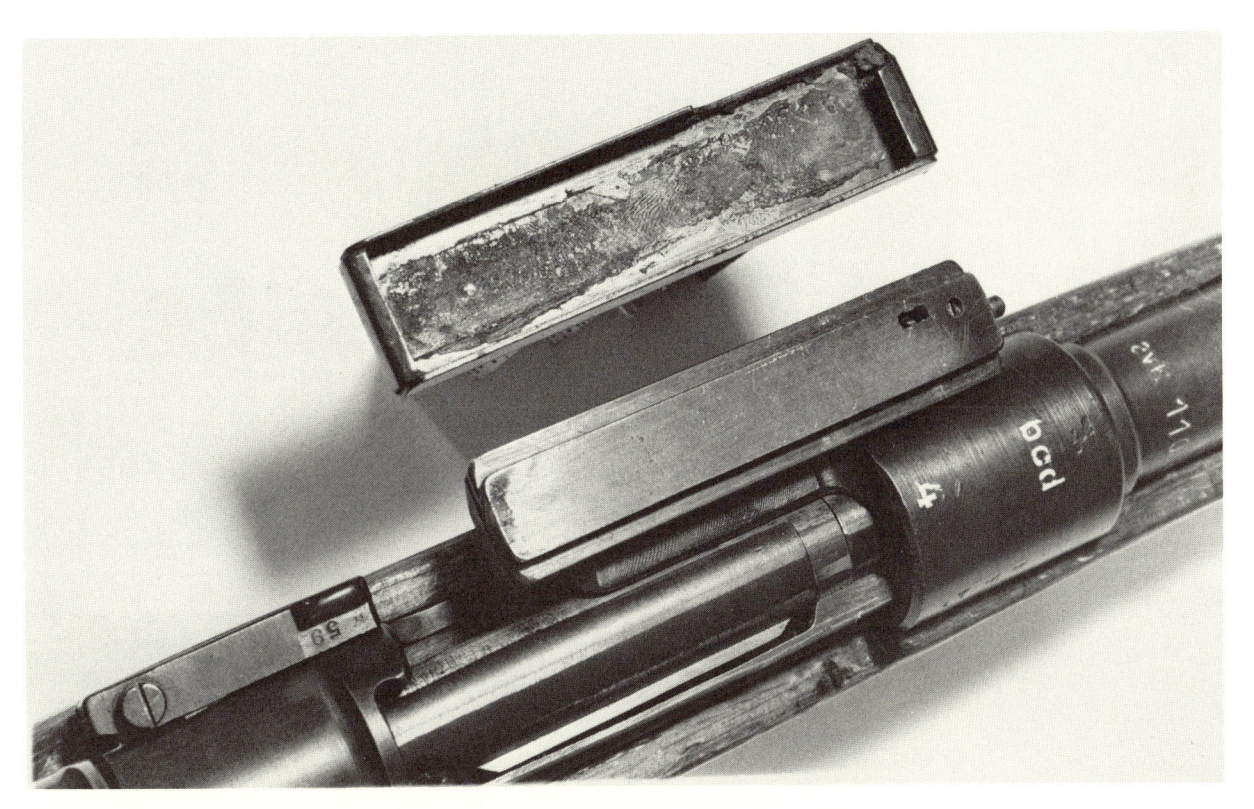

An underview of the sheet metal rail cover. Both the metal rail cover and telescope windage adjustment wrench are long side rail accessories that are rarely encountered.

The special windage wrench (key) used for zeroing the long side rail rifle. It was not intended for correction on individual shots. The wrench is 49.21mm long and 23.57mm at its widest point. The legend reads: "Rechts-Schuss" (Right-Shot); "R. Los." (Right Loose); L. nachstellen" (Left Adjust).

The "low" scope safety designed for the "bcd" sniping rifle. As originally issued they were numbered to each rifle. This particular piece bears the production mark (gnn) of Pyro-Werke G.M.B.H.

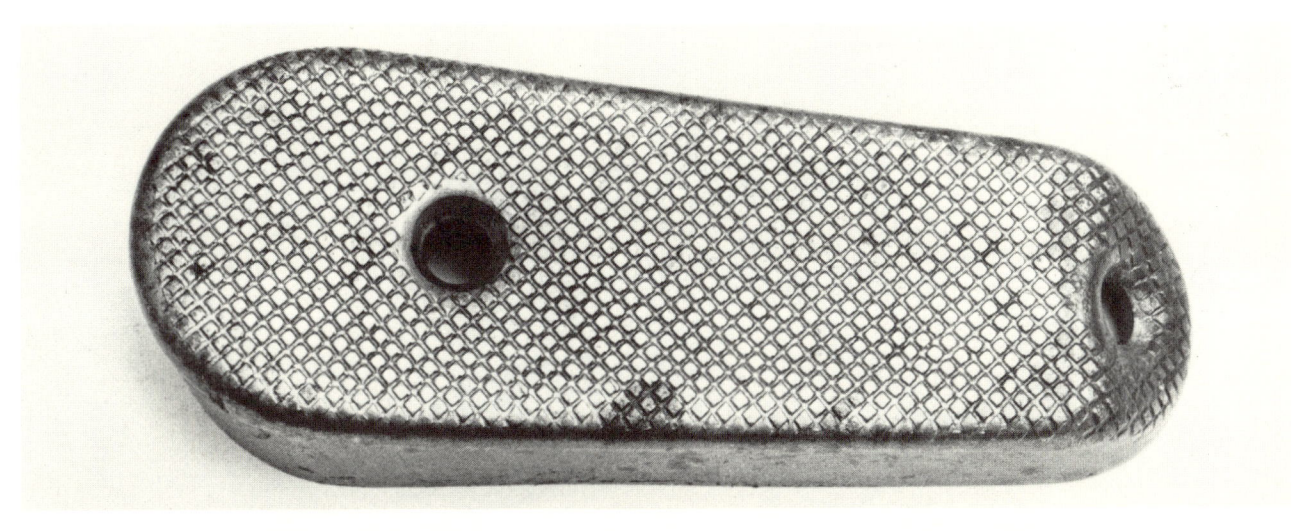

A checkered butt plate as originally designed for the long side rail sniping rifles. Its purpose was to provide a better "bite" on the marksman's shoulder than did the smooth plate of the standard 98k.

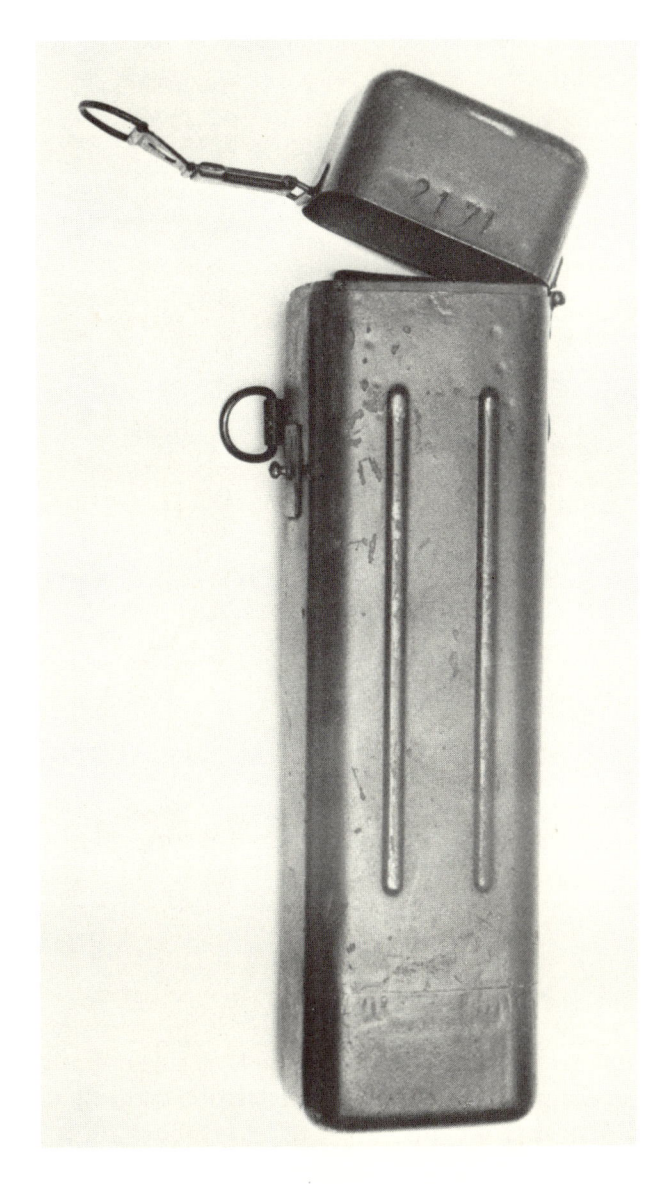

Formed metal container for carrying the long side rail scope assembly. A shoulder sling was attached to the "D" rings. Note the rifle number stamped on the lid.

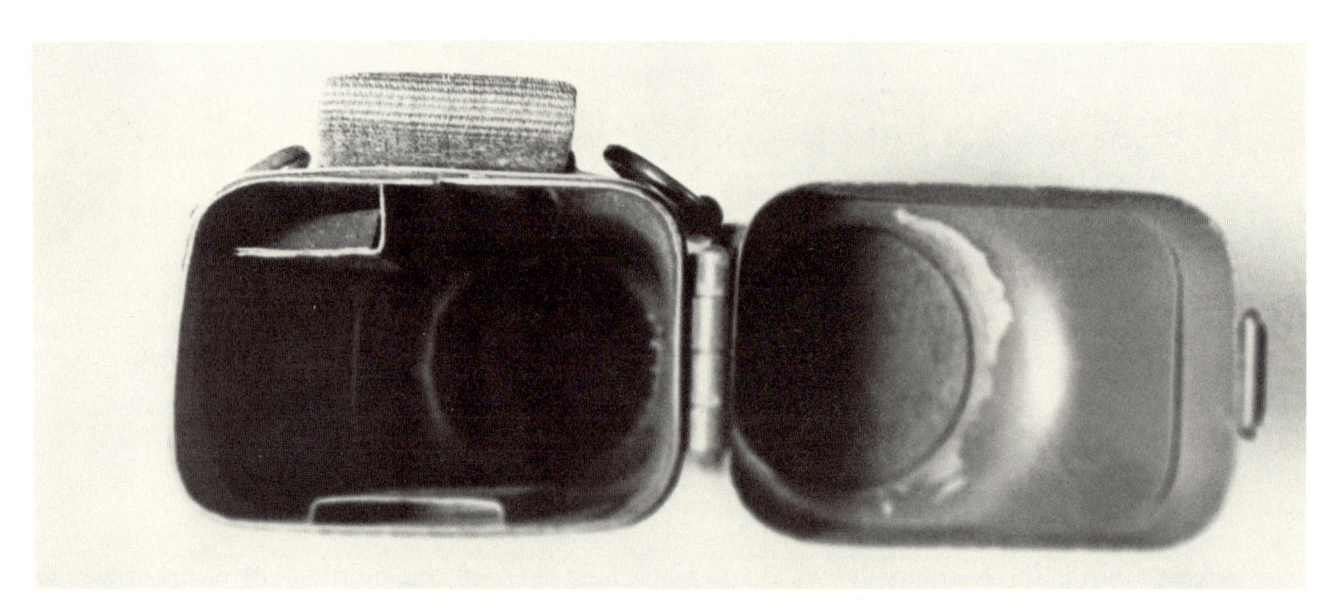

Inside view of a typical "bcd" long side rail scope carrying case. Metal pockets on either side were intended for holding the windage adjustment wrench and sheet metal receiver base protective cover.

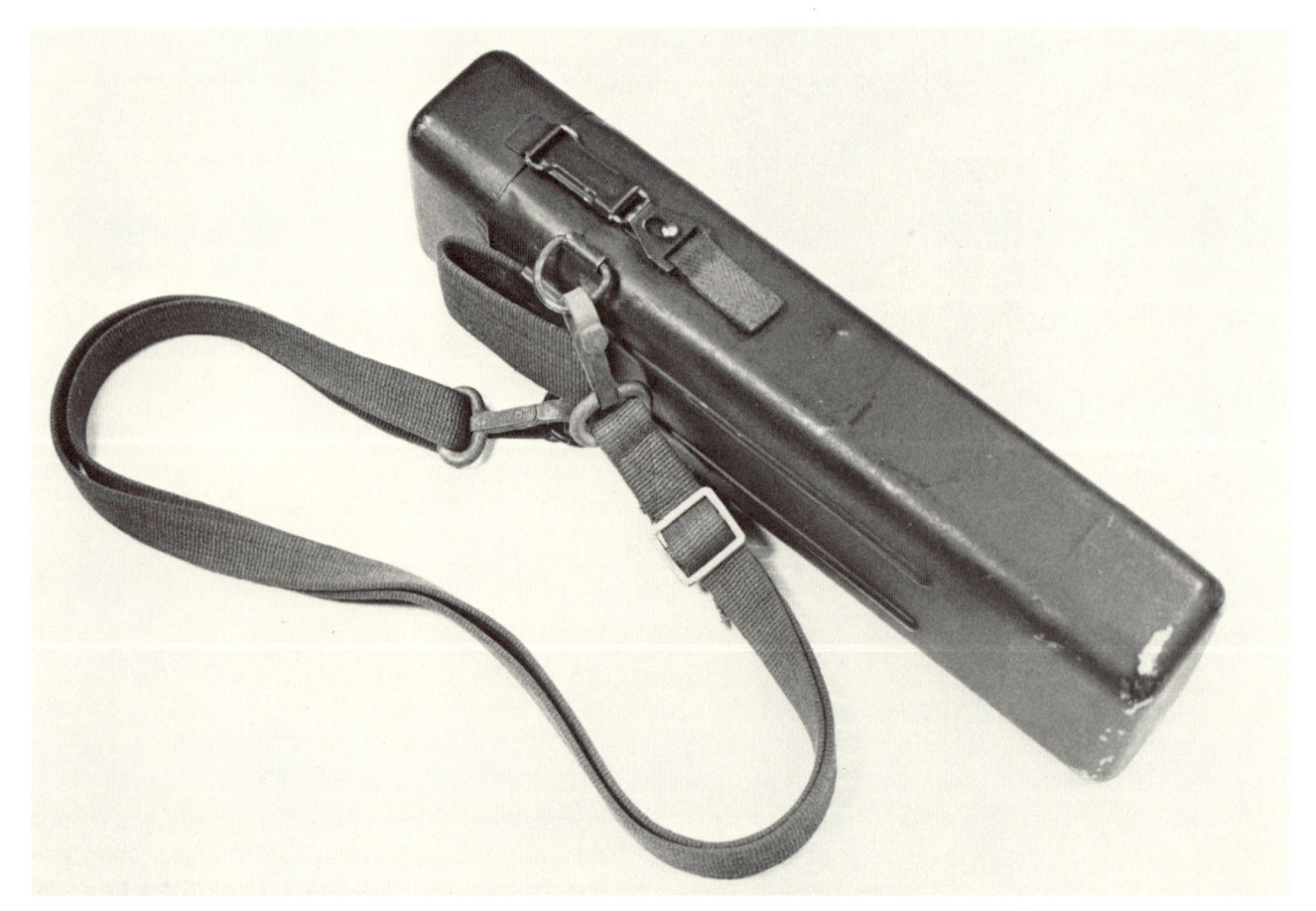

Metal scope carrying case with web belt loop and shoulder sling as issued with the long side rail sniping system. Made by Wessel & Muller (jvb). This firm was a principal supplier of telescope carrying cases for German military use during World War II.

A comparison of the turret mount (left) and long side rail scope carrying cases. Turret case dimensions are 314.32mm x 68.26mm x 49.21mm; dimensions for the rail variant are 314.32 mm x 79.37mm x 53.97mm.

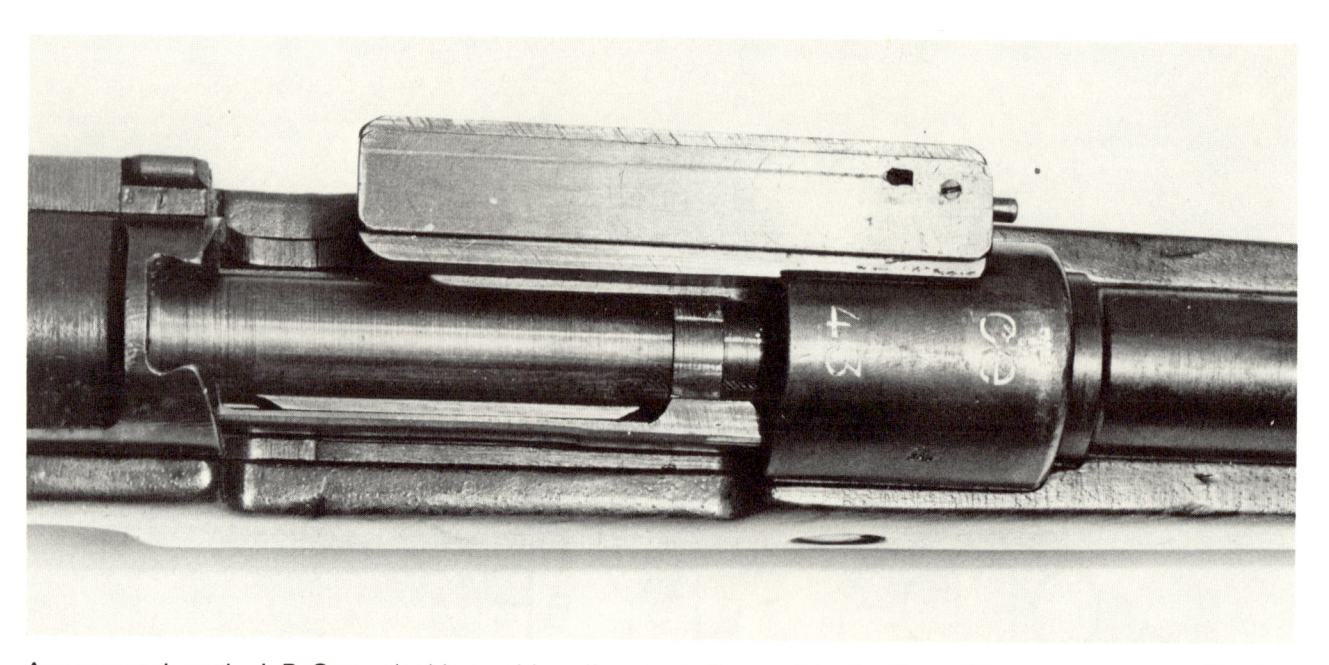

An unusual, early J. P. Sauer (ce) long side rail variant. From all indications, Sauer was responsible for the prototype work on the long side rail sniping system late in 1943. The Gustloff Werke, Werk Weimar (bcd) eventually received the production contract. Note the manner in which the base is fitted to the receiver. The number "118" is stamped on the base but no Waffenamt is present.

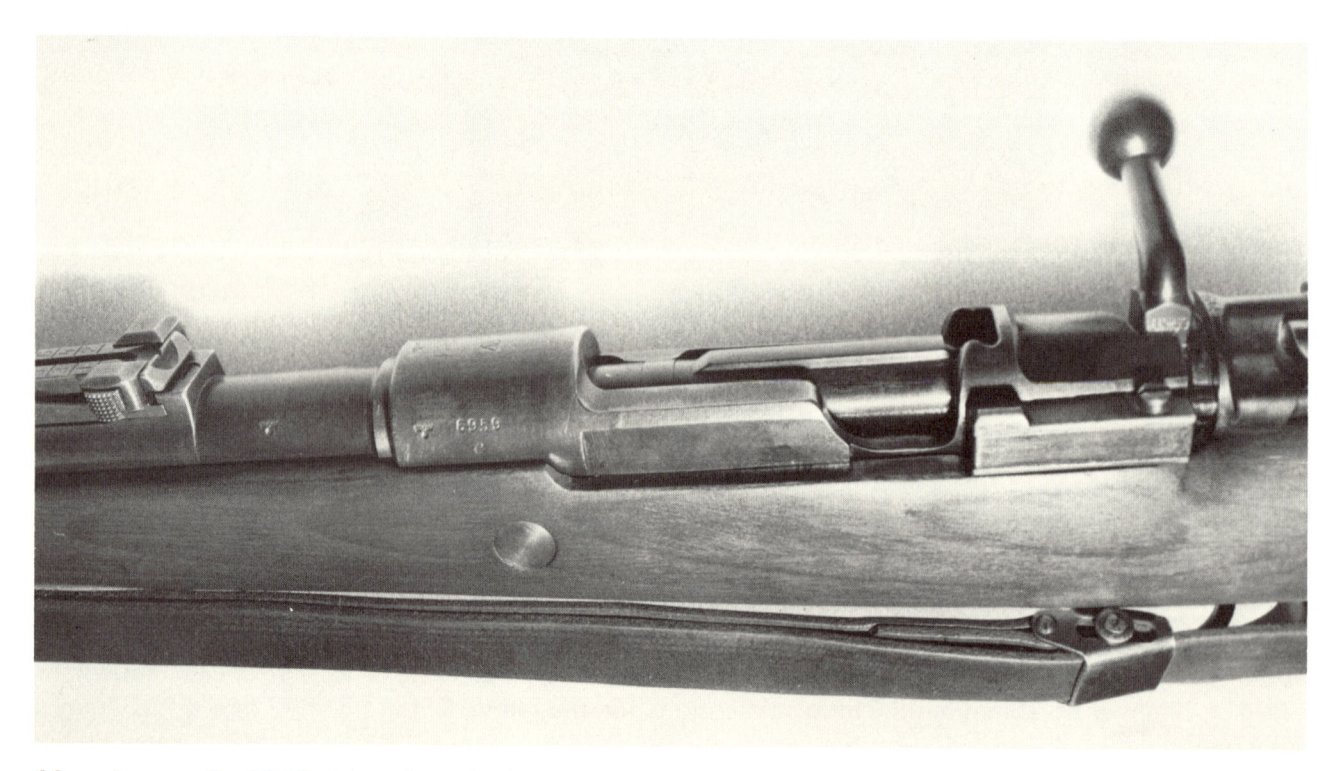

Manufactured in 1943, this unique "ce" code early long side rail variant has remained in a virgin state, no receiver base mounting holes.

Top view of Sauer (ce 43) special enlarged receiver markings. Note the thickness of the receiver wing; compare this with a standard 98k.

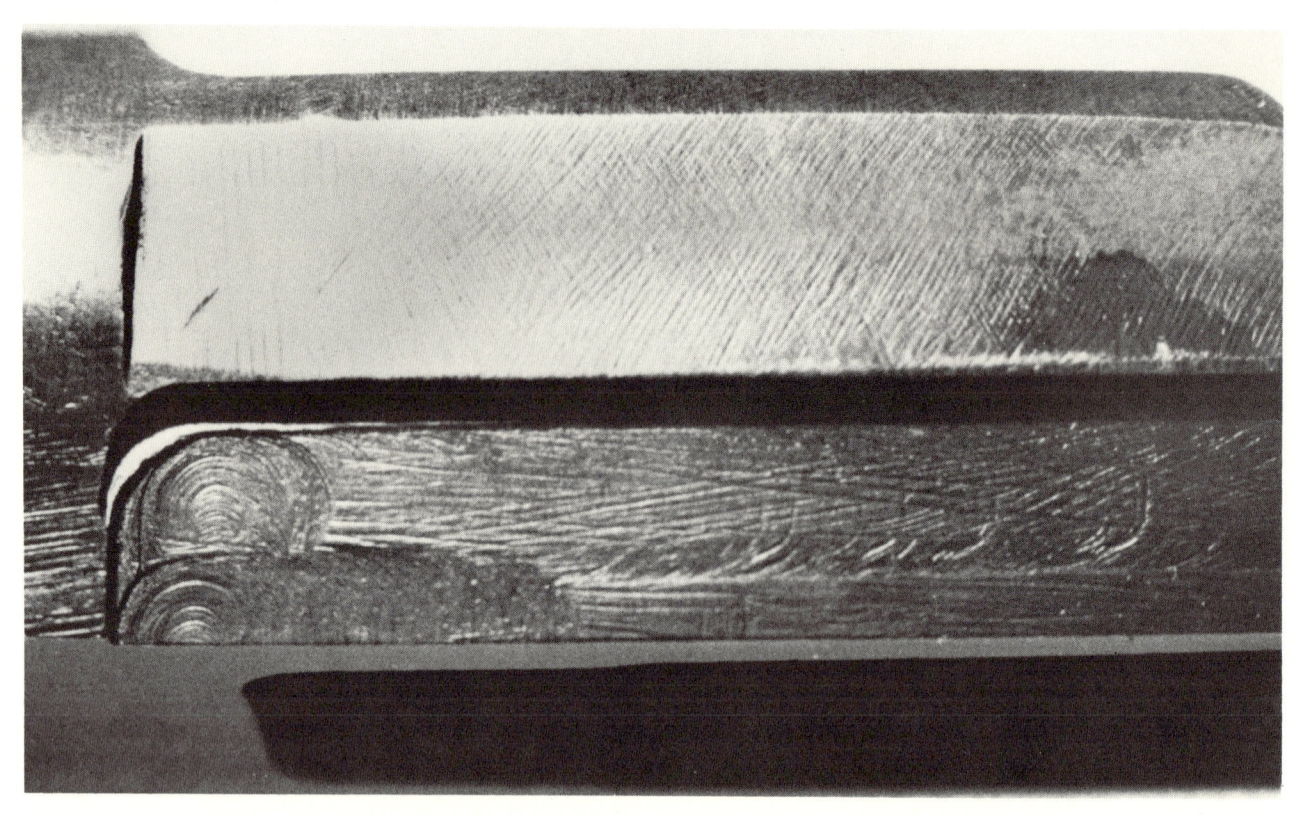

Close view of Sauer (ce 43) special enlarged receiver removed from the stock. Note the difference between the machining (flat) on this piece and the typical "bcd" long side rail receiver.

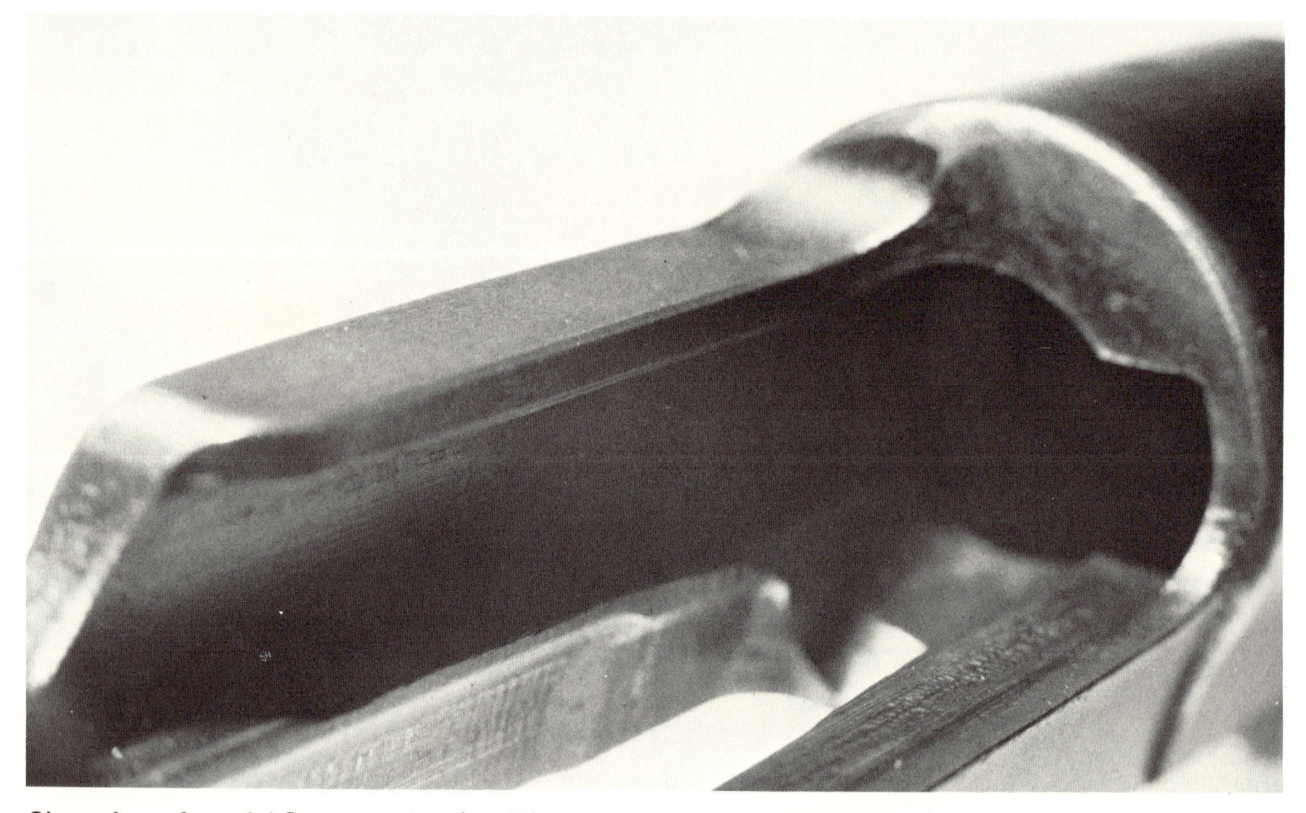

Close view of special Sauer receiver (ce 43) with bolt assembly removed.

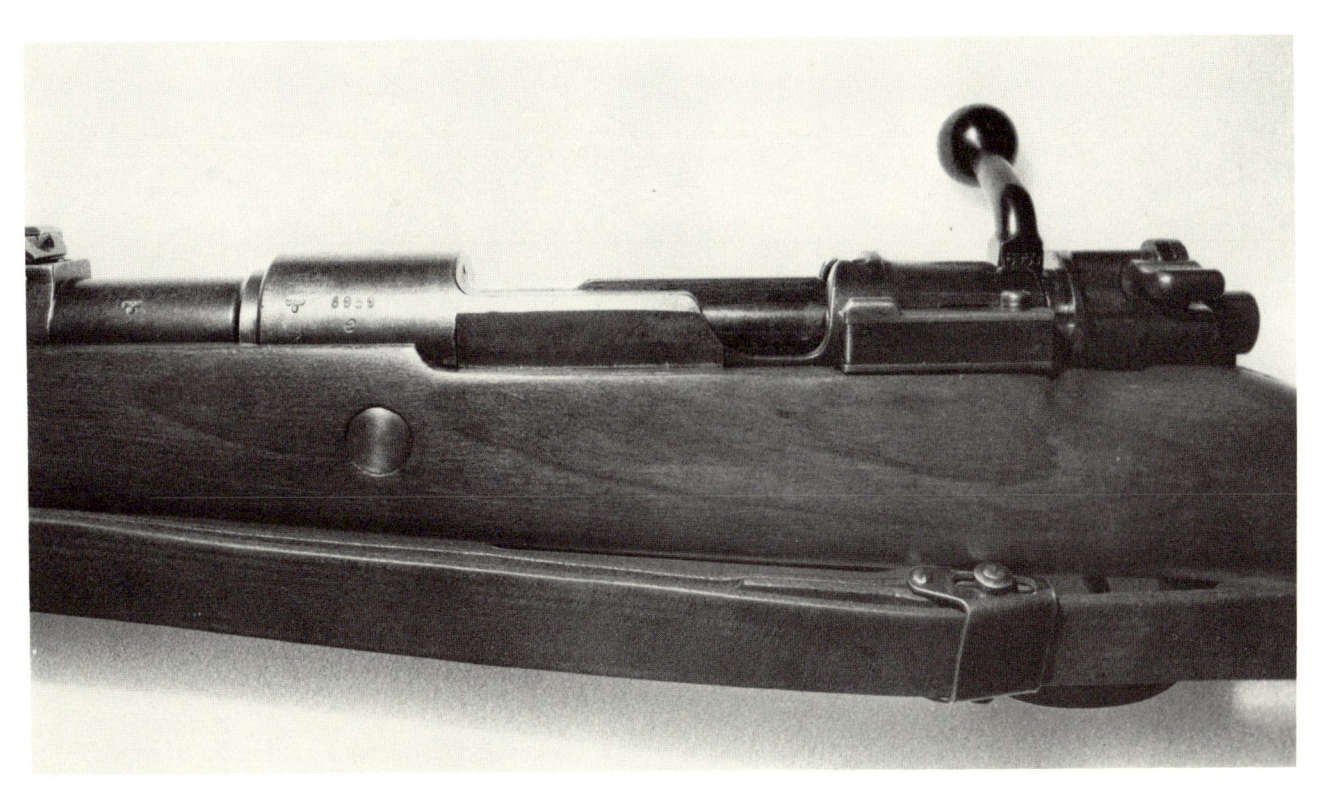

Another view of the original J.P. Sauer (ce 43) 98k rifle with special enlarged receiver.

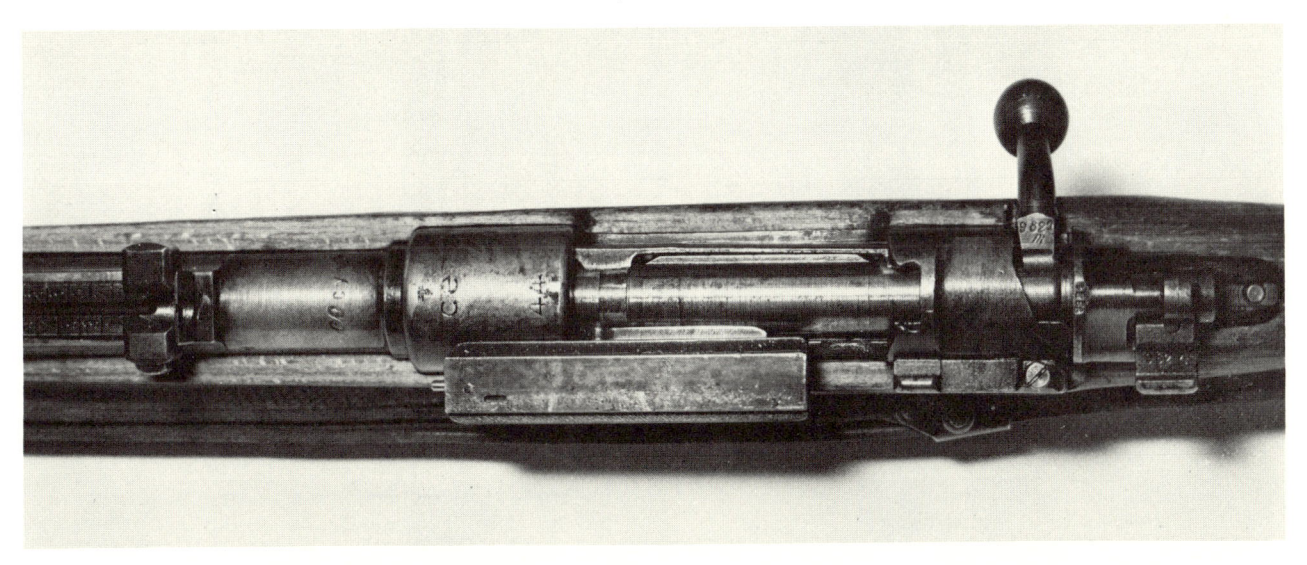

A very limited production 98k long side rail sniping rifle with special enlarged receiver as manufactured by J. P. Sauer. Note the "ce 44" markings on the receiver ring. Although the scope mounting base is identical to that found on the Gustloff (bcd) variants (Walther manufacture base with 359 Waffenamt), this receiver base bears a "37" Waffenamt instead.

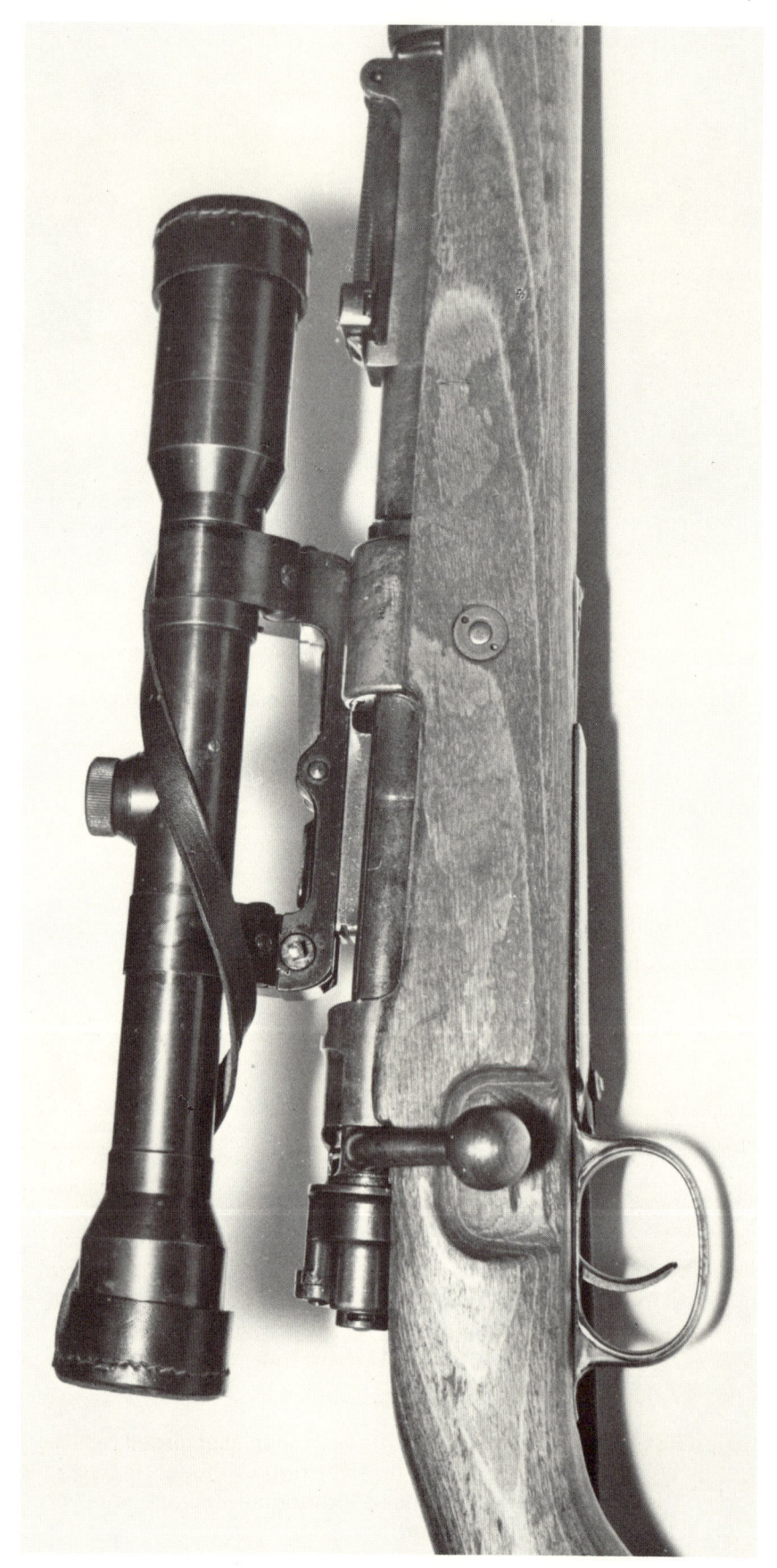

Right view of J. P. Sauer (ce 44) long side rail sniping rifle with "bek" code telescopic sight.

CHAPTER XI

Karabiner 98k— Claw Mount System

With the "bnz" code (Steyr-Daimler-Puch) 98k serving as the base weapon, approximately 10,000 claw-mount sniping rifles were reportedly produced in late 1943 and 1944 for German sharpshooter use.

While similar in appearance to the offset scope mounting system utilized with the World War I Austro-Hungarian Model 95 sniping variants produced at the Steyr Armory, in this case (bnz), a single claw foot was used on both the front and the rear scope mounts.

The front base is a massive two-piece machining consisting of a mounting block soldered to the left side of a curved steel base and contoured to the receiver ring. This piece is soldered to the receiver ring and covers its entire top section. The rear base is a two-piece machining soldered and screwed to the rear of the receiver. Its center section is a sliding wedge attached to a push button.

Scope rings are one-piece machined steel bands split at the top; the scope must be taken apart for the rings to be put on. There are two tightening screws facing opposite and located at the top of the rings. By design, the telescope mounts place the sight directly over the bore, even though the receiver bases are in-line and offset to the left to permit use of the conventional sights with the scope in place.

The scope is pushed into the front base assembly at a 45-degree angle until its foot meets the rounded pin on which it rests. The rear section of the scope is then pushed down, causing its front claw foot to pivot into place. A button at the rear is an integral part of the sliding center section of the rear base, and pushing it moves the entire center section forward, allowing the rear foot to fall into place. When the button is released, the sliding section moves rearward, wedging into the rear foot groove. The telescopic sight is then locked in position over the rifle bore.

Unlike other World War II German scope mounting systems that were based on 98k rifles of different manufacture regardless of the limits of their actual issue in some cases, the claw-mount variant is considered unique from the standpoint that its method of telescope mounting was only utilized with weapons bearing the "bnz" production mark.

Although 4-power telescopic sights of various manufacture have been noted on rifles of this configuration, judging from the weapons that have survived in an absolute "as issued state" (no contemporary tampering), the "bmj" code (Hensoldt, Wetzlar) scopes appear to have been a principal issue with this variant.

While hardly classified as rare, "bnz" sniping rifles are highly prized collection pieces, particularly in a matched state with correspond-

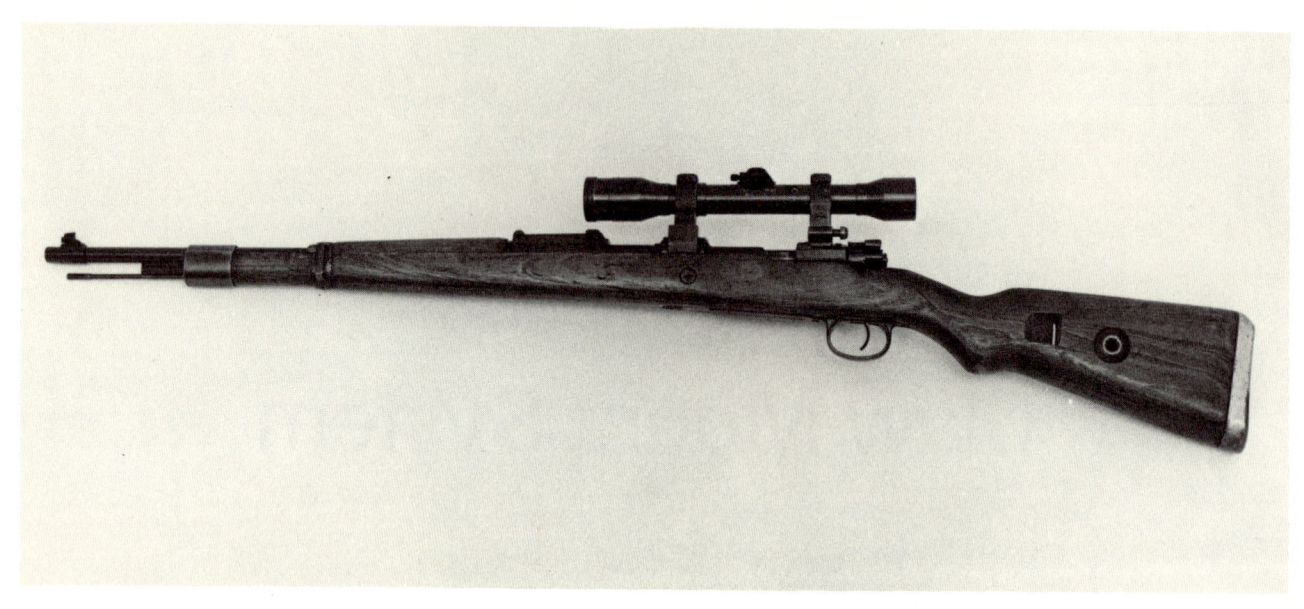

The "bnz" code (Steyr-Daimler-Puch) 98k claw-mount sniping system.

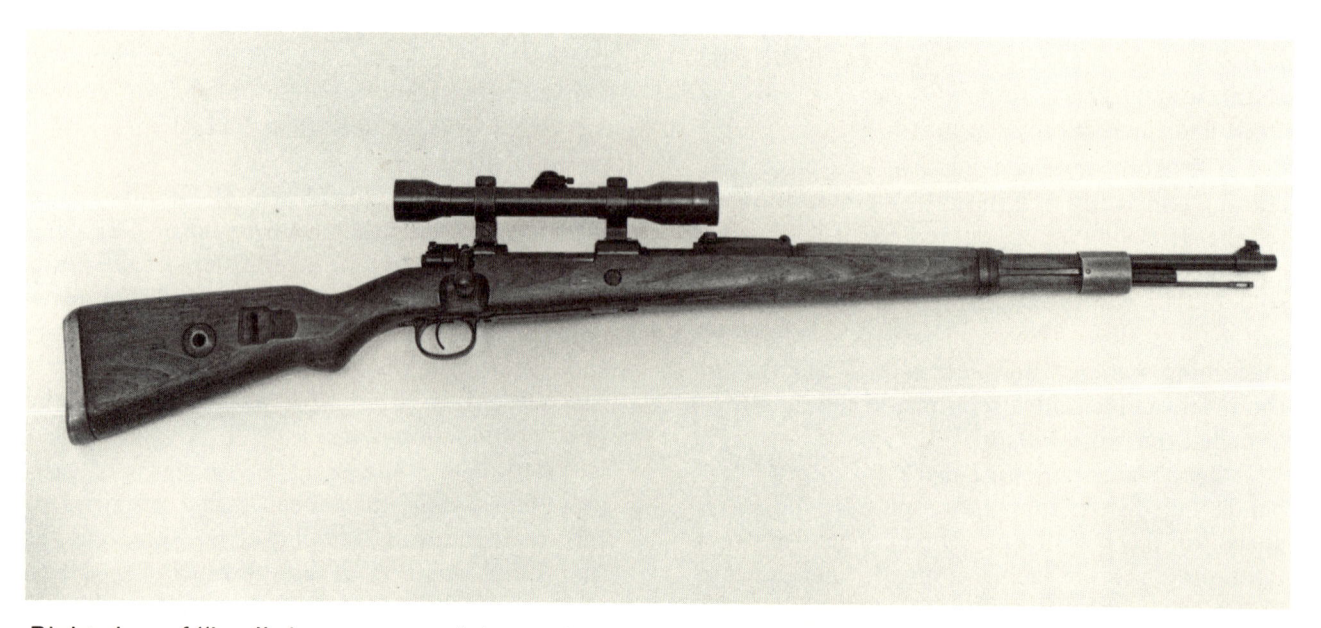

Right view of "bnz" claw-mount sniping variant mounting a 4-power military contract "bmj" (Hensoldt, Wetzlar) telescopic sight.

ing serial numbers on the scope mount, base, and rifle.

Even though the 98k "bnz" claw-mount system represented nothing more than an extension of the state of the art as it existed in 1918, these sharpshooting rifles were simple and rugged, a most desirable asset at the field level.

Left three-quarter view of "bnz" claw-mount sniper rifle. Note the recoil ring and sun shield fitted to the Hensoldt sight.

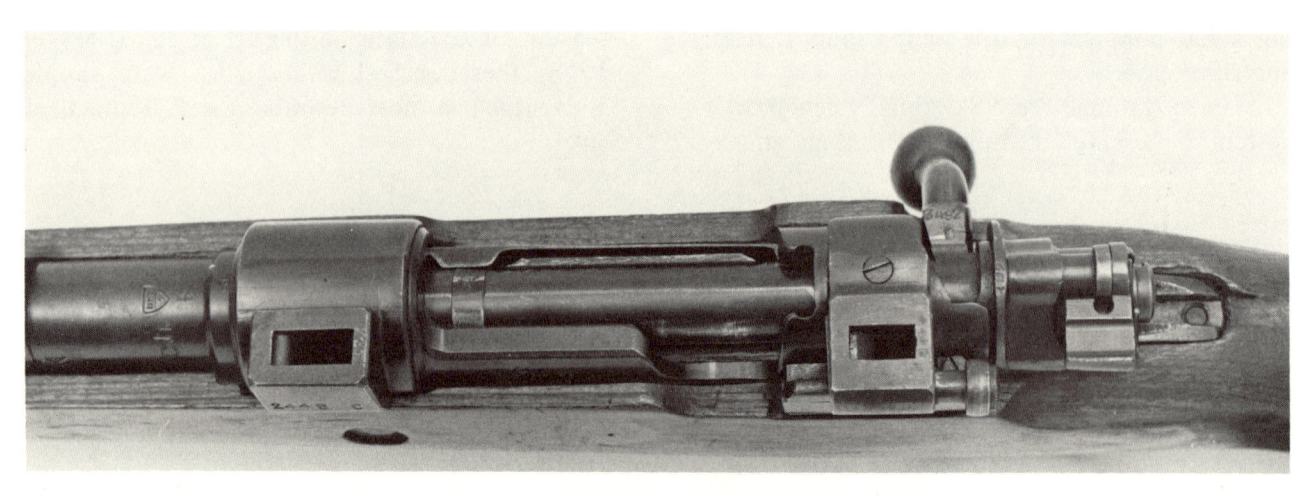

Top view of the receiver bases utilized with the "bnz" claw-mount system.

Right three-quarter view of "bnz" claw-mount sniper rifle.

Detail of front base of "bnz" claw-mount. Pivot pin is visible at front of rectangular hole. Note wrap around base contoured to fit the receiver ring, which covers the manufacturer's markings and reaches down to the stock on either side.

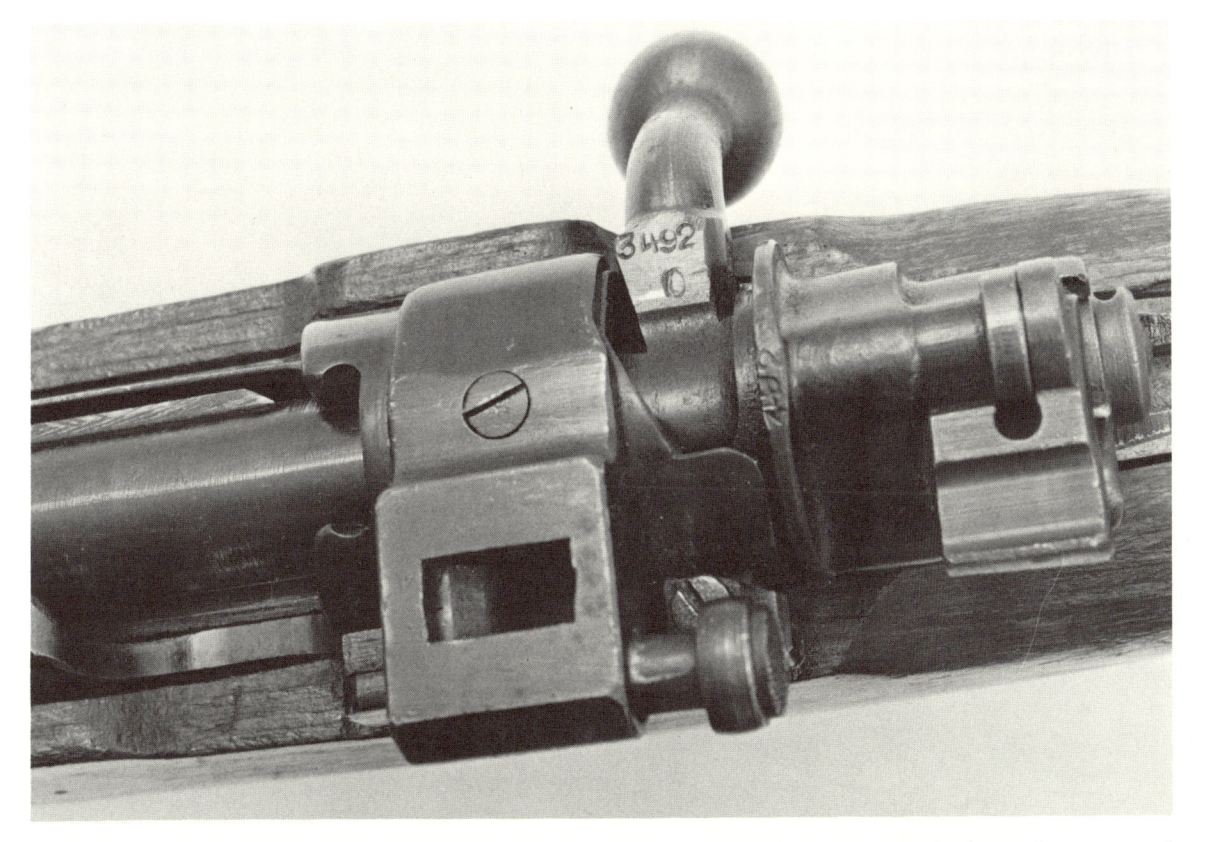

Detailed view of the rear base assembly, "bnz" claw-mount. Single screw is the only one used in either base. Wedge-shaped section inside the rectangular recess is the locking section of the center sliding ledge. The push button is plainly visible. Note that safety lever has been "ground-off" to provide additional clearance.

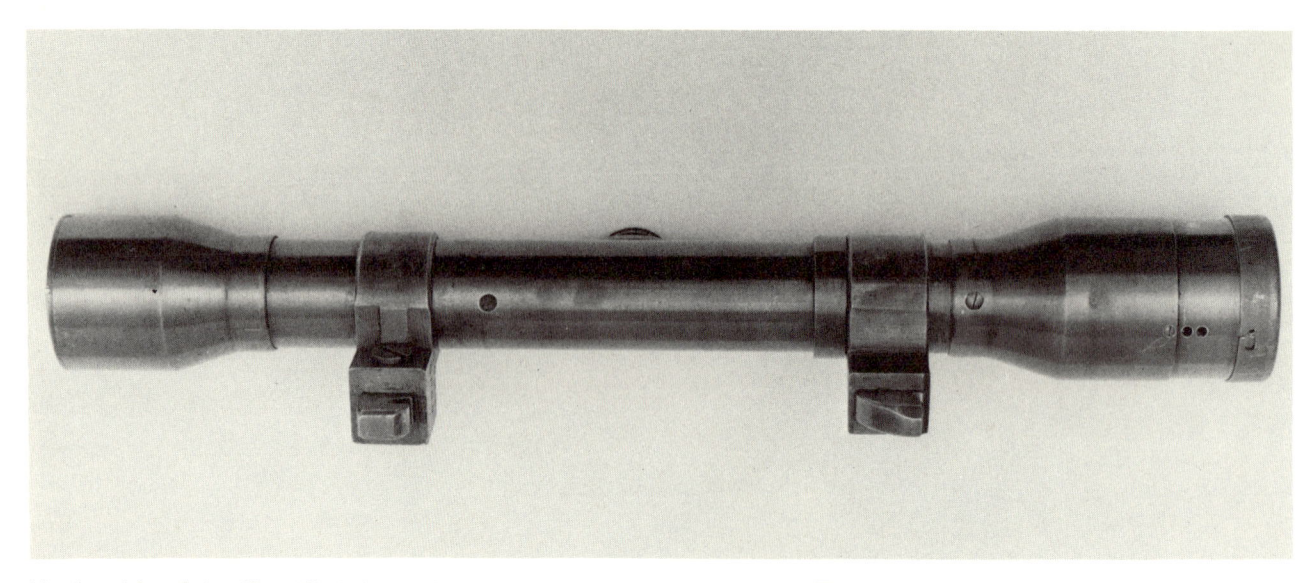

Underside of the "bmj" sight and single claw-mount assembly. The sun shield was intended to reduce the glare from snow during winter operations. Note the two small, circular openings provided for draining moisture from inside the special arctic attachment.

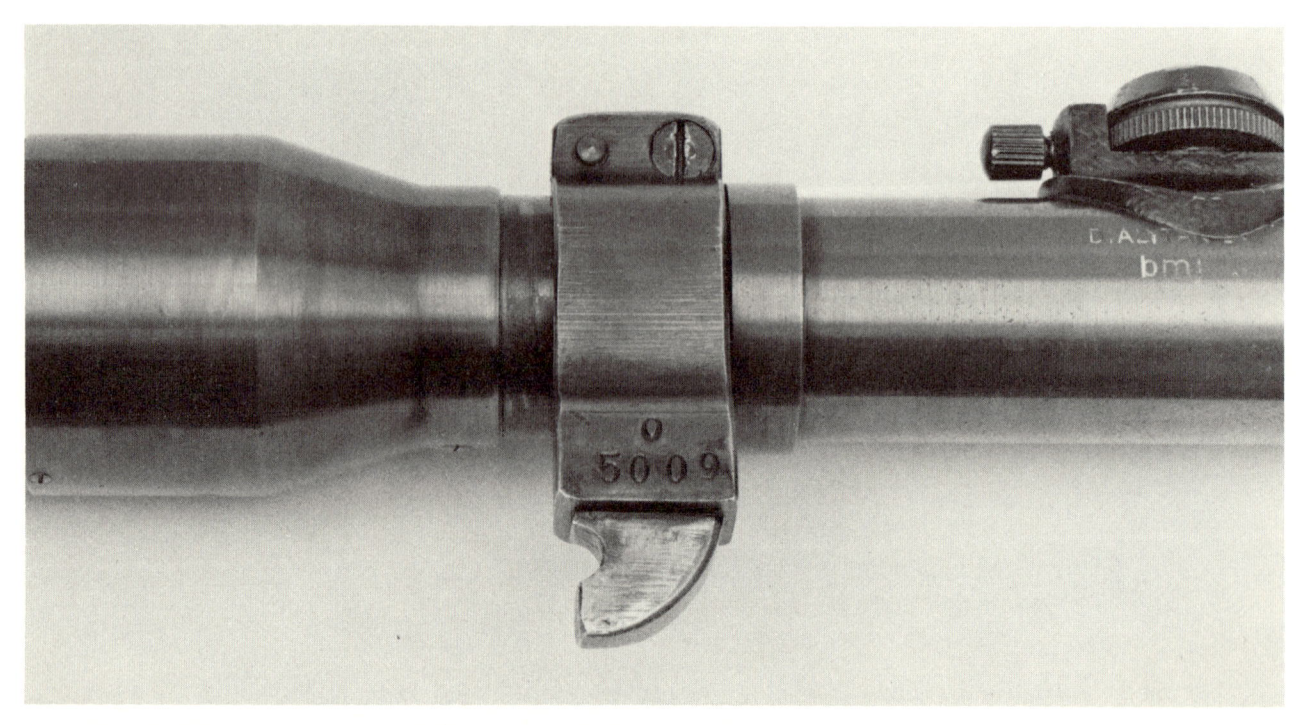

Detail of front ring and claw foot. Rounded pivot cut is at front of foot. The front receiver base and telescope mount were stamped with the rifle serial number as issued.

Detail of ring and claw foot. Locking wedge cut is at front of foot. Note the large windage adjustment screw and opposite facing ring tightening screws.

A "bnz" claw-mount 98k rifle (8847 d) without telescopic sight.

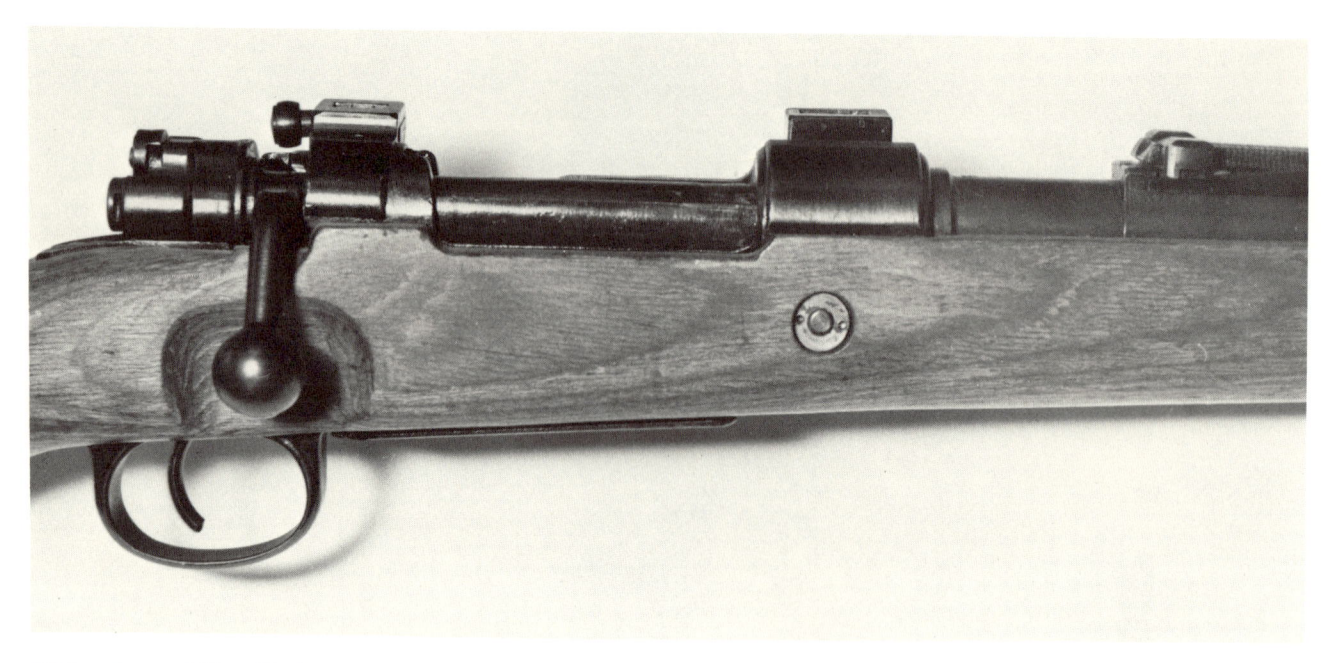

Right view of "bnz" claw-mount rifle (8847 d).

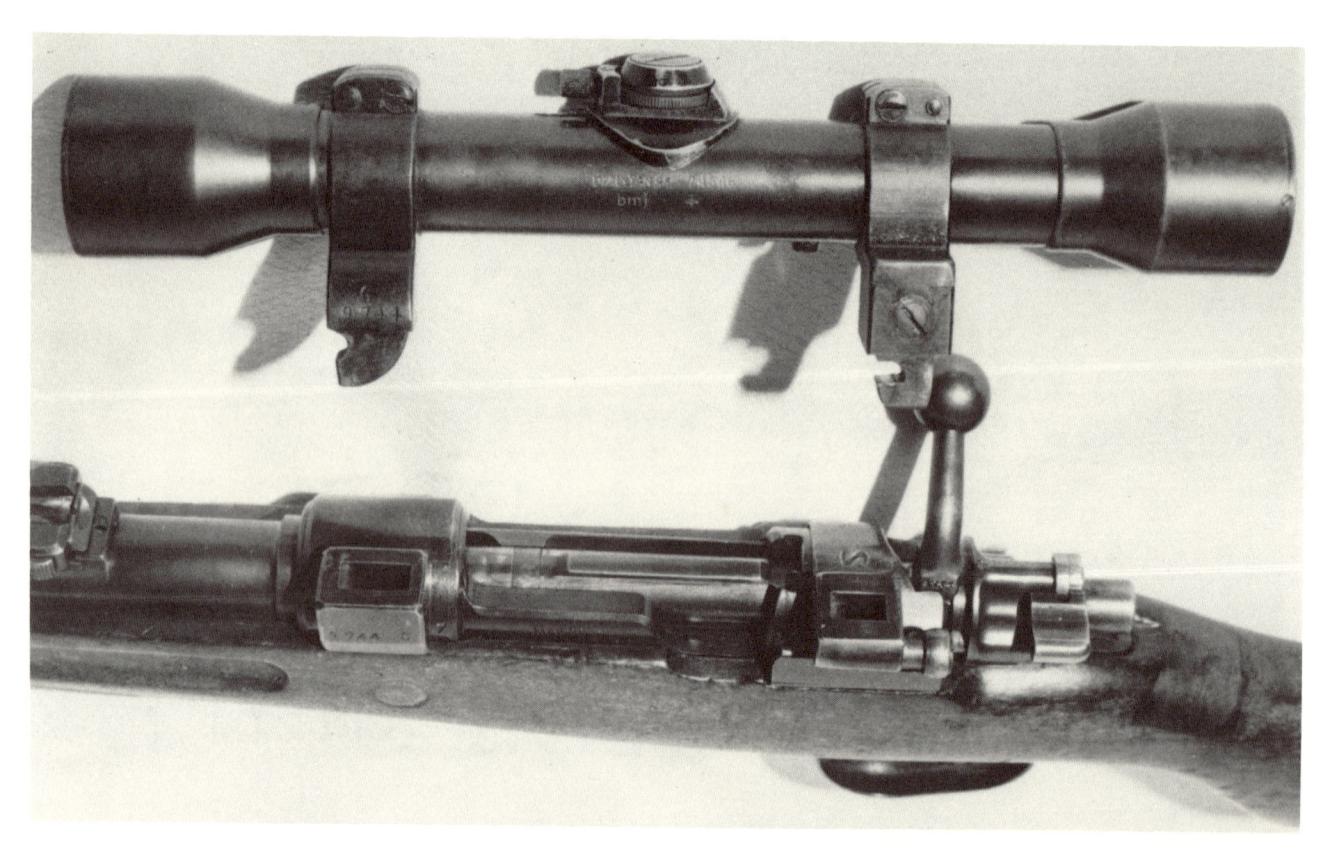

A "matching number" (scope mount assembly and rifle) "bnz" claw-mount sniping rifle (9744G) with a standard "bmj" code scope in this case (no extended objective shield or recoil ring) as "brought home" following the war. Although scope carrying containers (formed metal) were a part of the original issue, they are rarely encountered with surviving examples of this system.

KARABINER 98K — CLAW MOUNT SYSTEM

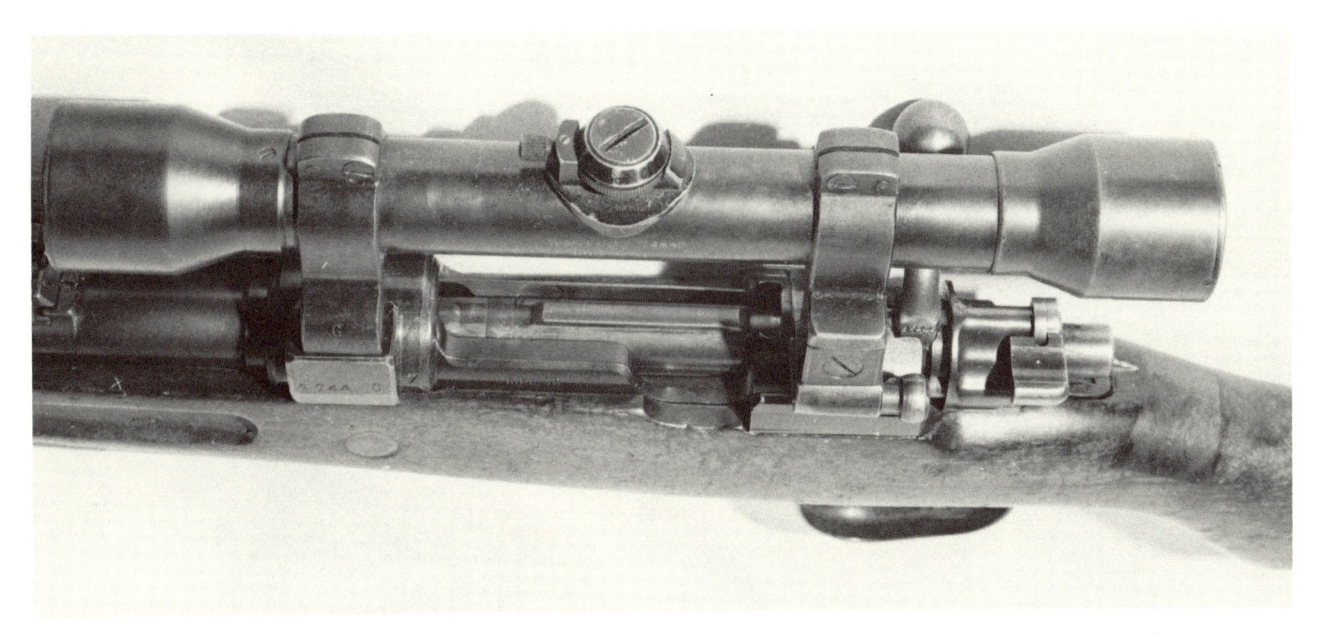

Top view of 98k claw-mount rifle (9744G) showing the upper ring assembly. The scope was taken apart to install the mounts. Although short objective "bmj" scopes appear to have been the principal model employed with this system, both the short and extended objective telescopic sights *were issued* with this sniping rifle. Once in place the scope was rarely removed from the weapon.

An additional "bnz" claw-mount rifle (9131 d) is shown for comparison.

An original issue, matching number scope carrying case, a highly desirable part of any claw-mount assembly. The metal case bears "bnz" rifle serial No. 3146.

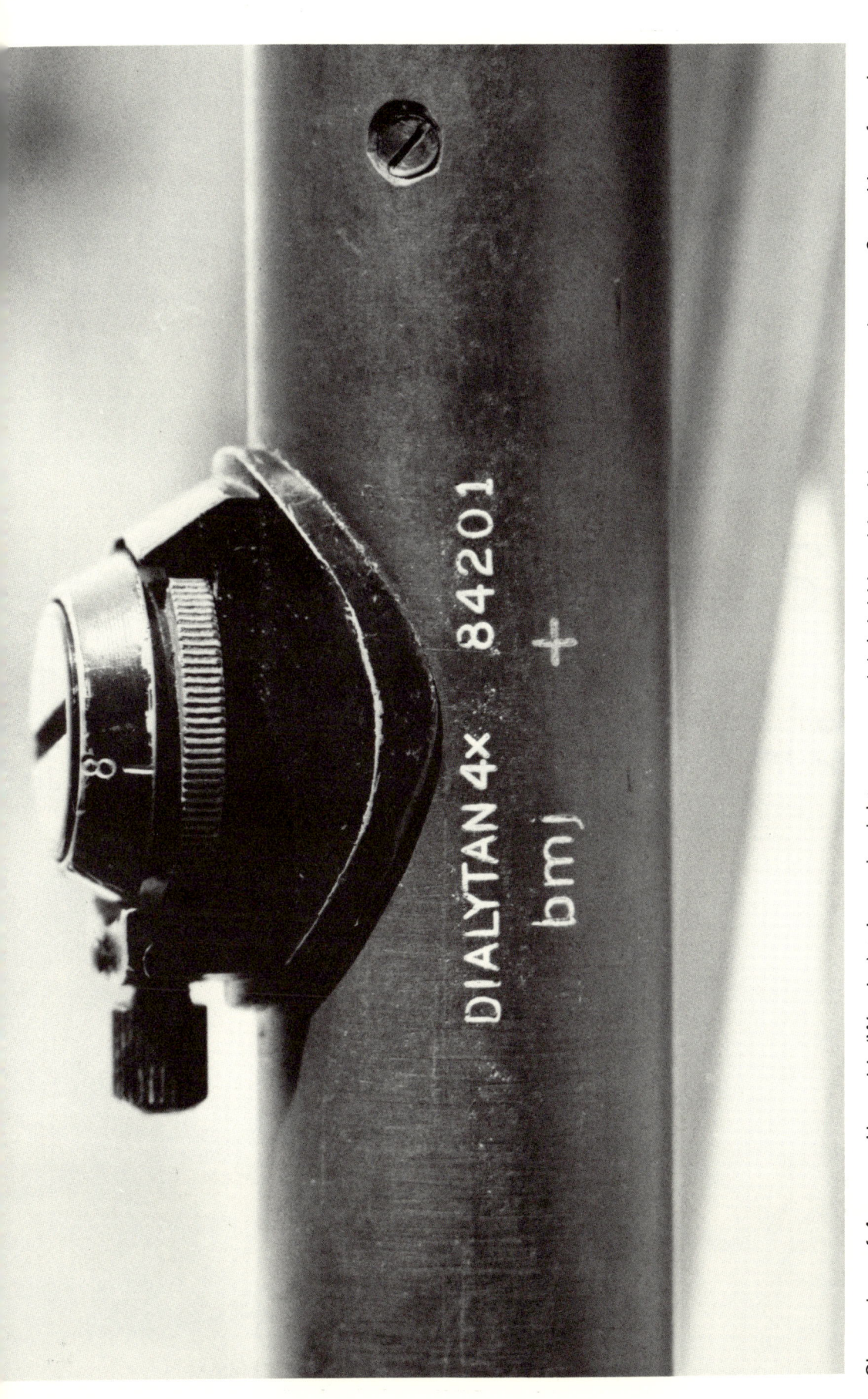

Close view of 4-power Hensoldt (Wetzlar) telescopic sight produced for German military use showing the model designation (Dialytan 4X), serial number (84201), code (bmj) and function mark (+). Inlaid in white, blue, or green, the + or O markings found on sights of this type indicated the reliable function limits in various climates.

Close view of "bmj" claw-mount system scope with front mount assembly bearing rifle serial No. 3146.

A prime example of senseless tampering: "bnz" claw-mount sniping rifle (3146) with the scope mounting bases removed from the receiver. A sad fate for a fully matching rig.

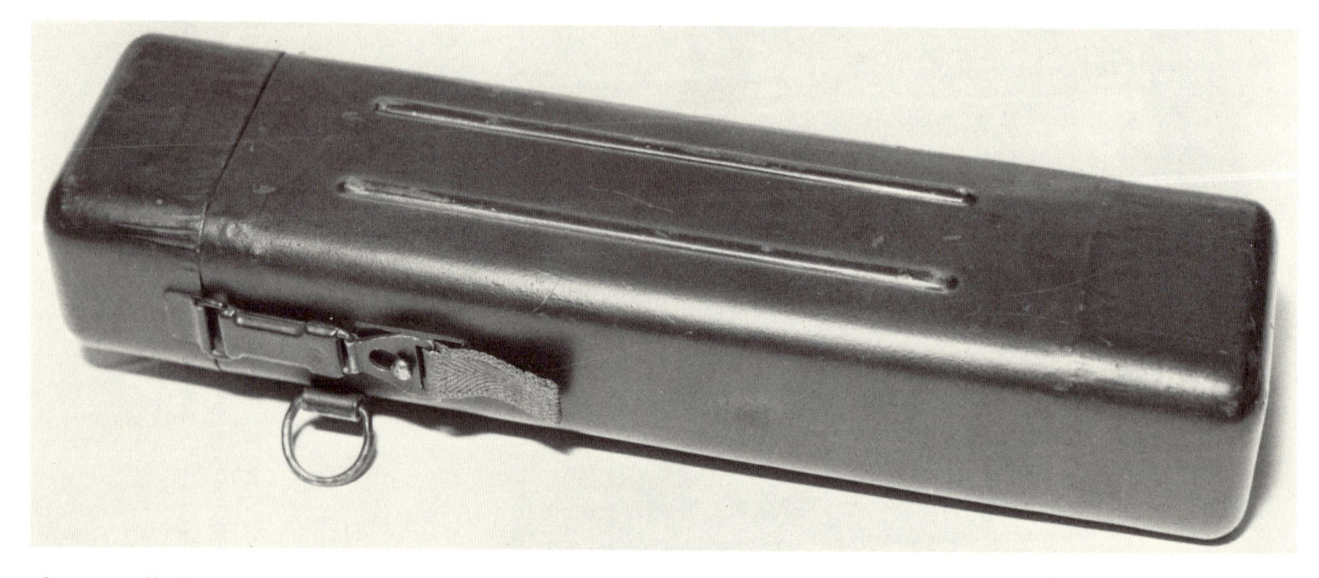

An overall view of the metal scope carrying case originally issued with rifle number 3146.

CHAPTER XII

Scharfschutzen — Waffen-SS

The Waffen-SS (SS-in-arms), which came into being at the onset of World War II, placed considerable emphasis on the use of sharpshooters beginning in late 1939. At this time, the value of telescopic-sighted rifles was considered important enough to warrant the acquisition of a large quantity of sniping rifles (98k-SS short side rail variant) for use by early SS combat units.

Even though a firm foundation had been laid, added impetus was given the organized use of expert riflemen following SS confrontations with Red Army snipers during the first months of the Russian campaign.

Various documents, including official correspondence, reveal direct personal involvement by the Reichsfuhrer of the SS (RF-SS) Heinrich Himmler in Waffen-SS sniping activities, beginning with a mild internal controversy concerning the distribution of "captured high-powered Polish rifles fitted with sniper scopes" in January 1940.

Parallel with the overall effect of Soviet sniping efforts, in what appears to be among the first high-level SS references to Russian combat use of telescopic-sighted versions of "das halbautomatische Tokarevgewehr" (the half, or semiautomatic Tokarev rifle), correspondence from Himmler's chief adjutant to the Chief, SS-Ordnance in September 1942 posed the question of when front-line troops could expect to be furnished "rapid fire carbines with telescopic-sights like the Russians."

Ancillary documents drew attention to plans for an MKb (Maschinenkarabiner 42) sniper variant with a 600-meter capability. But aside from fragmentary reference to contemplated application of a 2-power or a 3-power sight "superior to that in use by the Russians" mounted 300mm from the eye so as not to interfere with ejection, the extent and results of these plans have not been fully revealed.

It should be emphasized, however, that development of a sniping weapon based on the MKb 42 appears to have been conducted by the Heereswaffenamt (Army Ordnance Department) during 1942 as a part of the overall Maschinenkarabiner program rather than by SS-Ordnance on an independent basis. Nevertheless, SS documents did mention testing of a "belt-fed" variant of the MKb concurrent with Army evaluations.

Of further interest are periodic references to "demands" by the RF-SS concerning telescopic-sight and related weapon development noted in official correspondence directed to or exchanged between the various German Ordnance agencies from early 1943 through the early months of 1945. In line with obvious high-level SS pressure for adequate sniping equipment, the most repetitous reference to Himmler centers on the statement: "The Reichsfuhrer of the SS demands the training of sharpshooters at ranges at least to 1000 m."

A measure of Himmler's keen interest in Waffen-SS sniping activities can be drawn from the following letter forwarded to Albert Speer

Waffen-SS marksman shown with an early model German sharpshooting rifle. Many were again pressed into service in World War II.

by the RF-SS under date of 18 December 1944.

Dear PartyMember Speer:

Perhaps you have already heard that I'm encouraging and accelerating sharpshooting training in the Grenadier Divisions. We have already attained outstanding results. I have initiated a contest between all divisions of the army and the SS that are under my command. With the 50th confirmed sharpshooter hit—that is, when he has virtually eliminated a Soviet Infantry Company—each man receives a wristwatch from me and reports to my Field Headquarters. With the 100th hit, he receives a hunting rifle and with the 150th, he is invited by me to go hunting to shoot a stag or chamois buck. The heavy requirement of sharpshooting underlies my following representation: Per experience, it is entirely possible that a division can effect at least 200 hits in a month. I have several divisions which have attained 300 and 400 hits. Suppose there were only one hundred divisions on the entire eastern front—there are significantly more—that would mean 20,000 dead foes in one month. It should be taken into consideration that these fallen foes belong to fighting infantry, not to the supply lines, the artillery or the rear support services.

The Soviet rifle division today has 2 Infantry Regiments of 12 Companies with 50 men each; in all about 1200 men. 20,000 dead foes per month by means of sharpshooter hits means the elimination of the infantry of almost 17 rifle divisions, a result we cannot obtain more effectively, and —if you prefer, more inexpensively—with the employment of the least amount of armament.

For this, however, it is necessary that we obtain more sharpshooter rifles. I would be very grateful if you could step up the production of telescopic sights, rifles with telescopic sights and perhaps also machine carbines with telescopic sights as soon as possible.

Signed Himmler

Trained and equipped under almost the same circumstances as their comrades in arms, the Waffen-SS is known to have maintained sniper schools for Panzer-Grenadier, Panzer, and Gebirgs Divisionen through the course of the war.

Waffen-SS sharpshooting rifles were obtained through regular supply channels from overall sniping weapon production on a monthly basis along with those intended for use by the army and to a lesser extent, Luftwaffe ground forces. The weapons allocated for Waffen-SS use were diverted directly to the SS-Main Operational Office (SS-Fuhrungshaupamt or SS-FHA) responsible for their ultimate distribution to SS combat units.

According to knowledgeable individuals who served in the German Ordnance system during this period, efforts to supplement the ever increasing need for sniping equipment included the periodic random selection of suitably accurate conventional 98k rifles at various army ordnance depots for conversion to sniping arms. This was done especially in the latter months of the war when combat losses of telescopic-sighted rifles offset the supply of those produced in the normal manner. A portion of these aforementioned standard 98k army rifles were forwarded to Waffen-SS depots on a regular basis "for conversion to sniping rifles by their own personnel."

Even though SS marksmen were issued the same types of sharpshooting equipment employed by other Wehrmacht ground forces, in addition to the unique early war short side rail system based on reworked Gewehr 98 rifles, two particularly interesting sniping variations have been attributed to the Waffen-SS, the "double-claw" mount and the "objective-mount" 98k sniping rifles.

Reminiscent of World War I scope mounting practice, the SS double-claw system possessed characteristics identical to those found on prewar Czechoslovakian commercial sporting rifles produced by Ceskoslovenska Zbrojovka, A.S. (Czechoslovak Arms Factory, Ltd.) located in Brno, Czechoslovakia.

Although receiver modifications were required (machining), the relatively uncomplicated scope mounting bases were dovetailed directly into the 98k receiver ring and bridge. The front base, with two narrow rectangular openings, retained the claw-mount hooks,

```
                              A b s c h r i f t

Der Chef der Heeresrüstung und            Berlin, W 35, den 18.12.1944.
Befehlshaber des Ersatzheeres.            Matthäikirchplatz 6
  Chef des Ausbildungswesens              Tel. J2 49o1
       im Ersatzheer
Stab/Ia Nr.5112/44 geh.

Betr.: Scharfschützenausbildung.

        Nachstehende Abschrift eines Briefes des Reichsführers SS u. Be-
fehlshaber des Ersatzheeres an Reichsminister Speer zur Kenntnis.
                                   I.A.

                          gez. Unterschrift.

" Der Reichsführer SS Bra/H Nr.35/111/44 geh. v. 29.11.44.

Lieber Parteigenosse  S p e e r !

Vielleicht haben Sie schon davon gehört, daß ich den Scharfschützenabschuß
bei den Volks-Gren.Divisionen in besonderer Weise pflege und vorantreibe.
Wir haben schon ausgezeichnete Ergebnisse damit erzielt. Es hat bei den
ganzen mir unterstehenden Divisionen des Heeres und der SS ein Wettbewerb
eingesetzt. Jeder Mann erhält bei dem 5o. bestätigten Scharfschützenab-
schuß - wenn er also praktisch eine sowjetische Inf.Kp. erledigt hat - ei-
ne Armbanduhr von mir undmeldet sich in meiner Feld-Kommandostelle. Beim
1oo. Abschuß erhält er ein Jagdgewehr und beim 15o. wird er als Jäger von
mir zum Abschuß eines Hirschen oder Gemsbockes eingeladen.

Der starken Förderung des Scharfschützen schiessens liegt bei mir folgen-
de Überlegung zugrunde: Es ist nach den Erfahrungen absolut möglich, daß
eine Division im Monat mindestens 2oo Abschüsse tätigt. Ich habe schon
mehrere Divisionen, die 3oo und 4oo Abschüsse erreicht haben.

Wenn ich im Osten auf der gesamten Front nur einhundert Divisionen anneh-
men würde - es sind bedeutend mehr - so würde ein im Monat 2o ooo  tote
Gegner bedeuten. Hierbei ist noch zu berücksichtigen, daß diese Gefalle-
nen des Feindes nicht dem Troß,der Artillerie oder den rückwärtigen Diens-
ten angehören, sondern der kämpfenden Infanterie.

Die sowjetische Schützen-Division hat heute zwei Inf.Rgt. zu zwölf Kompa-
nien mit je 5o Mann; insgesamt etwa 1 2oo Mann.

2o ooo tote Gegner im Monat durch Scharfschützenabschüsse erzielt, be-
deutet die Erledigung der Infanterie von fast 17 Schützen-Divisionen;
ein Erfolgt, wie wir ihn nachhaltiger und - wenn Sie es wollen - billi-
ger mit Einsatz von geringsten Mitteln der Rüstung nicht haben können.

Dazu ist aber nun notwendig, daß wir mehr Scharfschützengewehre bekommen.
Ich wäre Ihnen sehr dankbar, wenn Sie die Produktion von Zielfernrohren
Zielfernrohrgewehren und vielleicht auch von Zielfernrohr-Maschinenkara-
binern möglichst bald stark fördern könnten.
                                             gez. Himmler"
```

An official "Abschrift" (copy) of correspondence from the Reichsfuhrer of the SS (RF-SS) Heinrich Himmler to Albert Speer requesting increased production of sharpshooting equipment (telescopes and rifles) in December 1944.

while the rear base, of similar configuration, was provided with projections on either side to engage the spring-loaded rear mount latch assembly of the telescope.

By design, the bases were tightly wedged into the receiver dovetails and soft-soldered. As a further precaution, a pointed-punch was used to "upset" the metal on the receiver next to the bases to prevent their moving.

Telescopic sights manufactured by Opticotechna in Prerau, Czechoslovakia (dow) saw extensive if not exclusive use with this variant.

In addition to the number "63" acceptance stamp located on the telescope mounts, the SS marking "SSZZA2" (SS Zentral Zeugamt A2 or SS Central Ordnance Depot A2) also invariably appears on the left side of the 98k receiver ring on weapons of this configuration. While subject to debate, it is believed that this marking represents a specific SS facility responsible for contract weapon acceptance or final distribution and had no connection with the conversion work for sniper use.

It is not known whether the Waffen-SS double-claw sniping rifles were produced on an as needed basis from existing stocks of miscellaneous manufacture 98k rifles from within the SS supply system, obtained on a secondary basis from the army, or based on SS allocations of newly manufactured rifles obtained from the manufacturers.

However produced, there is no mistaking the Waffen-SS controlled, double-claw mount telescope sighting system of Czechoslovakian origin.

The second and perhaps most enigmatic Waffen-SS sniping rifle was referred to as the objective-mount 98k which has survived in even fewer numbers than the double-claw variant.

Without question, the scope mounting assembly utilized with this unusual variant was commercial in origin. Judging from the complex machining necessary to produce the receiver bases, this system was adapted to select 98k rifles from existing components.

As mounted to the 98k rifle, the double-claw type front mount positioned the telescope objective directly over the receiver ring, producing a distinctive appearance with the sight resting further back on this receiver than on the other telescopic-sighted sniper weapons dating from this period. While similar to the SS double-claw method of mounting, provision for locking the rear mount was an integral function of the rear base assembly rather than a part of the telescope mounting.

Unlike the 98k double-claw system, which is considered a pure Waffen-SS sniping variant, the objective-mount sharpshooting rifle has not been confirmed as being used exclusively by SS marksmen as some experts contend.

Waffen-SS double-claw mount 98k sniping rifle (byf) with "dow" code telescopic sight as fielded for SS sharpshooter use.

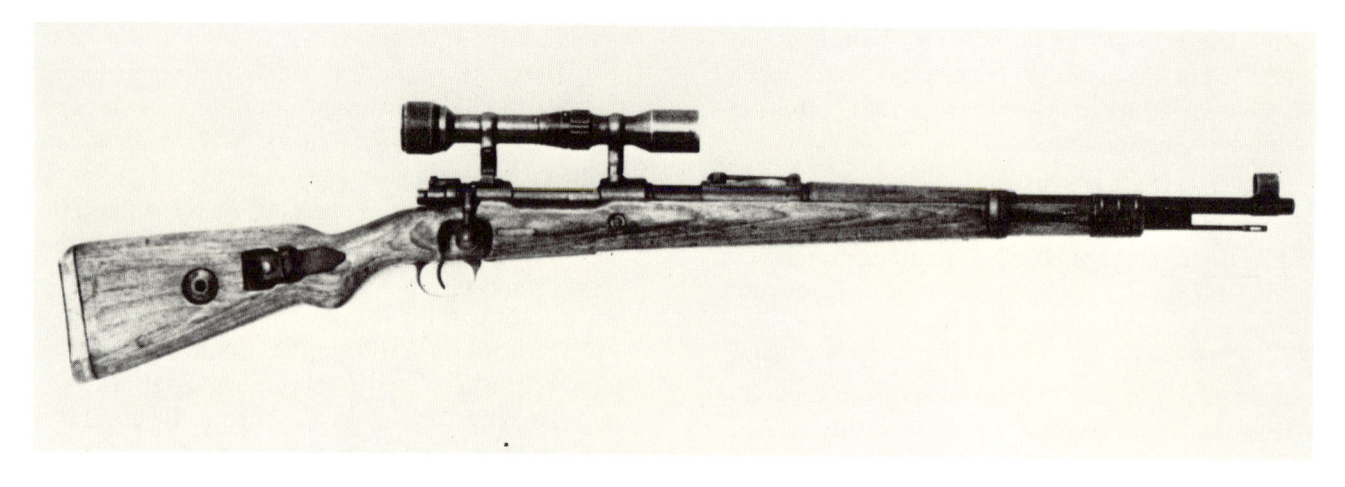

Right view of Waffen-SS double-claw mount sniping rifle.

Close view of SS marking (SSZZA2) as it appears on the left side of the 98k double-claw mount receiver ring.

SCHARFSCHUTZEN — WAFFEN-SS

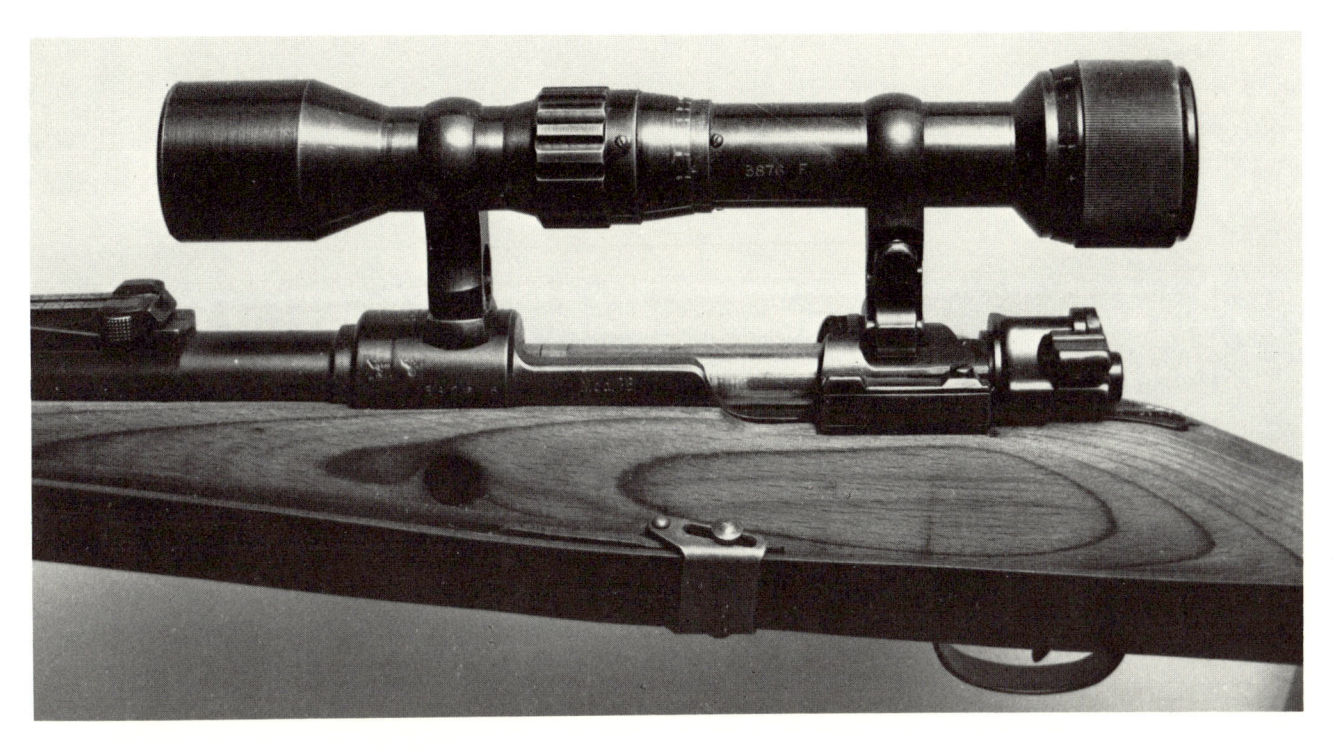

Gustloff manufacture (bcd 43) 98k rifle with SS double-claw telescope mounting system in original form (as issued) with matching serial number "dow" (Opticotechna) sight.

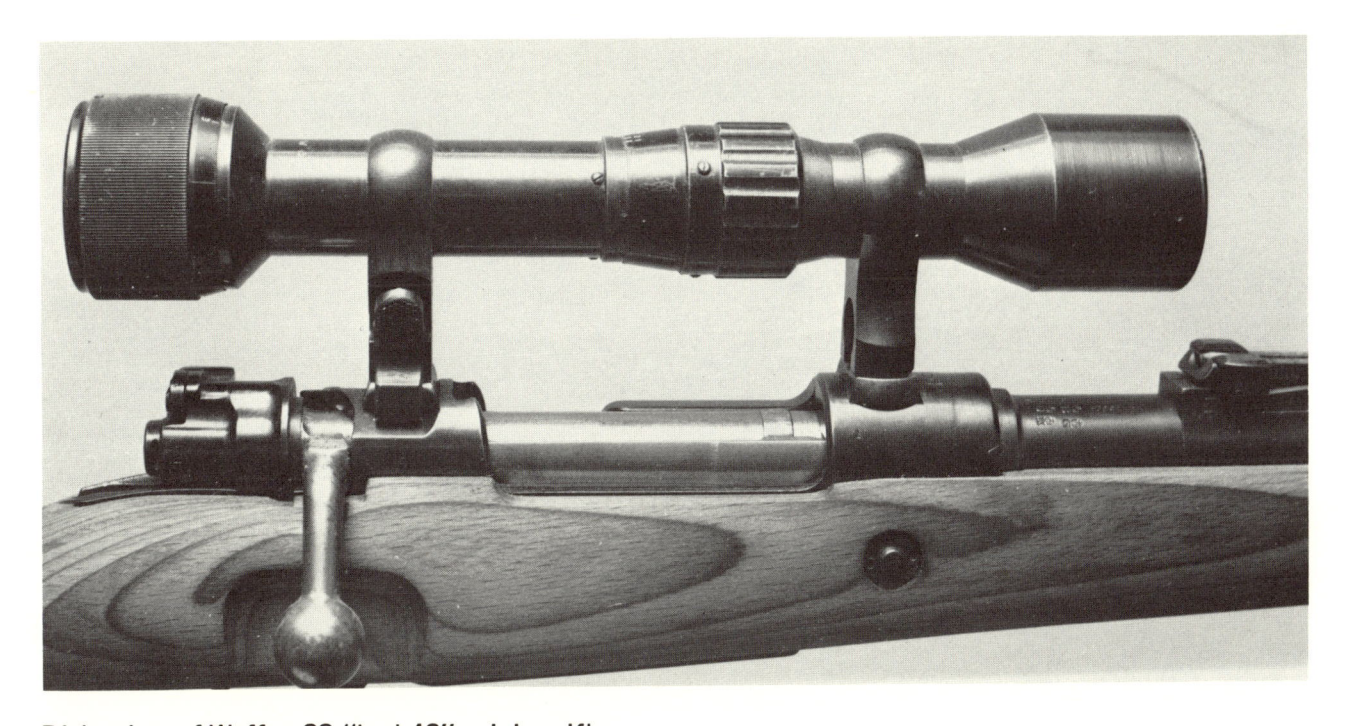

Right view of Waffen-SS "bcd 43" sniping rifle.

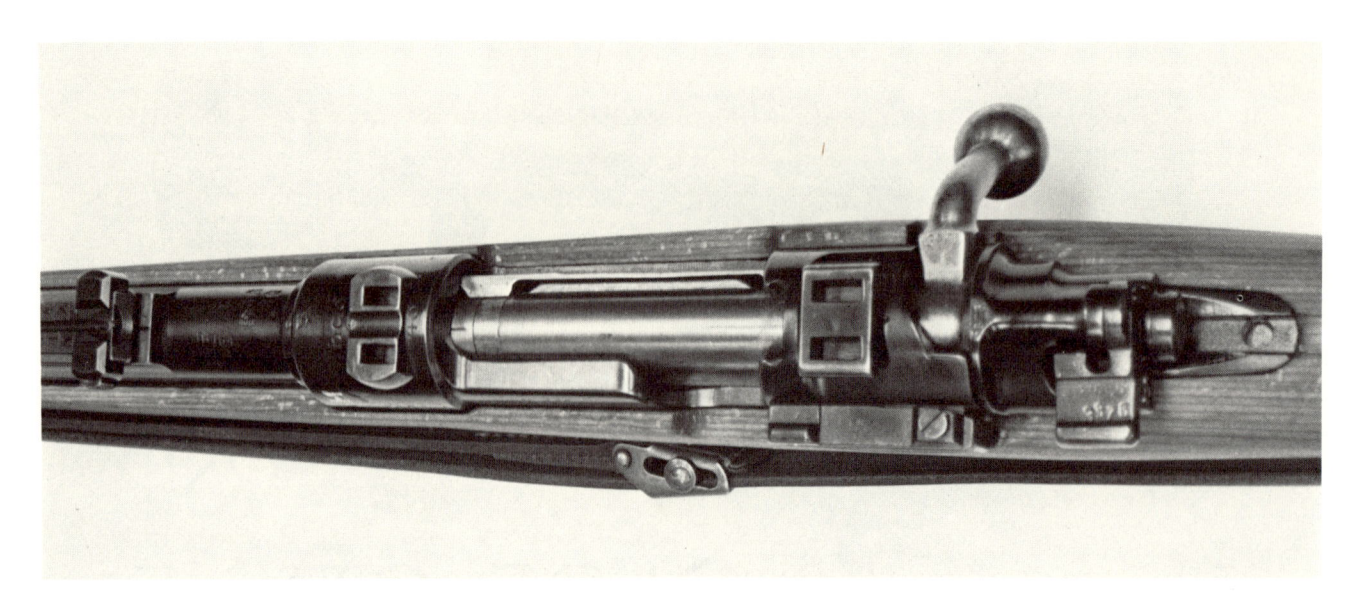

Top view of "bcd 43" SS double-claw sniping rifle showing front and rear scope mounting bases. The recess in the front base provided added clearance for use of the standard sights with the scope in place.

Close view of "bcd 43" SS sniping rifle. Note the manner in which the front base was dovetailed into the receiver ring. The "SSZZA2" stamp is clearly visible.

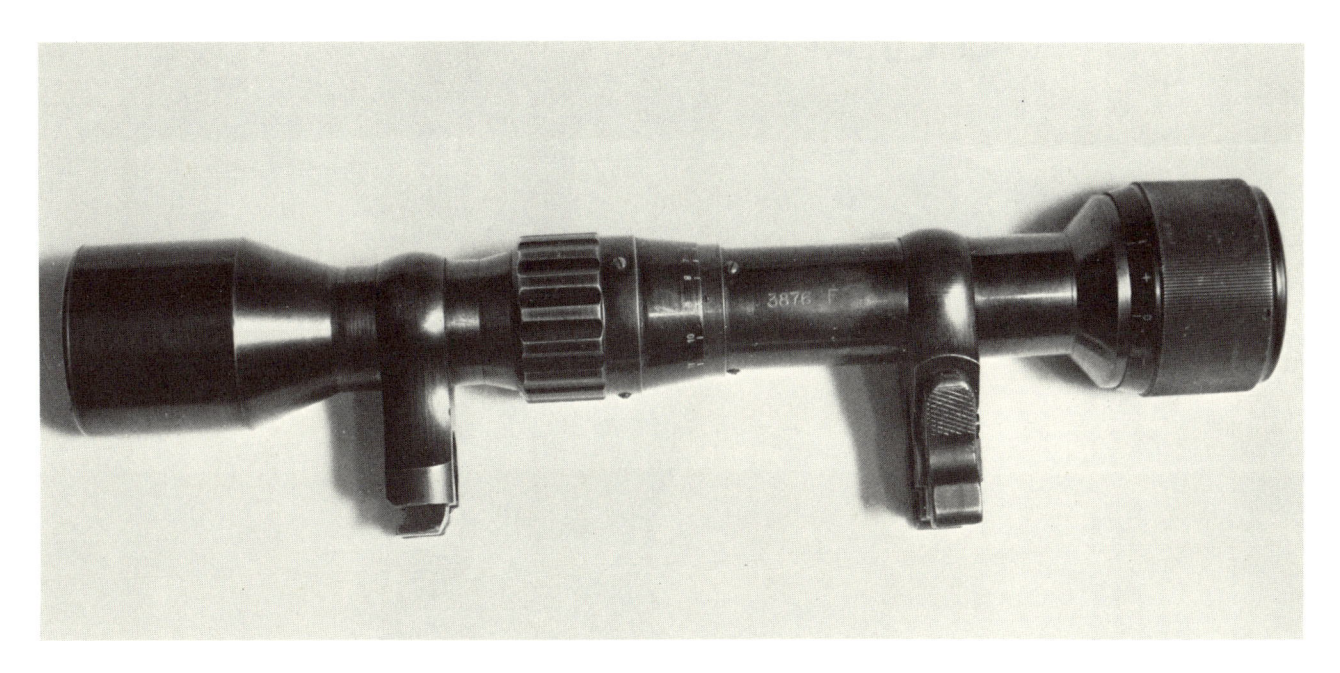

Opticotechna (dow code) telescopic sight and mount assembly used with the Waffen-SS double-claw 98k rifle. The elevation adjustment ring has provision for 1200 meter sighting (1-12). The knurled ring on the ocular end of the sight rotated right or left for proper focus. Note the claw-mount hooks which engaged a corresponding ledge in the front base.

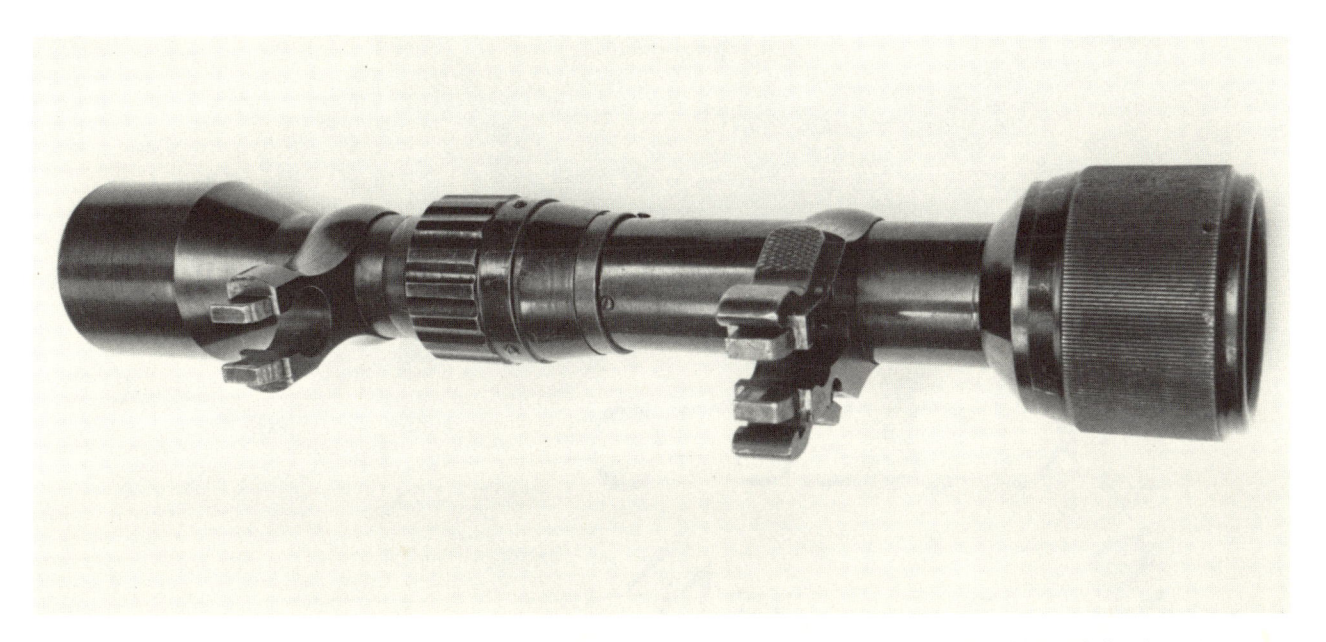

An underview of the "dow" scope showing the front claw-mount, latch assembly, and dual-projections which fit into the rear base openings. The rings were soldered directly to the telescope tube.

Close view of SS double-claw rear base and mount assembly. The rear mount attached to the receiver with spring-loaded latches closing over projections on either side of the receiver base.

Mauser manufacture (byf, serial No. 26045d) 98k rifle with SS double-claw system. The 4-power "dow" scope elevation adjustment ring is graduated to only 800 meters in this case (1-8). Note the fine knurl finish and latch configuration (rear mount latch assembly) as compared to others depicted in this section.

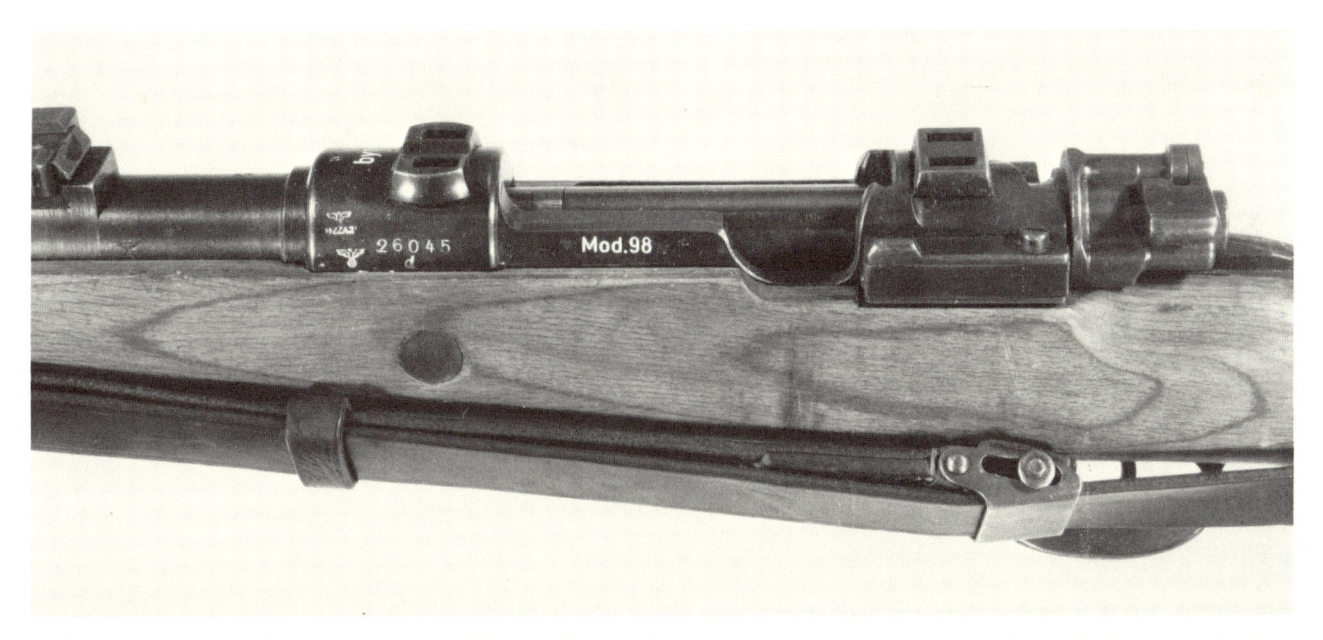

Left view of "byf" double-claw sniping rifle with telescope removed. Regardless of the rifle (manufacturer) used with this system, base mounting characteristics were consistent on weapons of this configuration.

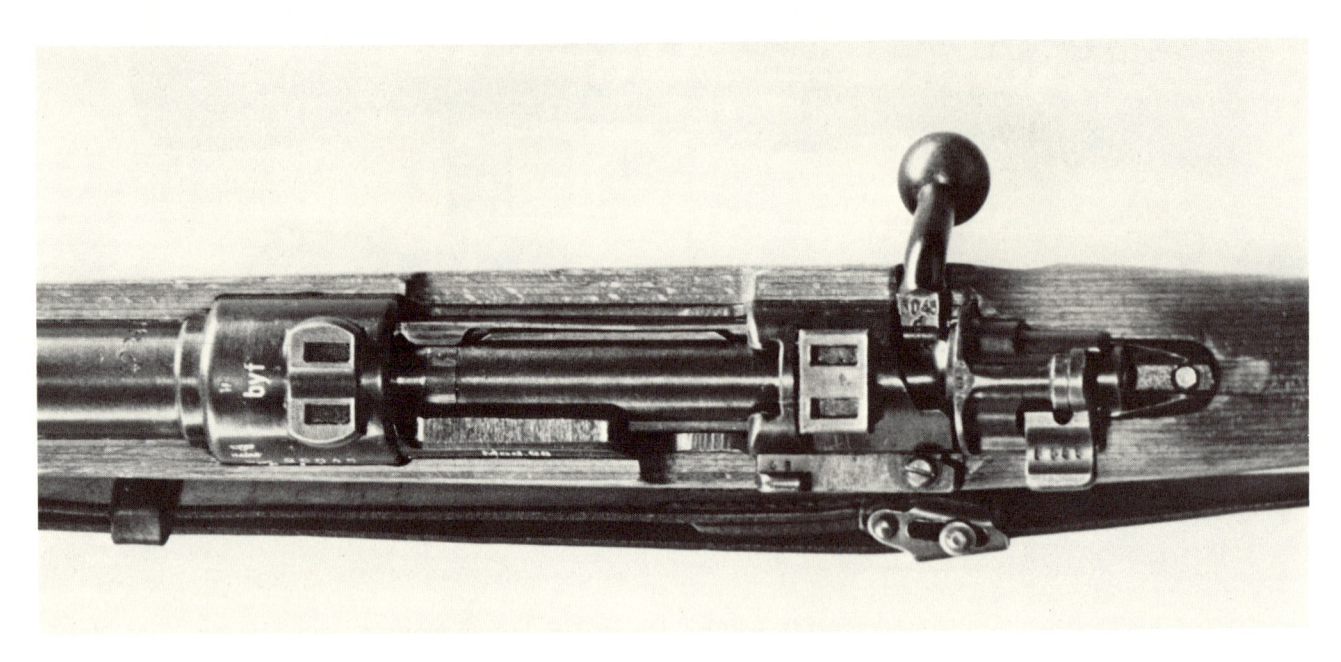

Top view of SS double-claw scope mounting bases fitted to the "byf" variant provides an effective comparison with those on the "bcd 43" rifle.

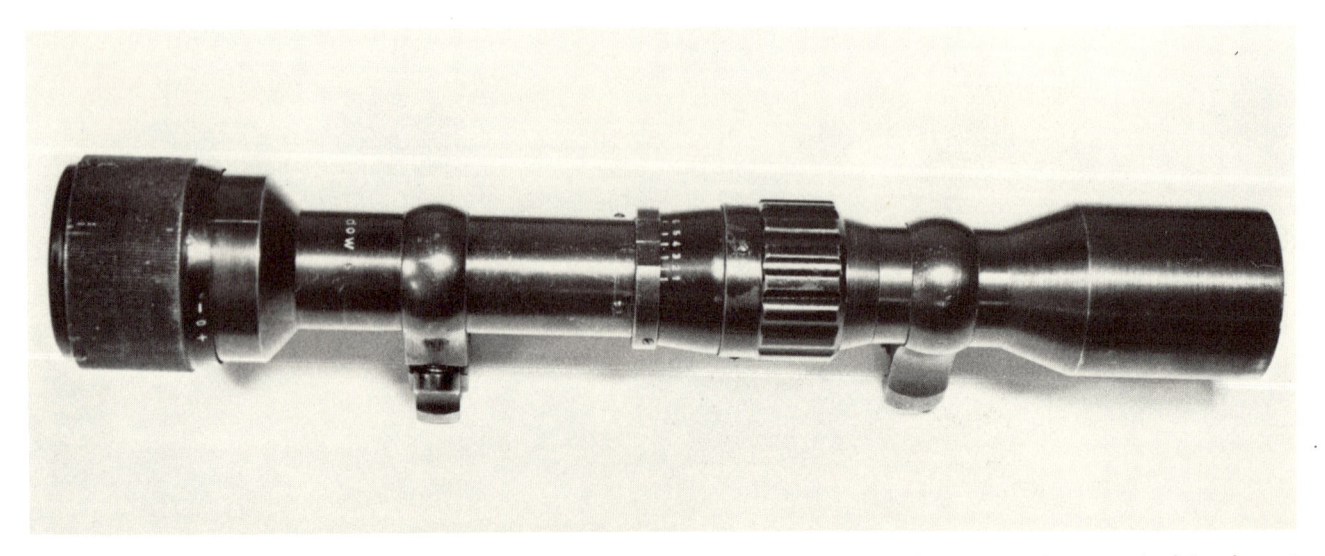

Top view of 4-power "dow" telescopic sight. Note the manufacturer's production mark (dow) and elevation adjustment ring graduations (1-8). Except for their range markings, scopes of this type were virtually identical in appearance.

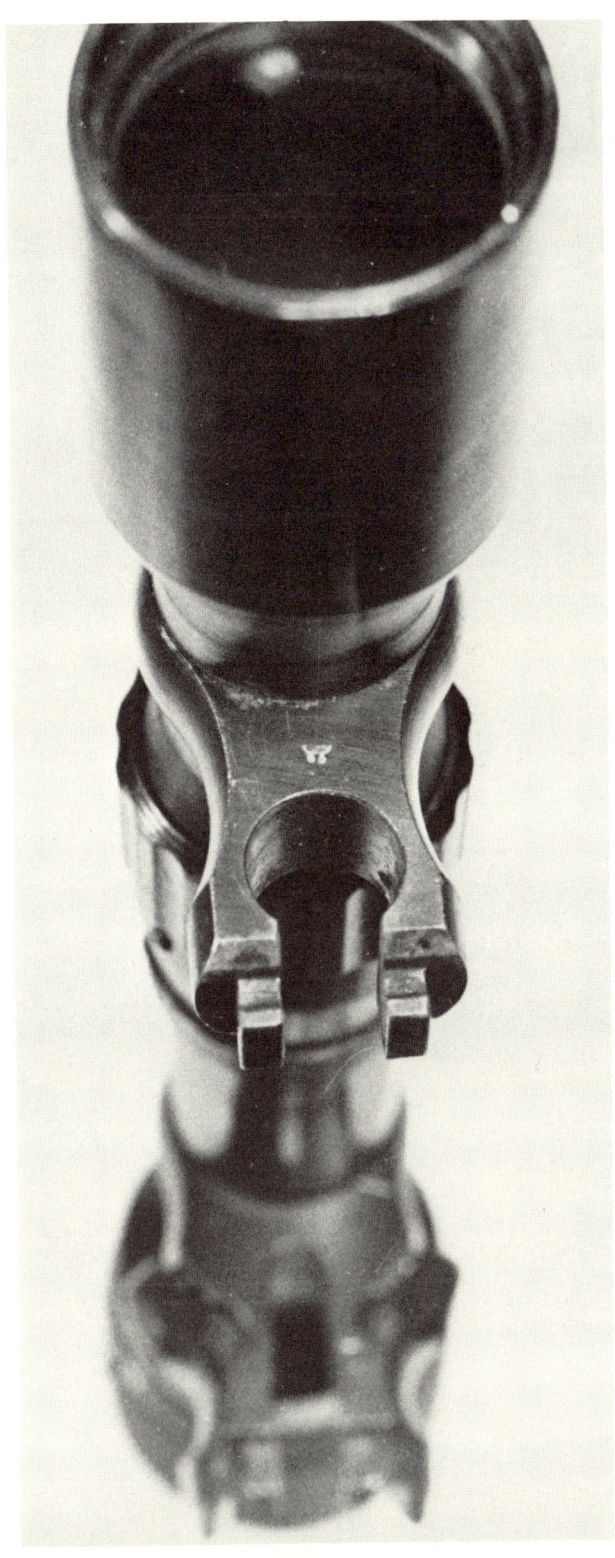

An underview of SS double-claw "dow" front scope mount assembly showing the claw feet and opening ("tunnel") which allowed use of the standard 98k sights with the scope in place. Note the "63" Waffenamt stamp.

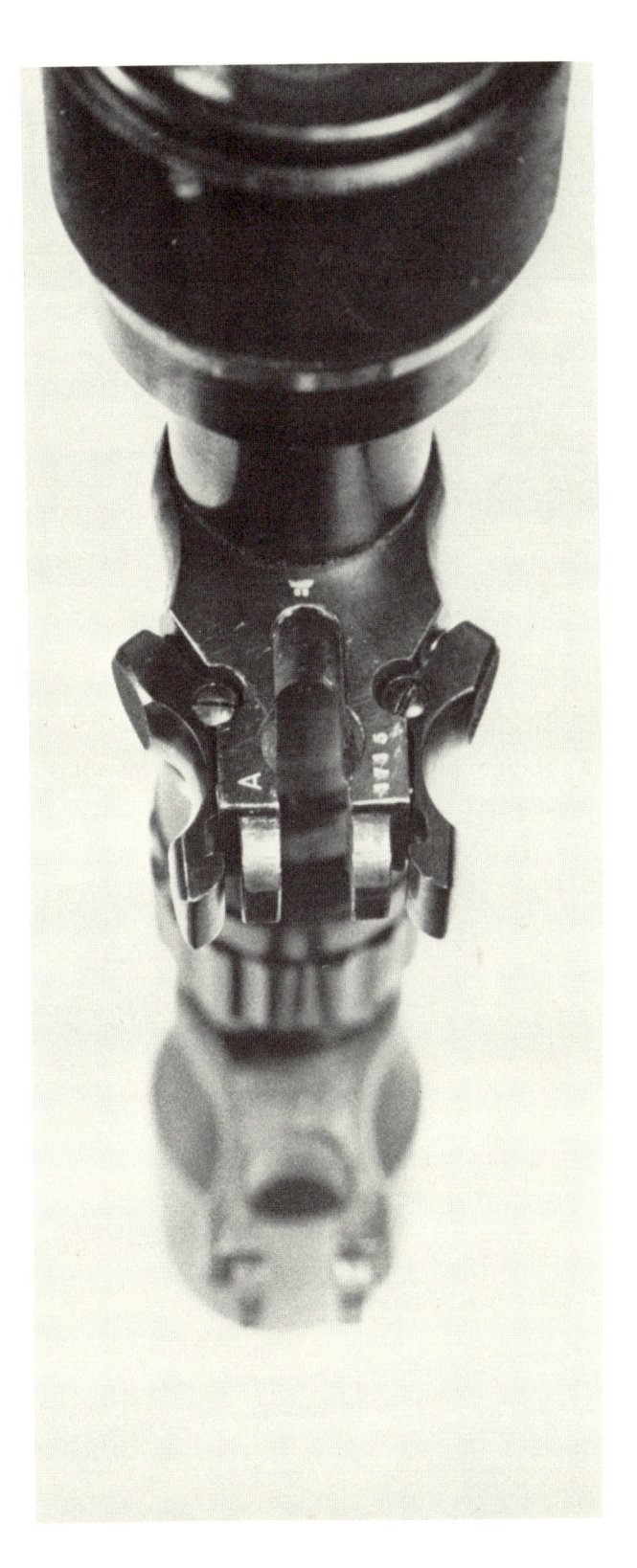

An underview of SS double-claw "dow" rear scope mount assembly showing the dual-projections, spring-loaded latches, and "tunnel" opening. Note the "63" Waffenamt stamp.

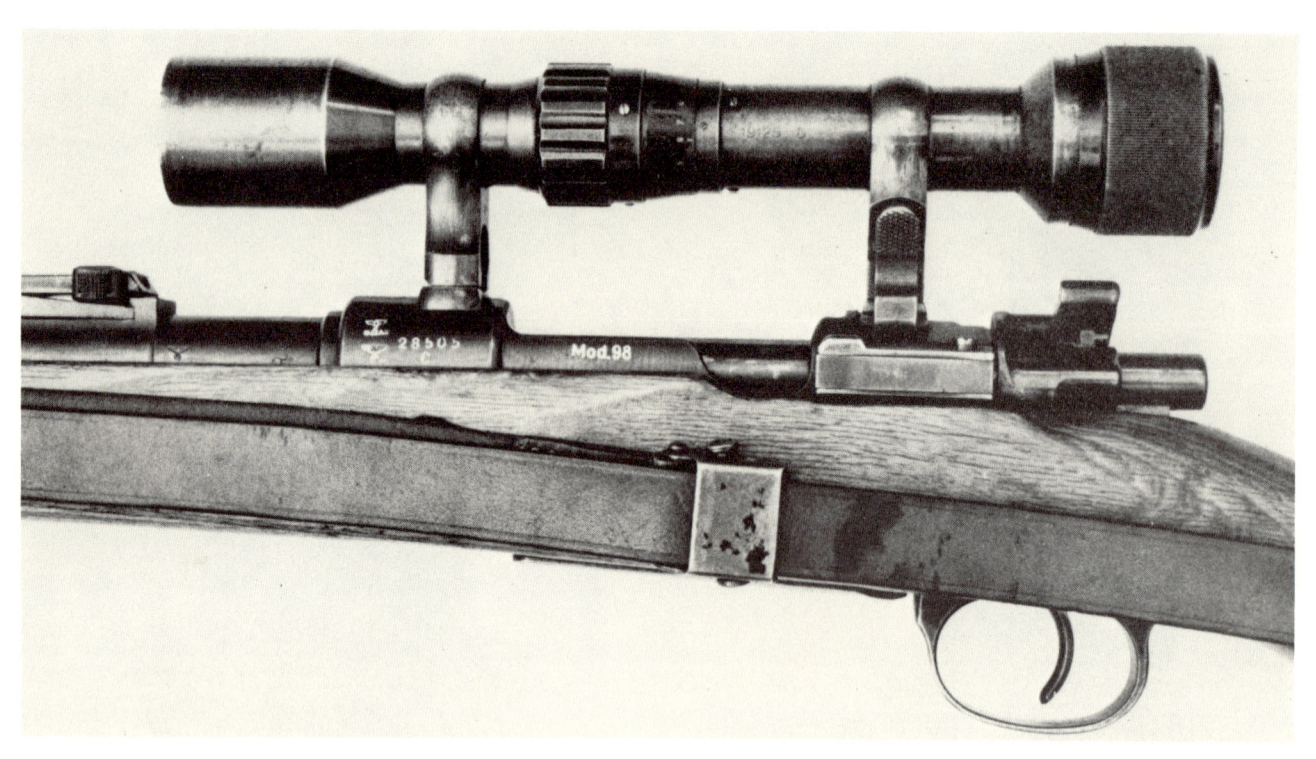

An additional Waffen-SS double-claw mount variant (byf, serial No. 28505c) with "dow" telescopic sight is presented for further comparison.

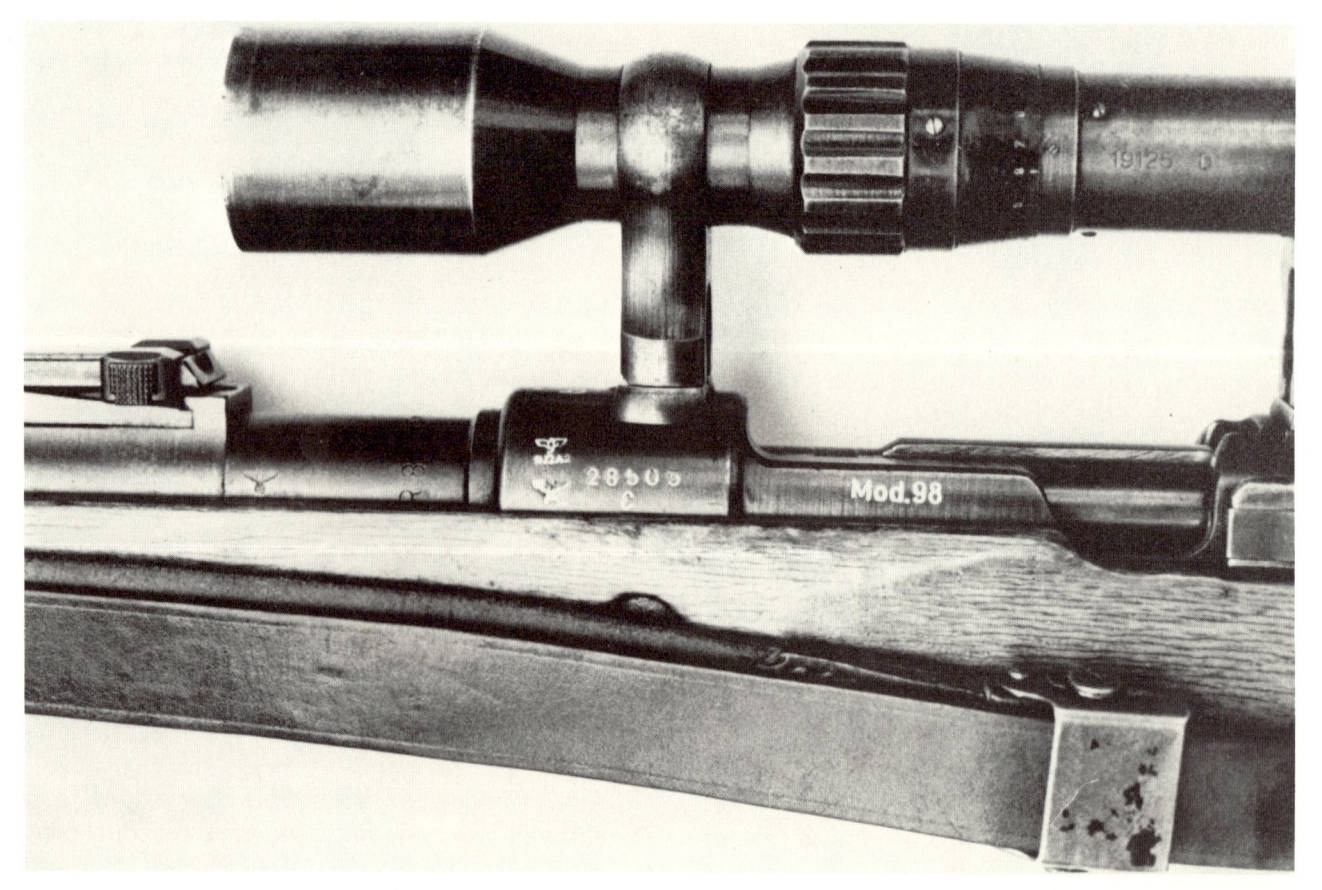

Close view of Mauser manufacture 98k rifle (28505c) front scope and receiver area with Waffen-SS marking (SSZZA2) clearly visible.

Close view of ocular end of "dow" scope showing knurled collar with diopter scale used to make adjustments for vision (focus).

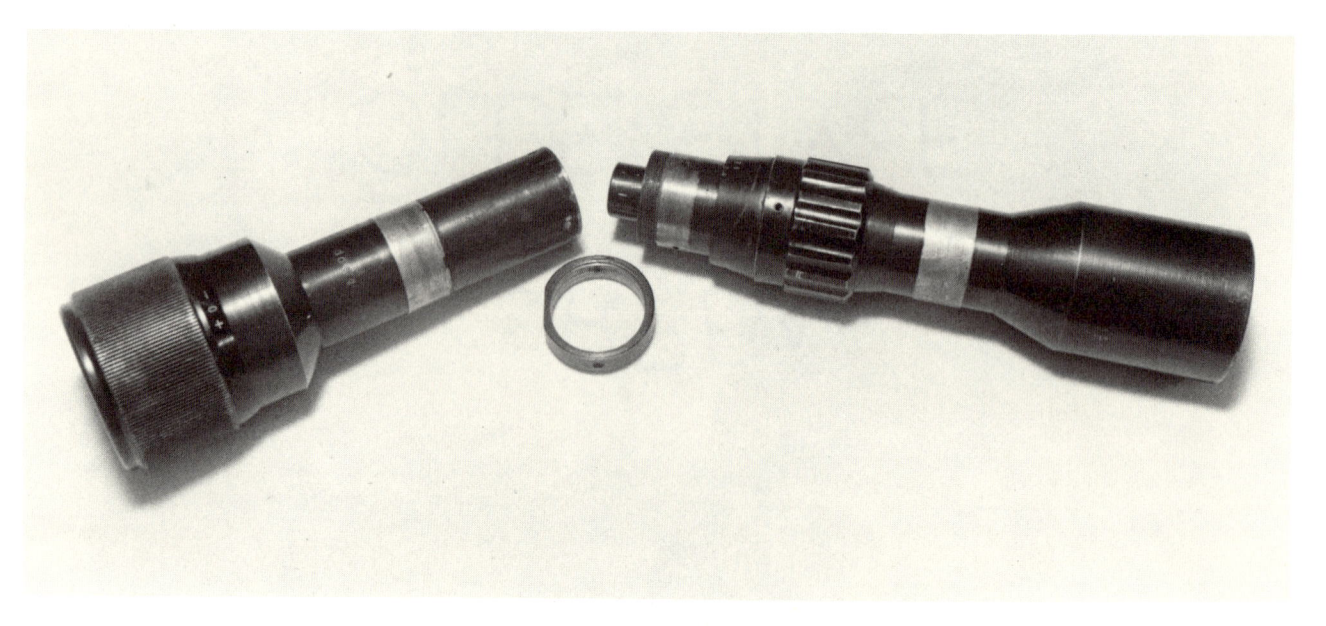

A partially disassembled "dow" telescopic sight. The mounts have been removed.

Close view of SS marking (SSZZA4) as noted on a standard 98k "bnz 43" rifle. In this case the stamp was applied to the left side of the barrel between the rear sight base and the receiver.

SS marking (SSZZA4) as noted on a standard Gewehr 98 rework rifle is shown for comparison. This stamp appears on the right side of the barrel between the rear sight base and the receiver. The marking is hidden by the stock.

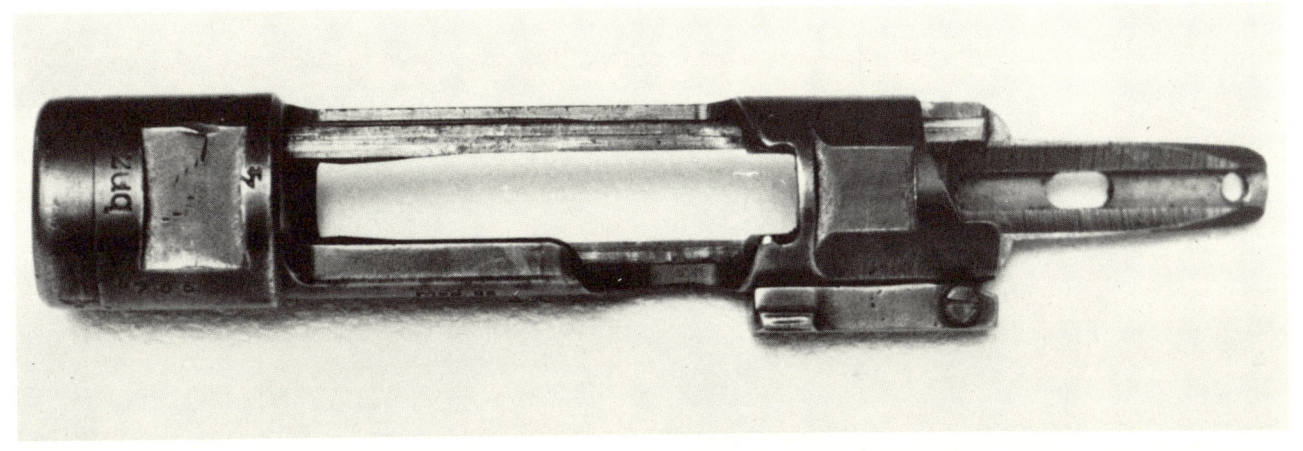

The remains of an SS double-claw sniping variation after a "basement gunsmith" has completed his work. Prior to defacing, this receiver (bnz) was a principal component of a unique World War II German sniping rifle. Note the manner in which the dovetail form was machined into the receiver.

The unique objective-mount sniping system typical of those depicted in numerous Waffen-SS combat photos. It is believed that commercial scope mounting components were simply adapted to select 98k rifles as needed by the SS.

Sighting from a defensive position, an SS sniper makes use of the distinctive objective-mount 98k sniping rifle during winter action.

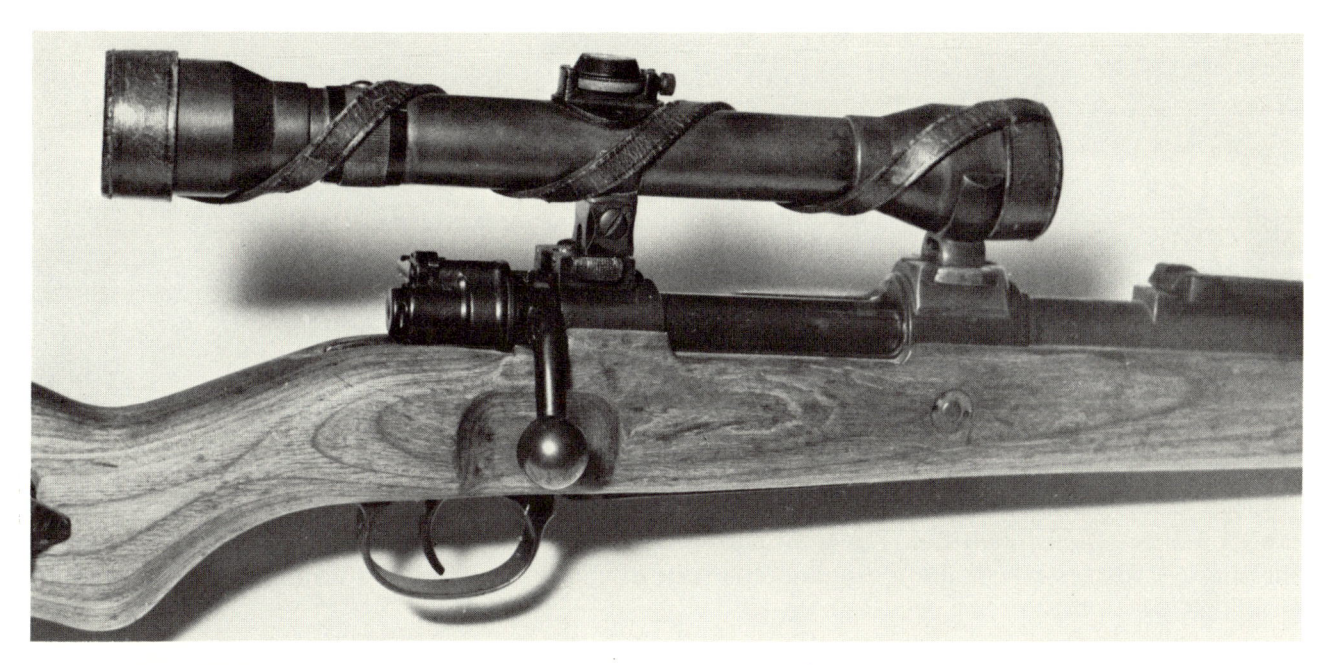

Right view of 98k objective-mount sniping system. The sight, a 4-power Hensoldt-Wetzlar device, has the early style focus adjustment ring and "half-rings" soldered to the tube. The front mount utilized double-claw feet with dual rear mount projections similar to those on the SS double-claw mount variation.

Waffen-SS armored force pausing to assess the situation. Note the objective-mount 98k sniping rifle in the foreground.

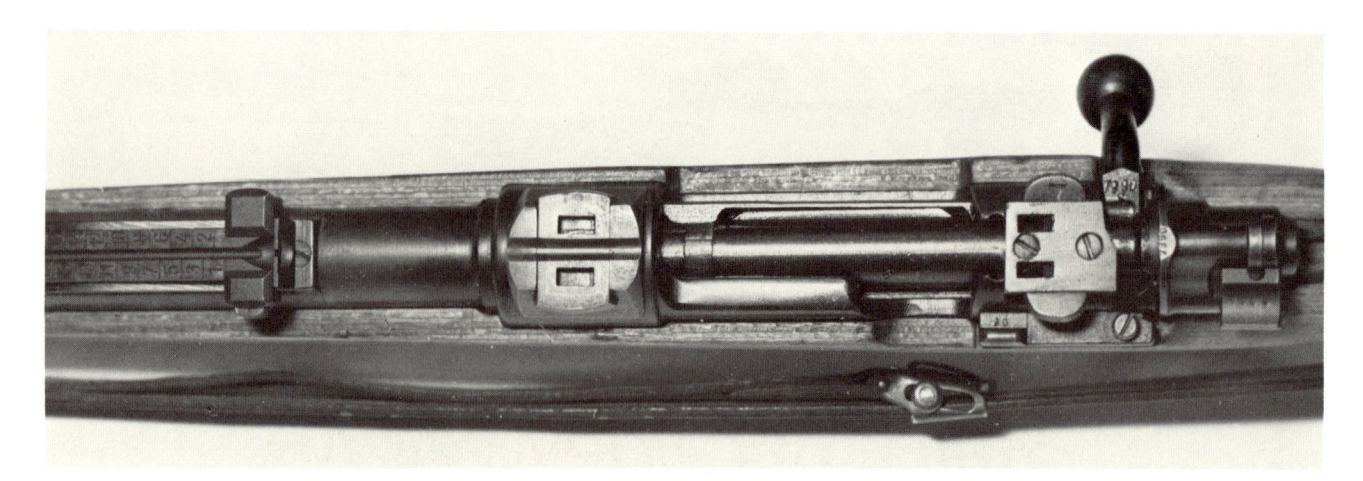

Receiver bases used with the 98k objective-mount sniping variant are typical of those found on many prewar European sporting rifles. The front claw mount engaged the base in the standard manner while the spring-loaded, semi-circular wings on either side of the rear base were drawn back when mounting the scope.

Front view of objective-mount sniping rifle showing provision for standard sight use (mount tunnel) with scope in place.

CHAPTER XIII

Selbstladegewehr 41

The original German Ordnance concept of a semiautomatic rifle resulted in the introduction of the Selbstladegewehr 41 (self-loading Rifle 41) design as affected by the Walther (Zella-Mehlis) and Mauser (Oberndorf) firms in 1941.

Concurrent with the overall plan of telescopic-sight use on service rifles, a course necessitated by events in Russia, at least three variations of optical sights were tested or utilized with these weapons, the ZF 40, the ZF 43(B), and the Gw ZF 4-fach. Only the 1.5-power Zielfernrohr 40 device has been established as having been an operational item, with field use extremely limited at best.

Earmarked for use with the Gewehr 41, 1.5-power sights were adapted to both the Mauser G41(M) and the Walther (W) variants by different means.

In much the same manner as those fitted to the 98k-ZF 41 rifles, the G41(M) telescope mounting base (rail) formed a part of the rear sight base assembly, and while similar in appearance, differed slightly from the 98k-ZF 41 configuration. When mounted on the Gewehr 41(M) rifle, the telescope rested directly over the rear sight base in line with the bore. Because the Mauser variation proved the less desirable of the two designs, its manufacture was reportedly terminated in 1942 with a total of 20,000 rifles produced at the Mauser Werke in Oberndorf.

Unfortunately, neither German Ordnance nor Mauser records have yielded any specific reference to the use of telescopic-sighted G41(M) rifles.

Whereas a thorough explanation and field application of the 1.5-power ZF 40 sighting system with Walther design rifles is referenced in the ordnance manual, *D191/1 Gewehr 41, Beschreibung, Handhabung und Behandlung* dated 2 February 1943, the description, operation, and handling instructions for the Gewehr 41(M) Mauser variant (dated 26 May 1941, published 1942) makes no mention of telescopic sights.

The Zielfernrohr 40, as utilized with Walther design rifles (ac and duv) was attached to integral machined rails located on either side of the rear sight base by means of a unique mounting. The mount, which straddled the standard rear sight assembly, slid on to the rails from the rear, positioning the scope directly over the receiver.

From a collector's standpoint, aside from the representative (replica) ZF 40 "straddle-mounts" currently produced for the martial arms collector, it should be emphasized that no original World War II German manufacture Gewehr 41 telescope mounts of this configuration are known to exist.

Originally the 1.5-power sight was intended to serve as a universal rifle scope, but limitations of the sighting system in combination with the

An early Mauser experimental selbstladegewehr mounting a 1.5-power telescopic sight. The receiver ring and various parts are marked "S/42 D-03." There are no acceptance stamps on this unusual weapon.

The extremely rare Mauser manufacture G41(M) semiautomatic rifle with 1.5-power telescopic sight and mount assembly.

Left view of telescopic-sighted Mauser Gewehr 41(M).

Close view of 1.5 power telescopic sight as mounted to the early "S/42" Mauser rifle.

Left view, early Mauser experimental Gewehr 41 with 1.5 power telescopic sight. Note the "S/42 D-03" marking on the magazine housing.

Gewehr 41 prompted the subsequent development of an unusual telescope and mount for use with the Gewehr 41. Bearing the designation Zielfernrohr 43(B), the unique 4-power sight is approximately 190.50mm in length with an ocular lens diameter of 18mm. The objective lens is 22mm in diameter with a field of view of 5 degrees. Elevation adjustments were made with a range drum graduated from 100 to 800 meters (1-8) located on top of the tube, while windage, a function of the mount assembly, was obtained by rotating a large knob on the right side of the mount.

The mount itself, noted as a "heavy forging machined on the bearing surfaces to provide for accurate alignment with the base," presented a most unusual appearance when in place on the rifle.

Even though the ZF 43(B) was reportedly subjected to army field trials, the fate of the rather complex sight assembly was parallel to that of the overall Gewehr 41 weapon system: a realignment of both development and manufacturing priorities with the introduction of the improved Selbstladegewehr 43 (self-loading rifle 43 or G43) and as followed, the Gw ZF 4-fach sight design brought to fruition by the Voigtlander firm during the latter part of 1943. The G43 and the Gw ZF 4 sight were a direct result of concentrated efforts to mass produce a simplified telescopic sight and a new semiautomatic rifle for general service use.

As a matter of interest, ordnance documents predating initial manufacture (regular production) of the Gw ZF 4 device cite testing of a telescopic "B" sight with the Gewehr 43. But it is not known if this reference had any direct bearing on or link with the ill-fated ZF 43(B) sight.

Without question, the least known variation of telescopic-sighted Gewehr 41 rifles was that fitted with a dovetail scope mounting base of the exact configuration that is found on the early G43 rifles, no mount locking recess machined into the base.

Intended ostensibly for mounting the 4-power ZF 4 sights, the base was fitted to the G41 receiver in approximately the same position as the integral Gewehr 43 receiver base, with the bolt slide cocking handle (G41) repositioned on the left side of the bolt assembly to provide adequate clearance for the ZF 4 scope mounting.

Believed to represent erstwhile attempts to field satisfactory scope-sighted Gewehr 41 rifles from existing production, original weapons of this type are considered quite rare.

On an overall basis, the scarcity of surviving scoped versions of the Gewehr 41 can be attributed to the limits of original use and to the fact that untold numbers of those weapons remaining operational were either destroyed or lost to the Russians during the four years of action on the eastern front. At this time, there remains nothing to substantiate the extent of actual G41 use in a sniping capacity.

Detailed view of G41(M) scope mounting base. Note the flat surface typical of those used with early manufacture "flat-roller" scope mounts.

SELBSTLADEGEWEHR 41

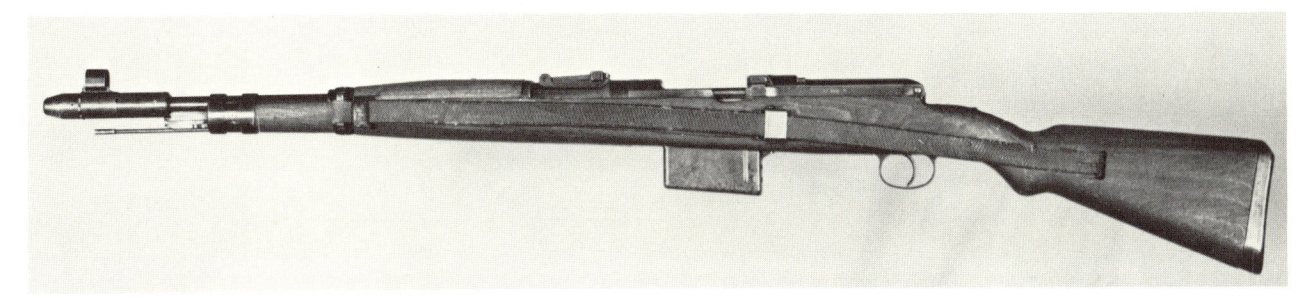

A complex design with components requiring "excessive machine work," a Mauser manufacture Gewehr 41(M) is shown in standard trim.

Receiver markings (typical) found on the Mauser Gewehr 41(M). This rifle has no scope mounting base.

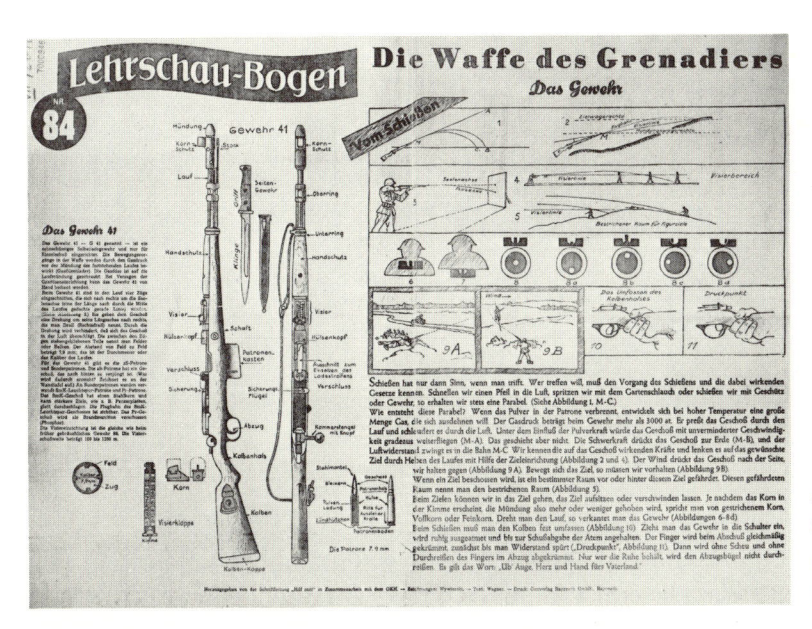

An early description of the Mauser Gewehr 41(M) weapon system. Original material covering this rifle remains very scarce.

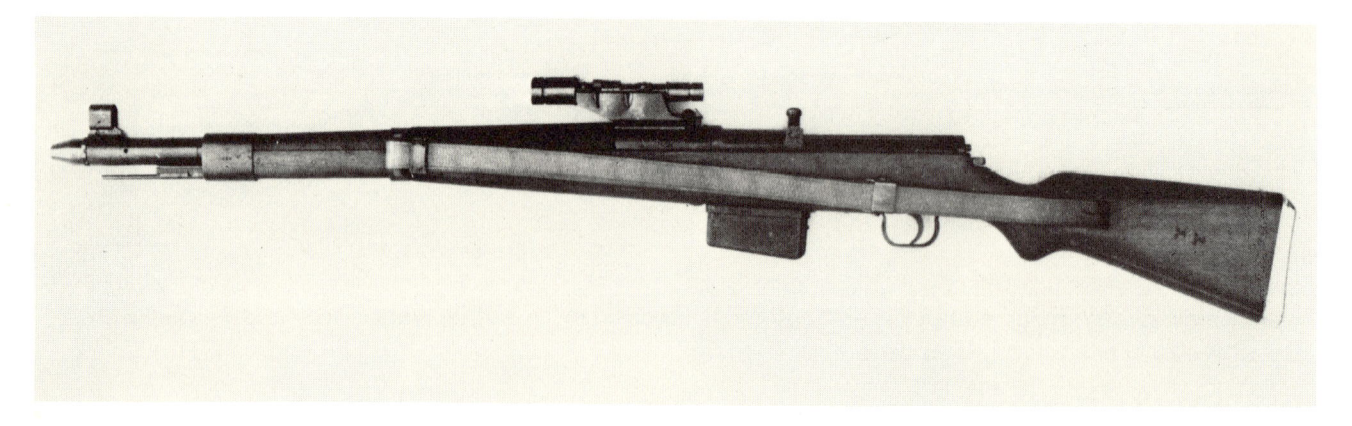

Original photograph taken from German Ordnance (Army Weapons Bureau) report RH II 1/53, 17 November 1942 showing a Gewehr 41 rifle with a 1.5-power ZF 40 sight and mounting utilized with Walther design rifles. This is the only known photograph of an authentic "straddle-mount."

Conventional rear sight of a Walther Gewehr 41 flipped forward to show the telescope mounting rails present on many of the variants produced by Walther (ac) and Berlin-Lubecker Maschinenfabriken (duv).

Early production Walther "G41(W)" marked rifle (5459) with the bolt release button found on those weapons bearing the "(W)" designation. There are no scope mounting rails on this *particular* rifle.

Close view, Walther (7062) Gewehr 41(W). An unusual variation with the number 4 stamp omitted from the legend. Original in every respect, this rifle has *no* bolt release button. Note the scope mounting rails.

Top view of Gewehr 41 rear sight assembly and integral scope mounting rails. Note the notch in the right rail for locking the mount.

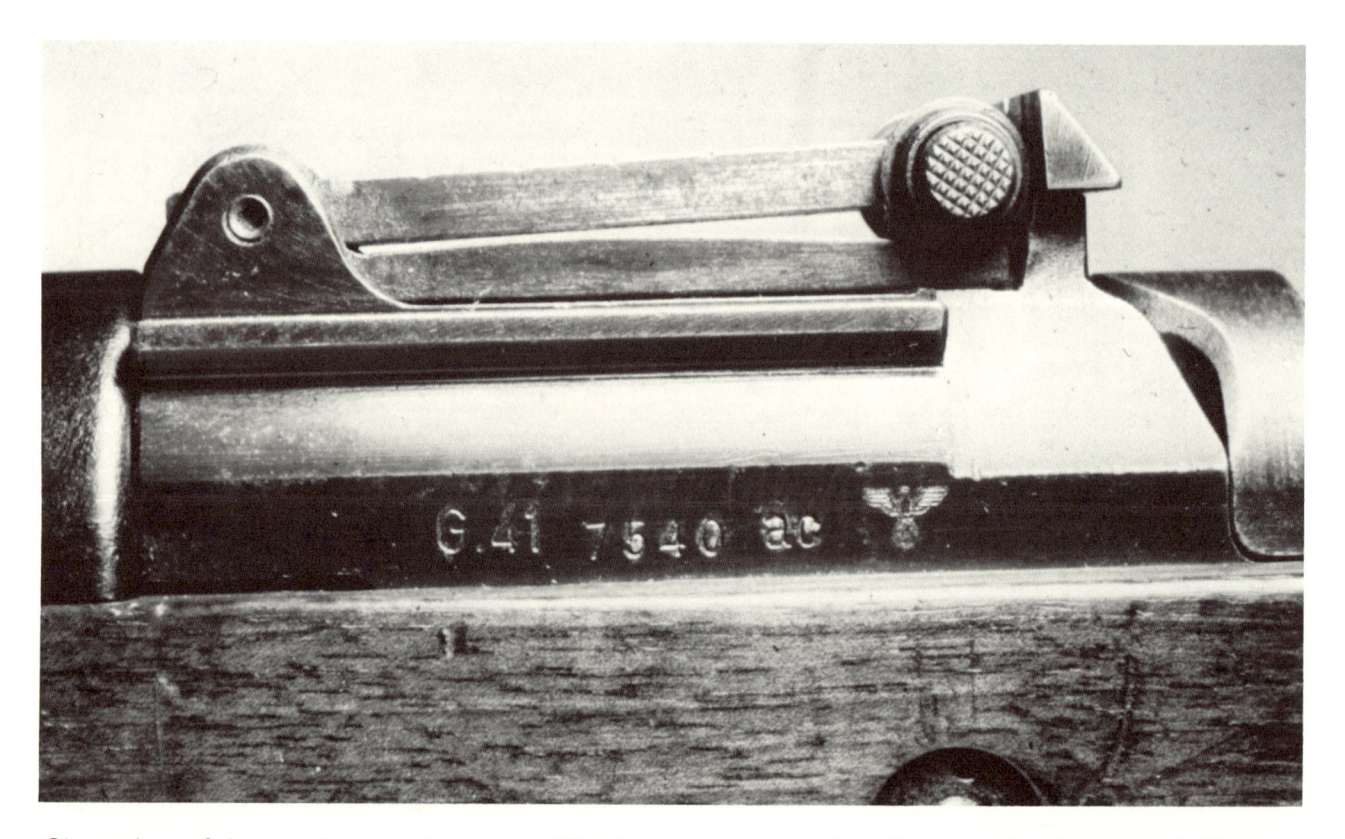

Close view of the receiver markings on a Walther manufacture (ac) Gewehr 41 rifle with ZF 40 scope mounting rails.

SELBSTLADEGEWEHR 41 309

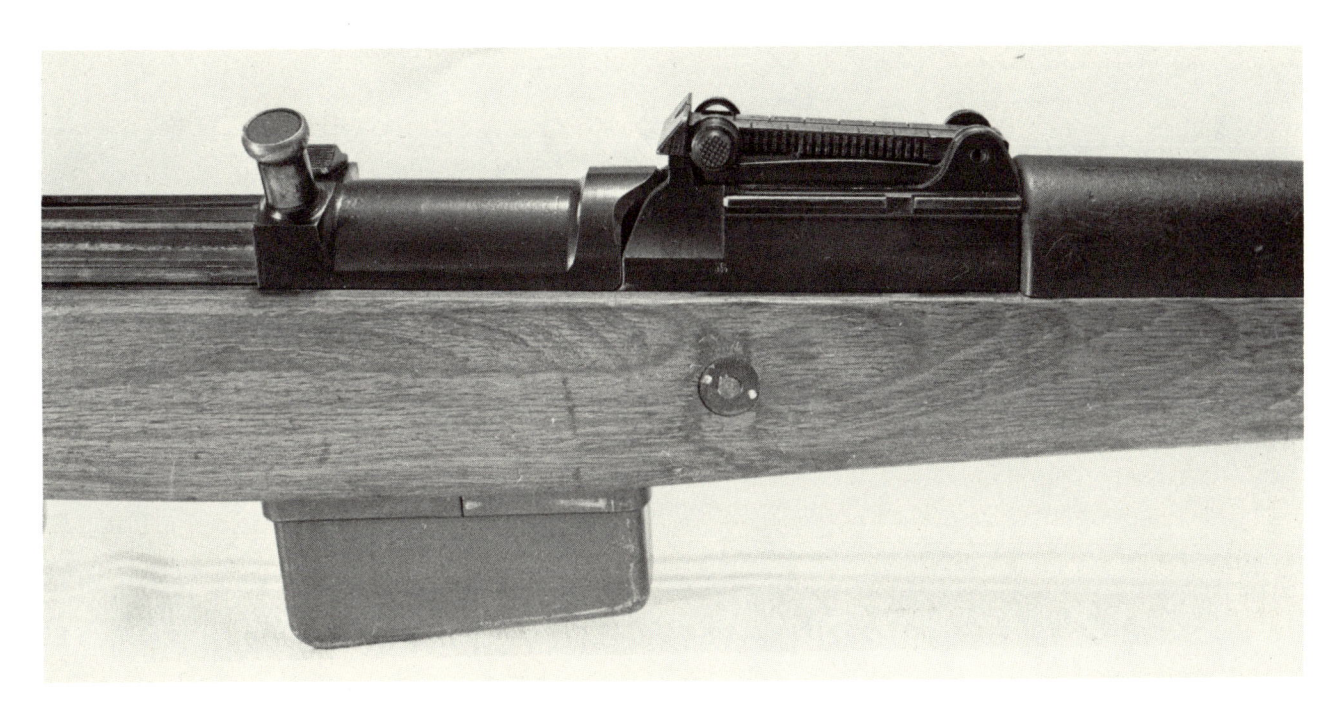

Walther manufacture G41 (ac 43) rifle rear sight and receiver area showing position of scope mounting rails.

D 191/1

Gewehr 41

Beschreibung, Handhabung und Behandlung

Vom 16. 2. 43

The front cover 191/1 of German Ordnance pamphlet D 191/1 Gewehr 41, 16 February 1943 which detailed the description, operation, and handling of the Selbstladegewehr 41 system (Walther design).

Gewehr 41 mit Zielfernrohr 40, 1,5fach, Ansicht von links

Gewehr 41 mit Zielfernrohr 40, 1,5fach, Ansicht von oben

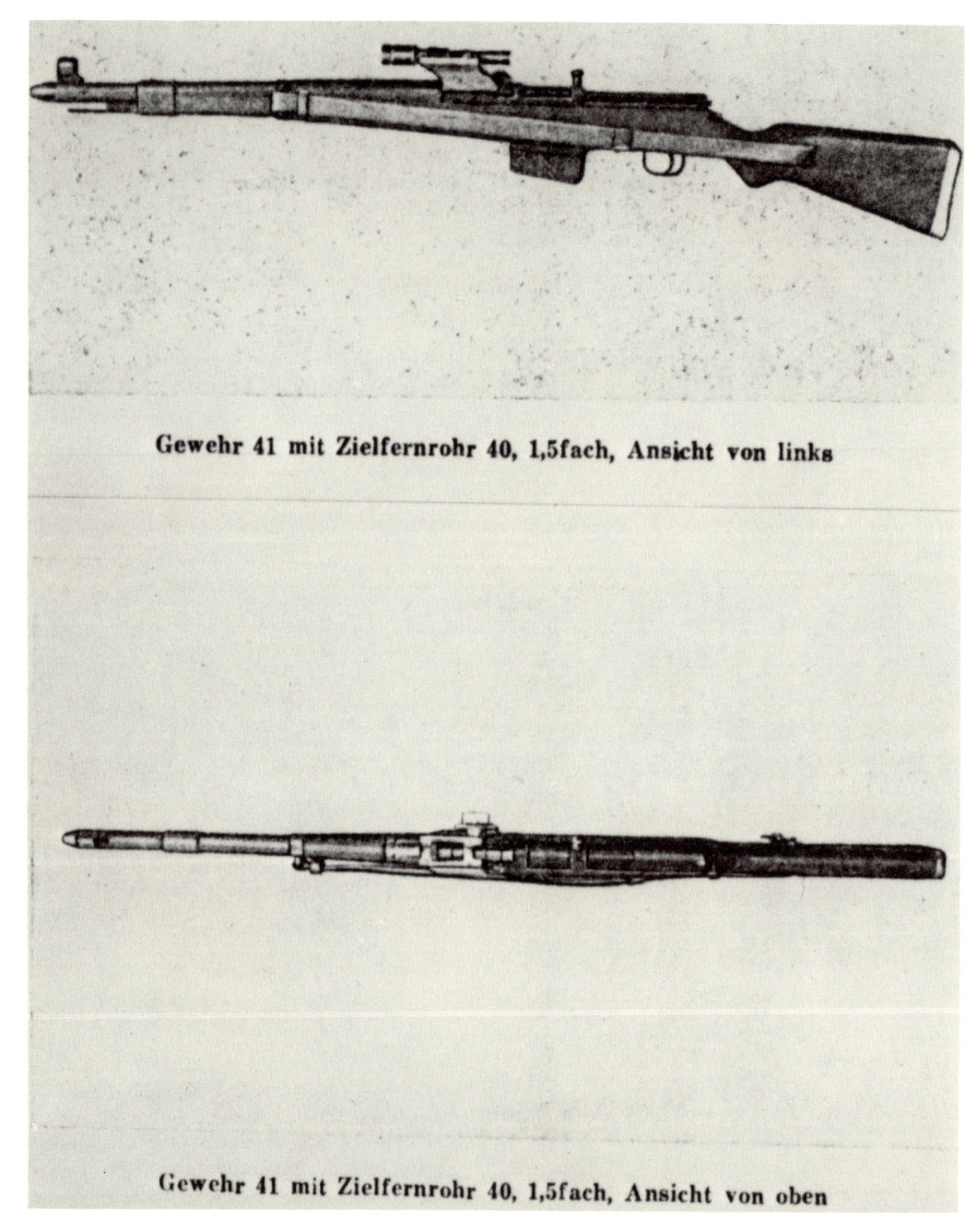

Illustration from D 191/1 showing the "Gewehr 41 mit Zielfernrohr 40, 1.5 fach" sighting system (Rifle 41 with telescopic sight 40, 1.5-power). Note the position of the telescope mount which "straddled" the rear sight assembly.

SELBSTLADEGEWEHR 41

Zielfernrohr 40, 1,5fach

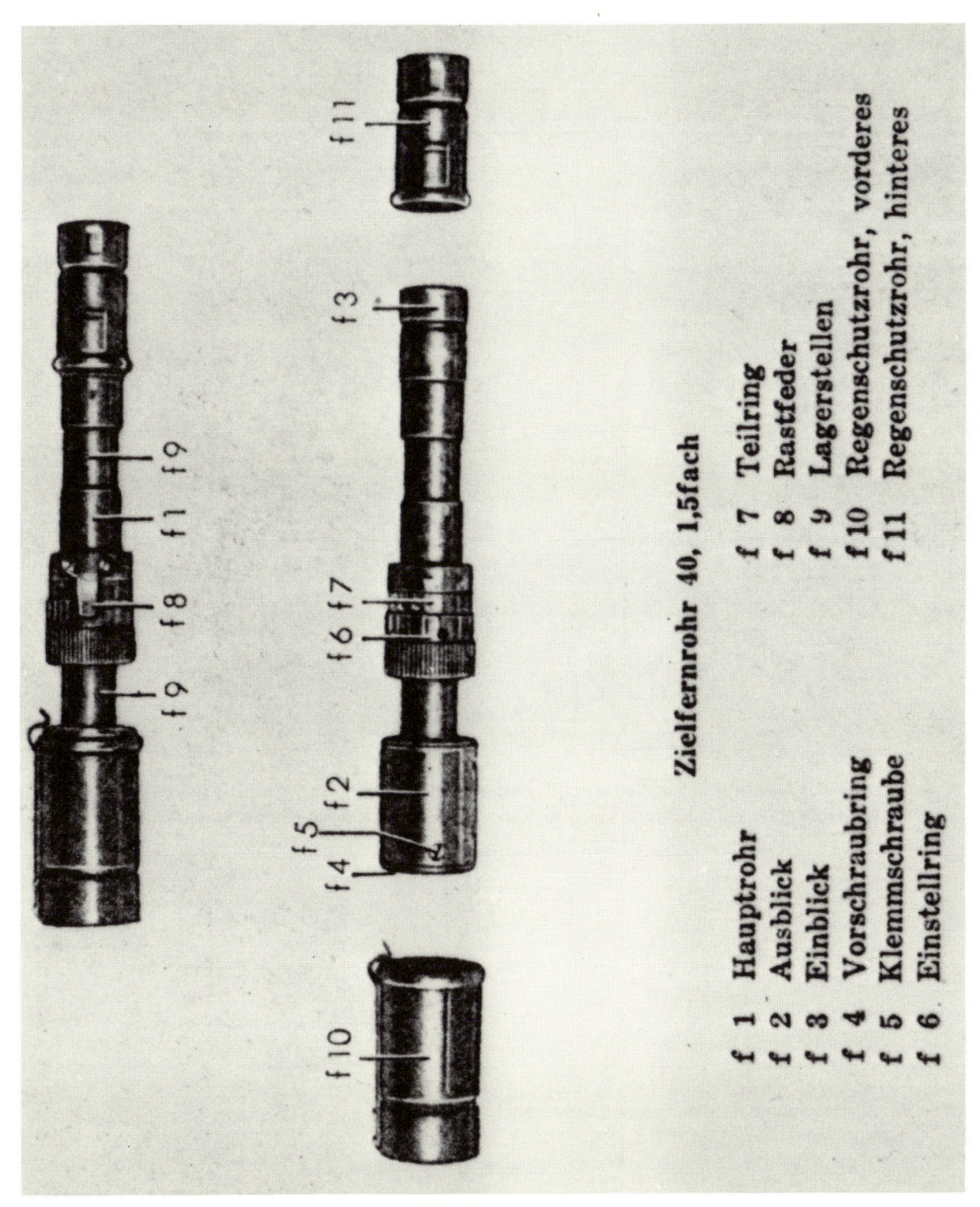

- f 1 Hauptrohr
- f 2 Ausblick
- f 3 Einblick
- f 4 Vorschraubring
- f 5 Klemmschraube
- f 6 Einstellring
- f 7 Teilring
- f 8 Rastfeder
- f 9 Lagerstellen
- f 10 Regenschutzrohr, vorderes
- f 11 Regenschutzrohr, hinteres

A view of the principal ZF 40 sight components as described in ordnance pamphlet D 191/1.

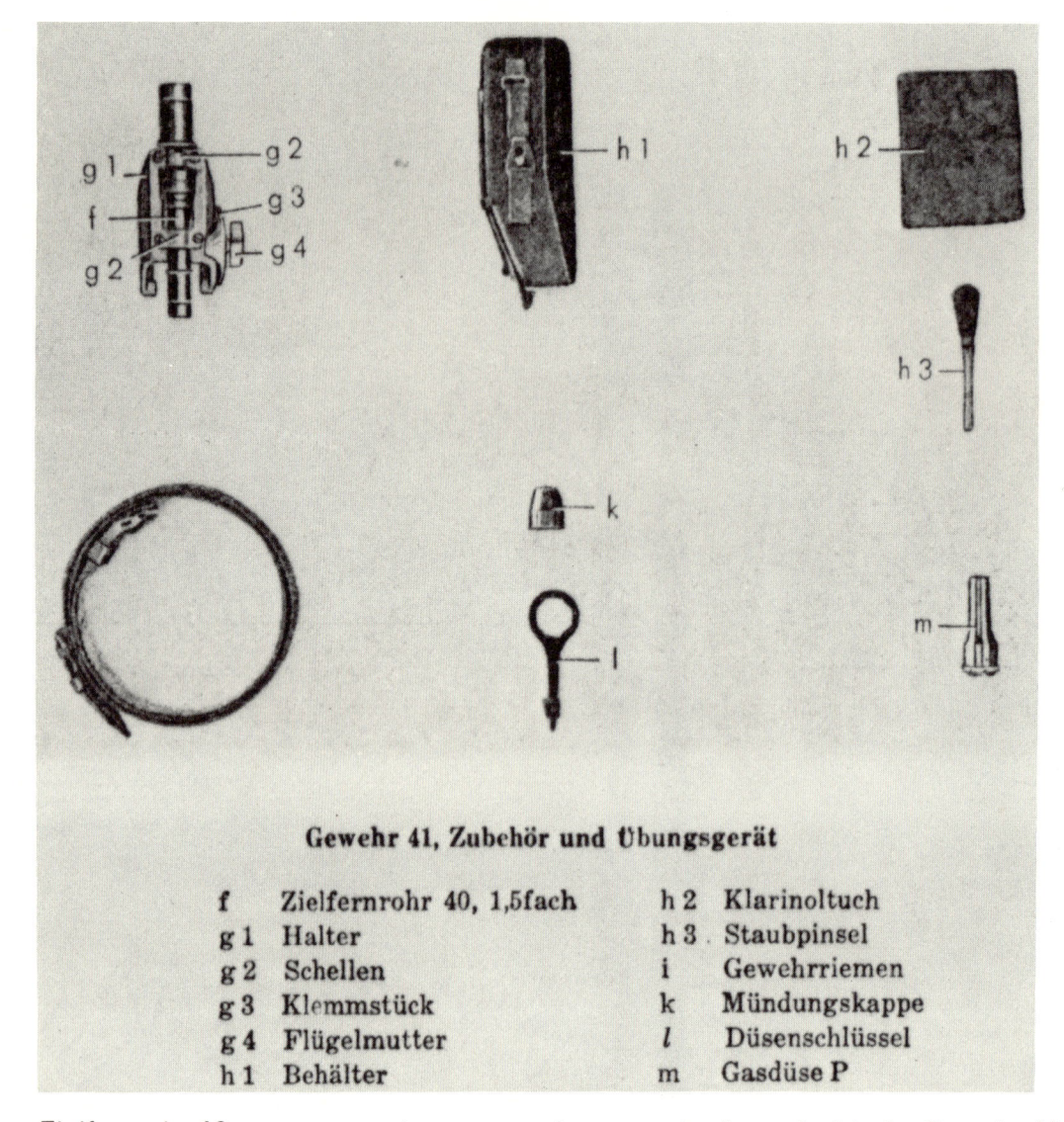

Gewehr 41, Zubehör und Übungsgerät

f	Zielfernrohr 40, 1,5fach	h 2	Klarinoltuch
g 1	Halter	h 3	Staubpinsel
g 2	Schellen	i	Gewehrriemen
g 3	Klemmstück	k	Mündungskappe
g 4	Flügelmutter	l	Düsenschlüssel
h 1	Behälter	m	Gasdüse P

Zielfernrohr 40, mount, carrying case, and accessories issued with the Gewehr 41 rifles. Note the configuration of the carrying case. None have surfaced thus far.

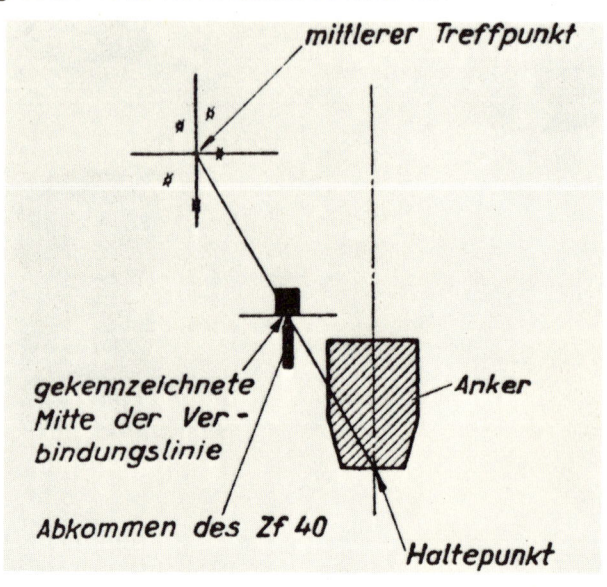

An illustration used in conjunction with an explanation for sighting-in the ZF 40 device with the G41 rifle as noted in ordnance pamphlet D 191/1.

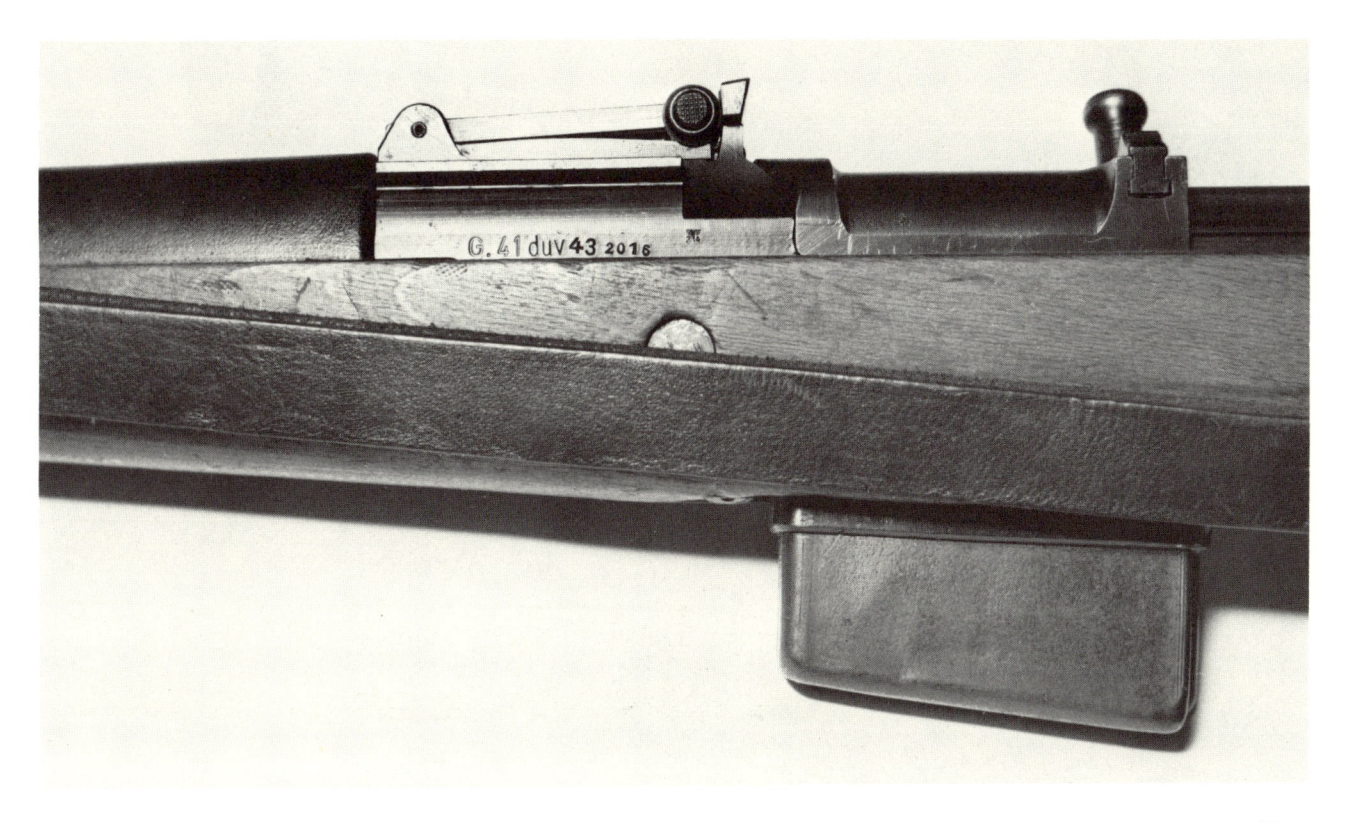

Berlin-Lubecker Maschinenfabriken (duv 43) Gewehr 41 rifle with ZF 40 scope mounting rails. The "214" Waffenamt stamp is typical of those found on "duv" production.

Gewehr 41 (duv 43) shown without the scope mounting rails.

Close view of the receiver markings on a Berlin-Lubecker Maschinenfabriken (duv 43) Gewehr 41 rifle with ZF 40 scope mounting rails.

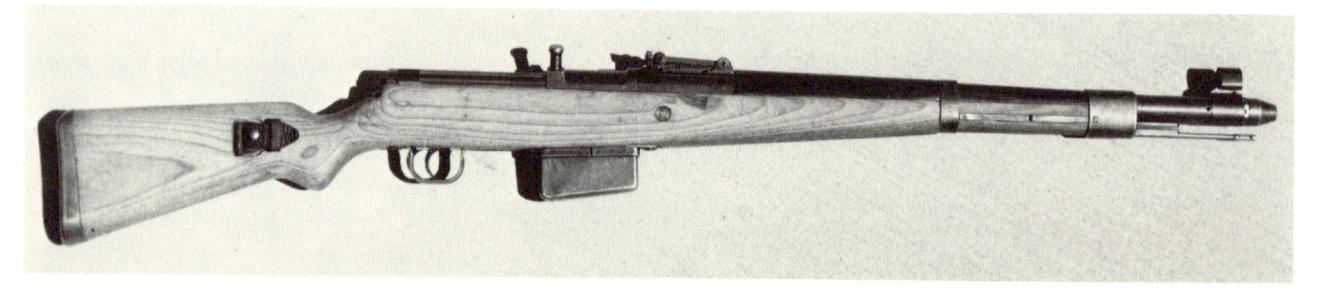

Despite the presence of scope mounting rails on many "ac" and "duv" G41 rifles, few were ever issued for sniping purposes. The weapon is a "duv 43" Gewehr 41.

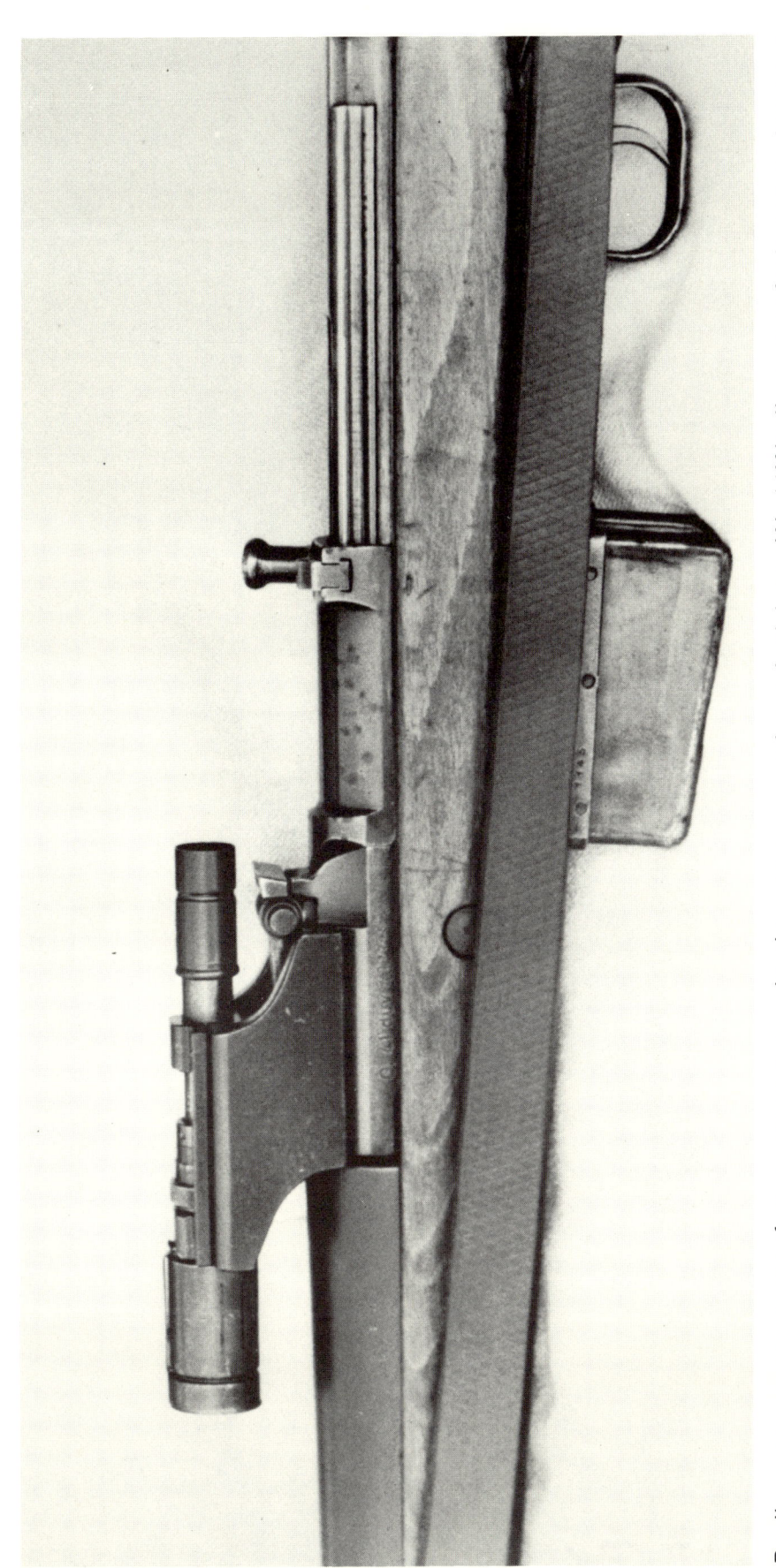

Replica, contemporary manufacture scope mount in place on a "duv 43" G41 rifle. No *original* (as first manufactured) ZF 40 telescopic sights or World War II mounts of this type are known to have survived the war.

Top view, contemporary manufacture "straddle-mount" for the Gewehr 41.

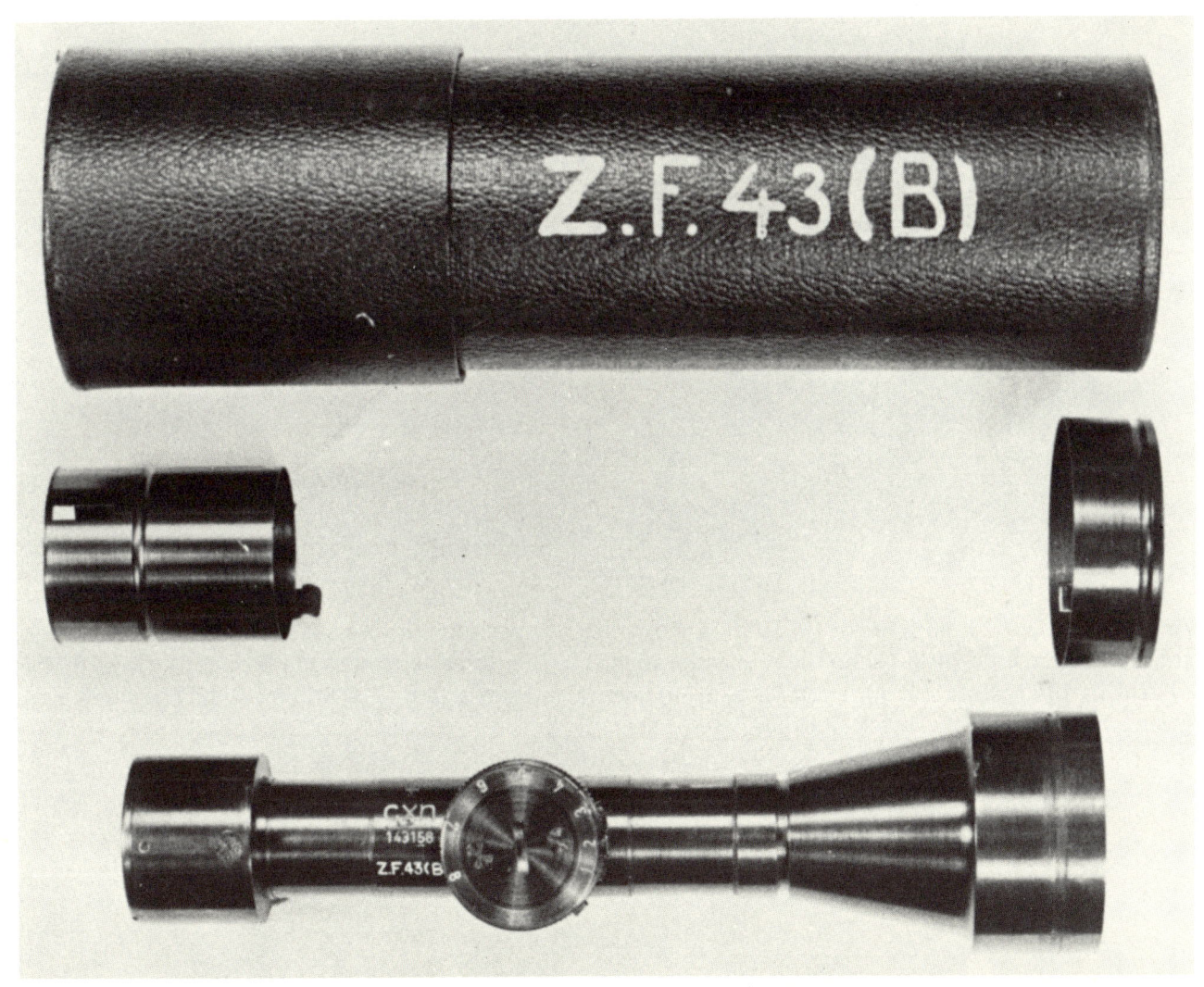

The ZF 43(B) telescope with original factory container. Fitted to a special mount which rested directly over the conventional rear sight, the eye-relief was cited as approximately 330mm. Note the rain shield (left) and "opaque" objective lens filter opposite. The unusually high serial number (143158) suggests that numbers were carried over from other scopes made at the "cxn" facility (Emil Busch A.-G.).

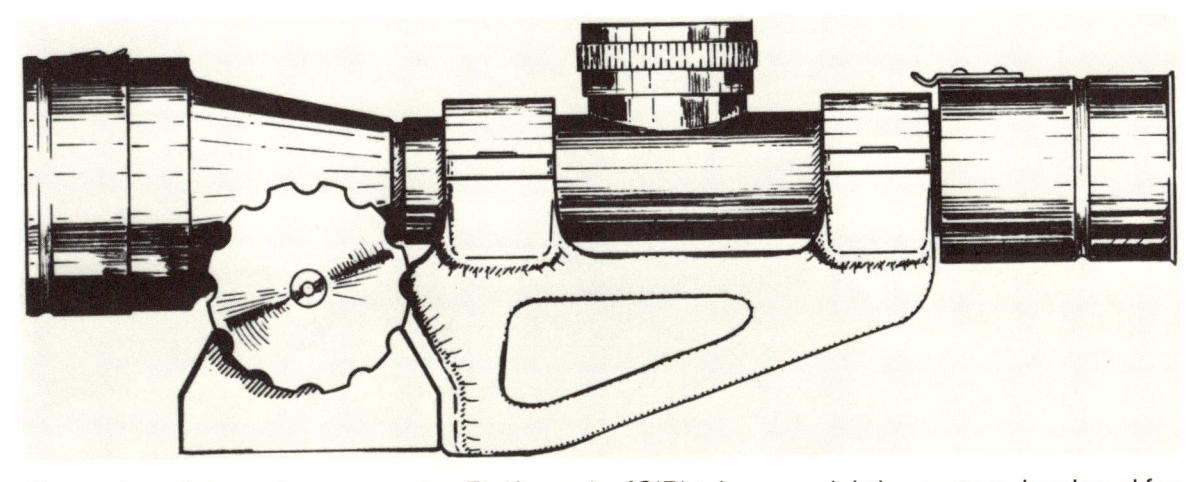

Illustration of the unique appearing Zielfernrohr 43(B) telescope sighting system developed for use with G41 series weapons. The large knob was rotated for windage adjustments. Copyright Niel Broky.

318   THE GERMAN SNIPER 1914–1945

Another 4-power ZF 43(B) sight (143161) is shown for comparison purposes. Although subjected to field trials, requirements for a "simplified" mass-produced semiautomatic rifle scope brought an end to this system in 1943. Note the elevation range drum graduated from 1 to 8 (100-800 meters). The sights were given a high quality blue-black finish to deter rust.

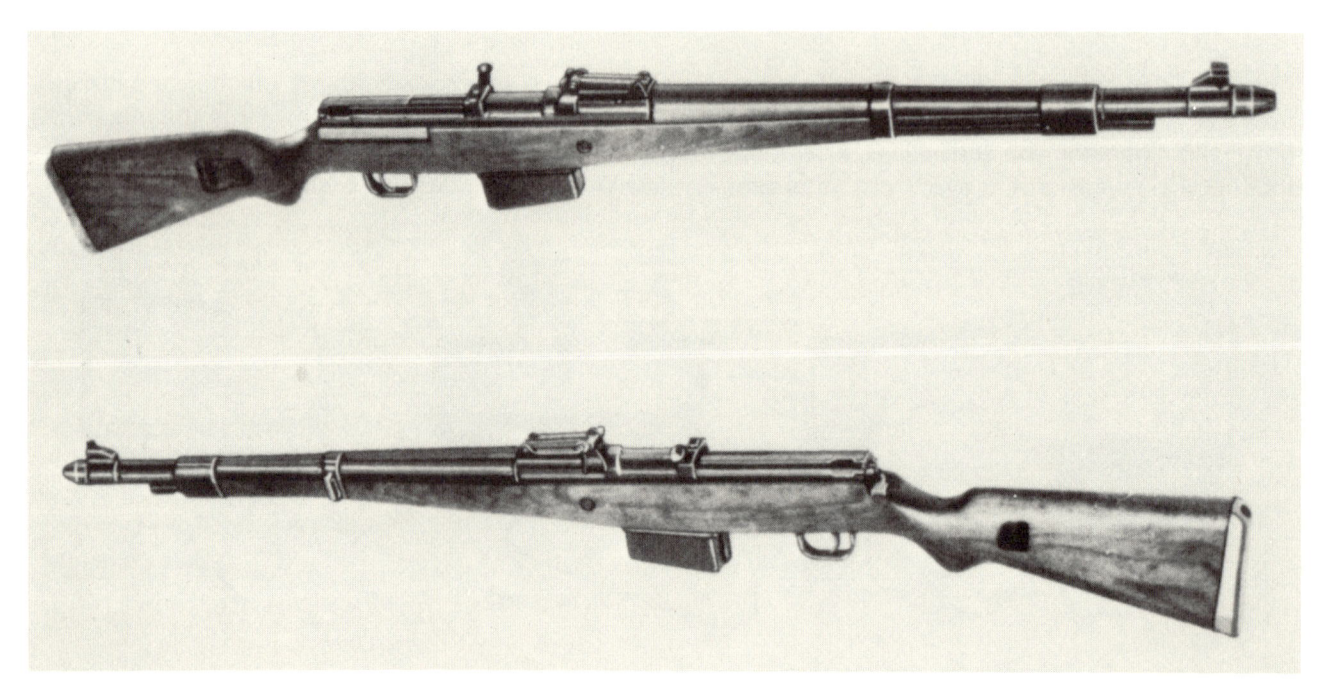

Original World War II German Ordnance illustration showing a right and a left view of a Gewehr 41 fitted with a dovetail receiver base for mounting ZF 4 sights.

SELBSTLADEGEWEHR 41

German Ordnance efforts to adapt a satisfactory rifle scope to Gewehr 41 rifles saw the use of G43 type dovetail receiver bases for mounting Gw ZF 4-fach sights (ZF 4). The rifle shown (ac 43) was fitted with such a base. Note the absence of a mount locking recess in the base, which is typical of those found on early Selbstladegewehr 43 rifles. The bolt slide assembly has the cocking handle positioned on the left side to provide clearance for the ZF 4 scope mounting.

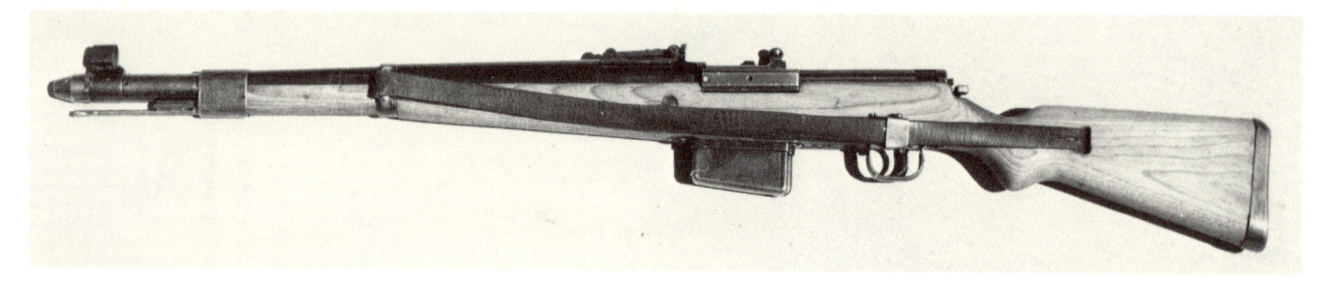

Walther manufacture Gewehr 41 (ac 43, serial no. 325a) with unusual telescope mounting base affixed to the left side of the receiver.

Held to the G41 receiver with solder and screws, the 88.09 mm long dovetail base bears the number 325. The lower front section of the receiver was machined to accept the base.

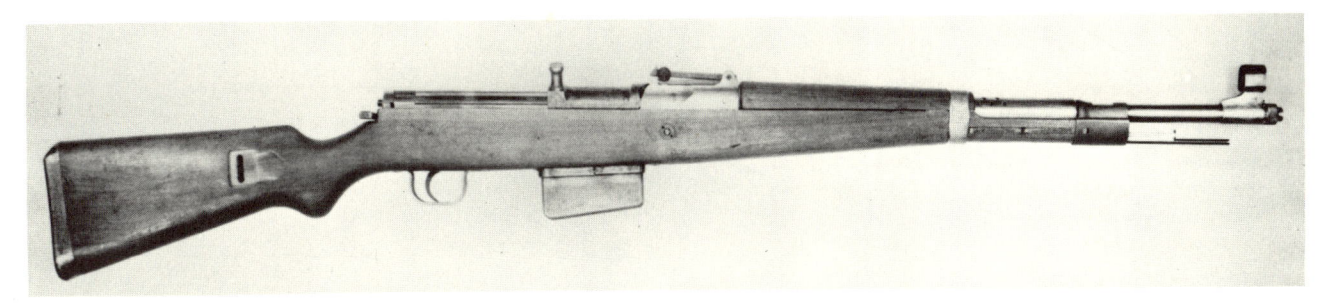

Experimental Gewehr 41 with transitional gas system. Although referenced as the "G41/43" it is not known if this was the official designation for this model. There are no markings on this rifle.

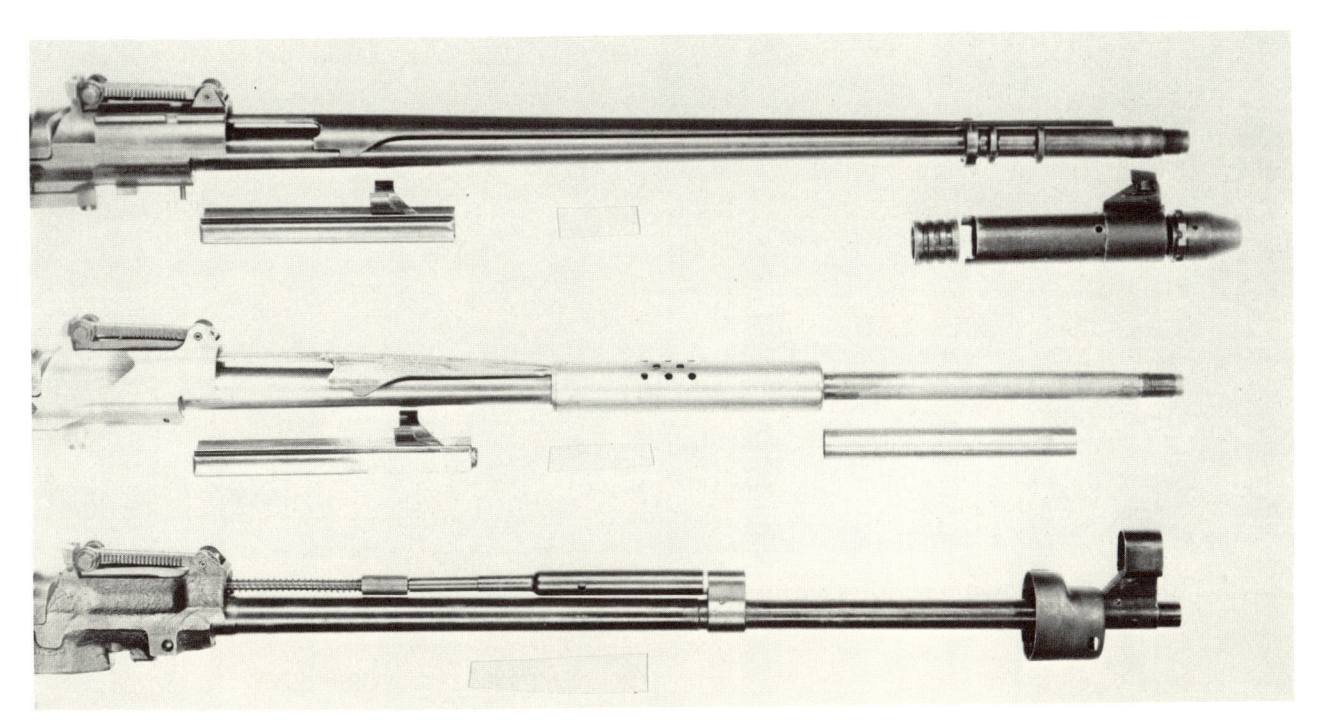

Comparison view of a typical Gewehr 41 gas system (top), the transitional model, and the G-K43 system as it evolved.

Canadian personnel shown test firing a captured German Gewehr 41 in Italy (1944).

CHAPTER XIV

Selbstladegewehr 43

From the very beginning, the Gewehr 43 semiautomatic rifle (Selbstladegewehr 43) was designed to be readily transformed into a sniping rifle by attaching a telescope assembly to an integral scope mounting base.

The G43 was equipped with a dovetail base (rail) located on the right side of the receiver. This feature was intended to permit optimum flexibility for fielding sharpshooting rifles by having two mass-produced, fully interchangeable components, a telescopic-sighting system and a weapon, come together following their separate manufacture to form a combat ready sniping rifle.

Even though this concept had originated with the Gewehr 41 series, the Selbstladegewehr 43 system was viewed by the German military as the ultimate solution to a problem that had confronted them since 1941.

When compared to the tedious, time-consuming practice of hand-fitting scope mounting components to the 98k, the idea of converting a rifle for sharpshooting use by merely sliding a complete scope assembly on to an existing receiver base was met with considerable enthusiasm.

So far as has been confirmed, the Selbstladegewehr 43 (Gewehr 43 and Karabiner 43) was manufactured by three German firms: Carl Walther Waffenfabrik, Zella-Mehlis; Berlin-Lubecker Maschinenfabriken, Werk Lubeck; and Gustloff Werke, Werk Weimar. In typical fashion, the origin of each weapon was indicated by a manufacturer's code (production mark) stamped on the receiver: "ac" "duv," and "bcd" for the respective firms.

The "qve" production mark, found on various K43 rifles manufactured in 1945 and reported as "Walther production" in numerous contemporary writings, has remained a subject of considerable controversy.

Both Walther and Berlin-Lubecker Maschinenfabriken were producing K43 rifles even as the Wehrmacht capitulated, with the latter firm responsible for a total of 15,000 pieces in the last month of the war. The presence of "ac 45" (Walther) marked K43 rifles bearing their typical "359" Waffenamt, the *total absence* of "duv 45" marked weapons even though this firm was in full production, and the presence of "qve" rifles bearing "214" Waffenamt which appeared on both weapons and hardware originating at Berlin-Lubecker Maschinenfabriken through the course of the war, supports the contention that the "qve" production mark was a replacement for the "duv" code and therefore represents rifles manufactured by Berlin-Lubecker Maschinenfabriken rather than by the Walther firm as cited in other sources.

Many noticeable variations were produced from early 1943 until the end of the war. From a standpoint of telescope mounting, the configuration of the receiver bases found on the vast majority of the G43 and K43 rifles remained virtually identical throughout production. Exceptions were those receiver bases with no scope mount latch recess machined into the rail as noted on experimental and early production

Gewehr 43 (G43) with 4-power Gw ZF 4-fach telescopic sight typical of those intended for German sharpshooter use during World War II.

Left view of Walther manufacture G43 (ac 44) semiautomatic sniping system.

G43 rifles.

Of parallel interest are variants without the scope mounting base thought to be the salvage from reclamation of the receivers that had been rejected for production defects (on the receiver base) of one form or another. At a time when weapon supply had become critical, as an expediency, it is believed that the defective rails were machined from the receiver leaving a functional component for use in the assembly of a standard semiautomatic rifle.

The telescope mount, a rugged steel casting, was machined to correspond with the dovetail form of the integral receiver base. To attach the telescope assembly to the rifle, the mount lever was rotated to the rear, the mount slid on to the rail with the lever, then pushed forward exerting pressure and thereby clamping the mount in place.

The locking lever, an extension of a screw positioned vertically through the center of the mount, is fitted to this screw with splines which allowed the lever to be adjusted to exert the correct amount of clamping force on the base. This design feature was intended to compensate for manufacturing tolerances in both base and mount.

To adjust for correct pressure, a small retaining clip above the lever is removed, the lever is reset on the splines as required, and the clip is replaced.

An auxiliary spring-loaded locking latch, independent of the pressure lever, engages the recess machined into the center of the receiver base, serving as an added measure of security to prevent the telescopic-sight assembly from falling from the rifle should the primary fastening device loosen or fail. In addition to rotating the pressure lever to the rear, it is necessary to depress the spring-loaded latch when the mount is removed.

Manufactured by Walther and Berlin-Lubecker Maschinenfabriken, Selbstladegewehr 43 telescope mounts bear the Waffenamt stamp "359" or "214," the numbers assigned to the inspection staff at each facility. Mounts made by Walther were the more common of the two.

The original slanted-strut scope mounting did not have the auxiliary locking latch as intended for the early G43 receiver bases without a corresponding machined recess. Regular production (parallel-strut) mount characteristics reflected minor dimensional differences, with the most obvious variance being the shape of the lower ends of the mounts; they were either rounded or square in form.

The Gw ZF 4-fach sight was held to the mount by two tempered steel spring bands working in conjunction with a screw and threaded dowel assembly for tightening purposes. Even though the bands proved more than adequate, a ridge located in the forward section of the mount cradle engaged a corresponding recess in the objective end of the telescope tube to prevent movement of the sight from the recoil action of the rifle.

The most commonly encountered scope retaining bands are of the continuous loop design with early variants spot-welded instead.

An additional modification, effected to simplify production or perhaps to allow the marksman greater access to the lever while wearing gloves, saw the elimination of the center bend in the pressure lever.

Rifle serial numbers were engraved or, to a lesser extent, stamped on the face of the mount and as such represented components issued under ordnance auspices for the G43 or the K43 system. Even though a fair quantity of unnumbered telescope mounts have been noted, the absence of a rifle serial number indicates that these were not issued for sniping purposes during the war. Regardless of the weapon system, all German sniping rifles and telescopic-sight assemblies were number matched prior to their issue.

The Gewehr-Zielfernrohr 4-fach (Gw ZF 4-fach) utilized on the Selbstladegewehr 43, when produced under carefully controlled conditions proved quite satisfactory. The sturdy lightweight scope eliminated the need for a complicated mounting system by having both elevation and windage adjustments made internally through the use of graduated drums located on the top and right side of the tube.

The scope itself, of unusual configuration compared to others of that era, was cylindrical at either end with the center section square in shape. The retaining bands wrapped around the cylindrical section with the adjustment drums positioned on the flats. The overall design, as noted in German Ordnance documents, was

Gewehr 43 (G43) with ZF 4 scope, rain shield, and variant leather lens caps.

Left view of G43 (ac 44) with ZF 4 scope and mount assembly.

Typical receiver markings from a Walther manufacture (ac 44) Gewehr 43. Both the method of marking and characteristics of the G-K43 series weapons were quite diverse.

to facilitate its rapid production.

Manufactured by Voigtlander & Sohn (ddx), Opticotechna (dow), and J. G. Farbenindustrie (bzz), the sights bore the manufacturer's production mark, the ordnance designation "Gw ZF 4," a white, green, or blue triangle to indicate the reliable function limits of each device in various climates, and a production serial number. Although the aforementioned markings appear the most frequently, specimens have also been noted bearing a "G43," "K43," "ZFK 43," and "KZF" designation as well.

Internal scope adjustments were made using the elevation range drum, which was graduated from 100 to 800 meters in 50-meter increments with approximately half-minute clicks and located on the right. The windage drum, located on top of the tube, was protected by an easily removed stamped sheet metal cover of which there are two distinct variations.

The most common reticle pattern consisted of the pointed vertical post with horizontal side bars. A limited number of ZF 4 sights, however, were also produced with a crosshair pattern as well as with a single vertical post having either a pointed or blunt end.

Of further interest are the unusual "Bu" marked ZF 4 scopes with a unique ranging style reticle pattern in place of that normally found in sights of this type.

When originally issued, the ZF 4 sighting system included a metal rain shield for the objective, rubber eye-guard, leather lens covers, cleaning cloth, amber-colored (haze) filters, clear glass disc to prevent fogging of the eye-piece and carrying case for holding the sight when removed from the rifle. There are at least three known variations of containers: wood and fiberboard with metal fittings, formed metal, and bakelite with fittings of metal. The latter variant is seldom encountered.

Despite an inordinate amount of production problems, the ZF 4 sight was considered both practical and efficient by all but the most proficient German snipers who, equipped with limited numbers of 6-power scopes, had compiled impressive scores with consistent results out to 1000 meters. These snipers had pressed for adoption of a sight possessing at least 6-power; ordnance studies, however, had indicated that the most effective sniping was conducted in the 400-600 meter range well within a good 4-power scope's capability. Consequently, higher magnification was never a serious consideration and the ZF 4 was considered more than adequate from that standpoint.

Late war efforts intended both to improve on the ZF 4 design and to supplement lagging production saw the development and introduction of the ZFK 43/1, 4-power telescopic sight brought to fruition by the Carl Zeiss firm (Jena) during the latter part of 1944.

Optical characteristics were cited as being the same as the ZF 4, however, the elevation adjustment drum, graduated from 100 to 600 meters in 50-meter increments, was positioned on top of the sight, while the windage drum, with graduated values of 2.5cm at 100 meters, was located on the right.

The new sight was cited as a departure from the "sheet metal design" and a return to the "more practical turning and milling method of telescope construction," an obvious reference to the production difficulties encountered with the sheet metal tube form of the Gw ZF 4 device.

Designed to fit the standard Selbstladegewehr 43 telescope mount with an eye-relief of approximately 80mm, the ZFK 43/1 was provided with a focus adjustment which permitted the marksman to adjust the sighting picture to suit his vision by means of a pressed metal drum located on the left side of the tube. An additional improvement was the inclusion of a glass reticle, the vertical post and side bars etched directly on to a glass plate, thus eliminating what had been a continuing problem source with the ZF 4 system.

Even though the overall design of the ZFK 43/1 had been established, and pilot production was in full swing early in 1945, a decision regarding the application of T-Schutzes (T-Protection), the use of magnesium fluoride coated lenses to increase light transmission and contrast of image, had not been made; the coated lenses were considered an impediment to fabrication of the sight.

Acknowledged as the most efficient of all German military telescopic rifle sights produced during World War II, ZFK 43/1 sights bear the

Early production Gewehr 43 integral receiver rail (base) without a machined recess as originally intended for the first ZF 4 telescope mounts (slanted-strut) that did not have an auxiliary spring-loaded locking latch. This rifle is a "bcd" variant.

An integral receiver rail with machined recess typical of those found on the vast majority of both G43 and K43 semiautomatic rifles. In this case a K43 (ac 45) made by Walther.

late war "rln" production mark assigned to the Carl Zeiss firm.

The ZFK 43/1 sights were produced in only limited quantities before the war ended. Despite considerable manufacturing difficulties, experts contend that approximately 150,000 ZF 4 sights (regular production) were produced from October 1943 through April 1945 with Voigtlander the principal source followed by Opticotechna and J.G. Farbenindustrie, the latter responsible for only a few thousand scopes at best.

The unique and relatively rare K98k "swept-back" mount sniping variation while hardly a part of the semiautomatic sniping system, also made use of the ZF 4 telescopic sight. Its origin can be traced to an official proposal from the Infantry Branch (Jn 2) to the Army High Command (OKH) under date of 22 July 1944. In anticipation of a readily available sighting system, the ZF 4 was recommended for general sharpshooter use with the 98k and with the K43. This recommendation would not see affirmative action until September 1944 when, as a result of ZF 4 production exceeding the availability of K43 rifles suitable for sniper use, it was decided to authorize use of the ZF 4 with the 98k "due to the great demand for telescopically sighted rifles at the front."

According to additional reference, the 98k in this particular configuration did not reach combat ready status until February 1945, at which time both the ZF 4 and the ZFK 43/1 were to be fitted to appropriate rifles as circumstances dictated.

As adapted to the 98k, the one-piece cast steel mount, very similar to that utilized on the K43, has a long dovetail slot which slides on to a base of mating configuration affixed to the left side of the receiver. A pressure lever and spring-loaded locking latch hold the mount securely in place, with forward and rearward motion tightening or loosening the mount as required, a method virtually identical to the mounting system employed with the semi-automatic sniping rifles. The mount was swept-back primarily to achieve proper eye-relief.

Sniping rifles of this type have been noted based on weapons manufactured by Sauer and Mauser. While the total quantity is acknowledged to have been quite small, the actual production reached or the extent of combat use is not known.

Original German Ordnance photograph made during comparative testing of a ZF 4 equipped MP 43/1 assault rifle and Gewehr 43 in October 1943 under the auspices of the Infantry School at Doberitz-Elsgrund.

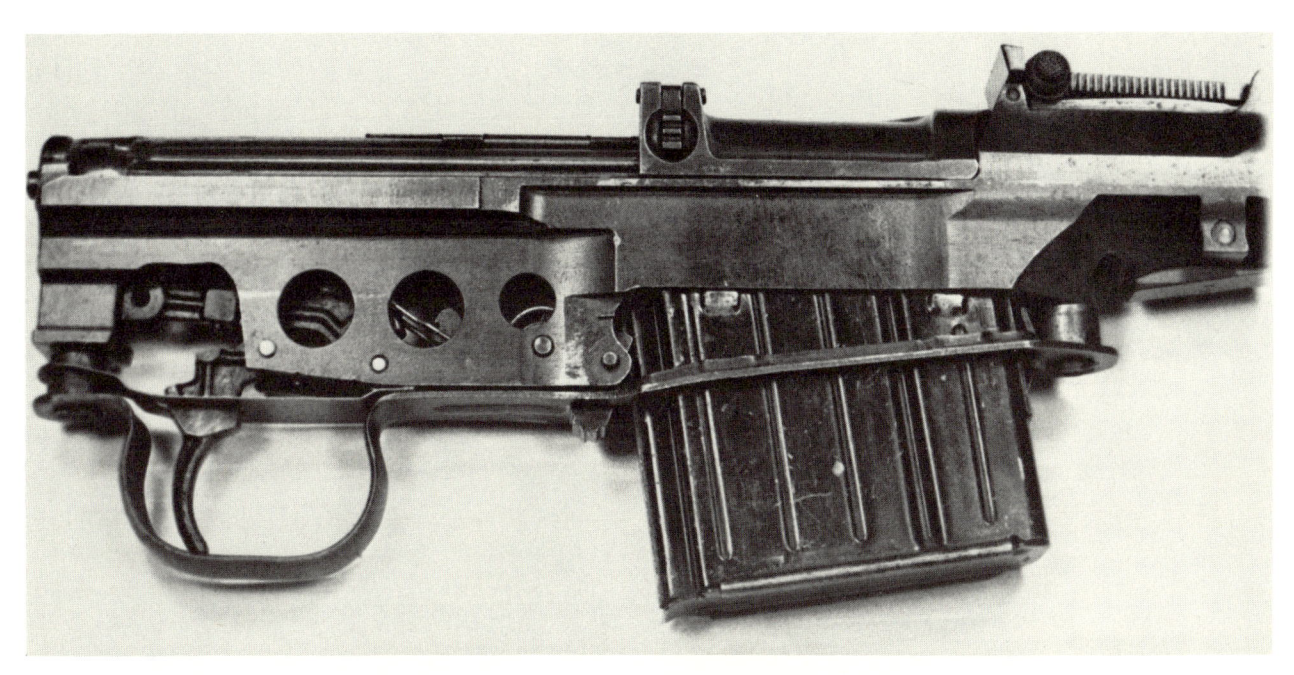

Rifles having the scope mounting base machined (removed) from the receiver are believed to represent late war efforts to reclaim receivers that had been previously rejected for production defects (rails) of one form or another and thus permitted the assembly of an otherwise functional weapon.

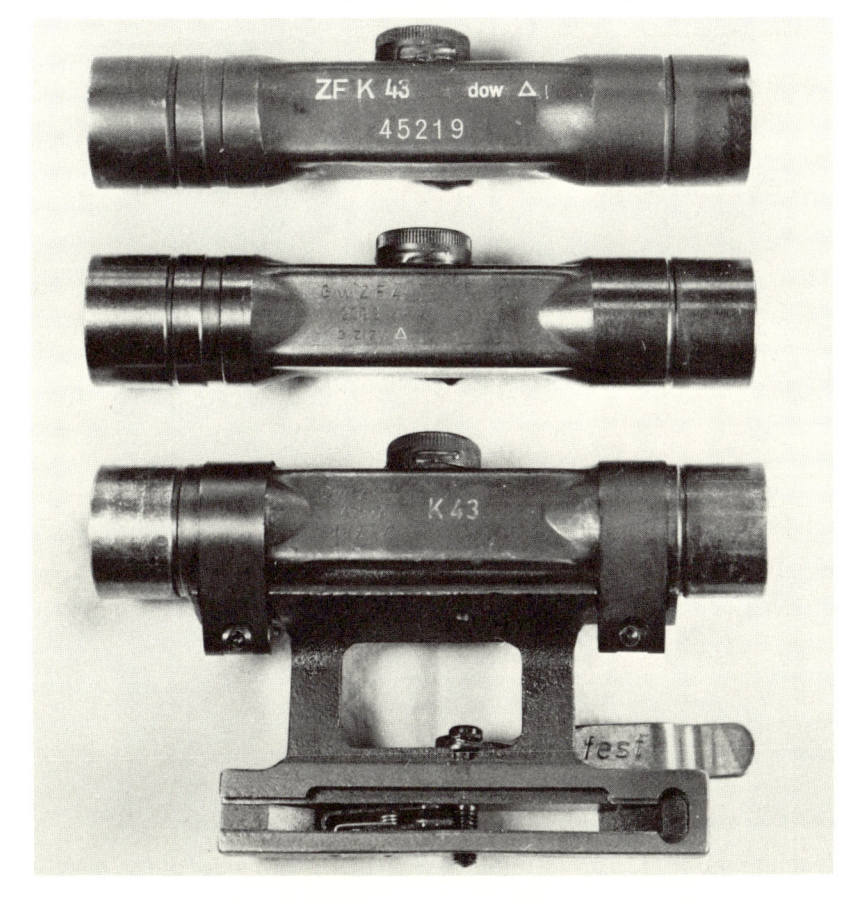

Referenced as the "sheet metal" design in World War II German Ordnance documents, Gewehr-Zielfernrohr 4-fach (Gw ZF 4) sights were manufactured by Voigtlander (ddx), Opticotechna (dow), and J. G. Farbenindustrie (bzz).

The reverse or receiver side of a typical ZF 4 telescope mount showing the method of machining, main pressure lever screw, and auxiliary spring-loaded locking latch assembly.

Walther manufacture ZF 4 telescope mount (359 Waffenamt) with a detailed view of the pressure lever, retaining clip, and auxiliary locking latch which engaged the recess machined into the center of the receiver base (rail). The lever was turned to the rear and the latch depressed to remove the scope from the rifle.

Early issue Gewehr 43 telescope mount with "center-bend" pressure lever (top) and the common variant (below).

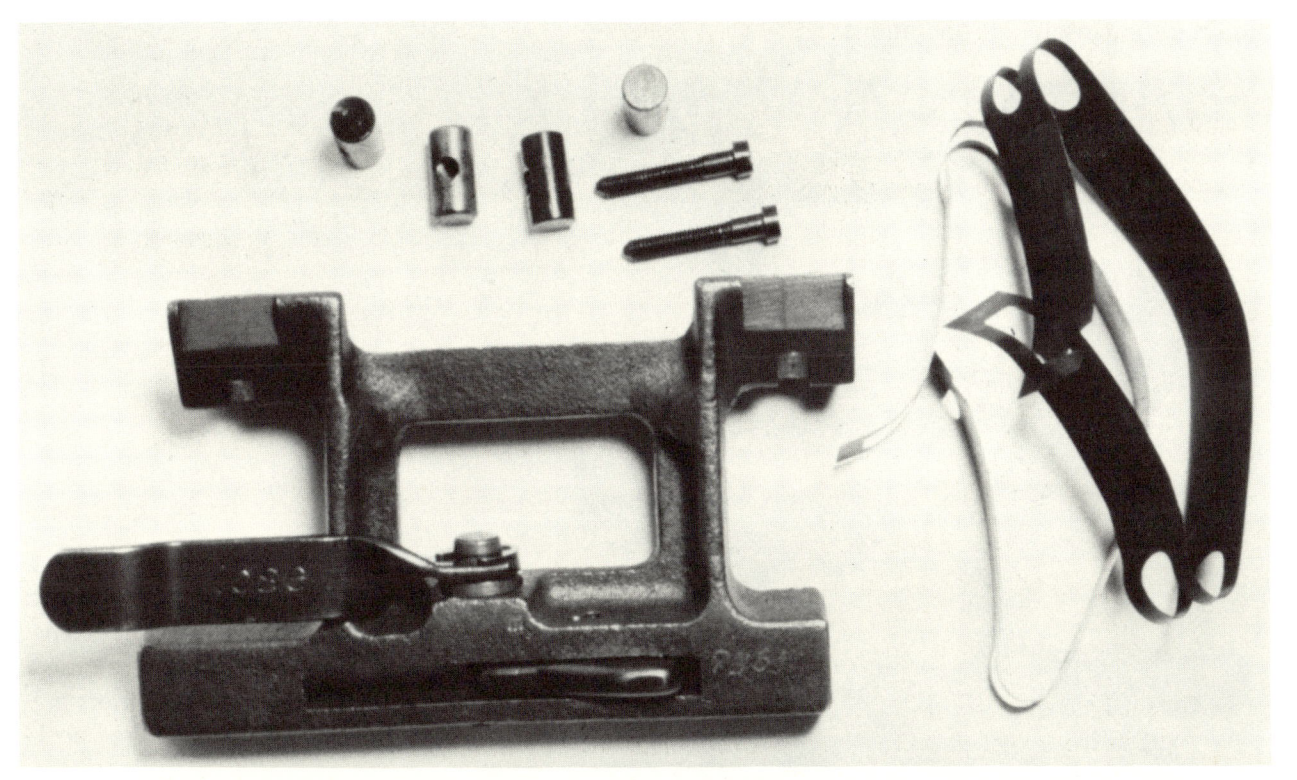

Tempered steel spring bands with screw and dowel assembly used for holding the ZF 4 sight to the mount.

Close view of tempered steel spring band and dowel assembly attached to the ZF 4 scope and mount. The small machined recesses in the vertical struts and beneath the scope cradle were used to index the mount during machining operations.

Although ZF 4 scope mounts reflect minor dimensional variances, the most obvious difference rests with the shape of the lower ends of the mount, rounded (top) or square. Both pressure levers are turned to the "lose," loose or free position.

A close comparison of the round and the square form ZF 4 telescope mountings. Both mounts bear the "359" Waffenamt.

The ZF 4 windage adjustment drum, located on top of the tube, was protected by an easily removed stamped metal cover that has two distinct variations. One, with slanted locking track approximately 12.70mm in length, the other about 9.52mm long with a slightly steeper angle. With few exceptions, ZF 4 scopes were given a blue-black finish for protection from the elements.

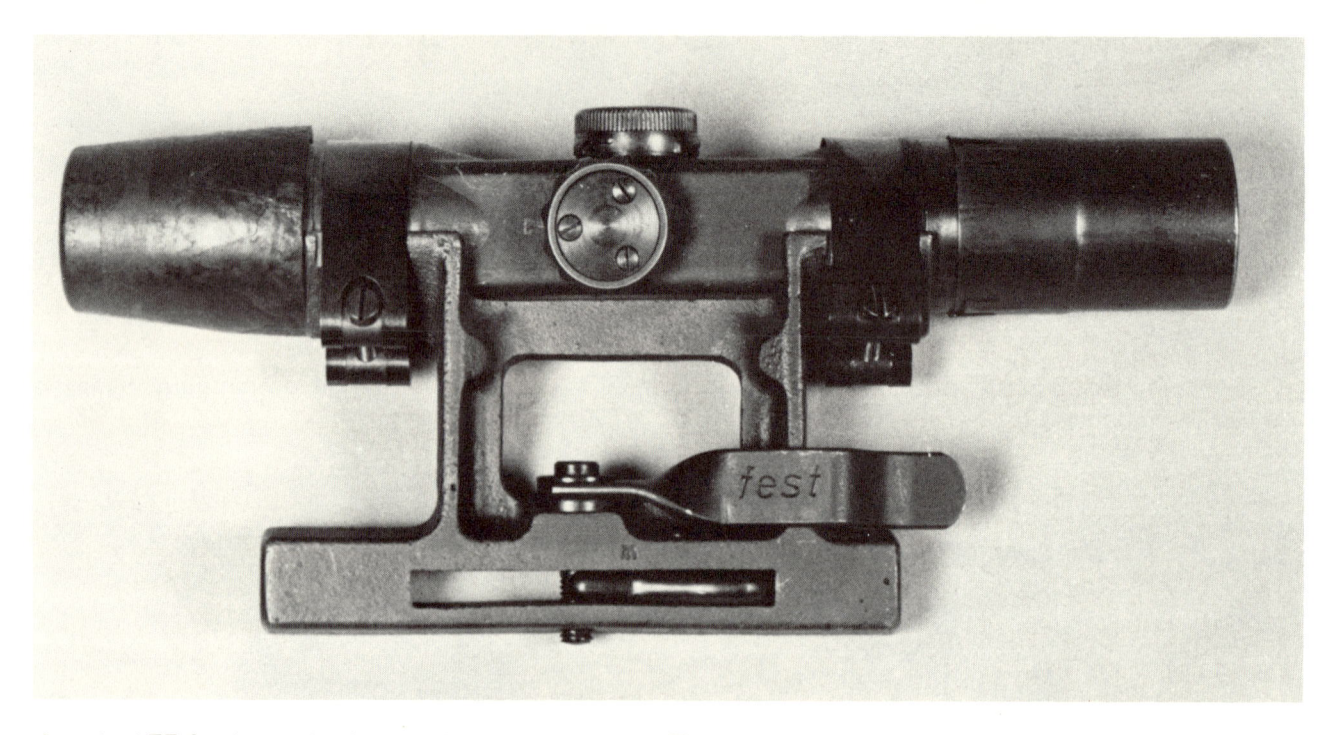

A typical ZF 4 telescopic sight and mount assembly. The scope carries a "reverse-taper" rubber eye cup believed to be the first style issued with the ZF 4.

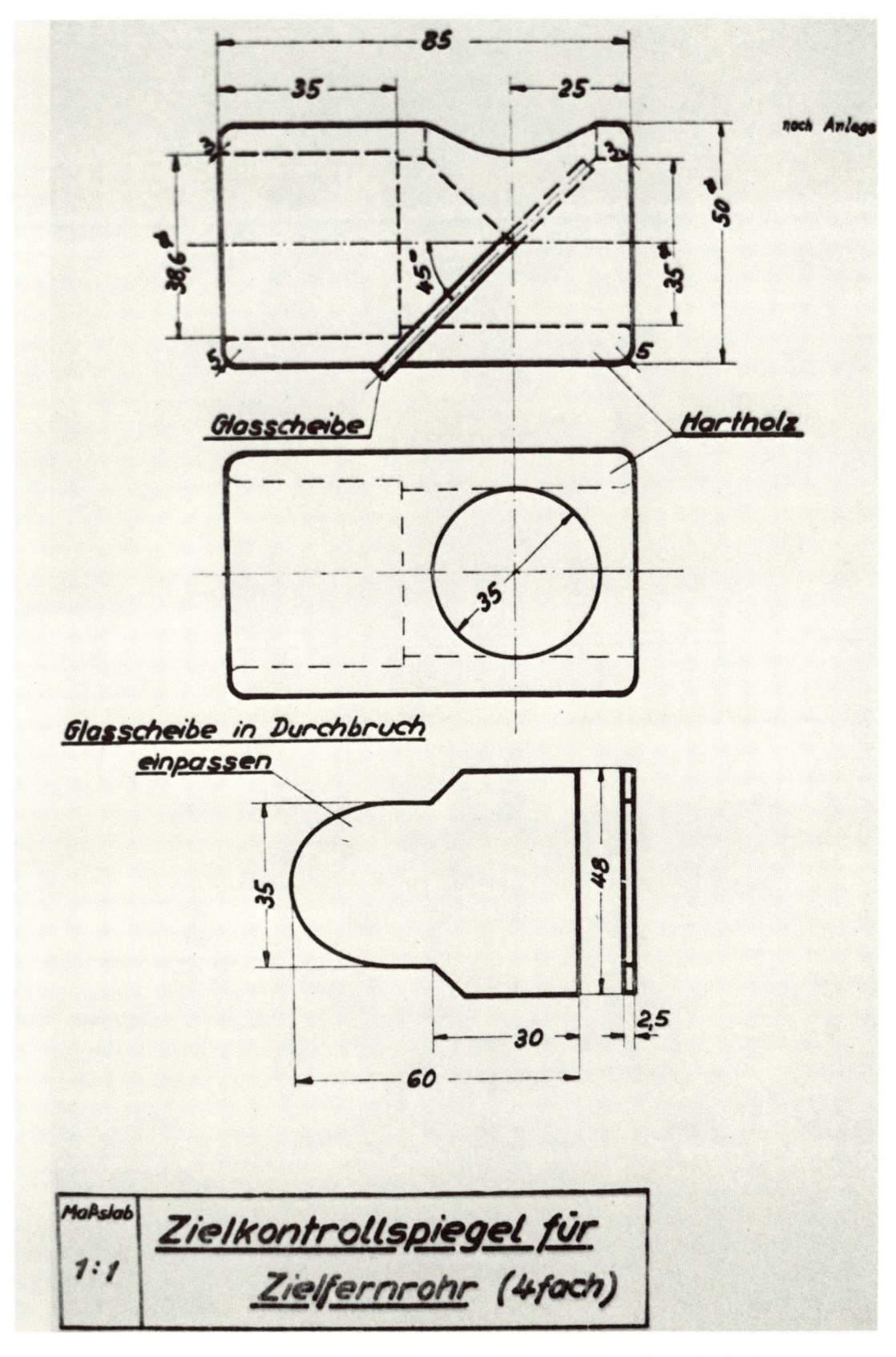

Mirror device employed for sharpshooter training purposes allowed the instructor to monitor the student's aim when sighting through a ZF 4 telescopic sight.

Unique view of Gw ZF equipped Gewehr 43 semiautomatic rifle, Circa October 1943.

Gewehr 43 with typical Gw ZF 4 telescope and mount assembly. The pressure lever is turned to the "fest" fast or tight position.

SELBSTLADEGEWEHR 43

Left view of G43 and ZF 4 sight assembly. Although referenced as a rain shield, the metal device was also used to minimize chance reflections from the objective glass. The rubber eye-guard was cited as a means to protect the shooter and to prevent damage to the sight should it recoil against the steel helmet.

ZF 4 telescope mount manufactured by Berlin-Lubecker Maschinenfabriken (duv). Note the "214" Waffenamt that invariably appears on "duv" hardware. Although ZF 4 mounts were produced by both Walther and Berlin-Lubecker Maschinenfabriken, the Walther variants with their characteristic "359" Waffenamt are encountered more frequently.

Berlin-Lubecker Maschinenfabriken G-K43 series telescope mounting. Note the "214" Waffenamt.

Walther manufacture ZF 4 telescope mount (359 Waffenamt). Note the rifle serial number which was placed on the mount (engraved or stamped) as issued for sniper use. While inconclusive, evidence suggests that rifle numbers were also placed directly on the scope tube (painted or engraved) as an alternate measure. In any case, the presence of such markings is not above suspicion.

Butt stock instruction manuals issued with both G43 and K43 rifles were revised in accordance with production modifications. Actual size: 101.60mm x 146.05mm.

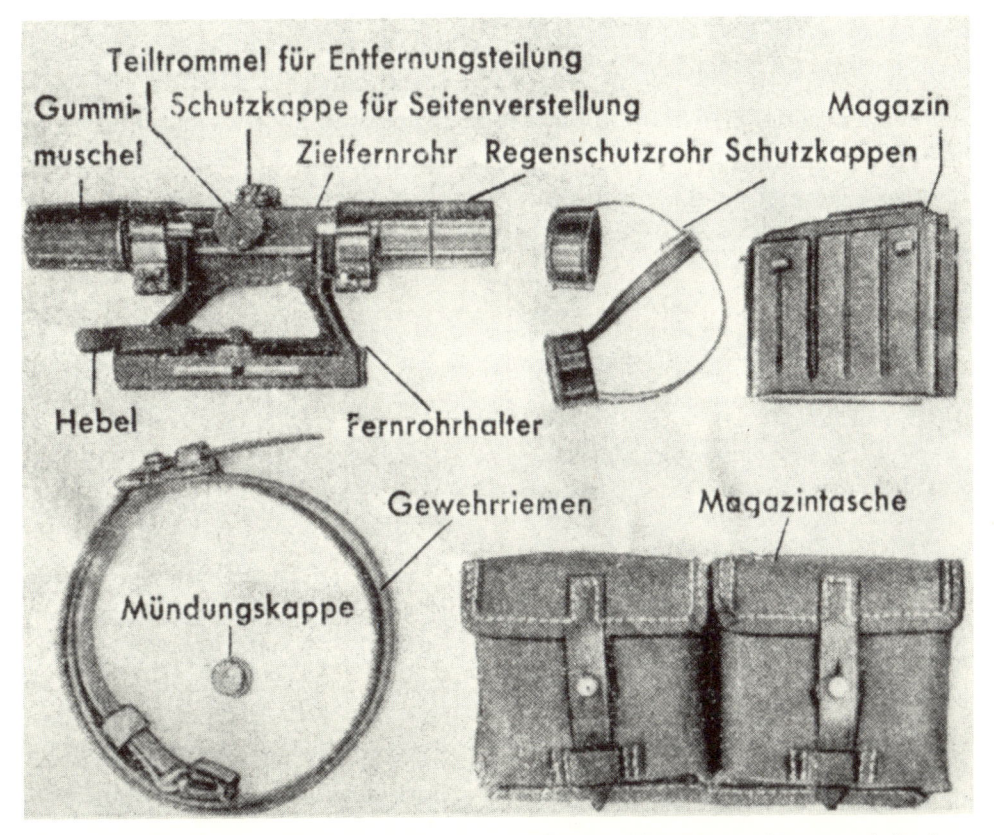

German Ordnance photo from an early Gewehr 43 butt stock manual depicts the original ZF 4 slanted-strut mount which did not have the auxiliary spring-loaded locking latch. Believed to be a prototype design that did not reach production, this illustration remains the only known evidence of the early G43 telescope mounting system.

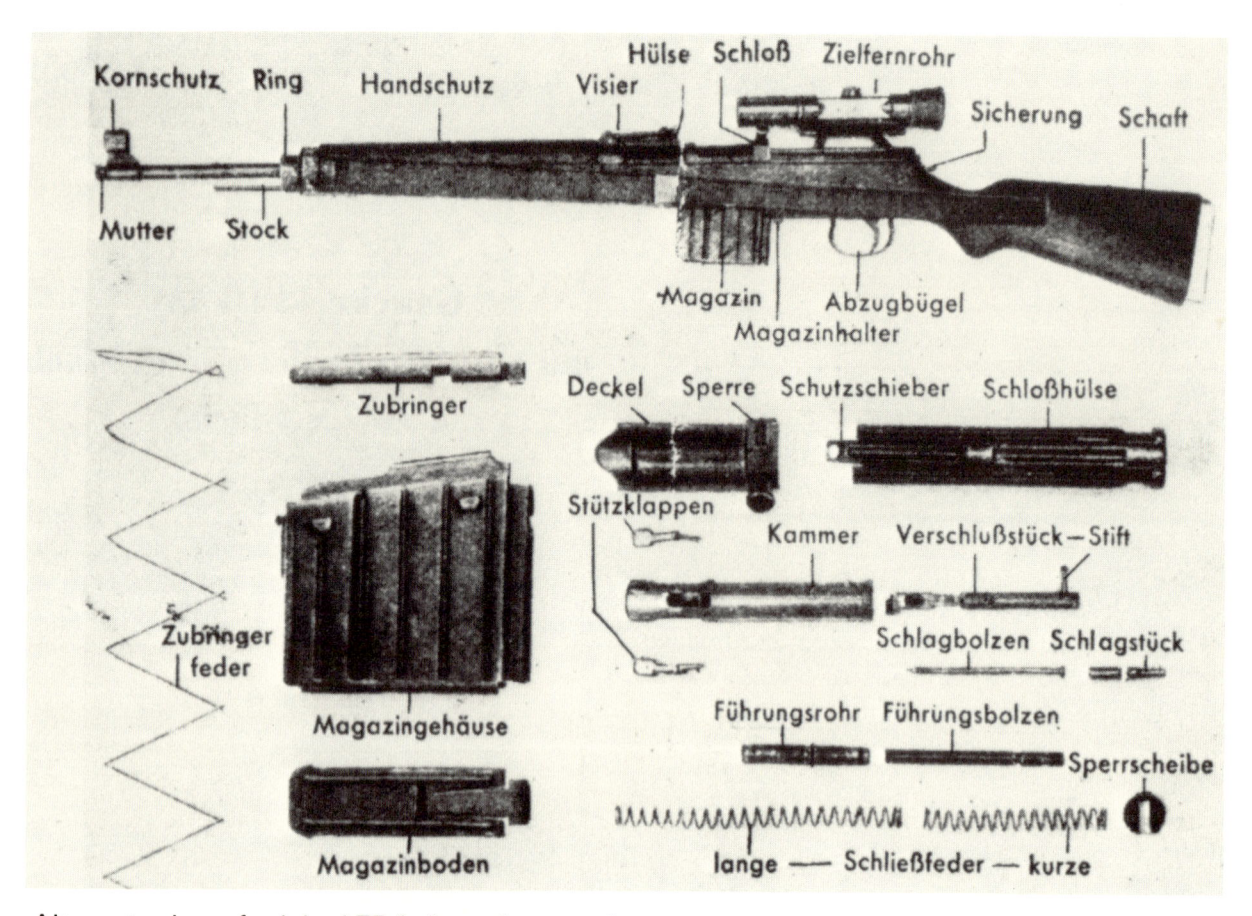

Alternate view of original ZF 4 slanted-strut telescope assembly shown with Gewehr 43 bolt and magazine components.

A butt stock illustration showing a rifleman inserting a magazine into the scope-sighted semi-automatic rifle.

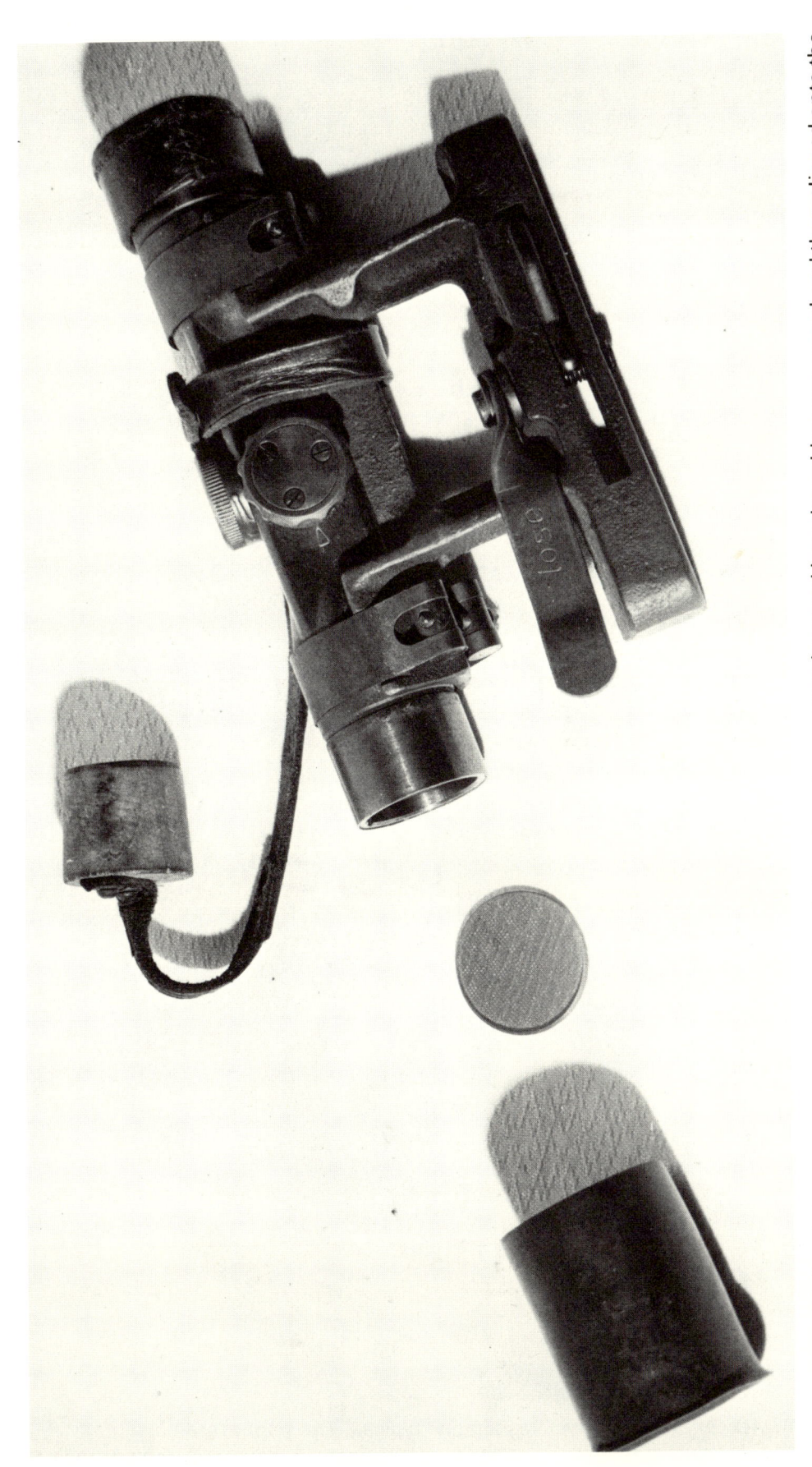

A part of the ZF 4 sighting system an amber-colored glass insert (disc), was provided for use on hazy days to improve vision. A clear glass disc was also included to prevent eye-piece fogging. The disc was inserted into the rubber eye-guard and then slipped onto the scope tube. Note the wooden plug, a simple form of protection for the ocular lens.

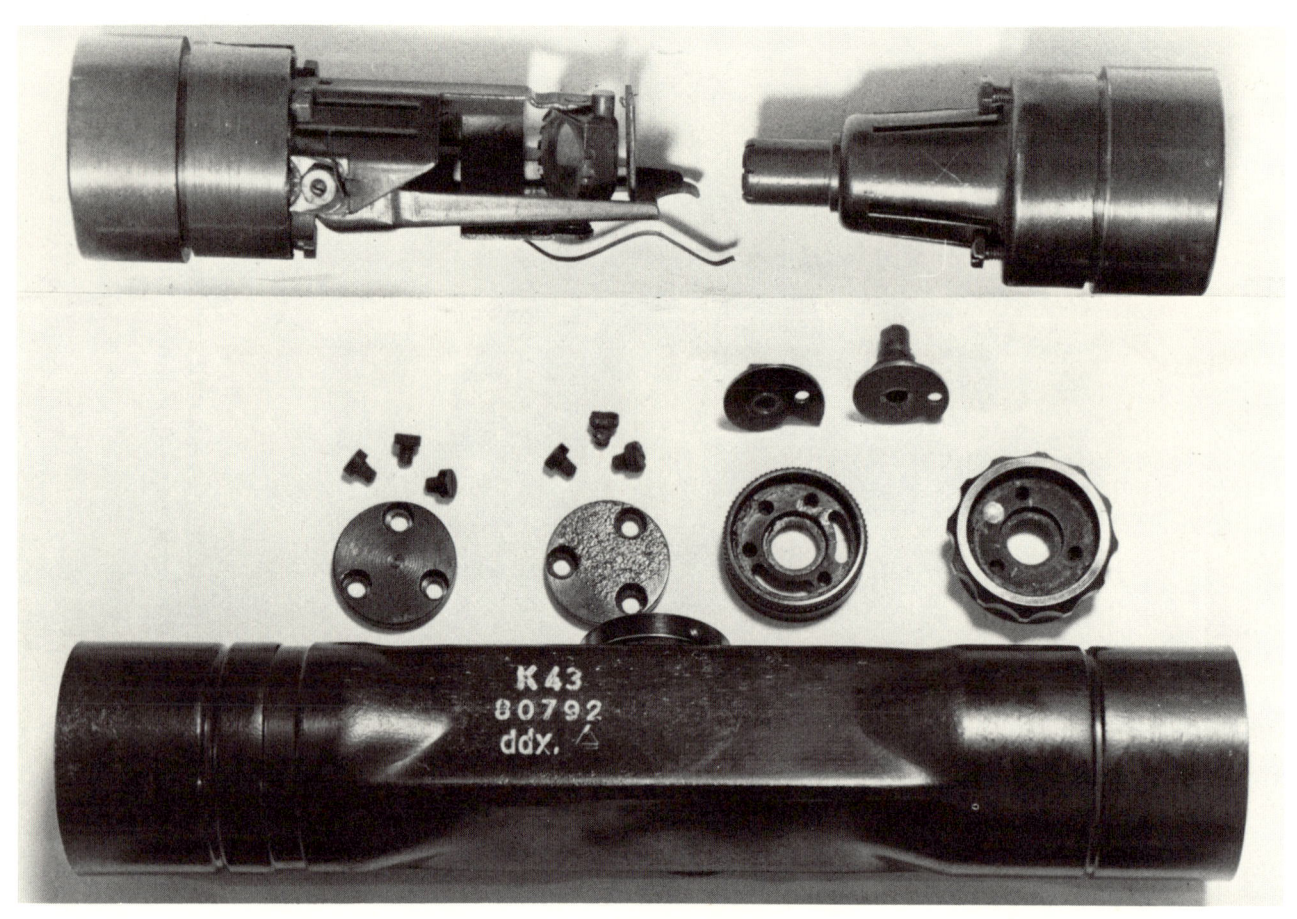

Disassembled ZF 4 (ddx, K43) showing principal components: adjustment drums, cover plates, objective lens, reticle assembly (center), and ocular (exit lens) assembly. Objective and ocular lenses were spin-crimped in their housings and not readily removed.

Close view of Gw ZF 4 reticle assembly (typical). The haphazard placement of the vertical post and side bars proved to be a constant production problem and a major field complaint.

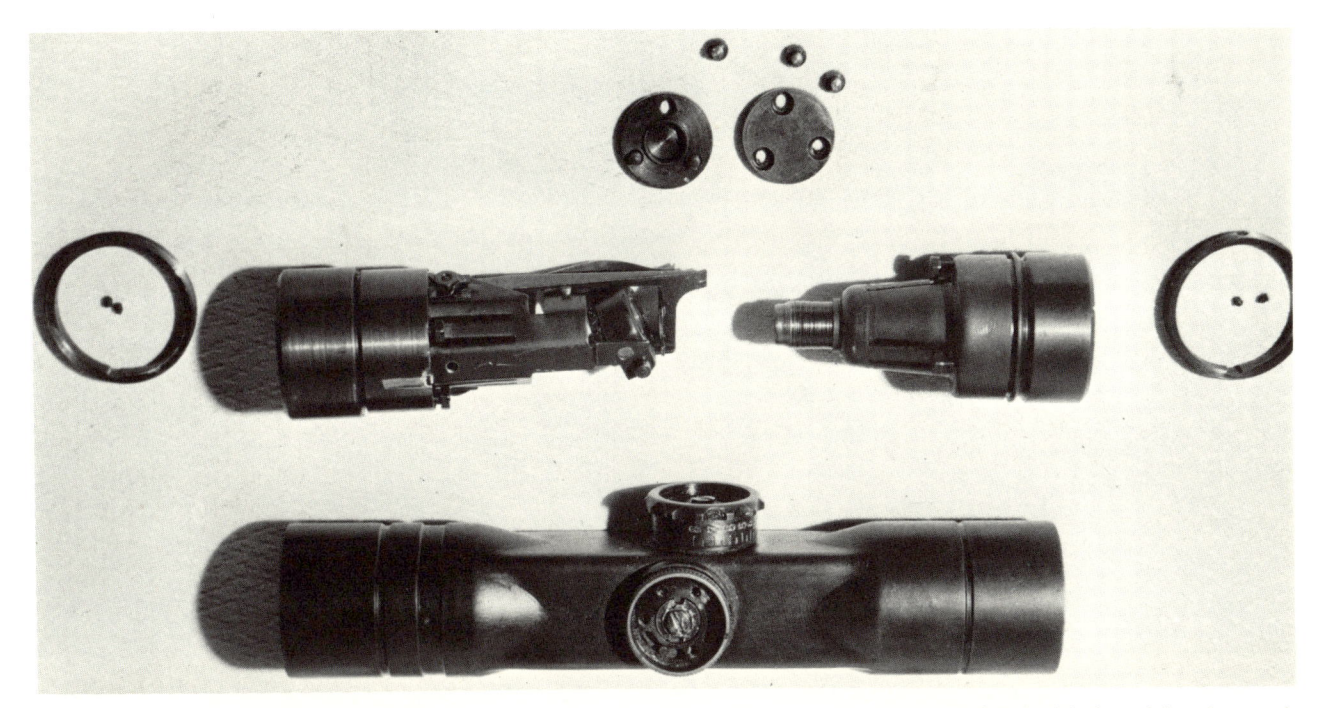

An alternate view of a disassembled ZF 4 sight showing the retaining rings which hold the objective and ocular assemblies in the tube. Most ZF 4 scopes have a retaining ring seal made from a material similar to plastic. Note the small set-screws (2) by each ring.

The most common Gw ZF 4 reticle pattern consisting of a pointed post with horizontal side bars. Sights were also produced with a cross-hair pattern, which proved too susceptible to shock, and a single vertical post having either a pointed or blunt end. The latter was deemed better for firing in subdued light.

Early Voigtlander manufacture (ddx) Gw ZF 4 telescopic sight (11545) showing typical scope markings.

Voigtlander Gw ZF 4 telescopic sight (37196) with variant "K43" marking.

Late manufacture "ddx" code ZF 4 sight (84779) bearing the "K43" designation. The Voigtlander firm was the principal source of ZF 4 production.

An early production Opticotechna (dow), Gw ZF 4 sight (12062).

Among the very last produced by Opticotechna (dow), the ZF 4 sight (49680) has a "ZF K 43" designation.

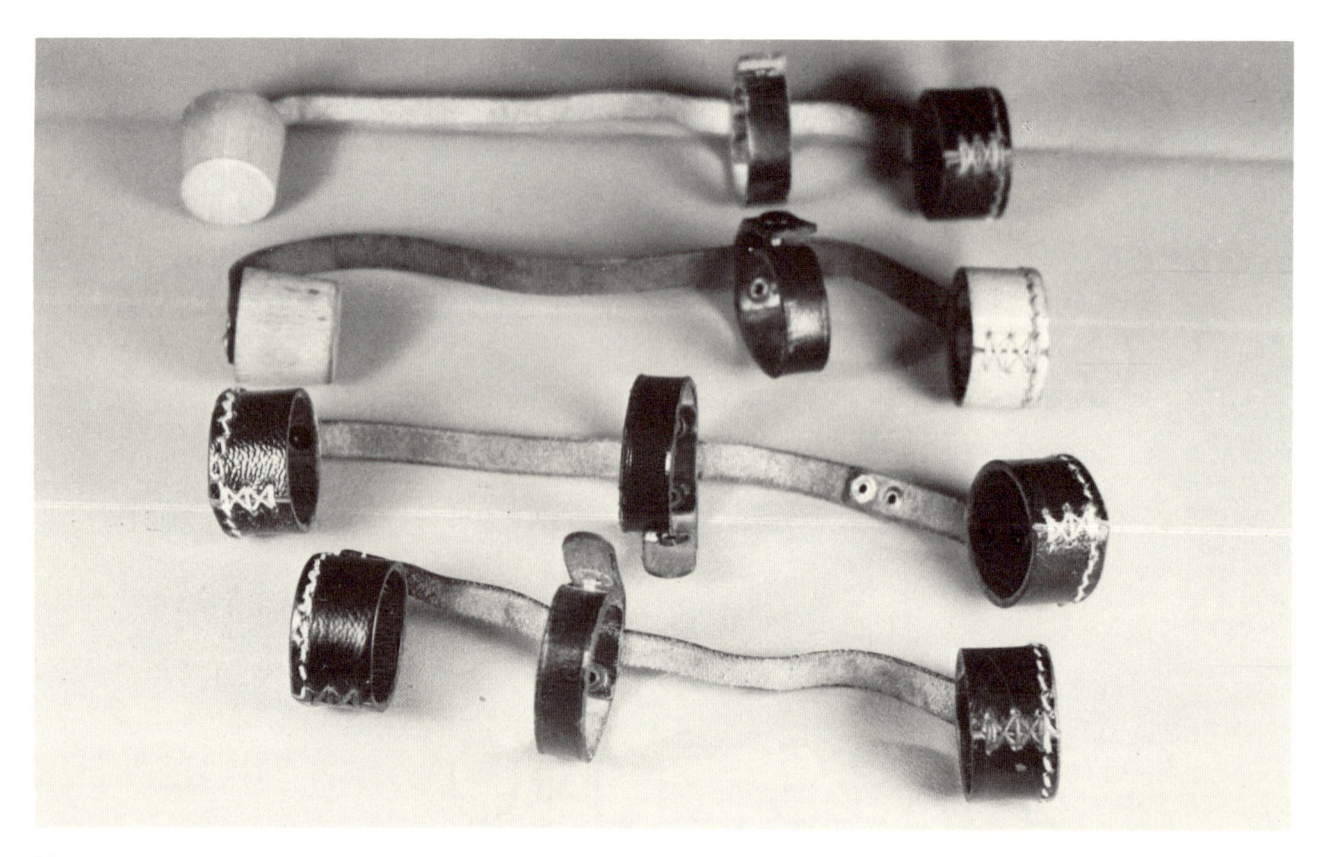

Variant leather lens caps issued with the ZF 4 sighting system through the course of the war. The wooden plug was inserted into the rubber eye-guard.

Formed metal case used for carrying the ZF 4 scope assembly when removed from the rifle. The exterior is finished in dark green with a special "damp-proof" paint (dull red) applied to the inside to deter the formation of rust from condensation. Note provisions for retaining the lens discs and rain shield.

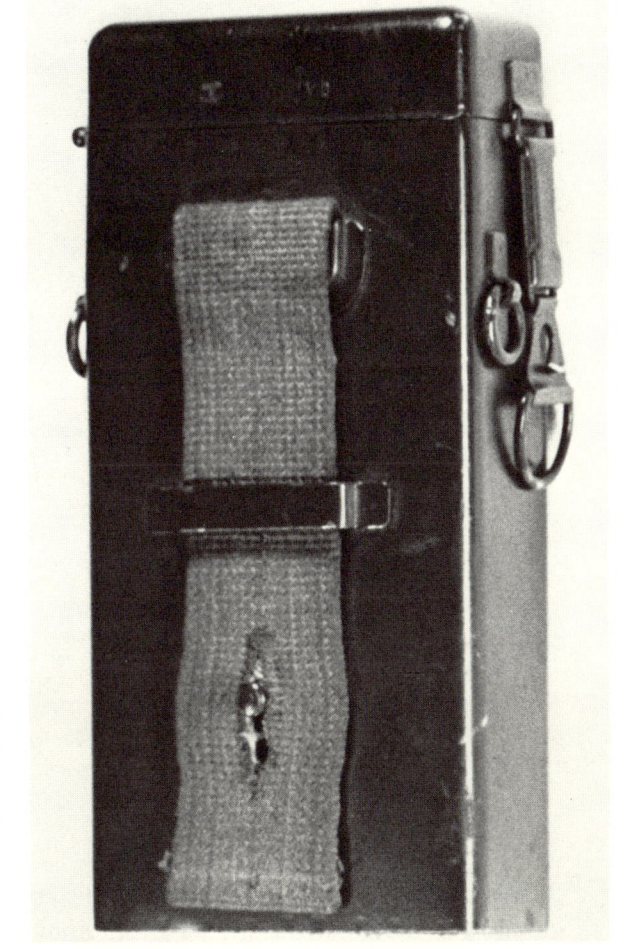

Reverse side of ZF 4 metal carrying case showing the web belt loop and "jvb" production mark on the lid. A shoulder strap attached to the small rings on either side of the case. Case dimensions are 200.02mm x 98.42mm x 46.03mm.

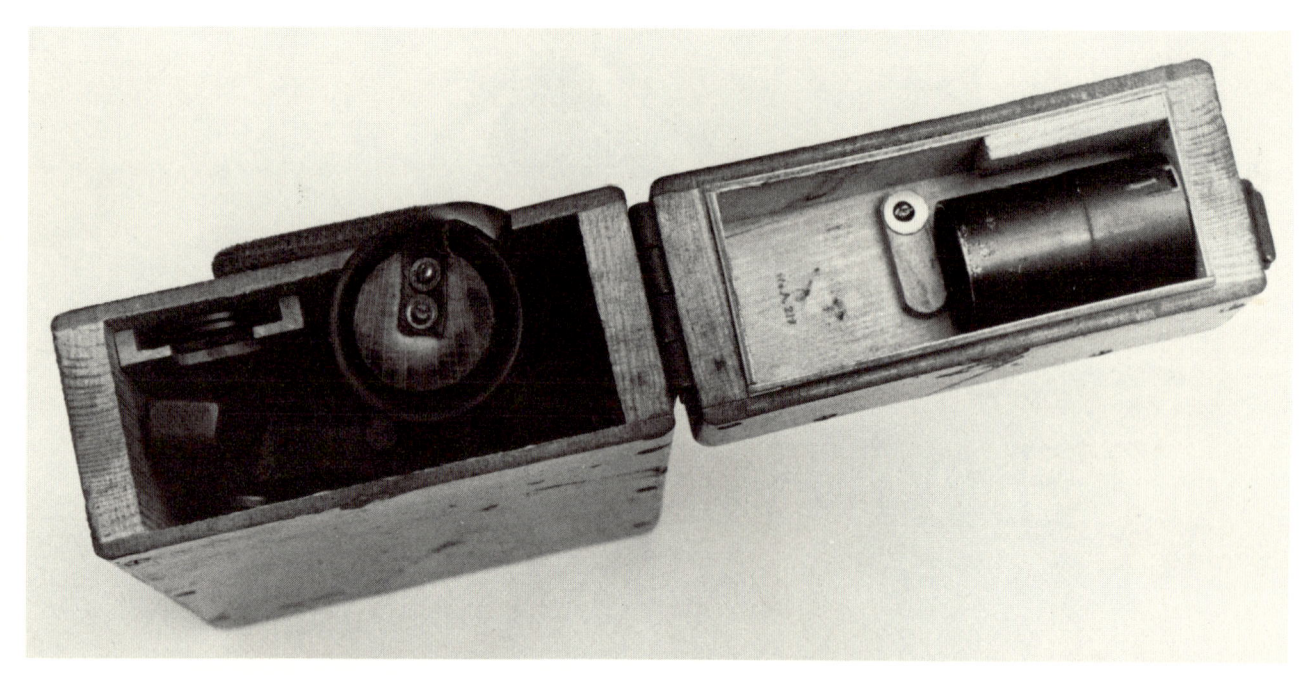

Carrying case fashioned from wood and fiberboard for the ZF 4 sight and accessories. The objective rain shield was retained by a large wooden dowel with the lens discs held in a small compartment.

Reverse side of wood and fiberboard ZF 4 carrying case showing the web belt loop and shoulder strap rings. Case dimensions are 215.90mm x 127.00mm x 50.00mm.

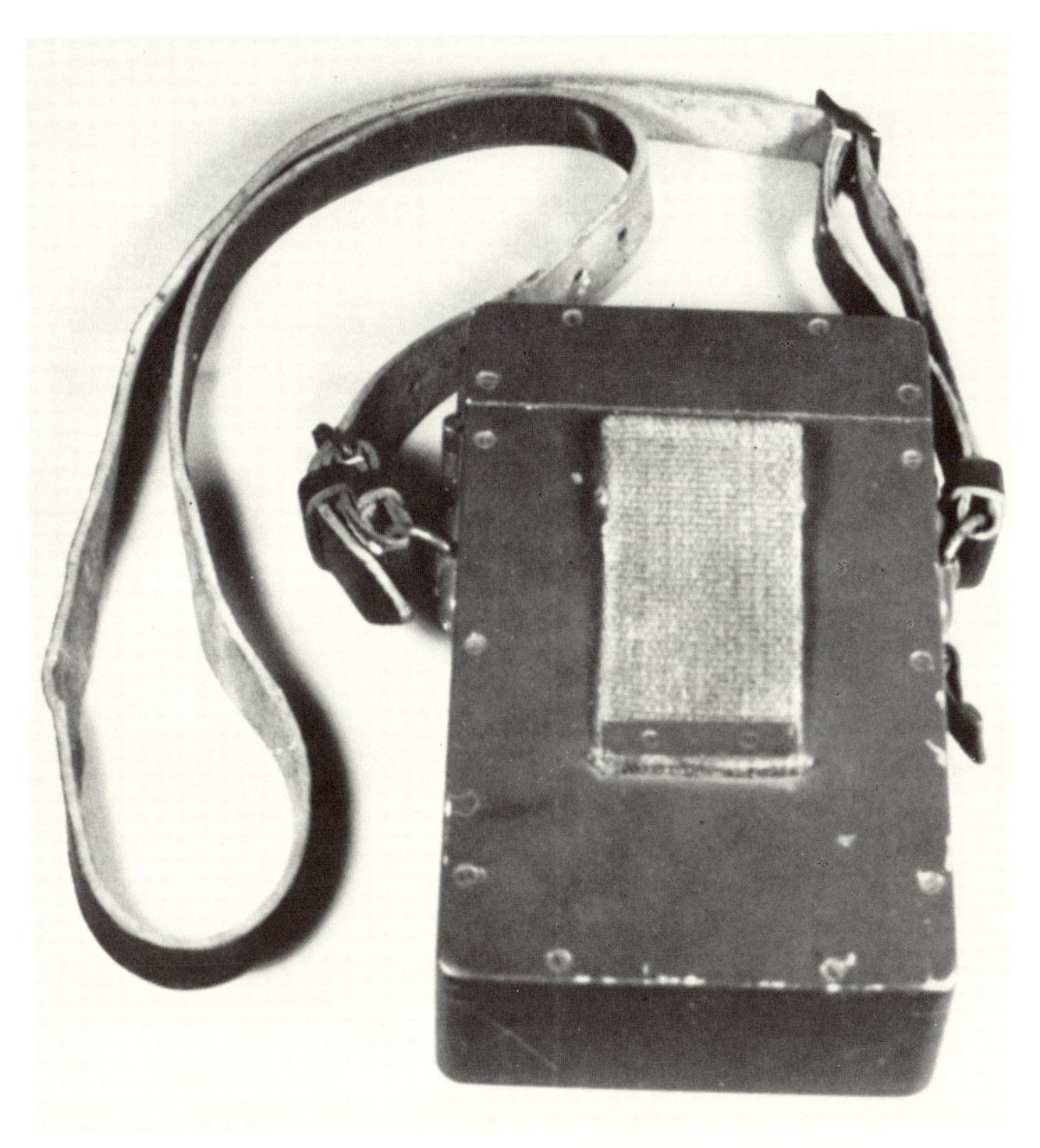

Wood and fiberboard ZF 4 carrying case with issue leather shoulder strap. Although cases are frequently encountered, original shoulder straps are not.

Comparative view of formed metal (left) and wood fiberboard ZF 4 carrying cases showing the latch devices used for holding the lids closed.

Unusual "Bu" marked Gw ZF 4 telescopic sight having a unique ranging style reticle pattern.

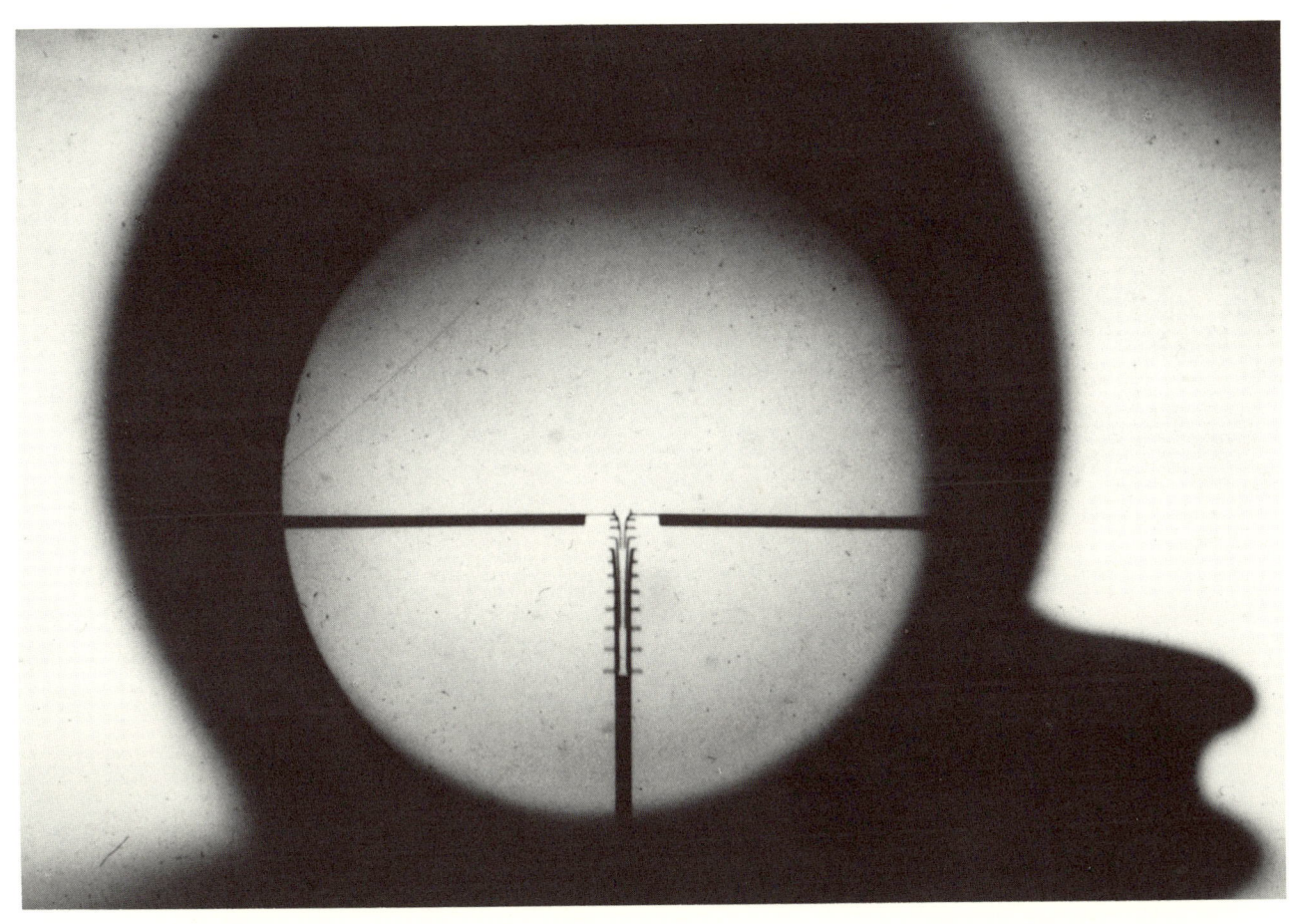

A view of the ranging reticle found in the Voigtlander (ddx) manufacture "Bu" marked ZF 4 scope.

The "Bu" marked ZF 4 sight with elevation adjustment drum identical to that used for windage (top). Note the absence of the usual range markings (graduations).

An extremely rare unconventional ZF 4 sight having a large threaded objective lens ring with a deep annular recess in the scope body.

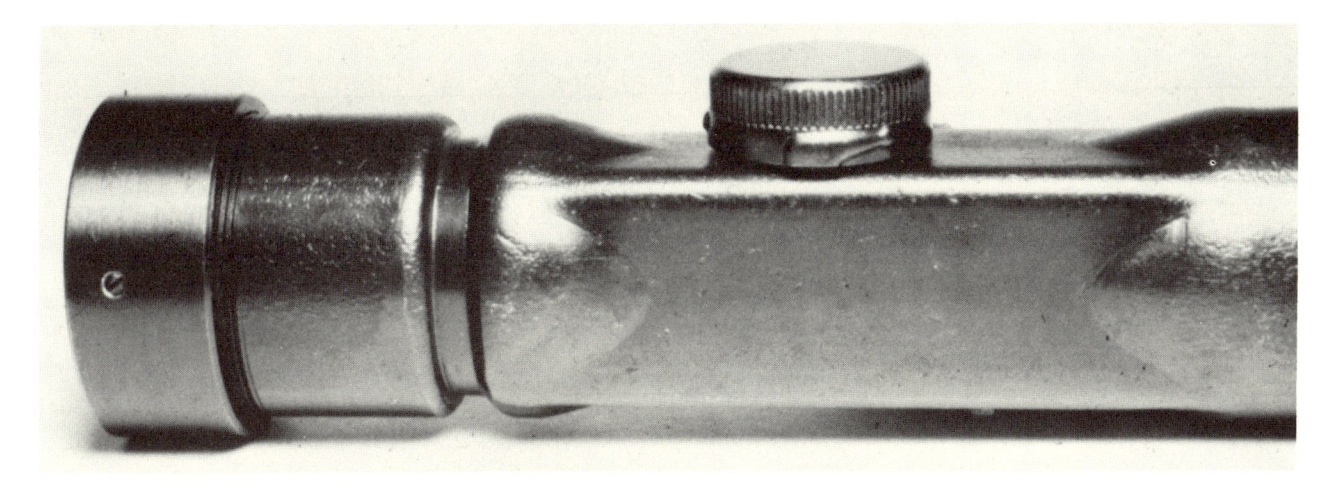

An experimental variant without question. Left view of the unique ZF 4 scope shows the absence of any markings.

Front view of unconventional ZF 4 sight. Although the threaded objective ring is much larger than in a standard ZF 4 tube, the lens diameter is identical.

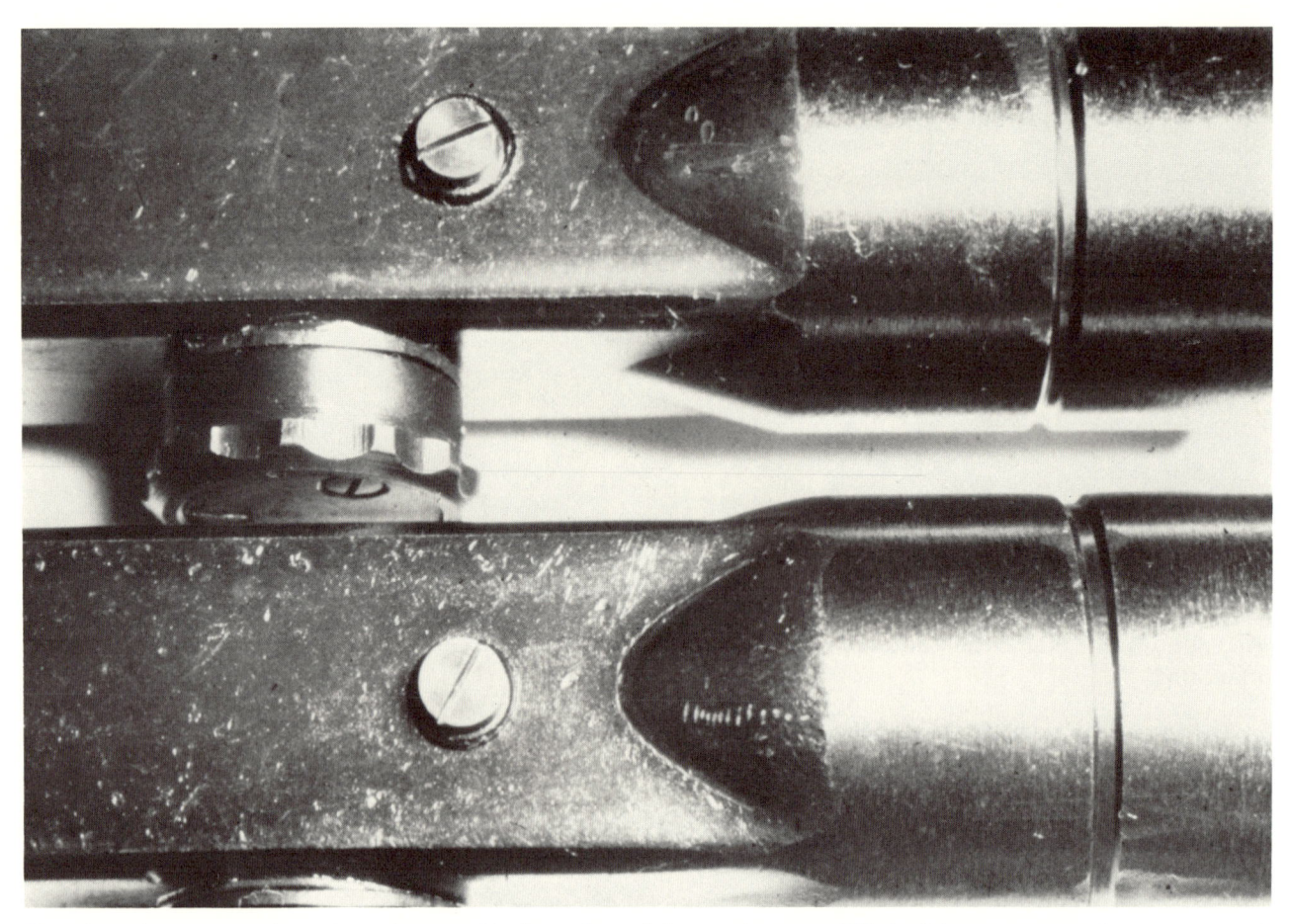

Despite the contention that the small plugs found beneath some ZF 4 sights were used on an experimental basis for filling the scope tube with dry nitrogen or other inert gas to provide "fogproof" internal lenses, or as otherwise suggested, to allow access for adjustment purposes, the specific reason for these sealed plugs has not been thoroughly explained.

Even though a considerable number of G43 and K43 rifles were eventually fielded during the war, combat photos showing these weapons in use, particularly for sniping purposes, are far from common. Note the semiautomatic rifle slung under the rifleman's shoulder.

Armed with a K43, a figure from a sniper training manual was used to show prospective sharpshooters an accepted firing position.

The ZFK 43/1, 4-power telescopic sight (no. 908895) as brought to fruition by the Carl Zeiss firm in late 1944. Designed to improve on the "sheet metal" ZF 4 device, the unique sights were to eventually replace the Gw ZF 4 for use with the semiautomatic sniping system. Cited as a return to the proven "turning and milling" method of telescope fabrication, ZFK 43/1 scopes were compatible with existing G-K43 series telescope mountings.

Left view of ZFK 43/1 "rln" code (Carl Zeiss, Jena) rifle scope showing the focus adjustment knob. Although similar in appearance to that on ZF 4 sights, the main body of the ZFK 43/1 was machined from a steel casting. The elevation drum was placed on top of the sight with the windage drum located on the right side.

Included in the collection of an obscure military museum in Germany, a unique telescopic sight believed to be a prototype version of the ZFK 43/1. Unlike the Zeiss production variant, in this case the windage adjustment drum was placed on the left side of the sight with the focus adjustment knob located on the right. While similar to those used with ZF 4 sights, this mount is distinctly different as a careful comparison will attest. Note the positioning of the vertical struts, machined clearance for the focus knob, etc.

Left view of unusual 4-power prototype telescopic sight, the only known example of its kind. There are no markings on the scope body, adjustment drums, or mount. Obviously intended for the semiautomatic sniping system, the origin of this scope remains obscure.

Another ZFK 43/1 sight (no. 908844) is shown for comparison. The recess in the objective end of the tube engaged the mount cradle in the same manner as on a ZF 4. The late war "rln" production mark also appears on binoculars made by the Zeiss firm.

Among the last manufacture in 1945, in this case the K43 (ac 45) rifle has the double-bolt guide rib design improvement. A number of modifications were made to the Selbstladegewehr 43 system during the span of production.

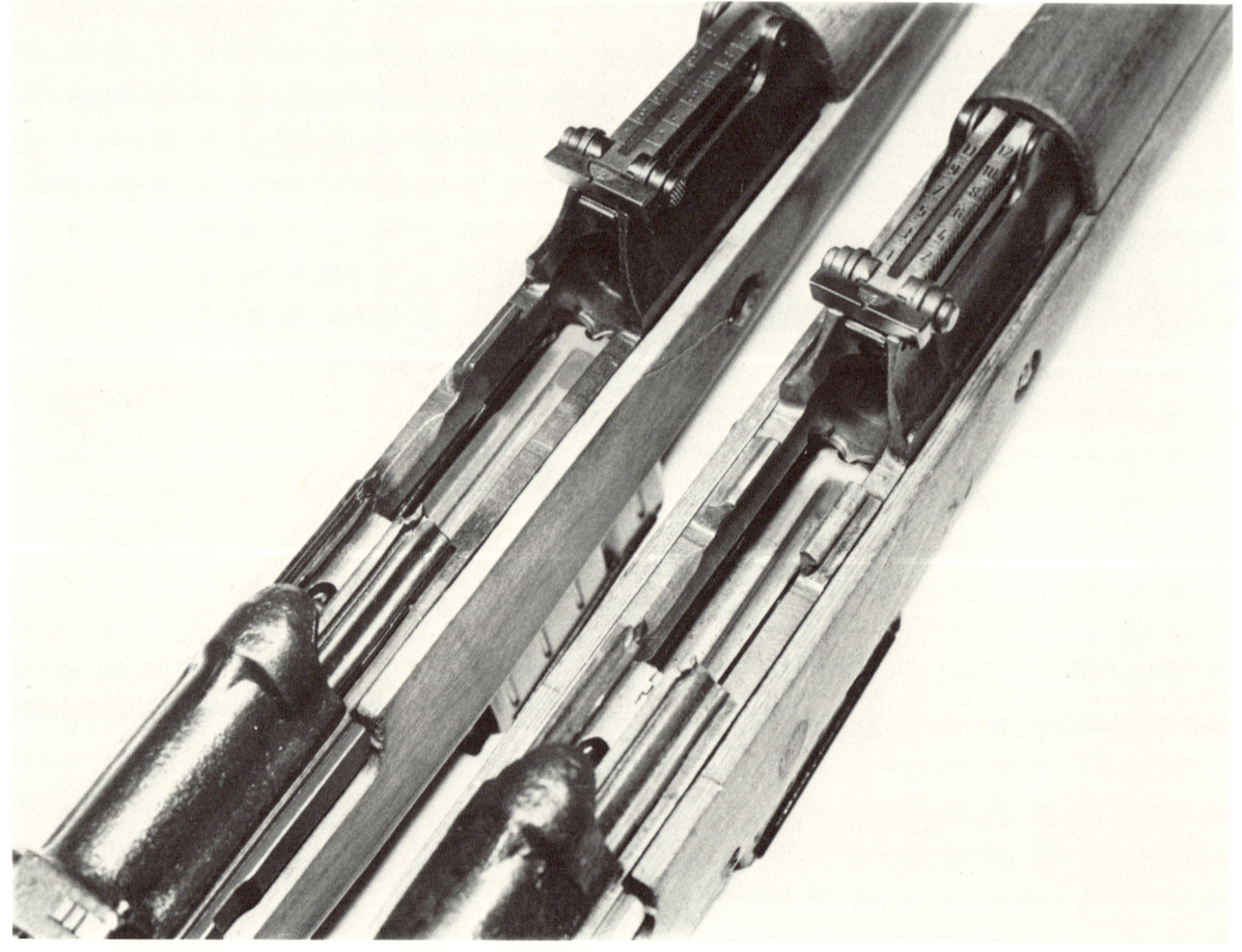

Comparison view of the common G-K43 single bolt guide rib (top), and the double guide modification found on some late war K43 production.

SELBSTLADEGEWEHR 43

Late war ZFK 43/1 (no. 908832) telescopic sight with Walther manufacture K43 rifle.

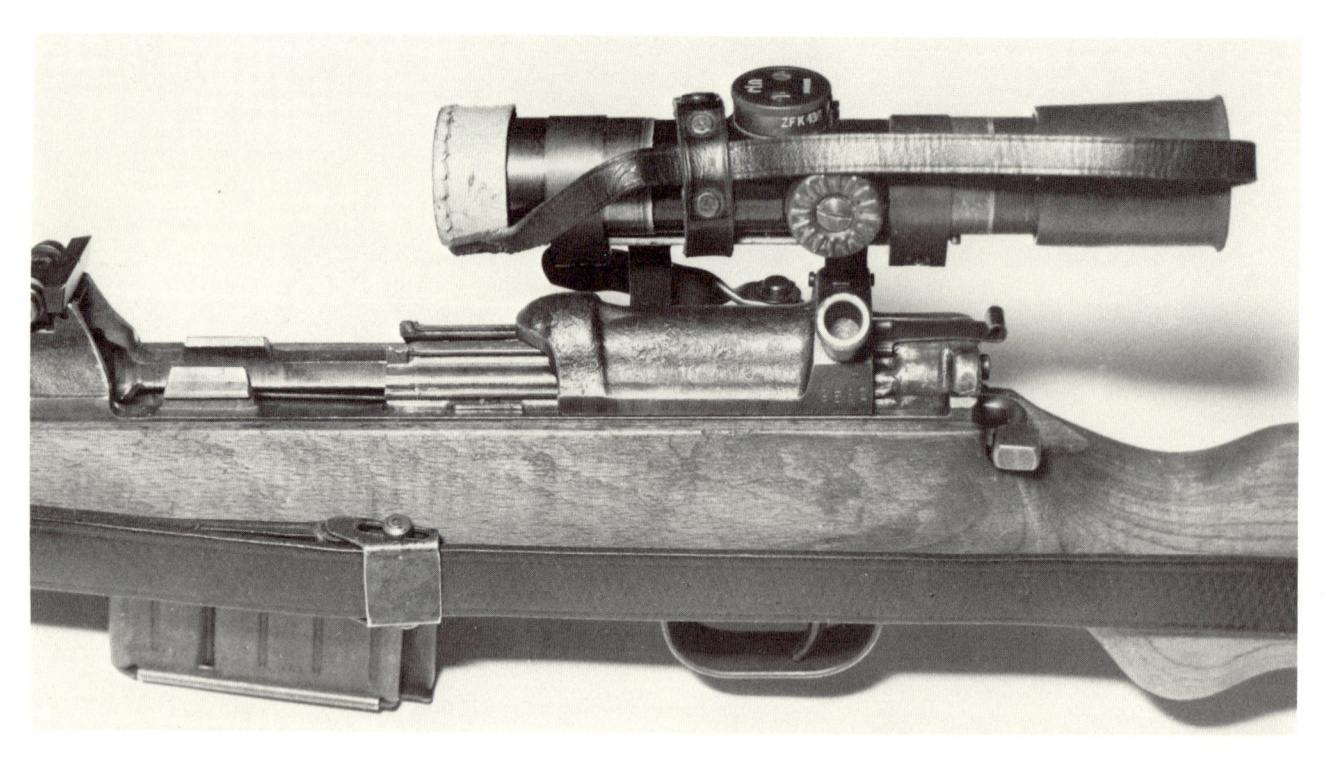

Left view of ZFK 43/1 and Walther (ac) K43 rifle. Note the double-bolt guide ribs, a late war design improvement.

Intended for the home front, a late war photo of Hitler Jugend (Hitler Youth) armed with "modern weapons." Although a scope-sighted K43, an MG42, and an Assault Rifle are obvious on the outside row, note the absence of additional weapons.

Das Ende, the capitulation of an overwhelmed and beleagured Wehrmacht: a frequent scene across Europe in 1945. In this case a German infantryman surrenders three semiautomatic rifles.

Even though bicycles moved a significant number of the Wehrmacht through the course of the war, this official ordnance proposal (1944) was quietly laid to rest. The weapon is a K43.

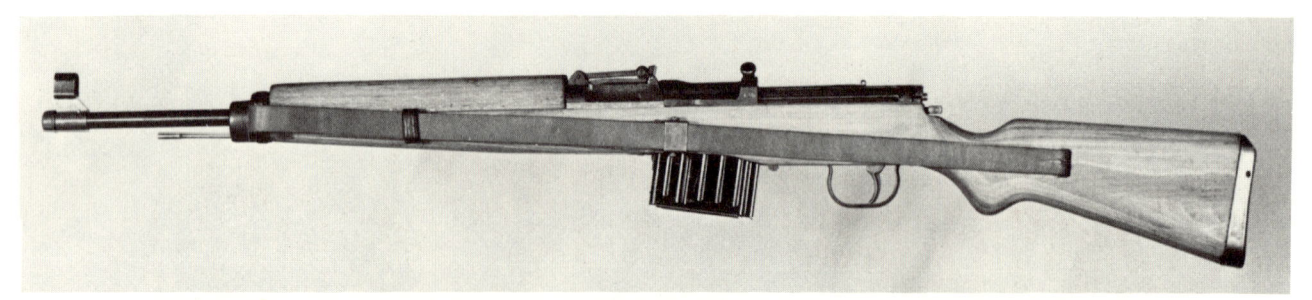

Karabiner 43 (K43) in standard trim. While not without flaw, conventional G-K43 rifles proved less of a problem in combat than those issued for sharpshooter use.

An unusual extended ZF 4 rain shield more than twice the length of the standard type. The shield is permanently attached to the scope and cannot be removed.

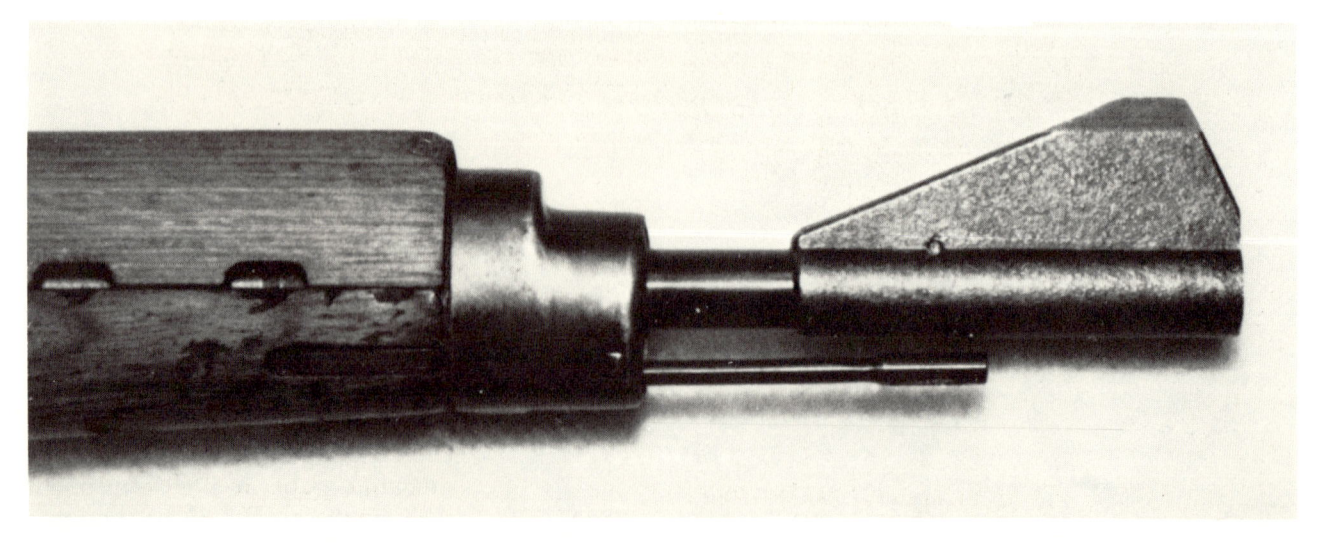

Integral flash hider and front sight assembly found on a late war K43 rifle (ac 45). Held in place by a small pin, the device is 76.20 mm in length with the barrel recessed 25.40 mm from the end.

Alternate view of a ZF 4 rain shield. Formed from sheet metal, the simple device slipped over the objective end of the telescope.

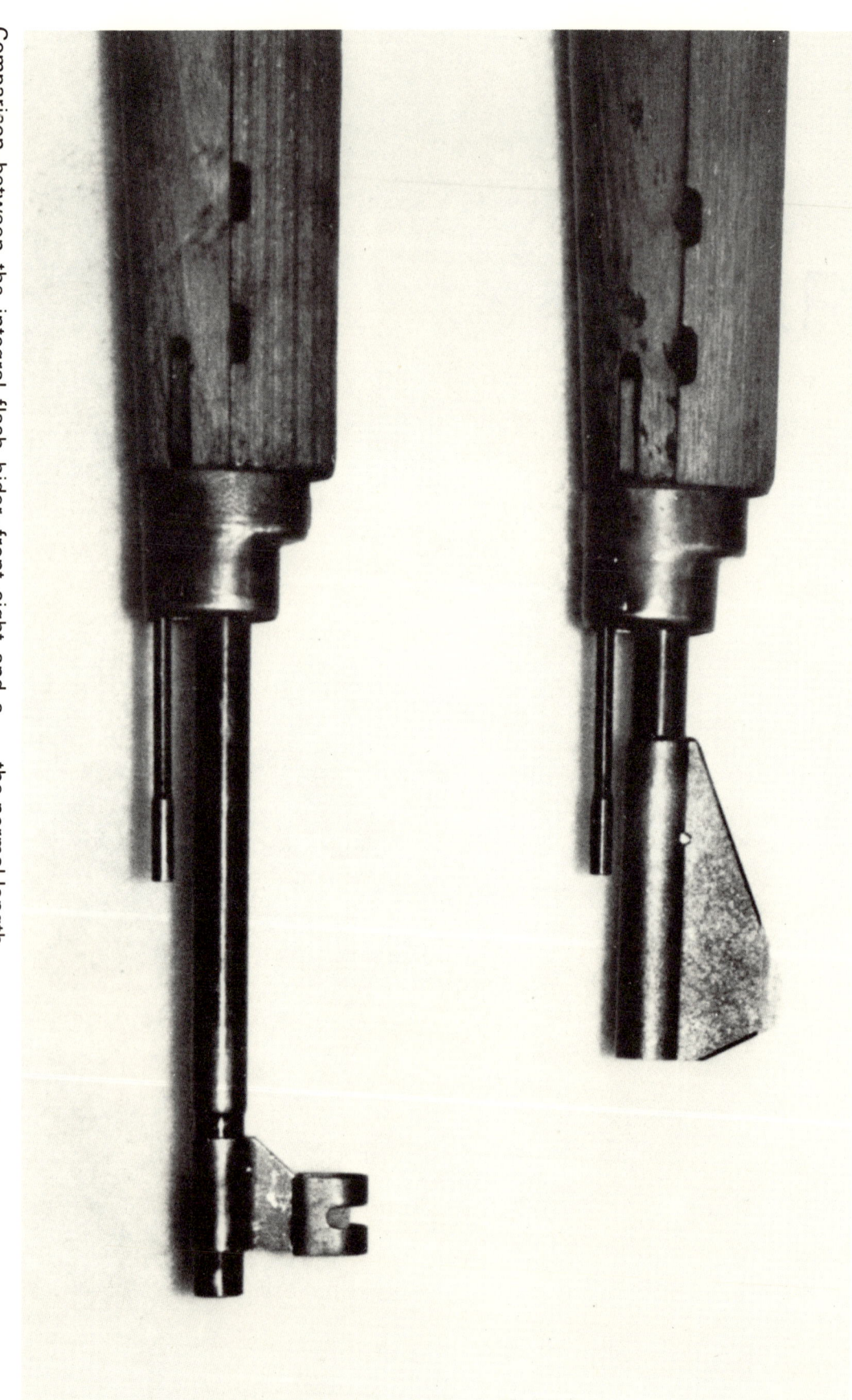

Comparison between the integral flash hider–front sight and a standard K43 assembly. The modified barrel is 76 mm shorter than the normal length.

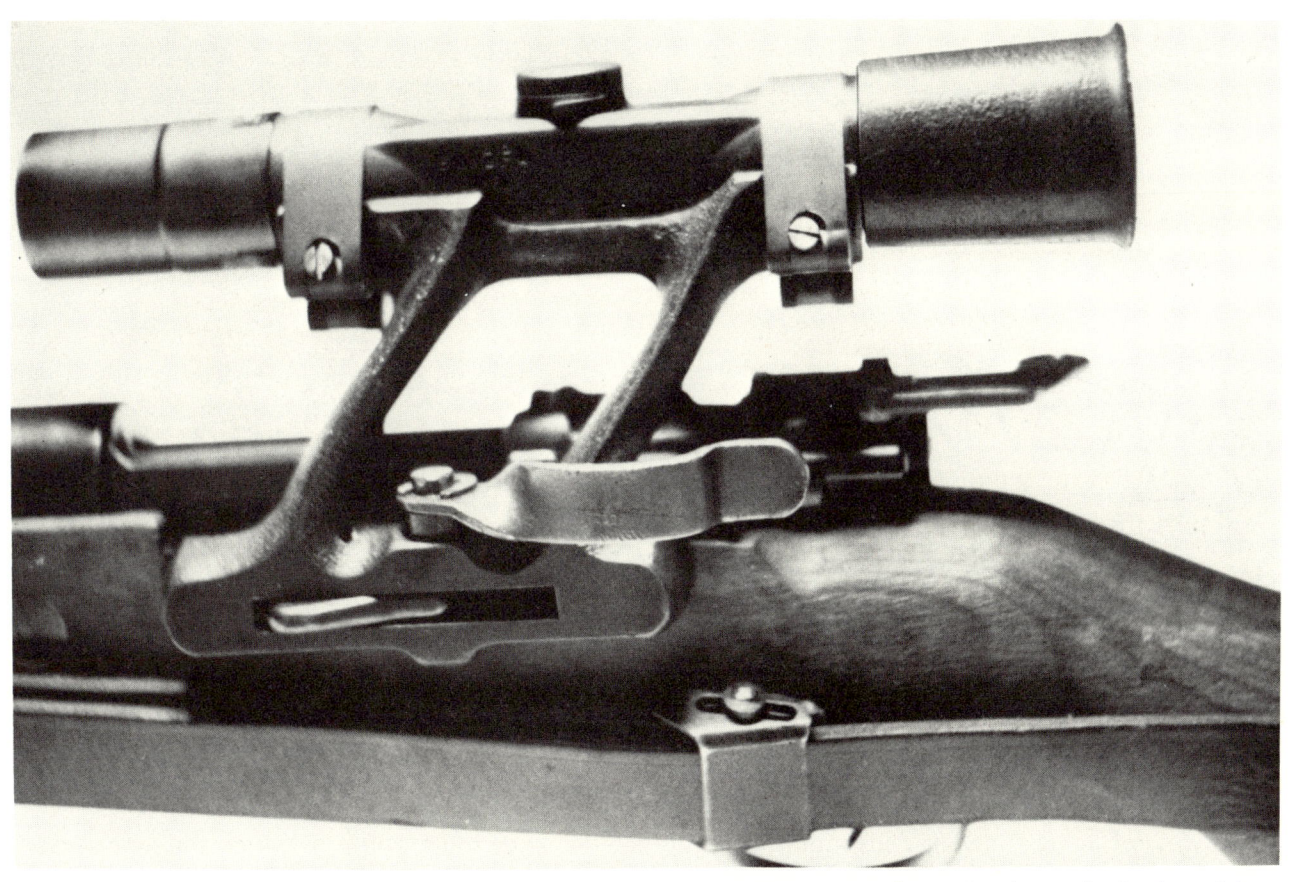

Detailed view of 98k sniping rifle with swept-back mount pressure lever in the unlocked position. Although "KZF" marked scopes have been noted on weapons of this configuration, this sight bears the regular "Gw ZF 4" designation. Note the mount characteristics similar to those employed with the semi-automatic sniping system. A "low" scope safety was used here.

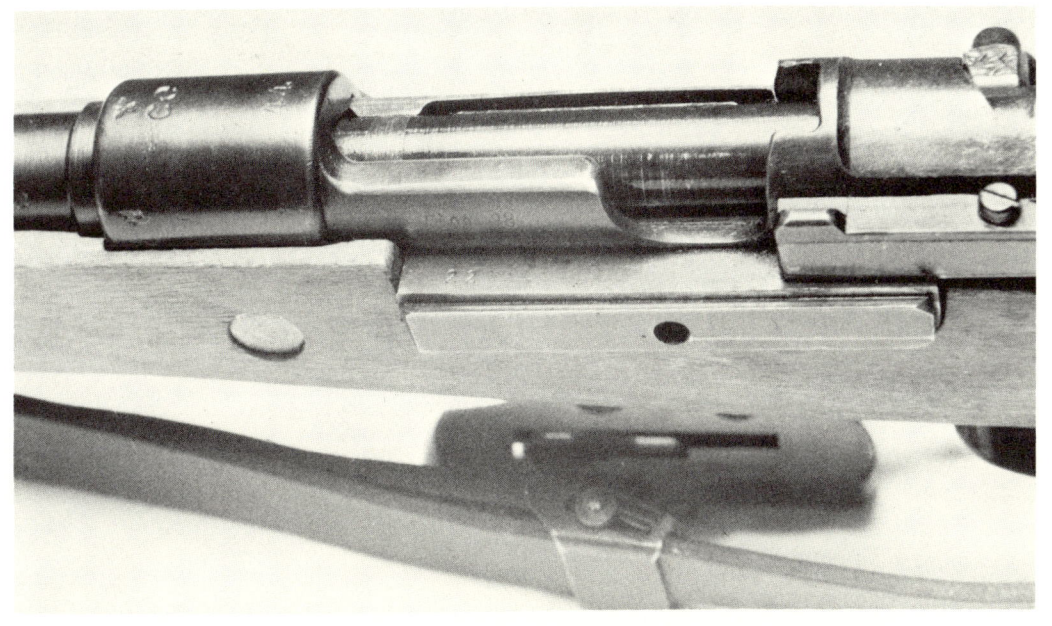

The telescope mounting base was affixed to the receiver with solder and two recessed screws passing through the inside of the receiver into the base. The circular recess engaged the auxiliary spring-loaded locking latch. Note the J. P. Sauer receiver marking (ce 44).

Side view of another swept-back mount 98k rifle (ce 44) showing manner in which the stock was modified to clear the receiver base. A standard safety was used here.

Late war issue 98k sniping rifle with swept-back scope mounting designed to accommodate both ZF 4 and ZFK 43/1 telescopic sights.

CHAPTER XV

Telescopic Sights— The Assault Rifles

Although no responsible claim can be made for German use of assault rifles for sharpshooting purposes, that is, for precise long-range firing, the study of weapons in this category indicates that significant design efforts were directed towards mounting both optical sights and eventually infrared devices to a succession of assault rifles from their inception through the end of the war.

At this time, complete records dealing with the design, development, and field application of telescopic-sighted assault rifles have remained obscure. Therefore, a presentation of this interesting facet of World War II German weaponry remains for the most part a pictorial chronicle of the various scope mountings and related hardware as noted on surviving specimens.

Neither the consideration nor the application of telescopic sights for assault rifles appears to have been a major prerequisite during early weapon development. But the Maschinenkarabiner 42 (machine carbine 42), as manufactured by Haenel and Walther for original service use (1942-43), the Mkb 42 (H) and the Mkb 42 (W), were provided with integral scope mounting rails on either side of the formed sheet metal housing which served primarily as the mounting block for the standard rear sight assembly.

Intended for fitting the 1.5-power sights, the telescope mount slid on to the formed rails and locked in place by means of a spring-loaded latch engaging a corresponding notch in the right hand rail.

As events transpired in early 1943, the Haenel design was deemed superior to the Walther variant, and as simplified for subsequent production as the Maschinenpistole 43 (machine pistol 43) series, was brought to fruition by the introduction of the MP 43 and the MP 43/1. Despite the fact that both weapons were the same in overall design and operation, they differed in detail at the muzzle (grenade launcher mounting).

Whereas the vast majority of the MP 43/I weapons appear to have been provided with integral telescope mounting rails in the same manner as on the preceding MKb 42 series, this feature was omitted on the MP 43 as well as on the later MP 44 and StG 44 variations.

The lack of official German reference to field use of 1.5-power telescopic sights in conjunction with assault rifles and the acute rarity of surviving mount assemblies suggests that few were ever issue items.

In addition to the obvious consideration of 1.5-power scope mounting to the MKb 42(H), MKb 42(W), and MP 43/1, a limited number of Maschinenpistole 43 series assault rifles, the MP 43/1, MP 44, and StG 44, were fitted with a machined steel 107.95mm long, V cross section base that was welded to the receiver on the right

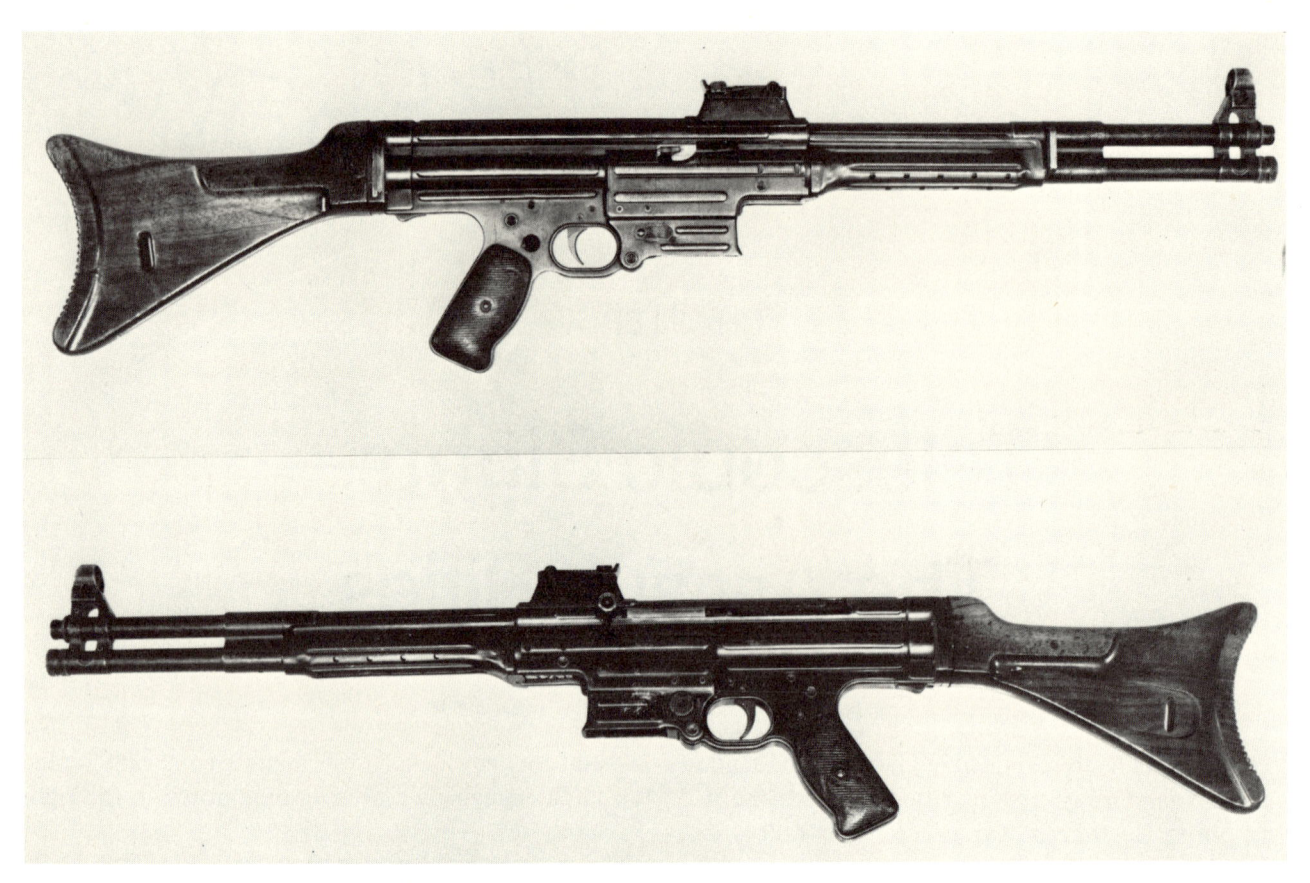

Prototype Haenel Maschinenkarabiner (machine carbine).

Early production Haenel MKb 42(H) with integral scope mounting rails.

German Ordnance photo (November 1942) of the Haenel MKb 42(H) machine carbine. Note the integral scope mounting rail on the rear sight base.

Haenel Maschinenkarabiner (machine carbine) MKb 42(H) with an extremely rare 1.5-power telescope mounting. The mount depicted is the only known example of its type.

side behind the ejection port cover. This was intended for mounting the Gw ZF 4-fach telescopic sight (ZF 4).

It should be emphasized that assault rifles fitted with the aforementioned scope mounting base were exceptions among weapons of this type.

Because the machined steel bases were almost identical in configuration to the integral G43 and K43 (Selbstladegewehr) receiver bases, the ZF 4 sight was readily adapted to the assault rifle by using the standard telescope mount.

Of parallel interest are examples of the Fallschirmjager Gewehr 42 (FG 42) Luftwaffe assault rifles, which also made use of ZF 4 sights. These rifles were used with a variety of telescope mountings designed specifically for them.

Original development of the FG 42 was based on the premise that a long-range, selective-fire weapon was necessary for parachute assaults. Luftwaffe Weapons Office specifications also had included the stipulation that the new rifle possess a capability for sniping. But at no point in successive development did the FG 42 attain the requirements necessary for efficient sniping.

In advance of the ZF 4, the 4-power ZFG 42 (Zielfernrohr-Gewehr 42), a limited production telescopic sight developed by the Voigtlander firm (ddx), was brought to fruition at the insistence of the Reich Air Ministry (Reichsluftfahrministerium or RLM) for use with the FG 42.

While the ZFG 42 was similar in appearance to the ZF 4, the principal difference between the two sights rested with the ZFG 42 having a full cylindrical tube with adjustment drums located on the top (for elevation) and left (for windage) of its somewhat square center section. Aside from the obvious configuration of the "sheet metal" tube assembly, adjustment drums on the ZF 4 were positioned on the top (for windage) and right side (for elevation) of the telescope tube. Although originally developed for Luftwaffe use with the Fallschirmjager Gewehr 42, the ZFG 42 sight is considered to have been the basis for evolving ZF 4 design which has also been attributed to the Voigtlander organization.

Perhaps the most enigmatic aspect of optical sight development associated with German assault rifles rests with the application of infrared night vision equipment.

Devices for seeing in the dark by means of electronic detection of invisible infrared radiation were developed to a high degree by German scientists during the war. Most of the equipment employed a picture-transformer tube called a Bildwandlerrohr that contained an infrared sensitive screen which emitted electrons, an electron lens system, and flourescent screen in which the image was formed by the lens.

In different forms, the unique tube was utilized in various types of German infrared equipment: for night-driving apparatus for regular vehicles, tanks, and armored cars, for the assault rifle system, and for aircraft applications.

Principal interest centered on the Zielgerat 1229 (ZG 1229) infrared sighting device developed specifically for assault rifle mounting.

A noteworthy account of this special system appears in the U.S. Army Ordnance School publication, *Submachine Guns*, Volume 1, July 1958.

In 1943 and through 1944, a program was started to develop a portable infrared "night sight" for the MP 43 series. A small electronic image tube (70-mm diameter) had been designed by the Forschungsanstalt der Deutschen Reichpost (RFP), Berlin. Ernst Leitz, Wetzlar, was contracted to develop the additional optical components and housings for such an infrared telescope. Limited manufacture at the Leitz plant seems to have been undertaken in 1944. The sight was designated Zielgerat 1229 (ZG 1229) and code named "Vampir". Leitz reported that 310 complete units of the night sight were delivered. The sight consisted of a 5" inch diameter, 35 watt transmitting lamp, a receiving telescope containing the 70-mm RFP tube, I:I 5f (8.5-cm) objective, and a high tension power supply. The transmitting lamp and receiving telescope were combined in a housing which could be mounted on the normal telescope sight bracket. The transmitter-receiver unit weighed 5 pounds, the power source was contained in a pack-type harness and gas mask container carried by the user. It weighed 30 pounds.

TELESCOPIC SIGHTS — THE ASSAULT RIFLES 373

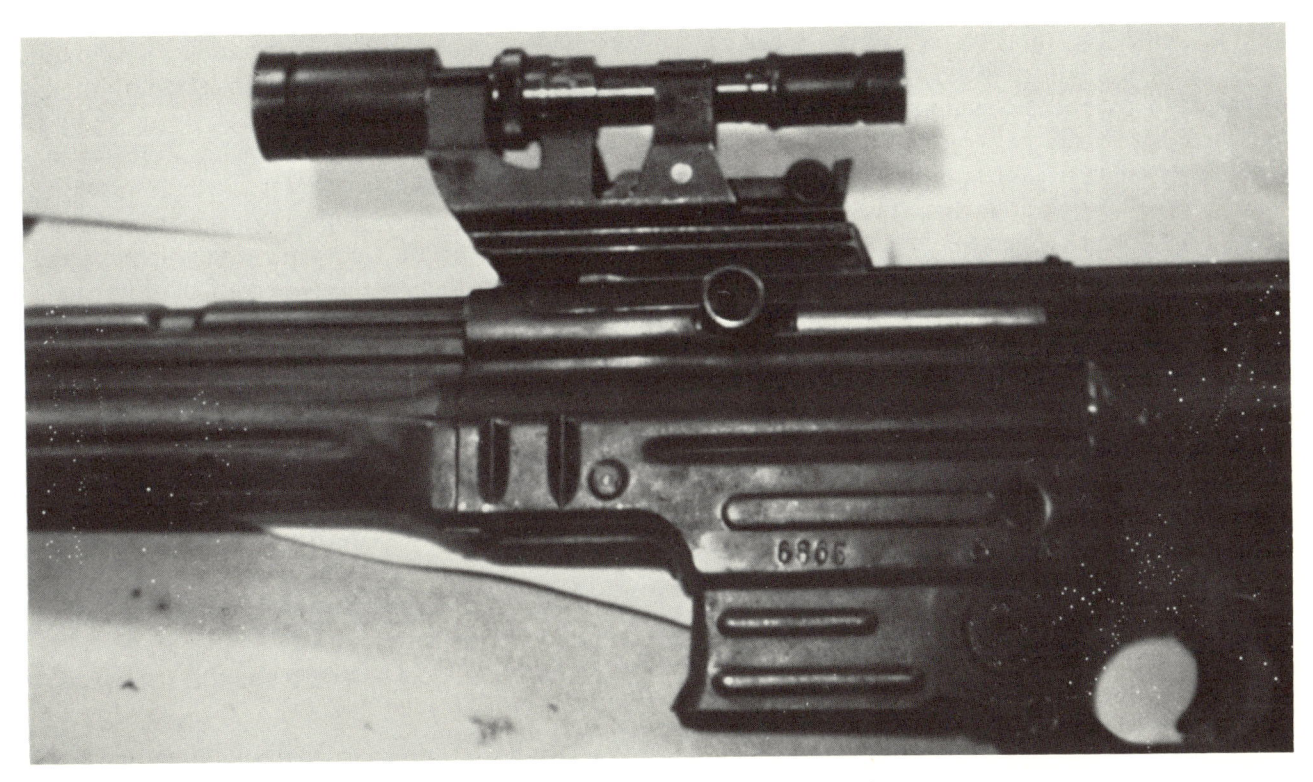

Haenel MKb 42(H) with 1.5-power telescopic sight and unique mounting.

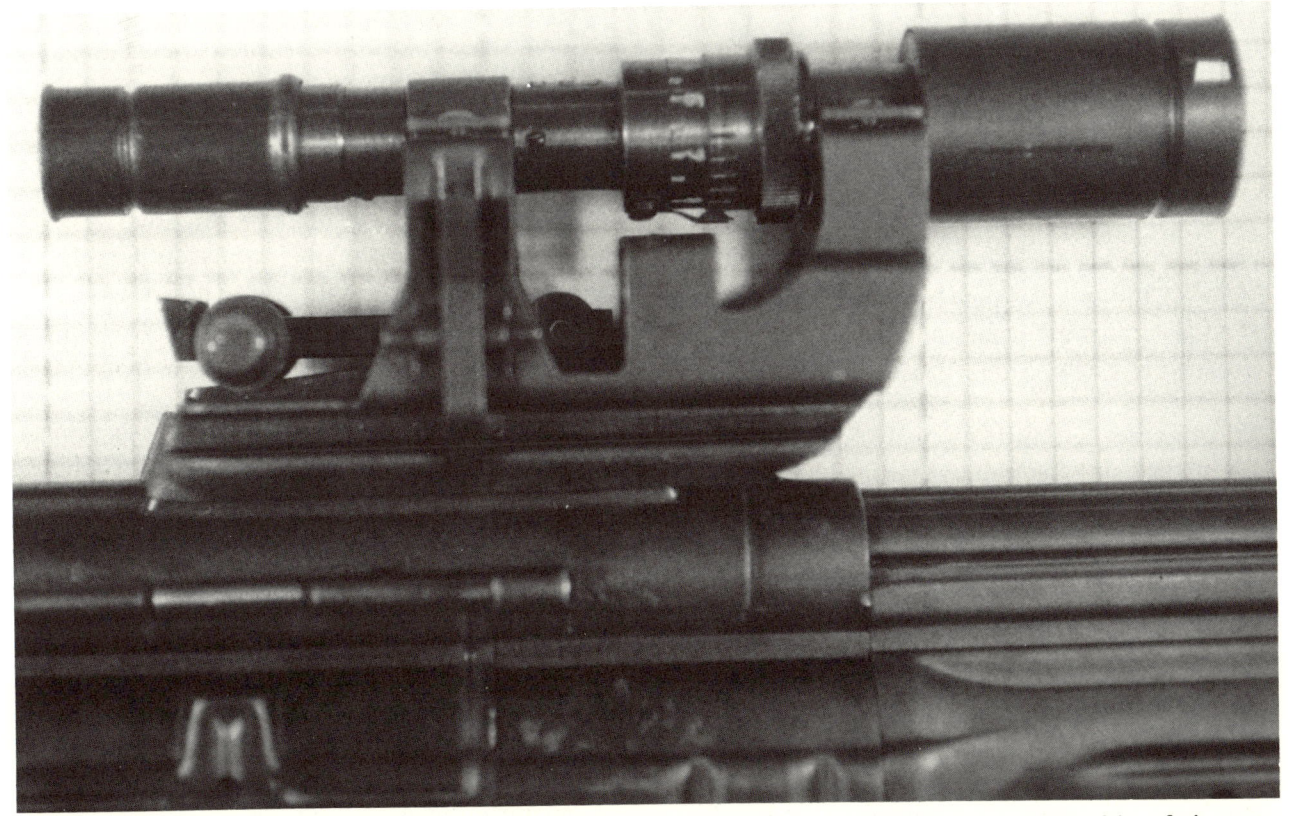

The formed metal MKb 42(H) scope mount slides on to integral rails located on either side of the rear sight base stamping and is locked in place by engaging a spring-loaded latch on the right side of the mount with a corresponding notch in the rail.

In 1943 the ministry of Albert Speer established a committee to coordinate research in the field of infrared among the army, navy, and air force. Until that point all three had been developing night vision equipment relevant to their particular requirements. The committee, headed by Dr. H. Gaertner, who was also the director of the optics branch (telescopic sights, etc.) in the headquarters of the Heereswaffenamt in Berlin, was charged with the responsibility of correlating the development of infrared devices that would benefit the Wehrmacht as a whole.

Dr. Gaertner, who in recent years served as a consultant to the West German Bundesministerium der Verteidigung, authored a review of German military infrared activities in advance of and during World War II in *Wehrtechnische Monatshefte* dated December 1961 and May 1962. With respect to the Zielgerat 1229 system Dr. Gaertner states.

> In October 1944 the military demand was made to develop a small I.R.-telescopic sight for the Sturmgewehr 44 to protect armored units against night attacks by tank-destruction-troops and to be in a better position to ward off the infantry attacking by night. The company AEG and the Reichspostforschungsanstalt participated in the development of this device. Several samples were introduced whose ranges showed only minor differences. One of the samples developed by the AEG had the following technical data:
> 1. Ancillary lens: = 7,5 cm, I:I, 5; visibility about 8°
> 2. magnifier: monocular, v = 10 fold
> 3. image-converter tube: type 128 (Diode) of AEG; line voltage 12 kV
> 4. over-all enlargement: v = 1,5
> 5. I.R.-floodlights for incandescent lamps: d = 10 cm; capacity 36 W
> 6. power-and high-voltage supply: especially light model for 12 kV, can be carried on back of marksman.
>
> Successful tests of the I.R. sighting device for hand fire arms were made by the Panzergrenadiere during the months of February and March 1945. Against single persons a range of approximately 70 m was established.
>
> The Panzergrenadiere, which were equipped with the telescopic sight, were transported with an armored car, which was also fitted with I.R. driving equipment. A machine gun with telescopic sight of the type used for the Panther was part of the armament. The armored car "Falke" was equipped in a similar manner and served as accompanying protection for the Panther-units at night.
>
> *Equipment of the Panther-units*
> With the previously mentioned devices—
> commander's device for the Panther observation vehicle "UHU"
> driving equipment for passenger cars and trucks
> I.R. telescopic sight for the Sturmgewehr 44

The I.R.-equipment of the Panther-units is described as it was planned for use by German armed forces at the end of 1944. Individual devices, like the commander's driving device for the Panther, were already available in large numbers at the end of the war. Other devices like the I.R. telescopic sights were only produced in small numbers. However all devices had been sufficiently tested for mass production.

The basic mode of operation for the assault rifle infrared unit was cited in the June 1946 issue of *Electronics Magazine* in an article dealing with World War II German night vision equipment.

> The zielgerat receiver operates on the same principles as the receiver used with night driving equipment for tanks, with the additional feature that it can still be used for about fifteen seconds after the power supply is turned off. The reason for this is that a larger value of capacitor is used for a greater time constant, so that the value of the applied voltage gradually decreases. All that is needed is a high potential to be built up, since there is practically no current drain.
>
> When a small button on the bottom of the power supply is depressed, the power is

TELESCOPIC SIGHTS — THE ASSAULT RIFLES

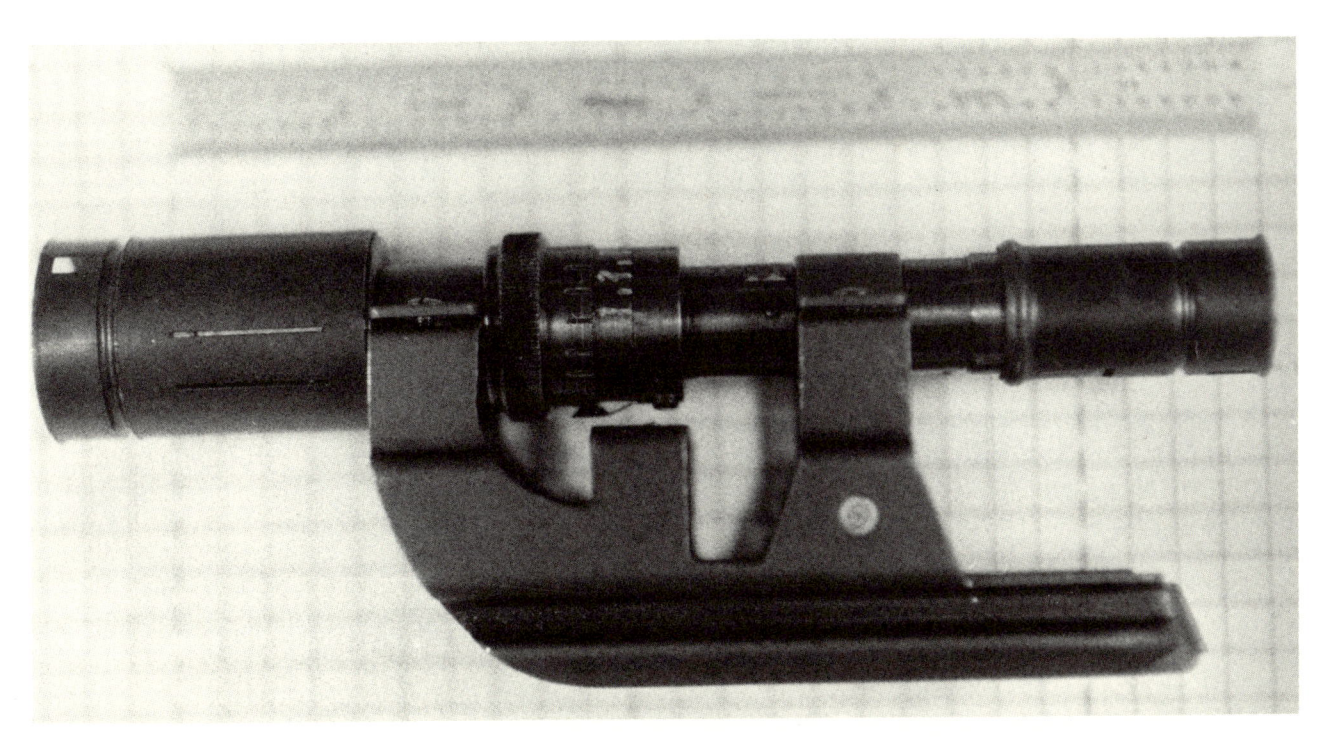

Left view of Maschinenkarabiner telescope mount removed from the weapon. A set screw, similar to that used with 98k-ZF 41 series mounts, is located in the bottom section of the forward ring assembly. Note the small Haenel proof mark on the lower rear section of the mount. This assembly *will not* fit the receiver rails found on Gewehr 41 rifles.

Haenel Maschinenkarabiner MKb 42(H) with integral rear sight base scope mounting rails as fielded for German service use early in 1943.

on and the image appears on the flourescent screen. When the view begins to fade, the button has to be depressed again. In this way the rifleman must stop periodically to apply high voltage, but he has ample time to sight and aim. The infrared light beam is worked separately and was probably kept on all the time. It is fitted so that the light beam is thrown on the same point at which the rifle is aimed.

Aside from confirmation that the Haenel firm (Suhl) had been furnished with infrared sights for fitting to the StG 44 in the final months of the war, specific details concerning their combat application remain unknown.

Reports do indicate, however, that the Vampir units were extremely delicate and considered altogether too cumbersome in an attack situation.

Infrared light, while invisible to the naked eye, irradiates objects so that they can be seen with the viewer. The main objection to infrared use was that the user could readily be spotted by a foe similarly equipped with infrared instruments or detection equipment.

In anticipation of Allied use of infrared, the Germans produced a simple device to detect its presence. The device consisted of a small paper tube with an infrared window and lens on one end and a sensitive screen on the other. The screen possessed a remarkable property: after exposure to strong sunlight it became so sensitized that for a long time afterward it would respond by emission of red visible light when illuminated by invisible infrared light. While in a sensitized state, the detector permitted German troops to pinpoint the source of any infrared device directed against them. Approximately 10,000 detectors were produced for the Wehrmacht.

Although British and American forces did employ infrared searchlights in Europe during World War II, owing to priorities already established for weapon mounted instruments in the South Pacific, the use of such devices following the Allied invasion of the Continent was nonexistent or extremely limited at best.

Original German Ordnance photo (November 1942) of the Walther manufacture MKb 42(W) machine carbine. Note the absence of scope mounting rails on the rear sight base.

TELESCOPIC SIGHTS — THE ASSAULT RIFLES

Walther Maschinenkarabiner MKb 42(W). Even though the integral scope mounting rails are present on the rear sight base, note the absence of a mount locking notch on the right rail.

Maschinenpistole 43/1 (MP 43/1) German assault rifle provided with telescope mounting rails on the rear sight base.

MP 43/1 assault rifle with rear sight base scope mounting area identical in configuration to that found on the early Maschinenkarabiner 42 series.

Left view of MP 43/1 showing the integral scope mounting rail.

TELESCOPIC SIGHTS — THE ASSAULT RIFLES

An illustration, one of a series, made during "comparative shooting between a MP 43/1 with telescopic sight and Rifle 43 with telescopic sight" as conducted by the Infantry School (Infanterieschule, Doberitz-Elsgrund) in October 1943 to determine the suitability of the assault rifle for sniping purposes. Note the unique ZF 4 telescope mounting.

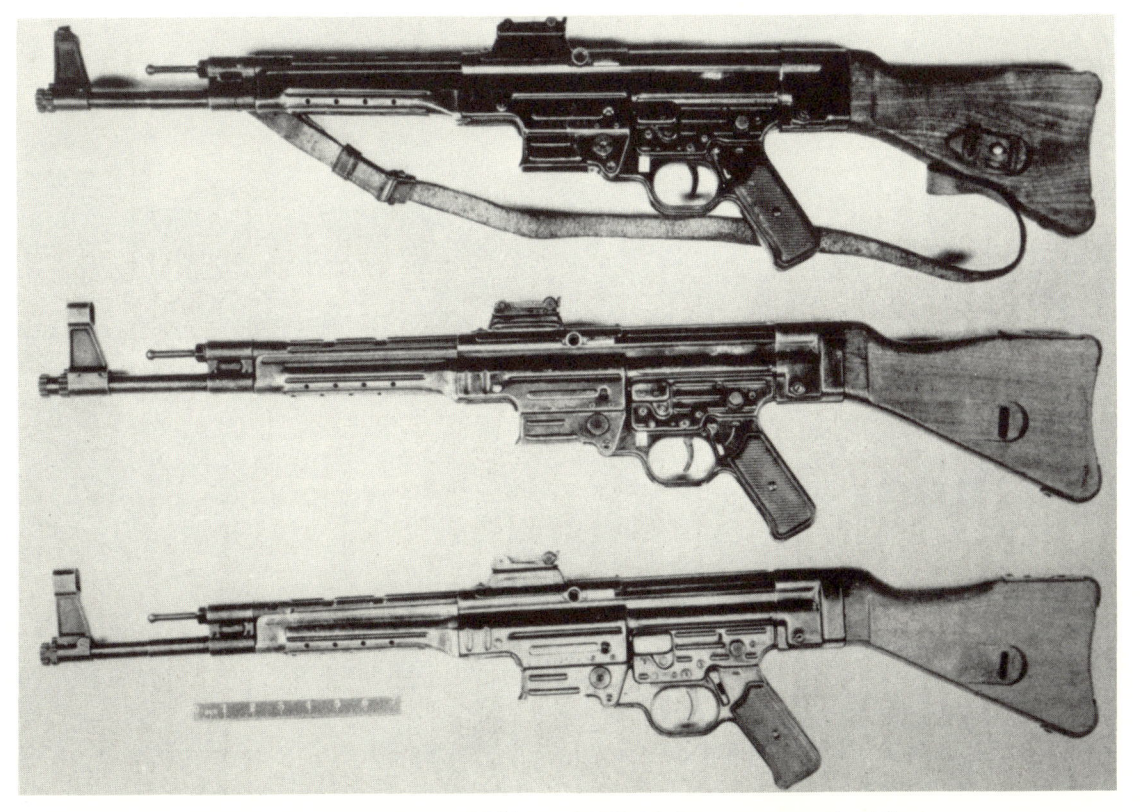

Comparative view of the MP 43, MP 43/1, and MP 44 German assault rifles.

Front view of MP 43/1 with ZF 4 telescopic sight as tested in October 1943.

Maschinenpistole 44 (MP44) with 4-power Gw ZF 4-fach telescopic sight assembly that attached to a machined steel dovetail rail (base) welded to the receiver behind the ejection port cover.

TELESCOPIC SIGHTS — THE ASSAULT RIFLES

Another view of the ZF 4 equipped MP 43/1 as tested in October 1943. While deemed "unusable as a sniper rifle because of great shot dispersion," it was the view of the Infantry School that the telescope mounting was responsible for the inaccurate shooting.

Maschinenpistole 44 with nonstandard machined steel dovetail base welded to the right side of the receiver for mounting ZF 4 and infrared sights.

An original German Ordnance photo of a prototype Zielgerat 1229 (Vampir) infrared night vision sight developed by the Reichspostforschungsanstalt, Berlin for use with the StG 44 assault rifle. The transmitting lamp (infrared light source) mounted directly on top of the receiving telescope assembly.

A British officer demonstrates a captured StG 44 assault rifle with the ZG 1229 infrared night vision sight. The batteries, necessary for operating the infrared assembly, were carried in a modified gas mask cannister, and the electrical power supply (power pack) was contained within a small wooden chest. Although such devices did prove beneficial when utilized in a defensive position, fog, haze, and smoke reduced the effective sighting range to only a few meters.

Close view of StG 44 infrared sighting system. Note the receiver mount with characteristics virtually identical to the lower section and locking lever assembly of a ZF 4 telescope mount used with G43 and K43 rifles.

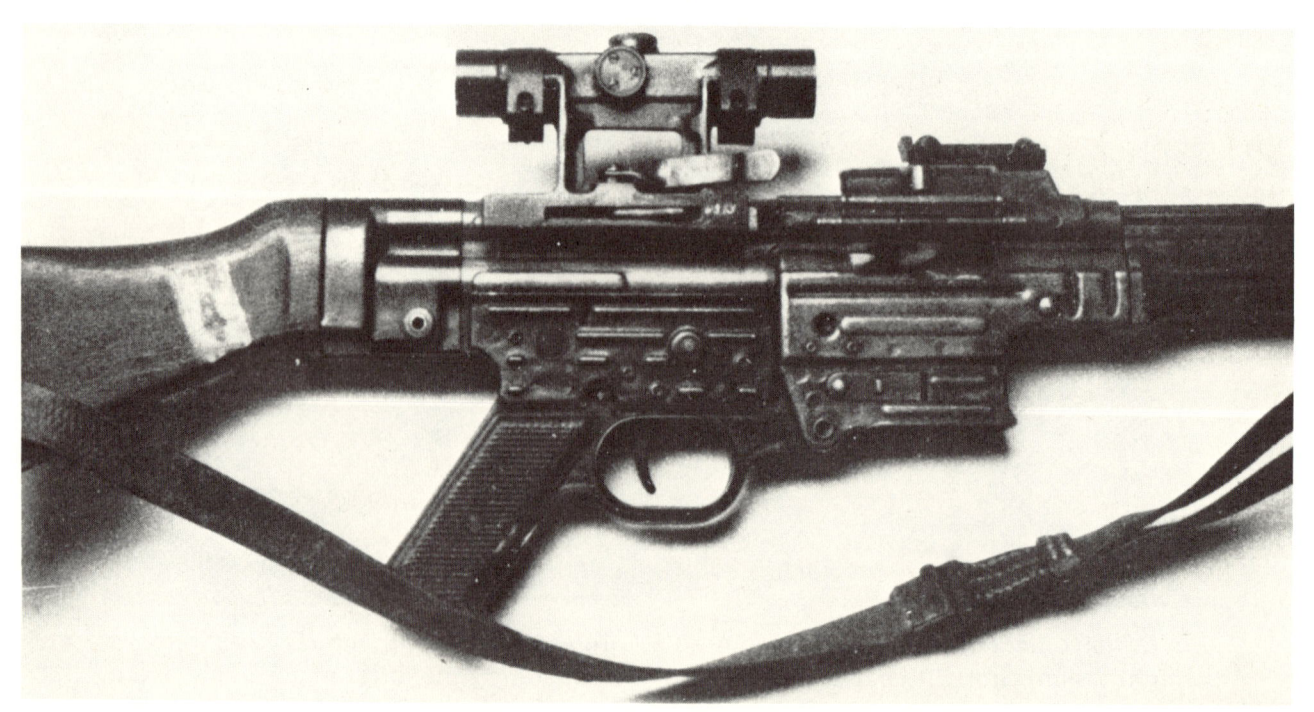

Sturmgewehr 44 (StG44 assault rifle) mounting a "ddx" code (Voigtlander) ZF 4 sight. The Haenel firm was reported to have been responsible for fitting both ZF 4 and infrared sights (ZG 1229) to the assault rifles.

TELESCOPIC SIGHTS — THE ASSAULT RIFLES

German paratroop assault rifle Fallschirmjager Gewehr 42, early version.

Top view of early FG 42 receiver, serial No. 1700, "fzs" code (Krieghoff), with integral dovetail form on both sides of the receiver (top) for attaching a telescope mount. The shallow machined recess behind the standard sight engaged a corresponding clamp on the scope mount.

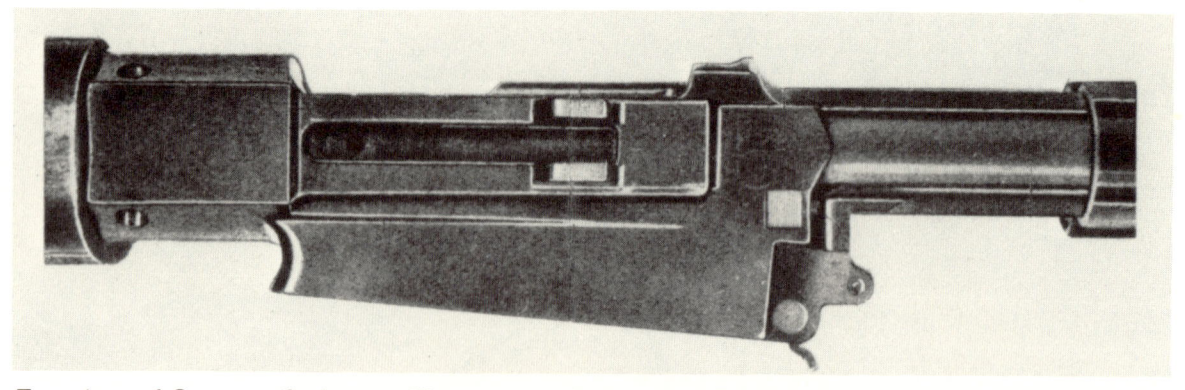

Top view of German Ordnance illustration of an early FG 42 receiver. Note the absence of a machined recess for the telescope mount clamp. The standard rear sight was removed from the receiver.

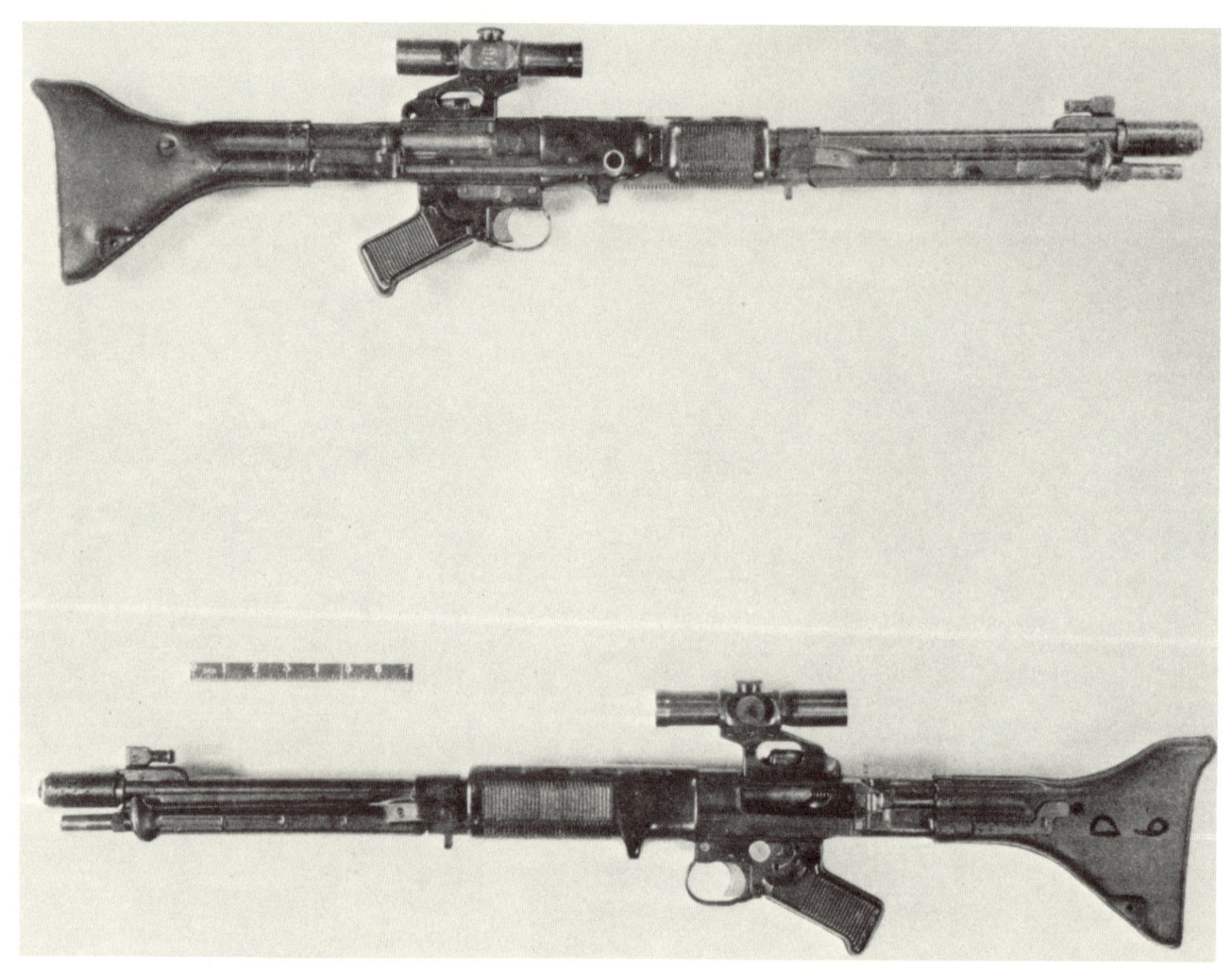

Early FG 42 mounting the limited production ZFG 42 telescopic sight. The scope bears the designation:
ZFG 42
FL   152973
ddx
104049
In addition to the model designation, code, and production number, the "FL 152973" is a Luftwaffe stock number.

TELESCOPIC SIGHTS — THE ASSAULT RIFLES

Improved production version (late model) Fallschirmjager Gewehr (FG 42) as issued for German paratrooper use during World War II. In much the same manner as on the original version, a small dovetail on either side of the receiver (top) was provided for telescope mounting.

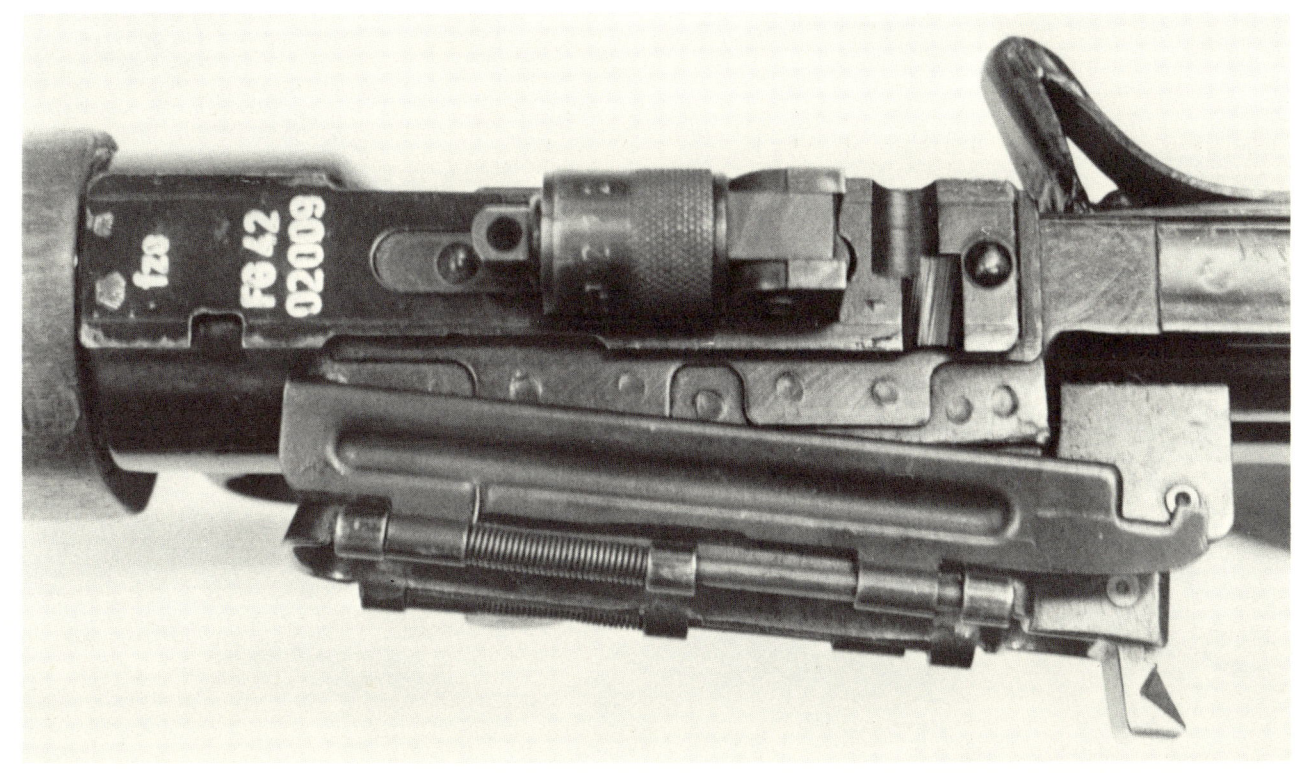

Close view of late model FG 42 receiver area with the standard rear sight folded forward. Close examination will reveal the small dovetail on both sides of the receiver (top). In this case, two shallow machined recesses were provided to engage locking clamps on either side of the scope mount. The "fzs" production mark was assigned to Heinrich Krieghoff, Waffenfabrik.

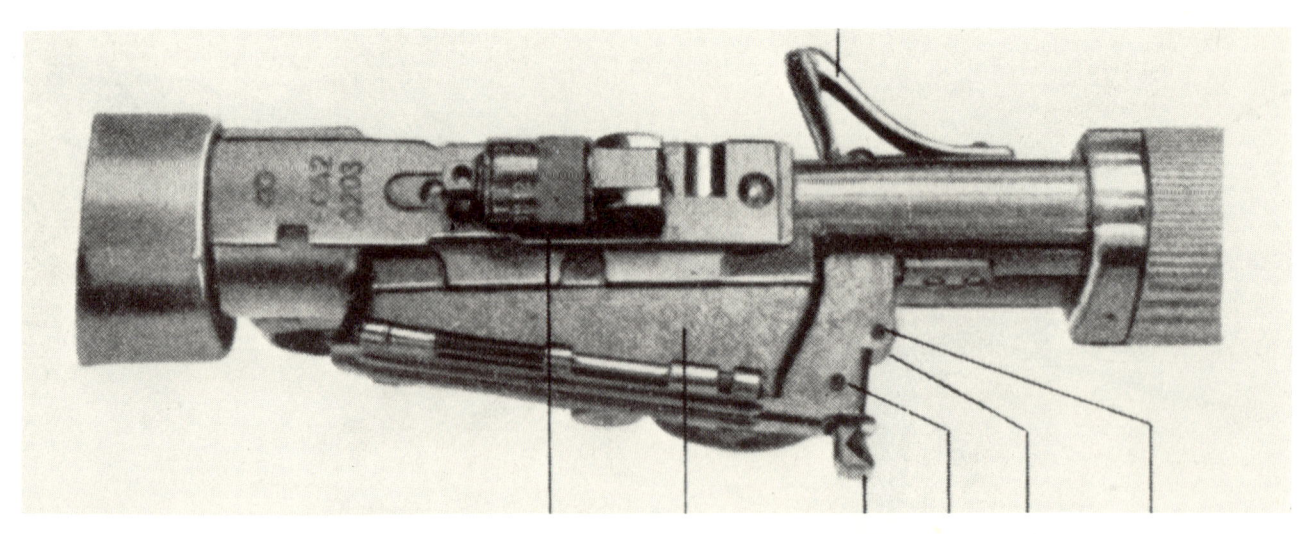

German Ordnance illustration of an improved FG 42 assault rifle (serial No. 0203). Note the single machined recess behind the rear sight.

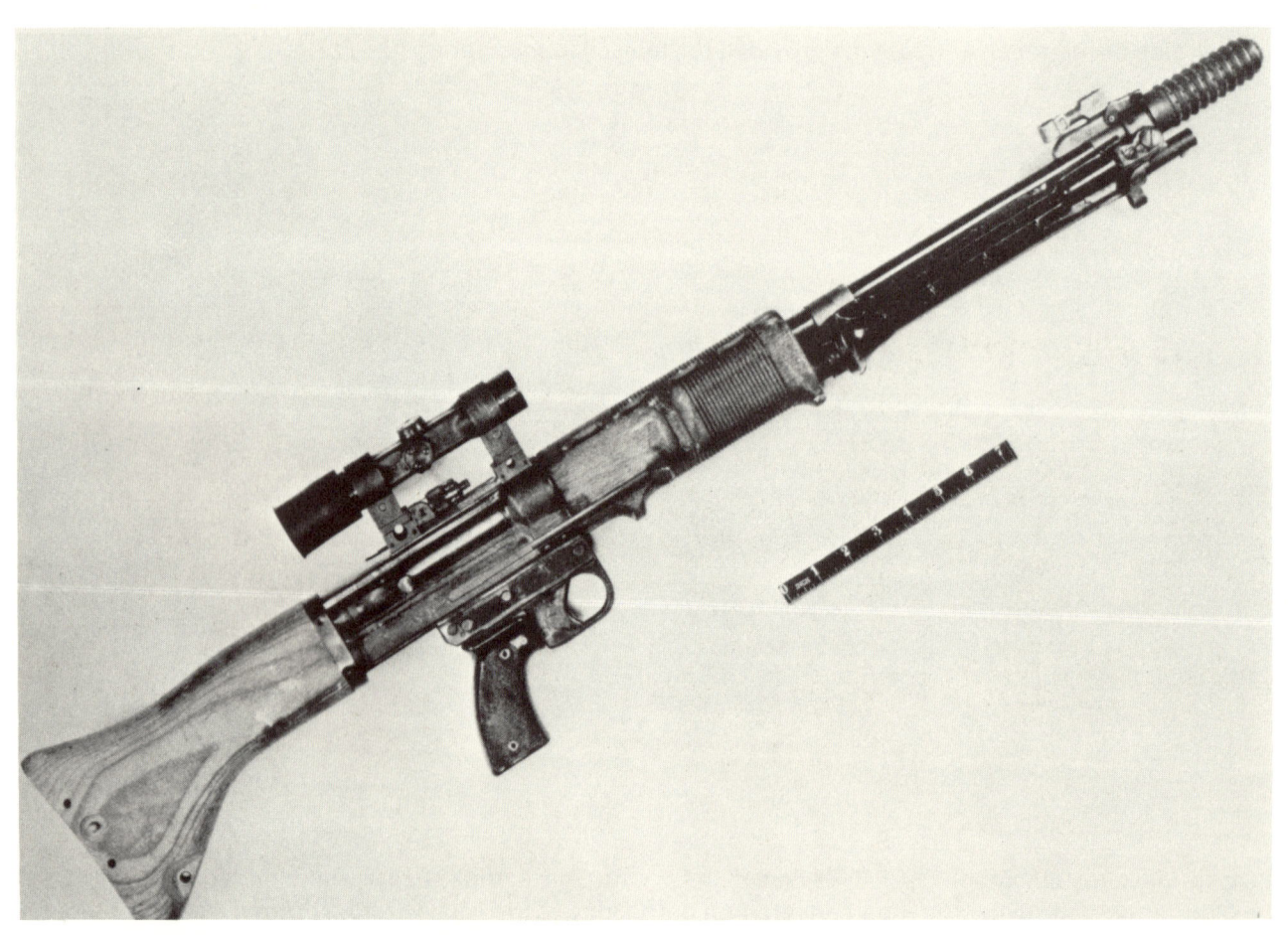

Right view of late model FG 42 with ZF 4 telescopic sight and variant mounting. The main mount locking screws are missing.

TELESCOPIC SIGHTS — THE ASSAULT RIFLES

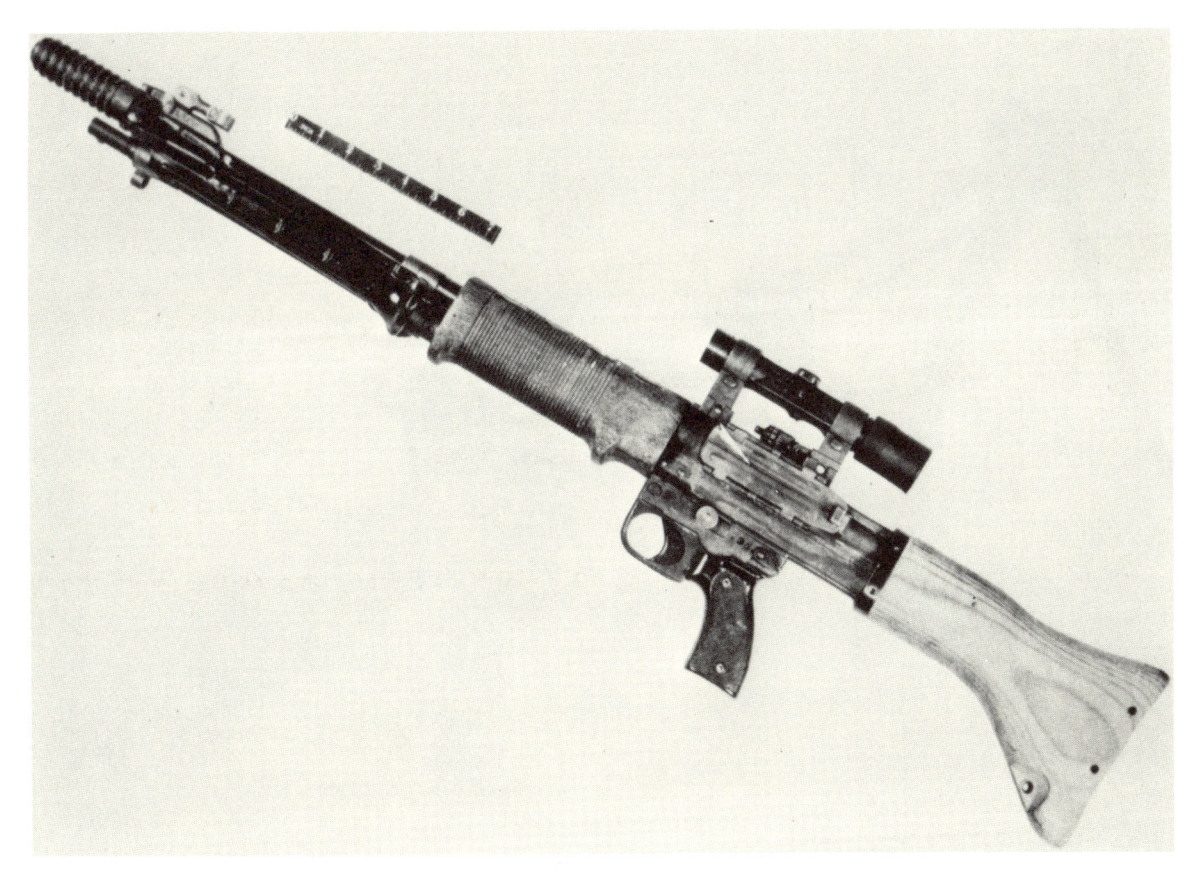

Left view of FG 42 Luftwaffe assault rifle. Note the prominent "L" marking on the ZF 4 telescope.

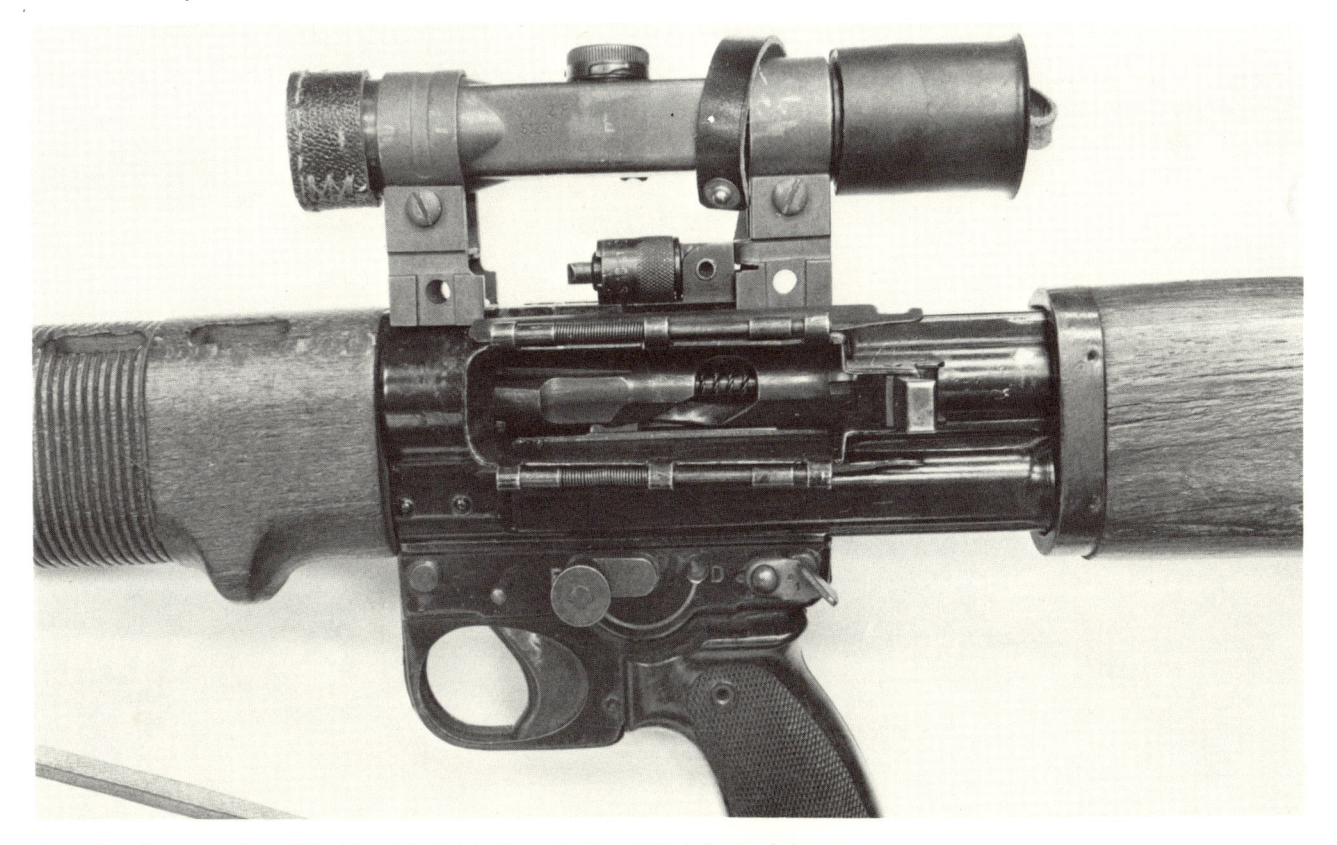

Another late version FG 42 with "ddx" code Gw ZF 4-fach sight.

FG 42 (late version) with ZF 4 sighting system. The main mount locking screws used for clamping the telescope assembly to the receiver are present.

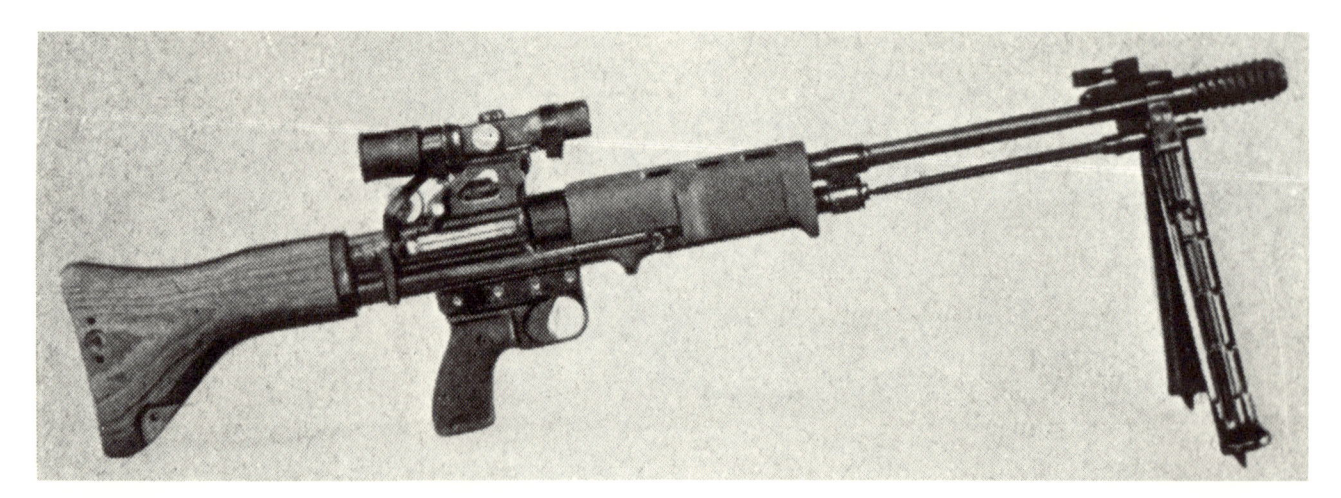

Variant ZF 4 telescope mounting as noted in the 1944 Luftwaffe FG 42 manual.

TELESCOPIC SIGHTS — THE ASSAULT RIFLES 391

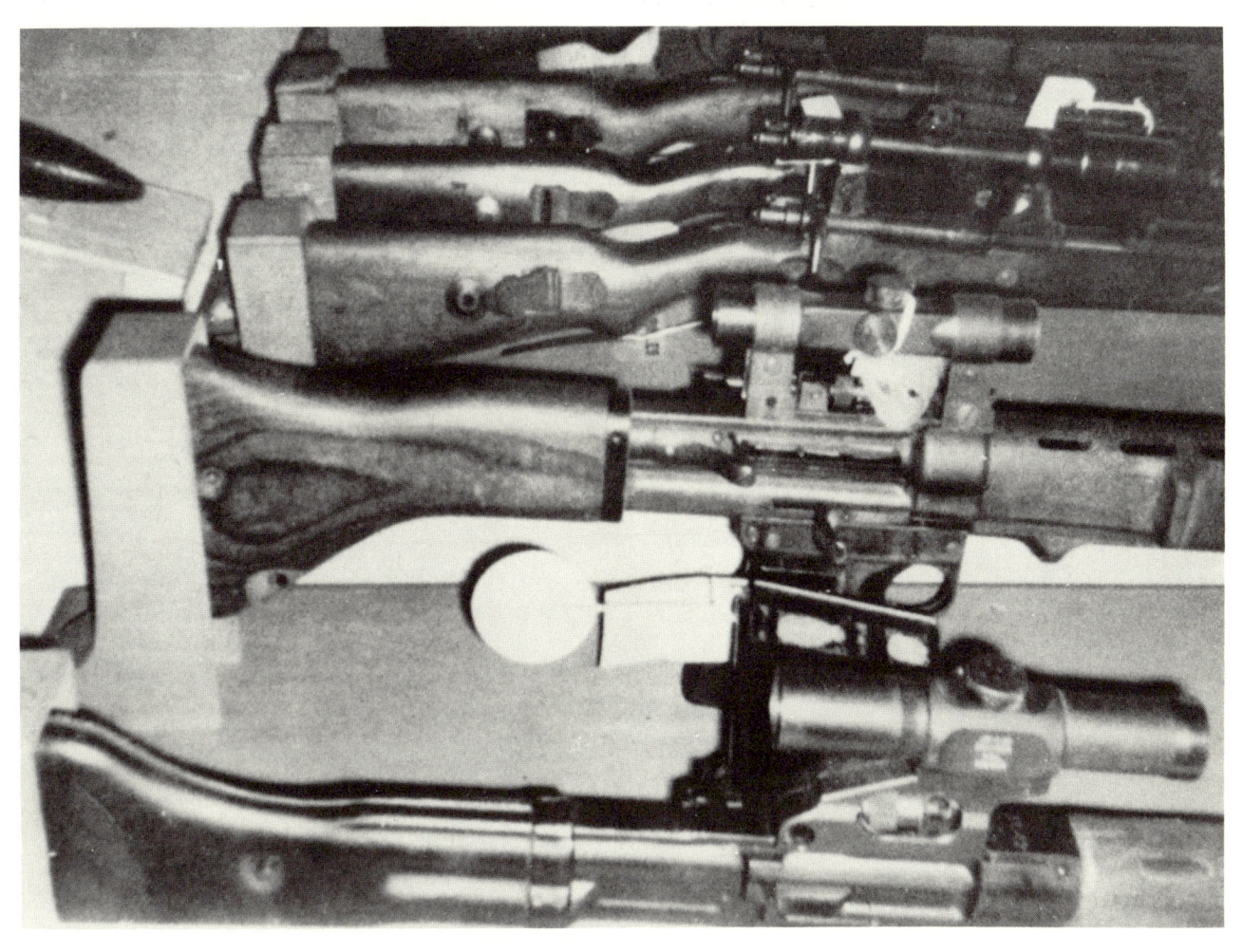

A part of the vast U.S. Army Ordnance Museum collection, the improved FG 42, serial No. 04189 (center), mounts a ZF 4 sight, while the partially obscured early model FG 42, serial No. 1831, is shown with a ZFG 42 telescope. Note the locking clamp missing from the lower rear corner of the scope mount.

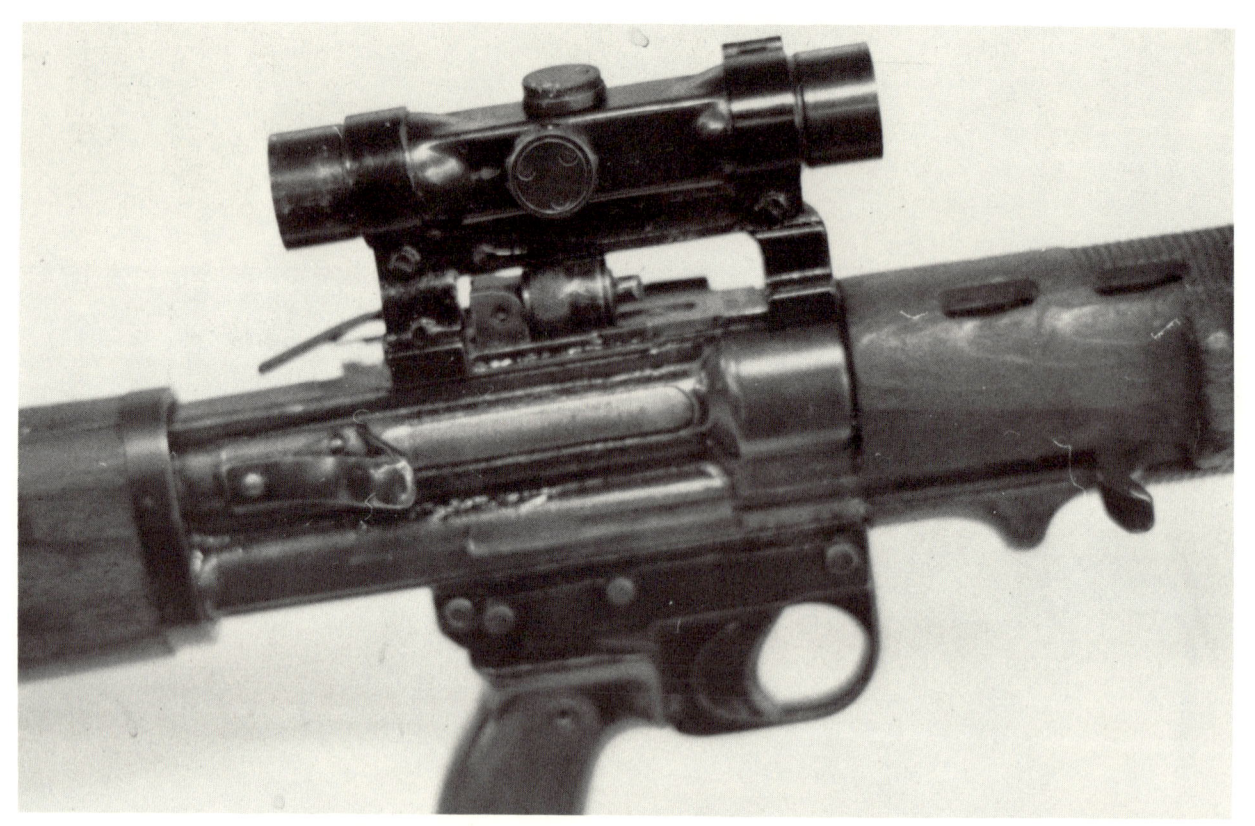

Variant ZF 4 telescope mounting as noted on FG 42, serial No. 02500, in the Armamentarium, Delft, Holland.

Left view of FG 42, serial No. 02500 and unique ZF 4 sight which bears no markings other than an unidentified logo—similar to that found on some post-war Czech scopes, and the legend 4X 4.5°. The lever served to lock the mount on the receiver.

Another unidentified ZF 4 scope with markings identical to that mounted on FG 42 serial No. 02500. The number (272635) was lined-out during postwar Czechoslovakian use.

Left view of ZF 4 telescopic sight showing the serial number and "ddx" code. The letter "L" indicates Luftwaffe issuance, and it is believed that scopes so marked were intended for use with the FG 42. The small colored triangles appearing on all ZF 4 sights represent the reliable function limits of each device in various climates. A white triangle indicates acceptable use under normal weather conditions encountered in central Europe; green, the standard German military color for tropical reference; and blue for extreme cold weather that confronted the Wehrmacht on the eastern front.

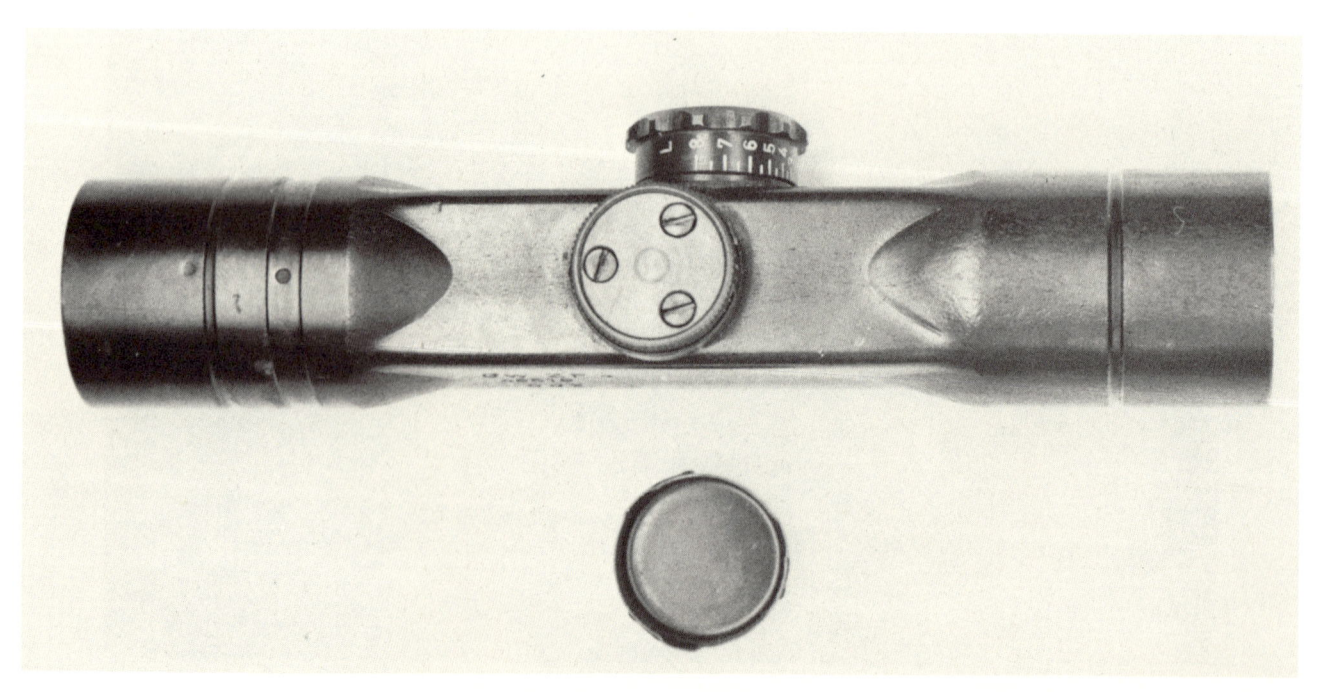

Top view of ZF 4 scope with protective sheet metal cap removed from the windage drum. Note the letter "L" above the 800 meter mark on the range drum.

TELESCOPIC SIGHTS — THE ASSAULT RIFLES

German paratrooper removing the ZF 4 scope from its carrying case prior to mounting the sight on his FG 42. This very rare case contained a side pocket which held the objective rain shield. An additional pocket in the lid carried the lens discs and cleaning cloth. (from 1944 Luftwaffe manual.)

German paratrooper mounting the ZF 4 sight assembly to his FG 42. (from 1944 Luftwaffe manual.)

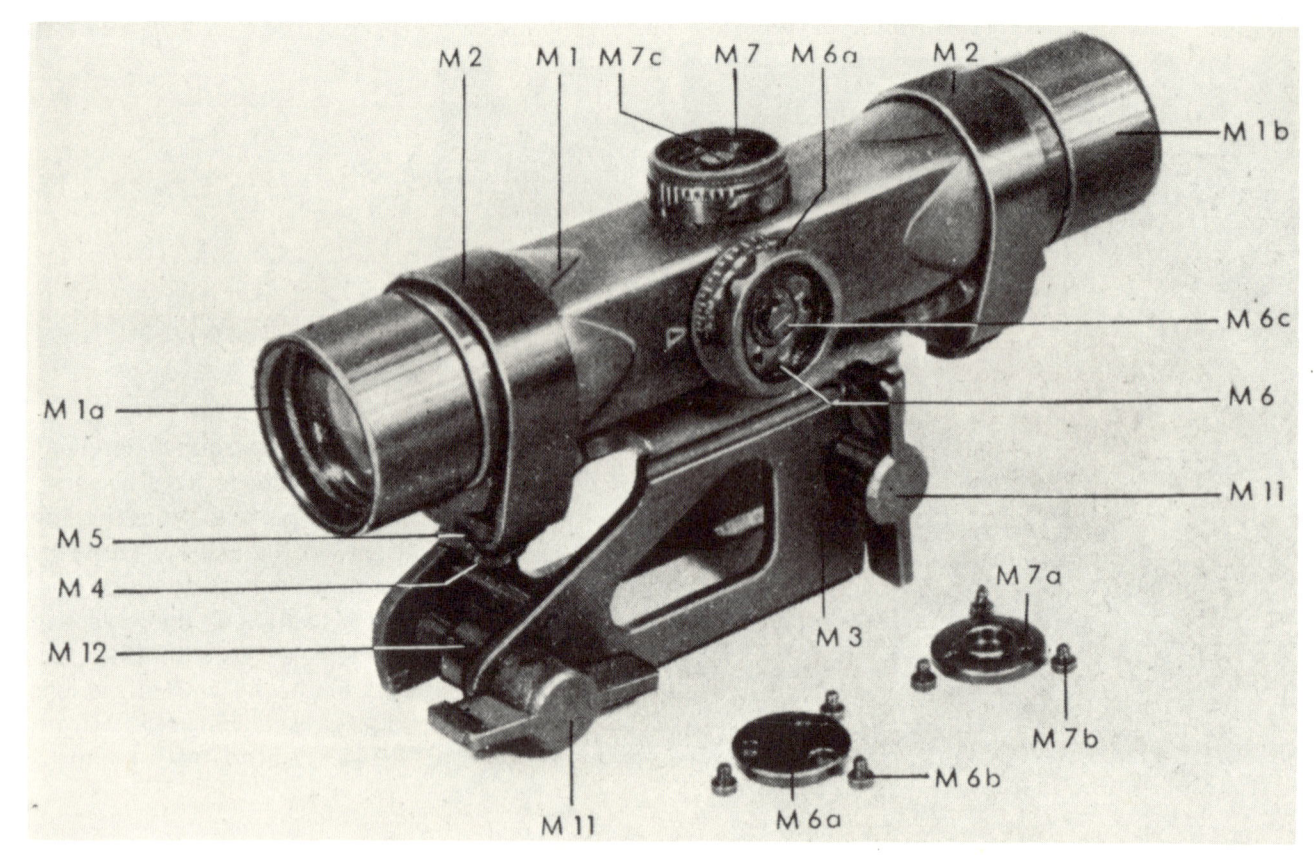

A description of the ZF 4 sighting system as translated from the 1944 FG 42 manual. "The telescopic sight (M1) is mounted by means of two holding screws to the supporting base (M3) and is clamped by means of two wedges (M5) and two locking screws (M4). The screws use a lockwasher to prevent loosening. The telescopic sight (M1) has a height adjustment (M6) on the right side and on top in the middle is the side adjustment (M7). To prevent injury to your head a eye guard (M8) made of rubber is slipped over the eye piece (M1a). Over the lens (M1b) a rain guard (M9) may be slipped. Eye piece (M1a) and lens (M1b) are protected against dirt by the lens guard (M10). This guard consists of a wooden plug (M10a) and a cap (M10b) that are tied together and are fastened to the telescopic sight (M1) with a looped leather belt (M10d). The mounting plate (M3) is formed like a swallow tail, is slipped on the gun sight plate of the FG-42 and locked at the rear by means of a half round bolt (M12) in a recess of the gun sight plate. Two toggle clamps (M11) lock the holding foot and thereby securely hold the complete ZF in position."

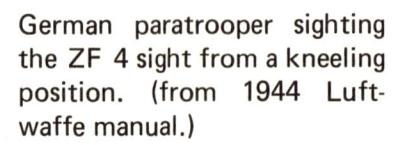

German paratrooper sighting the ZF 4 sight from a kneeling position. (from 1944 Luftwaffe manual.)

Intended to demonstrate the correct firing positions, the German paratrooper supports his weapon against a tree. (from 1944 Luftwaffe manual.)

398 THE GERMAN SNIPER 1914-1945

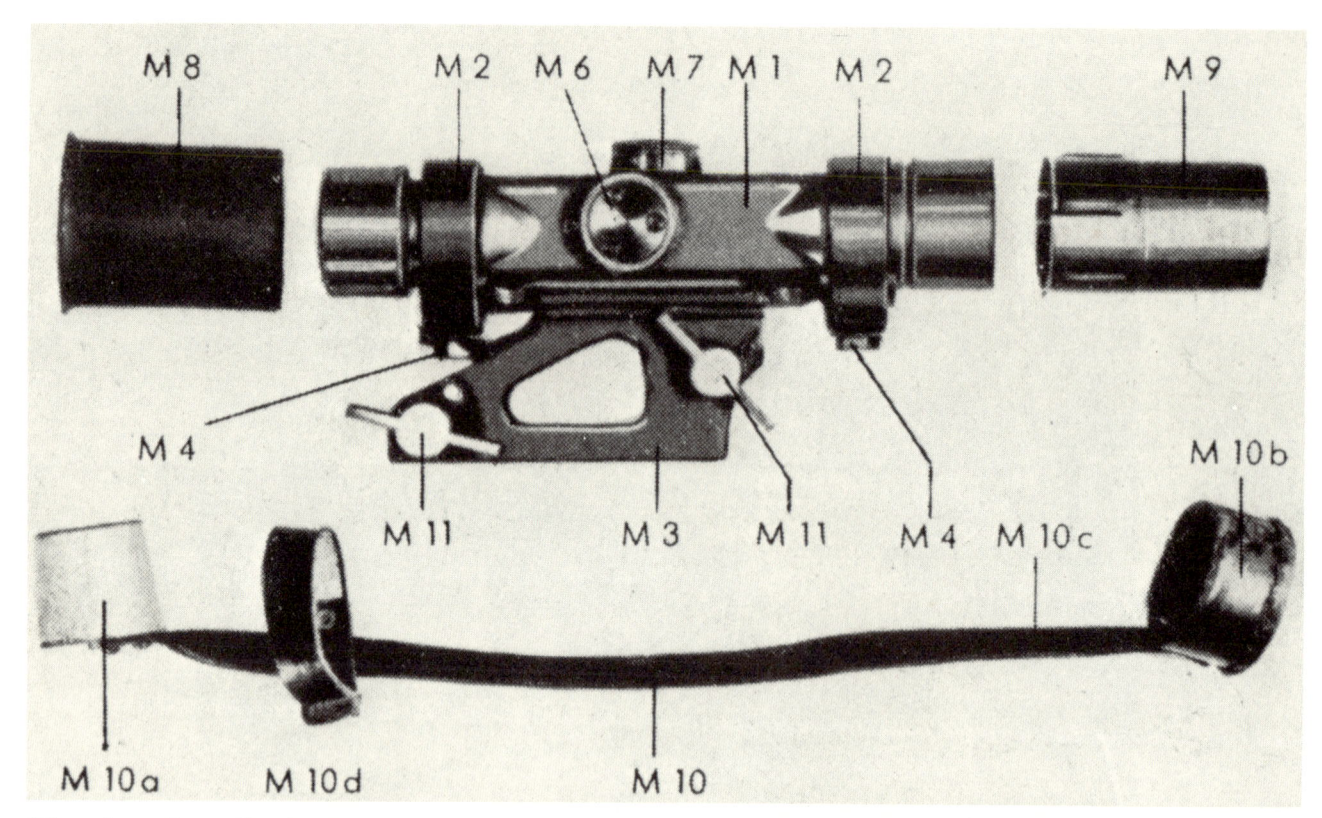

Directions for adjusting the ZF 4 sight as translated from the 1944 FG 42 manual. "To adjust the ZF slip the unit onto the sight finder of the rifle; Tighten the clamps (M11) by hand. Remove the protective cap for the side adjustment. Remove the holding screws (M6b & M7b) of the cover plates for side adjustment (M7a) and height adjustment (M6a). When the cover plates are removed, the adjustment screws can be seen (M6c & M7C). By turning these screws either to the left or to the right the target can be centered in. The adjustment is made after the height adjustment knob is set to 100 meters and the target plate is set at a distance of 100 meters also. On this adjustment target disk or plate, a triangle 10 cm high is marked. (Triangle at bottom) The adjustment is made that the triangle is right side up. Single shots are fired, the hit on the target is verified and corrections made until the sight is adjusted perfectly, while the triangle of the sight is adjusted perfectly, while the triangle of the sight is perfectly matched up with the triangle of the target disk. After the adjustment is completed the cover plates (M7a & M6a) are replaced and the guard caps for the side adjustment are replaced."

German paratrooper firing from the prone position using the bipod assembly (from 1944 Luftwaffe manual).

TELESCOPIC SIGHTS — THE ASSAULT RIFLES

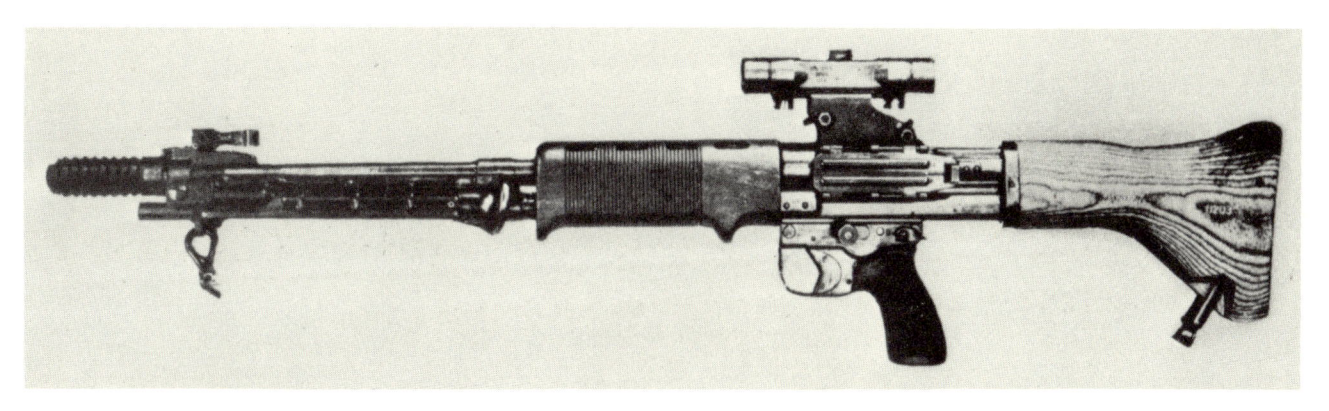

Variant ZF 4 telescope mounting representing one of a variety utilized with the FG 42 weapon system.

The exhilaration of parachute assaults long past, this paratrooper reflects the bitterly contested action of the Italian campaign. Garbed in "splinter" pattern smock and helmet cover, a lull in the fighting is used to clean the ZF 4 sighted FG 42.

CHAPTER XVI

Sniper—The Soviet Approach

In the months following the invasion of Russia, initial basic measures were implemented to offset the distinct advantage held by Soviet sharpshooters armed with their telescopic-sighted, bolt-action Mosin-Nagant and semi-automatic Tokarev rifles. The Armed Forces High Command (OKW), confronted with the almost insurmountable task of fielding both satisfactory telescopic-sighted rifles and competent marksmen, were hard-pressed to formulate an effective sniping program in the shortest possible time.

With no precedent other than what had been established by the Imperial German Army during the Great War, the OKW was forced to assimilate and improve on the state of the art then practiced by the Red Army.

Without question, the effect of Soviet sniping was the predominant influence determining the course of action taken by the High Command in 1942. A candid appraisal pertaining to this early situation appeared in the German publication, *Hamburger-Fremdenblatt*, May 1944, written by Captain Borsdorf of the Reserves. "The Russian soldier was and is, an individual fighter, a factor that must not be underestimated and indeed, the Russian sniper has for a long time been a very real factor which has influenced the training of our own modern infantry."

By mid-1943 a practical, unified doctrine had been established, and the German sniping program finally moved into high gear. By war's end, there was little difference between the sniping techniques practiced by both factions.

An effective comparison can be drawn from the following document, which appeared in a Soviet military publication in October 1943.

> The principal task of the sniper is the destruction of the most important enemy targets he can find. Officers, observers, scouts, liaison officers, enemy snipers, crews manning guns, trench mortars and machine guns, antitank riflemen and motorcycle skirmishers are his chief targets. He blinds enemy armored car and tank drivers by firing at their visors. He is capable of independent action under most difficult conditions in battle.
>
> As a rule snipers are given assignments in offensive action by their platoon or company commander, who at the same time tell their snipers of the immediate objective of the platoon or company. This is helpful in that he must clearly understand what the task is. When the attack begins, but before the rifle sections open fire, snipers advance to the front and to the flanks of the platoon under direct cover of the company. Moving independently from one firing position to another, and carefully adapting themselves to the terrain and making use of natural conditions of the particular locality, snipers fire at targets which hamper the advance of their units. Their operations, particularly after the company or platoon opens fire, must be closely synchronized with the fire of

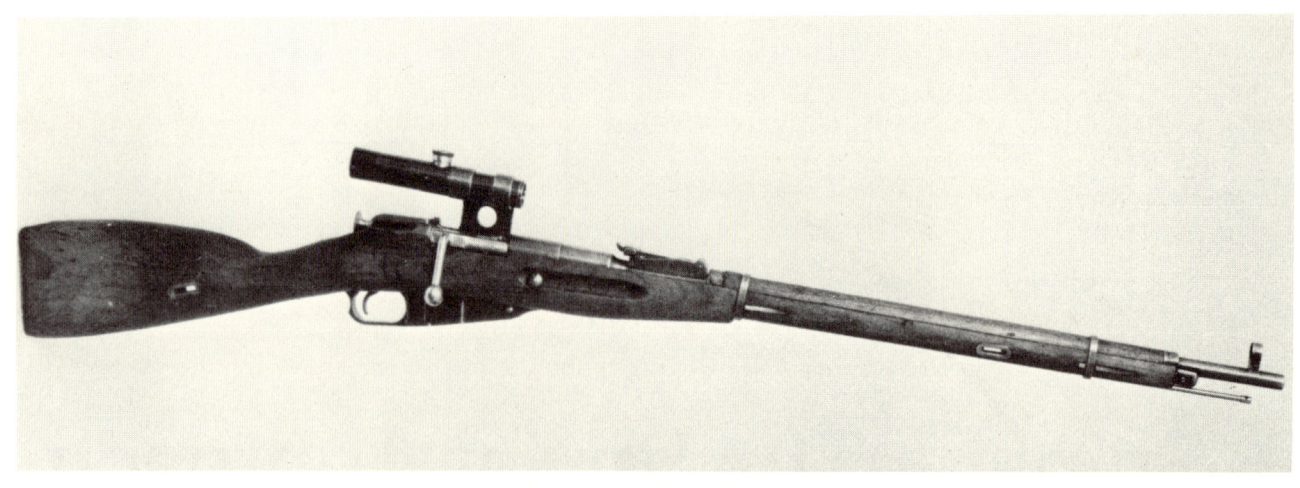

Typical Soviet Mosin-Nagant M91/30, 7.62mm sniper rifle with 3.5-power PU telescopic sight. In this case, manufactured at the Izhevsk Arsenal in 1944. This type of rifle served as the Red Army's principal sniping issue during World War II.

Right view of the 167.39mm long, 3.5-power PU telescope provided a field of view of 4° — degrees — 30' -minutes with an eye-relief of approximately 71mm. A turned-down bolt-handle was used to clear the sight. This particular scope was manufactured at the optical works in Leningrad.

mortars, machine guns and even tank forces.

When platoon or company fire begins, snipers move to the flanks to cover the concentrated attack. Their principal function at this time is to destroy the machine gun and mortar crews.

Snipers also participate in preparations for an attack by spotting and destroying targets which might be detrimental to the success of the impending attack. As artillery and mortar fire is shifted deeper into enemy defenses, snipers direct their fire at enemy fire points which reappear after the artillery barrage. They also fire on the enemy's main line of resistance, and when their own tanks appear and attack, snipers shift their fire to enemy antitank guns and crews, men with rifle grenades, bottles of combustible liquid and flame throwers. During subsequent action snipers remain at the point of concentration to support the infantry attack, their main targets now becoming pillbox embrasures, camouflaged machine gun nests, enemy flanks, antitank cannon and machine gunners.

When the enemy's main line of resistance has been pierced, snipers advance to positions in advance of their respective platoons and their principal targets are now officers and non-commissioned officers who are leading enemy counter attacks. If the enemy attacks with tanks, snipers use AP bullets and shoot at tank visors.

When the enemy begins to retreat, snipers seek out the riflemen of covering troops and also the officers attempting to halt the retreat and to retain a foothold. During action deep in enemy defenses, snipers march on the flanks of their units to observe and secure the unit against sudden counter attacks from ambush and silence enemy fire which harasses their infantry. Existing enemy firing points are the targets of snipers who precede troops in the second line of advance units.

Light and heavy machine gunners cooperate with snipers to destroy camouflaged fire points. When second line troops are ordered into battle, snipers cooperate with other means of fire to insure deployment. Snipers usually operate in pairs. One such pair must always be within close reach of the company commander who gives them orders to destroy particularly important targets that appear. These snipers conduct uninterrupted observation and help the commander to define enemy plans. They also insure the safety of the company as he directs the fight.

In defensive fighting a rifle platoon or company must firmly cling to the sector it holds by inflicting the heaviest possible losses on the enemy and frustrating his attempts to attack. Snipers operating and fighting with units on defense operate in groups and singly, picking the most important targets appearing in their unit zone of fire. The best positions for snipers in defense are on platoon flanks, at the main line of resistance and in front of it, as well as in the depth of the company's defenses. They may also advance with cover troops interspersed between fire points on both flanks and in front in a zone of obstacles. In this case a few well disguised positions must be prepared before hand in order to keep the enemy under fire from different directions.

As a rule, snipers cover the retreat of their rear guard troops and seek out enemy artillery observers who usually follow close behind. At the same time the company commander conducts observation reconnaissance, dispatching additional scouting detachments and groups with instructions to delay and destroy attempts of the enemy to overtake his soldiers and officers. Sometimes snipers deliberately lag behind cover troops and permit the enemy to pass. They then take up camouflage positions and pick off enemy officers and other important targets from the rear.

When the enemy is preparing an attack the task of the sniper is to harass his operations with well-aimed fire. For this purpose it is advisable for snipers armed with submachine guns to take up positions in front of the MLR. In one instance a company commander after occupying a defense

Left view of the elevation adjustment drum located on top of the PU sight, graduated from 1 to 13 (100 - 1300 meters). The windage drum has 10 graduations in either direction, beginning with zero; plus markings for corrections to the right; and minus markings for corrections to the left. There was no focus adjustment with these sights.

A unique photo taken through the Soviet PU telescope illustrates the typical reticle pattern.

district placed three pairs of snipers in front of his main line. He supplied them with two submachine guns and sappers (engineers) who helped the snipers prepare their positions and connected them by communication trenches with the main line of resistance. When, under cover of artillery fire, the Nazi troops began to concentrate on our main lines, the Soviet snipers and submachine gunners outside of the zone of artillery fire poured in well-aimed fire, picking off officers, gun crews, and advancing artillery observers. The fire of the snipers and machine gunners was so effective that the enemy was prevented from advancing with his machine guns and trench mortars. Moreover, the fire misled the enemy as to the location of our main line of resistance and when the Hitlerites finally opened fire our snipers had moved their positions. Every pair of snipers must have at least two spare positions in order to enable them to keep the approaches to the main line of resistance under fire as well as other important directions. As enemy tanks pass through their own lines of infantry, snipers along with other members of the company fire armor-piercing bullets at the visors of the tanks.

When the enemy infantry swings into action the snipers pick off the officers and leading soldiers. At this stage of the defense the task of the snipers is the same as that of the other members of the defense force: to cut the infantry formations away from the tanks they are following and pin them to the ground in front of the MLR until later, when in close synchronization with other means of fire it is possible to bring about the demoralization and destruction of the enemy infantry. In the meantime, snipers of the second line cooperate with other means of antitank fire to fight the enemy tank advance. The second line of snipers either destroys enemy troops who may have infiltrated into the defense zone or pins them to the ground with well-aimed fire. In this way snipers prepare for our eventual counter attack.

Darkness restricts the activity of the snipers. Therefore, whenever possible they should be allowed to rest so as to be able to function efficiently the next day. Generally speaking though, snipers, may be used as sound locators and observers as well as for reinforcing night-scouting parties.

In operations on stable sectors, Soviet snipers have missions similar to those they have in defense sectors. Their main targets are officers, observers with periscopes, liaison officers, gun turrets, embrasures, and low flying aircraft.

German recruits as a rule are not cautious and frequently present splendid targets. A seasoned German soldier, however, behaves differently and will not abandon his hole until he is certain that there is danger; he will resort to a thousand and one precautions and tricks to mislead the Soviet snipers.

A sniper must be able to outwit and mislead the enemy. One able Soviet sniper habitually goes on a mission carrying grenades and bottles of combustible liquid. He puts these on the ground near his dugout and then creeps away to another covered position from which he fires into the explosives himself. This diverts the enemy machine gun and mortar fire and enables him to locate enemy dispositions and fire points. He then moves to the best position from which he can go to work on crews he has seen.

Here is another example of skillfull Soviet sniper operations. One black night this sniper crept close to the enemy trenches and found a good hiding place with good observation. The next morning when an enemy officer appeared the sniper killed him with one shot. He then stayed where he was all day, carefully plotting every Nazi movement and keeping the Germans dug in; when darkness fell he returned safely to his own unit. In stable sectors of the front snipers may also be used for listening and observing at close range to the enemy camp. Each slight rustle of sound has meaning for a sniper, a number of sounds betraying to him the strength of the enemy concentration.

Snipers of a platoon detailed for re-

Soviet sniper team. Both are equipped with the telescopic-sighted M91/30 rifle.

Mosin-Nagant M91/30 sniper rifle with a 4-power PE telescopic sight. The 271.27mm long scope provided a field of view of 5° — degrees — 30' - minutes with an eye-relief of approximately 82mm. The elevation adjustment drum, located on top of the sight, is graduated from 1 to 14 (100 - 1400 meters). Windage adjustment is comparable to that of the PU device.

connaissance operations work either with the platoon or independently in pairs. If they work in pairs, they receive their assignments from the platoon commander. On one sector of the front a Soviet scouting detachment saw the Germans bringing up a cannon carriage. The platoon commander ordered a sniper team to destroy the detail bringing up the cannon. The snipers first picked off the officer and the two horses. The crew then abandoned their project and scuttled for cover, leaving the gun in the road.

In some particularly complex conditions snipers may work independently as scouts in groups of three to five men. Here they are usually on observation missions. They creep close to enemy positions or penetrate behind enemy lines. However, it may be advisable at times to include snipers in scouting patrols for they may be needed for more serious tasks which arise during reconnaissance operations. Crack marksmen protect a rifle platoon against surprise attacks while on patrol and when they approach an objective they select favorable positions from which they can observe the action of the patrols and open fire in its support if need be. In an unexpected encounter with an enemy patrol, snipers watch its operation and if the Germans attempt to slip away they cut off their retreat. It is advisable to take a number of prisoners if possible. Should this fail, the snipers protect the withdrawal of their patrol in order that it may continue its mission intact.

When patrols are attacked from ambush, snipers occupy positions on the flanks and cover the attack of their platoon. A reconnaissance patrol may sometimes be compelled to fight with stronger enemy forces, and in such an event the marksmen perform as they do in a defense or retreat, taking into consideration the specific conditions under which they are operating.

When detailed for night reconnaissance of permanent enemy outpost troops, snipers penetrate through to the main line of resistance of the Nazi and then organize an ambush. When scouting detachments attack, snipers map the disposition of enemy fire points and wipe out enemy cover troops when these troops retreat to their own main lines.

As a matter of interest, when compared to the almost complete lack of German preparations for fielding sharpshooting equipment during the 1930s, according to Soviet archives, a total of 54,160 Model 91/30 Mosin-Nagant sniping rifles were manufactured between 1932 and 1938, and an additional 53,195 telescopic-sighted weapons of the same type were reportedly produced during 1942 alone.

Even though "Das russische Scharfschutzengewehr Mosin-Nagant Modell 91/30" (the Russian Sharpshooter Rifle Mosin-Nagant Model 91/30) served as the principal Soviet sniping arm, the Germans were apparently more impressed with the conventional and telescopic-sighted versions of the M1938(SVT) and M1940 (SVT) semiautomatic Tokarev rifles. (SVT is the abbreviation for Samozaryadnaya Vintovka Tokarev, semiautomatic Rifle Tokarev.)

The firepower capability of a self-loading weapon made it eagerly sought after by Wehrmacht personnel during the early stages of their involvement in Russia when German troops made considerable use of captured Tokarev rifles. Nevertheless, the Russian weapons proved entirely too complex and fragile, with field maintenance and repair extremely difficult. When the severity of the winter conditions rendered these weapons unreliable, they were quickly discarded.

The telescopic-sighted bolt-action Mosin-Nagant, on the other hand, saw sustained use by German sharpshooters even after their own sniping equipment became available. This weapon proved extremely durable to both Russian and German specialists.

Although Soviet forces did utilize the Model 1938 and 1940 Tokarev rifles for sniping purposes against the Germans with reasonable effect, mechanical problems resulted in their eventual withdrawal from field service.

The original use of an operational semi-automatic weapon for combat sniping must be credited to the Russians, a factor which served as impetus for German Ordnance to develop their Selbstladegewehr sniping system along the same lines.

Red Army marksman with a M91/30 Mosin-Nagant sniping rifle mounting a 4-power telescopic sight.

Mosin-Nagant M91/30 sniper rifle with a variant 4-power scope. Vertical pointed post and horizontal side bar reticle patterns were typical of those used in Soviet telescopic sights. Leather lens caps were issued with all sight assemblies. Early war Russian documents cite their 4-power scopes as an "improvement" of the Zeiss design.

The Russian press made extensive use of Red Army snipers as a propaganda weapon. In this case a staged photo was intended to show a marksman bringing down a German soldier.

Soviet sniper armed with the bolt-action Model 91/30 Mosin-Nagant rifle.

Although telescopic-sighted, semiautomatic rifles were employed against the Germans with some success, sniping versions of the bolt-action M91/30 rifles proved extremely durable and saw extensive combat use throughout the war.

Although photos of this type served the Soviet propaganda machine, Russian women, many of whom actually functioned as snipers, were an integral part of the Red Army during World War II.

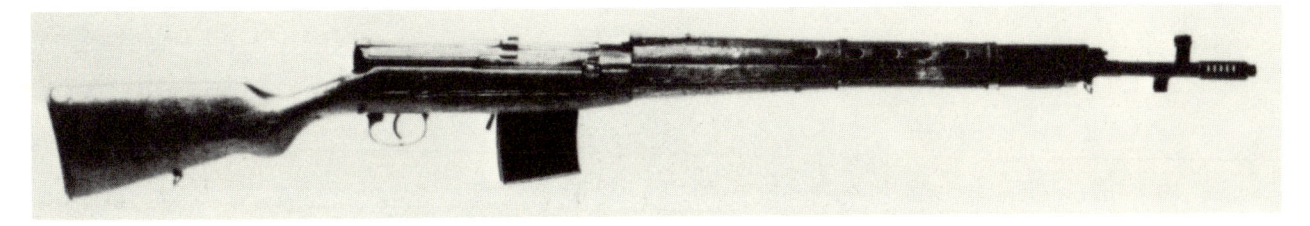

Tokarev Model 1938 (SVT) semiautomatic rifle in standard configuration.

Soviet sharpshooters sighting their Model 1940 Tokarev sniping rifles.

Typical World War II Soviet propaganda photo: a woman sharpshooter is shown with the M1940 Tokarev sniping rifle.

SNIPER — THE SOVIET APPROACH

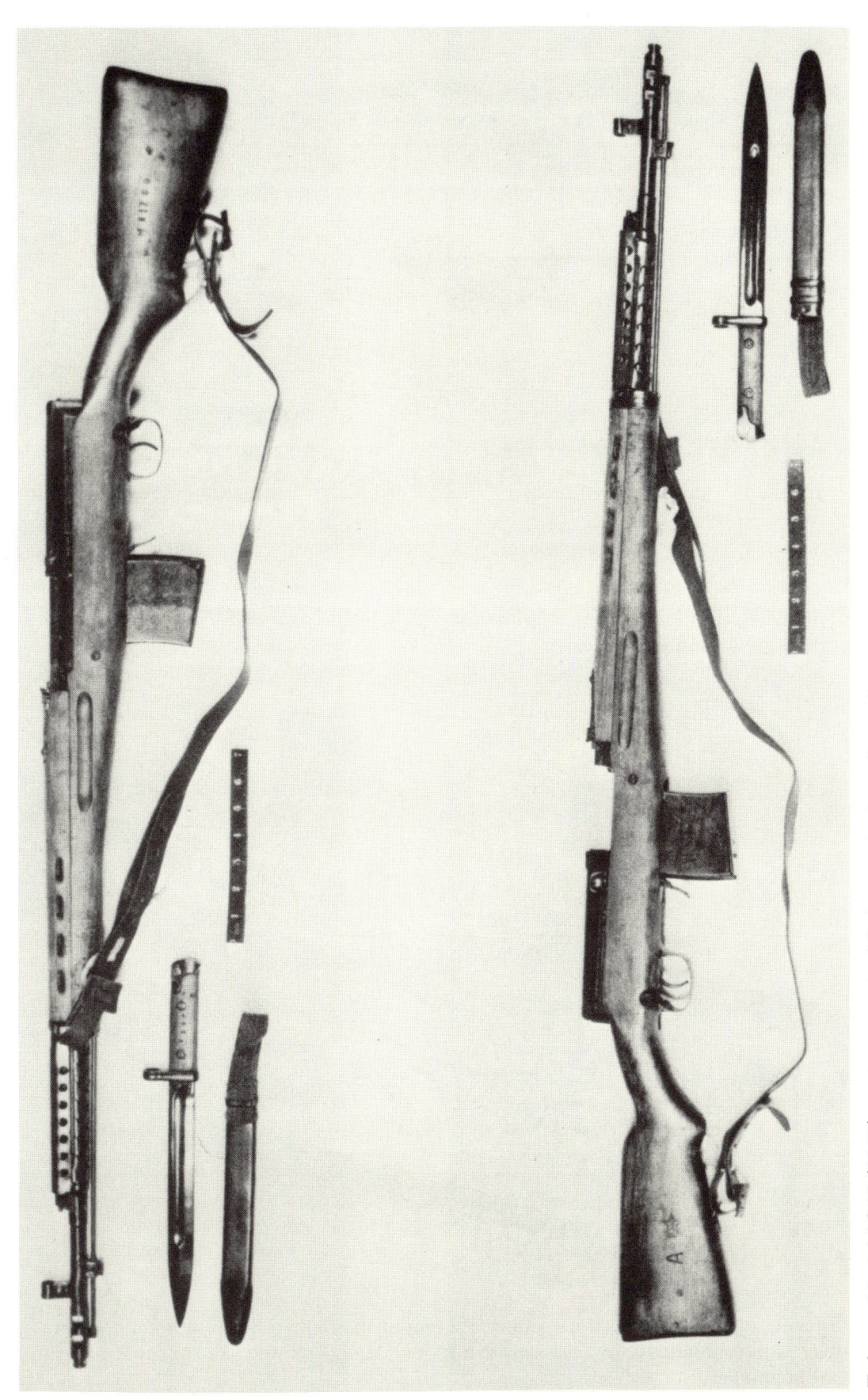

Tokarev Model 1940 (SVT) semiautomatic rifle in standard configuration.

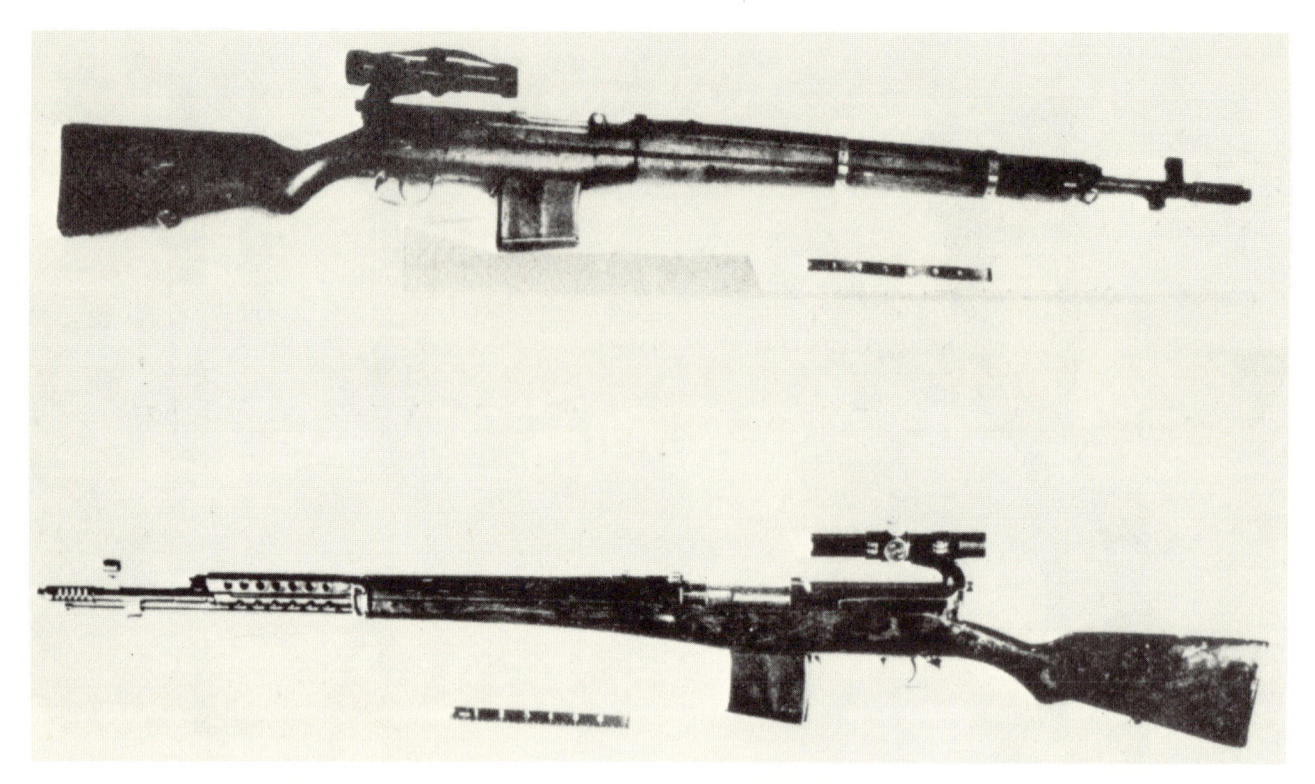

Sharpshooting versions of the Soviet M1938 (top) and M1940 Tokarev (SVT) semiautomatic rifles with 3.5-power telescopic sights.

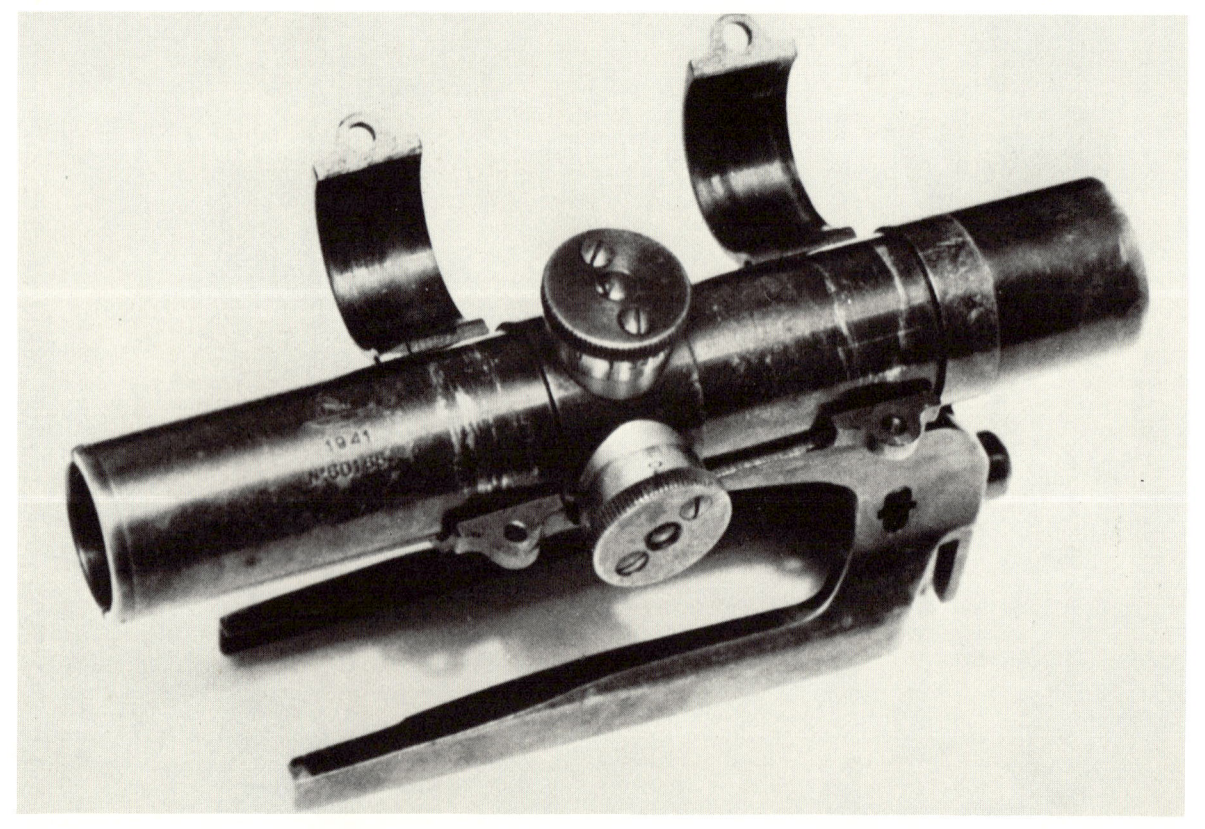

While essentially the same as the standard PU model, the main tube diameter of the back section on those sights, intended for use with the Model 1940 Tokarev (SVT) scope mounting, was somewhat smaller.

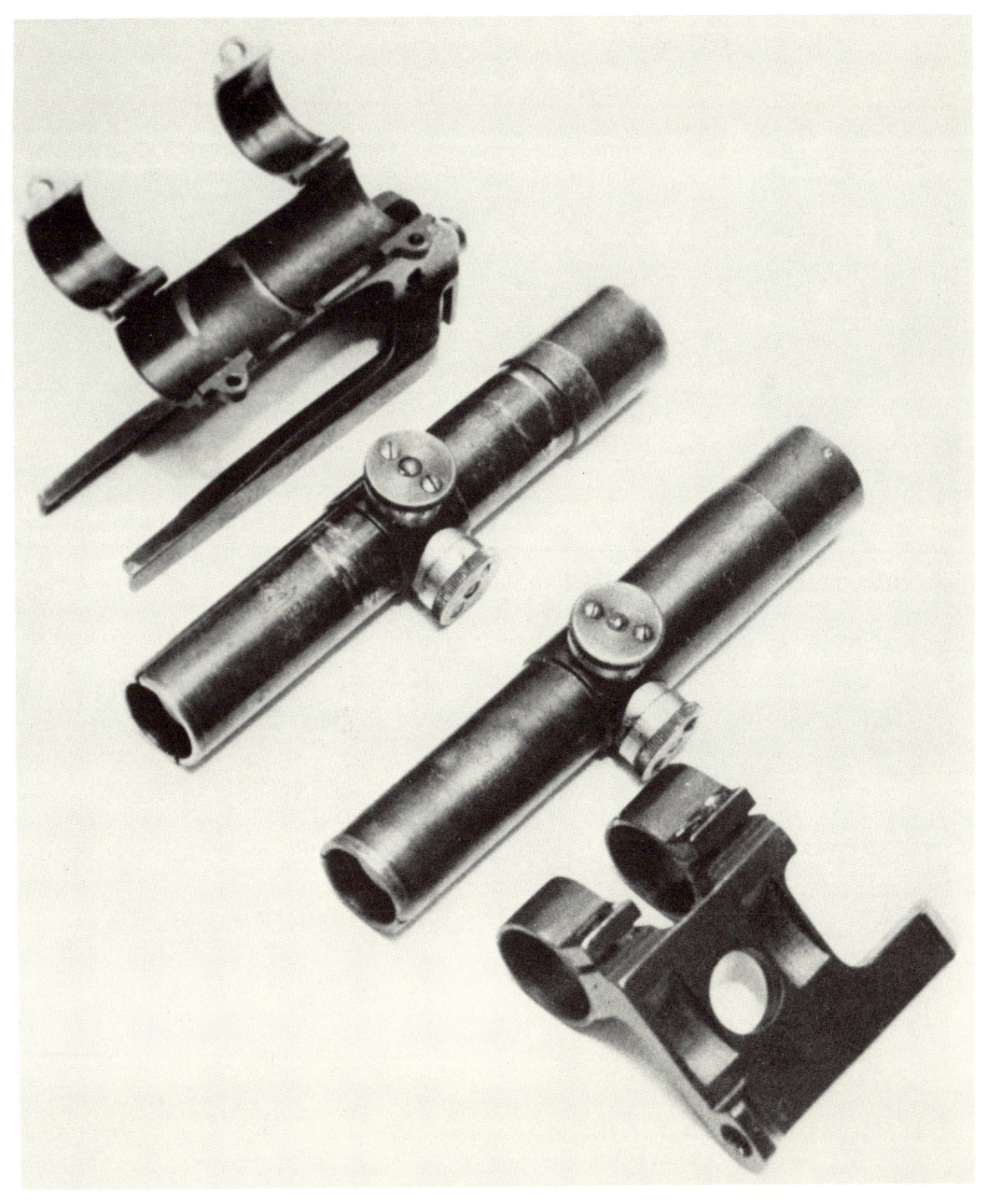

A comparison of the telescopic sight and mount intended for the Model 1940 Tokarev (SVT) rifle (top) and the standard PU scope and mount used with the M91/30 Mosin-Nagant.

A working Soviet sharpshooter with a telescopic-sighted M1940 Tokarev rifle.

German personnel at the Eastern Front. The officer (*left*) is armed with a captured Soviet Tokarev semiautomatic rifle.

CHAPTER XVII

Sniping Weapons Miscellany

The illustrations included in this chapter represent segments of German sniping hardware that only favorable circumstances have permitted to exist to this time. Although significant pieces of operational and prototype sniping miscellany were destroyed or discarded as worthless junk, some may still exist amongst the unidentified objects possessed by collectors and shooting buffs.

Unfortunately, wholesale transformation of German military rifles into hunting and sporting arms since the war's end has included countless numbers of valuable sniping variations. Evidence of this lies in the acute scarcity of original weapons of this type amongst collectors throughout the world.

Judging from the myriad pieces observed over the years, many of the existing German sniping weapons were "made up" for the collector trade long after the Wehrmacht fired its last shot. There exist no foolproof methods to keep serious collectors on the proverbial "straight and narrow" other than to assimilate as much factual material relevant to the subject of German sniping as interest permits. In the same manner as other fields where counterfeits are produced and sold for exhorbitant prices, martial arms collecting has proved to be no exception.

The prevalent practice of "upgrading" an existing weapon is accomplished by providing matching serial number stamping or engraving on the base, telescope mount, and related hardware to increase the overall value of a particular rifle. In addition to these deceptions, considerable numbers of conventional 98k rifles are being transformed into sniping weapons by using authentic parts from substandard original pieces as well as newly manufactured scope mounting components currently produced on both sides of the Atlantic Ocean. The new components bear appropriate but equally as bogus Waffenamt markings. With most German sniping weapons in the four-figure price range, the motivation for this sort of activity is quite obvious.

At some future point the contemporary manufacture of bogus German sniping equipment will undoubtedly attain a highly advanced state. At present, this level of sophistication has not been achieved in most cases. Therefore, careful diligent study by a knowledgeable individual will still reveal a counterfeit sniping rifle.

Aside from the well-known military sniping variations attributed directly to German Ordnance, the unexplained telescopic-sighted rifles that were fielded for Wehrmacht use either in advance of or during the war have thoroughly perplexed all levels of the collecting fraternity. Even though many of these unusual weapons have proved to be bonafide original specimens dating from this era, all weapons in this category should be viewed with considerable suspicion.

In the interest of a complete presentation of German sniping equipment, a cross section of the hybrid telescopic-sighted weapons fielded for undefined military use before the war, and those brought to fruition as an expendiency during the war, have been included in this section.

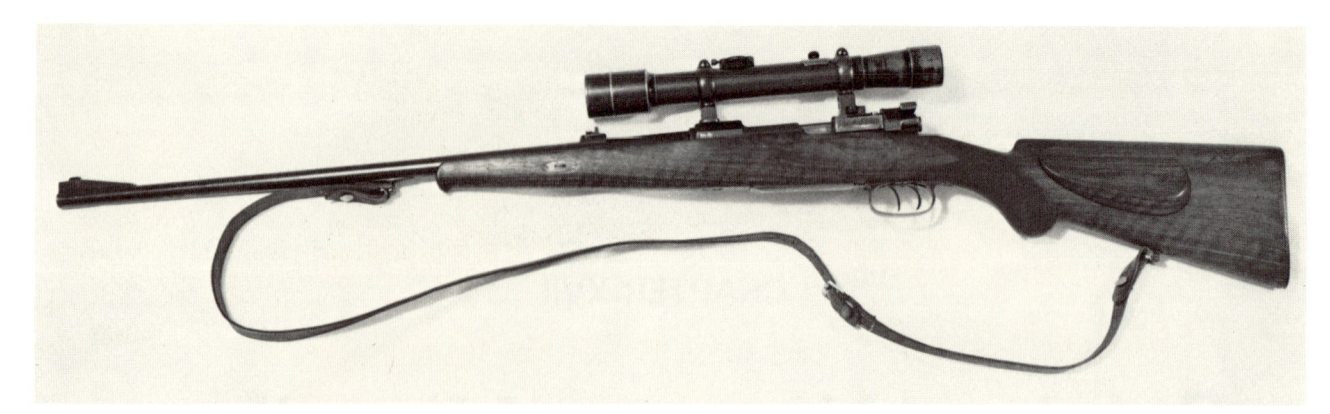

A prewar telescopic-sighted commercial sporting rifle typical of those pressed into German military service through the course of the war. In some cases a lug was fitted beneath the barrel for attaching a bayonet. While hardly regulation issue, rifles of this type have been noted bearing military acceptance stamps. Note the double-set triggers found on many European sporting arms.

A unique and relatively rare early issue item: a web cover designed to protect the 98k action in adverse conditions. While not intended as a sniping rifle accessory, the cover could be used with ZF 41 equipped rifles with the scope removed.

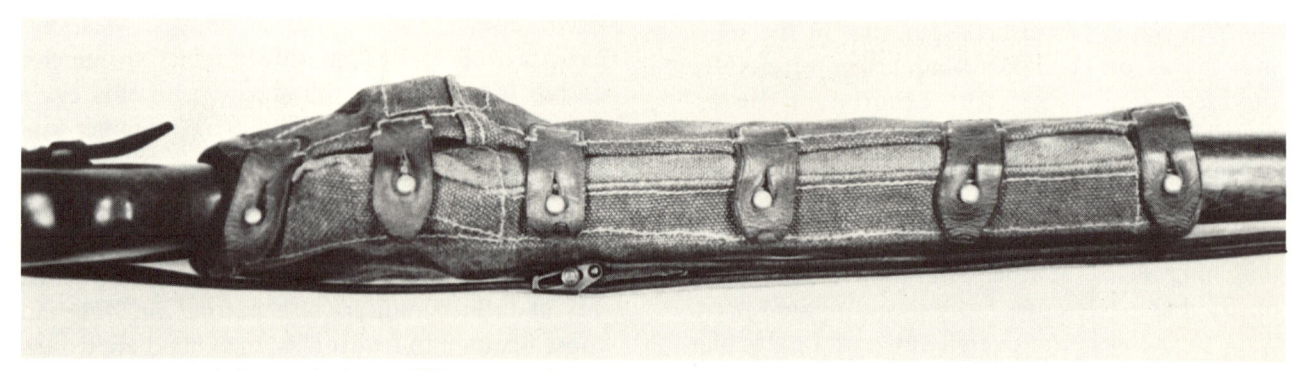

An underview of the early issue 98k protective web cover.

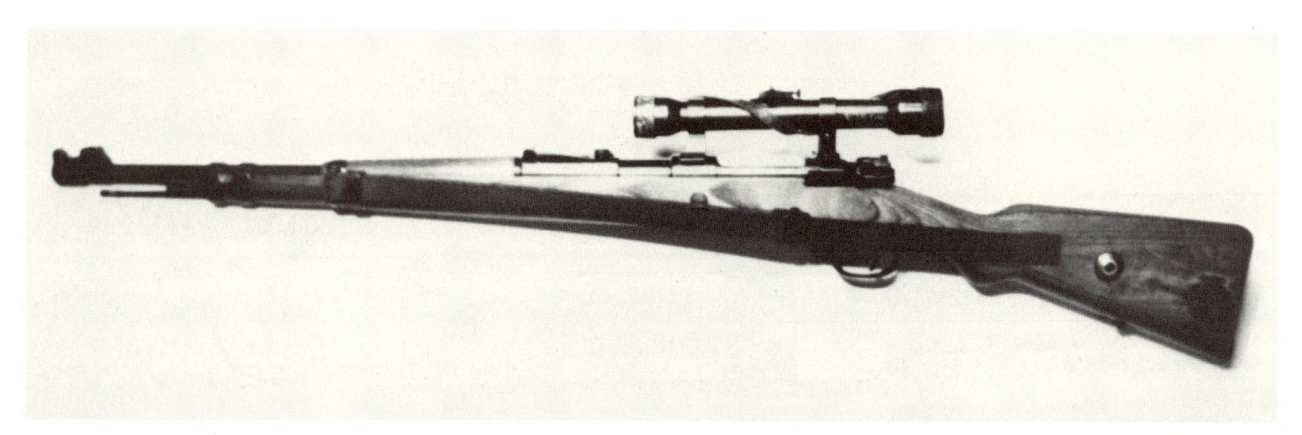

A most unusual prewar Mauser manufacture sporting type rifle in full military trim with 4-power Zeiss (Zielvier) telescopic sight and commercial mount assembly. Bearing the legend, "MAUSER-WERKE," A. G. OBERNDORF a.N.," this unique rifle fires the 5.6 x 61 Super Express cartridge introduced by E. A. vom Hofe in 1937 for deer and varmint hunting.

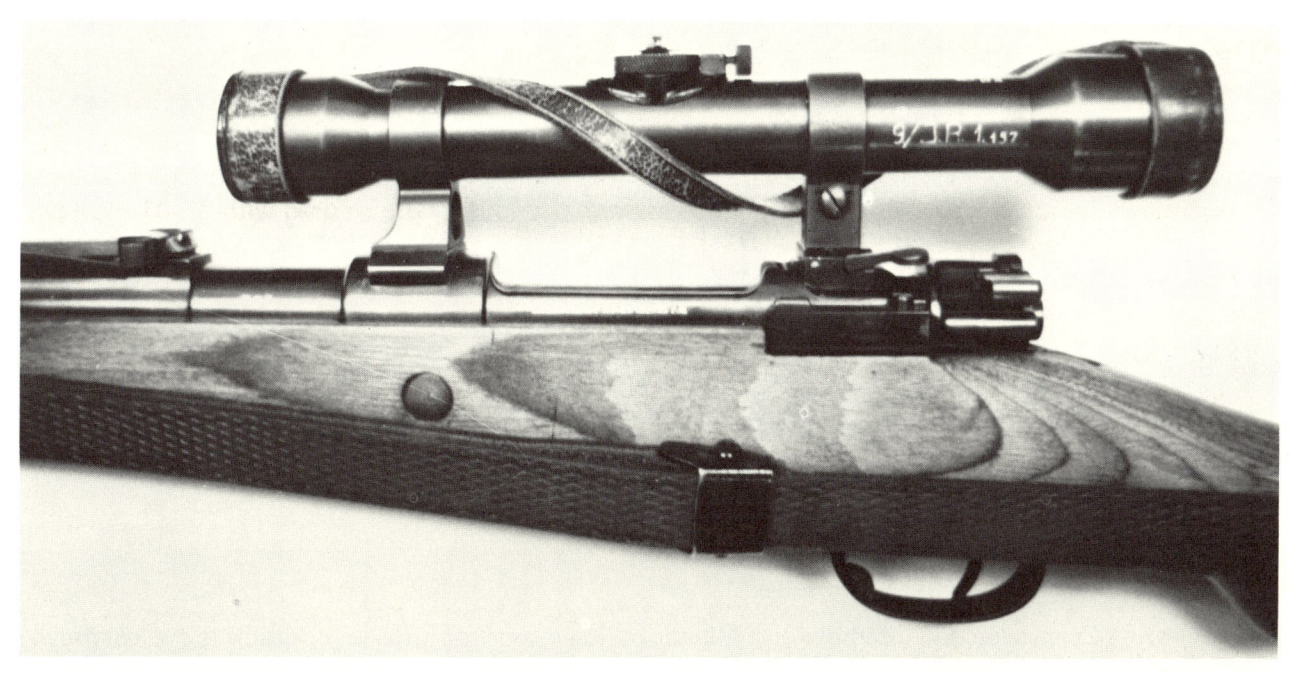

Close view of Mauser military sporter showing the Zeiss telescope and method of mounting. Note the original leather lens caps and unit markings engraved on the scope tube. Found on various prewar German scope-sighted military rifles, such markings served as identification and as a deterrent from theft by other units.

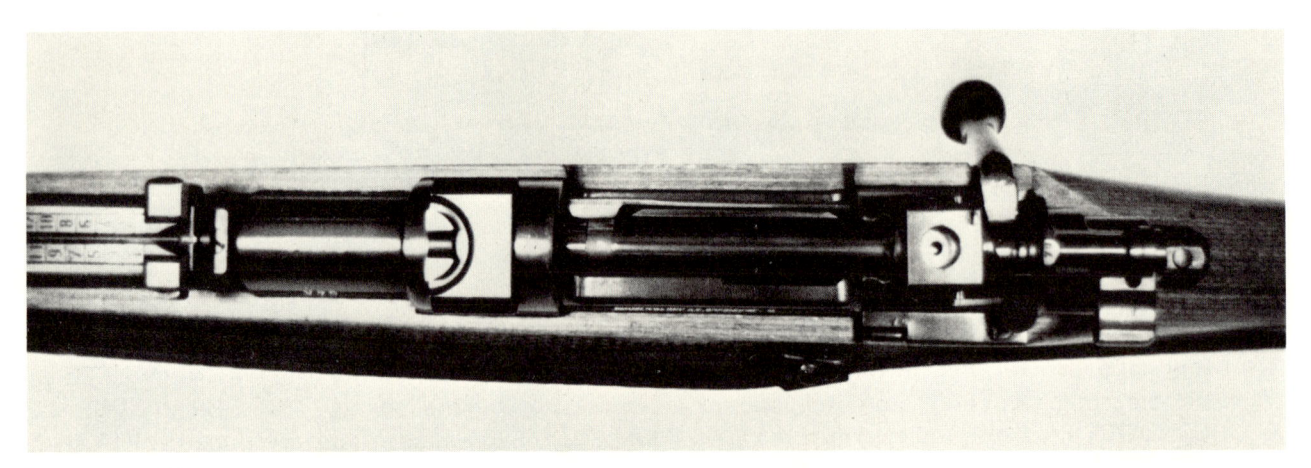

Top view of Mauser military sporter with telescope removed. Note the absence of a "thumb-cut" in the receiver wing, corresponding stock line, and laminated stock.

Right view of Zeiss telescope assembly and Mauser military sporting rifle. In addition to typical prewar commercial proofs, various components bear a "655" Waffenamt inspection stamp. While definitely not an "experimental sniping rifle" as some contend, this weapon was undoubtedly pressed into combat service early in the war as were most German telescopic-sighted rifles.

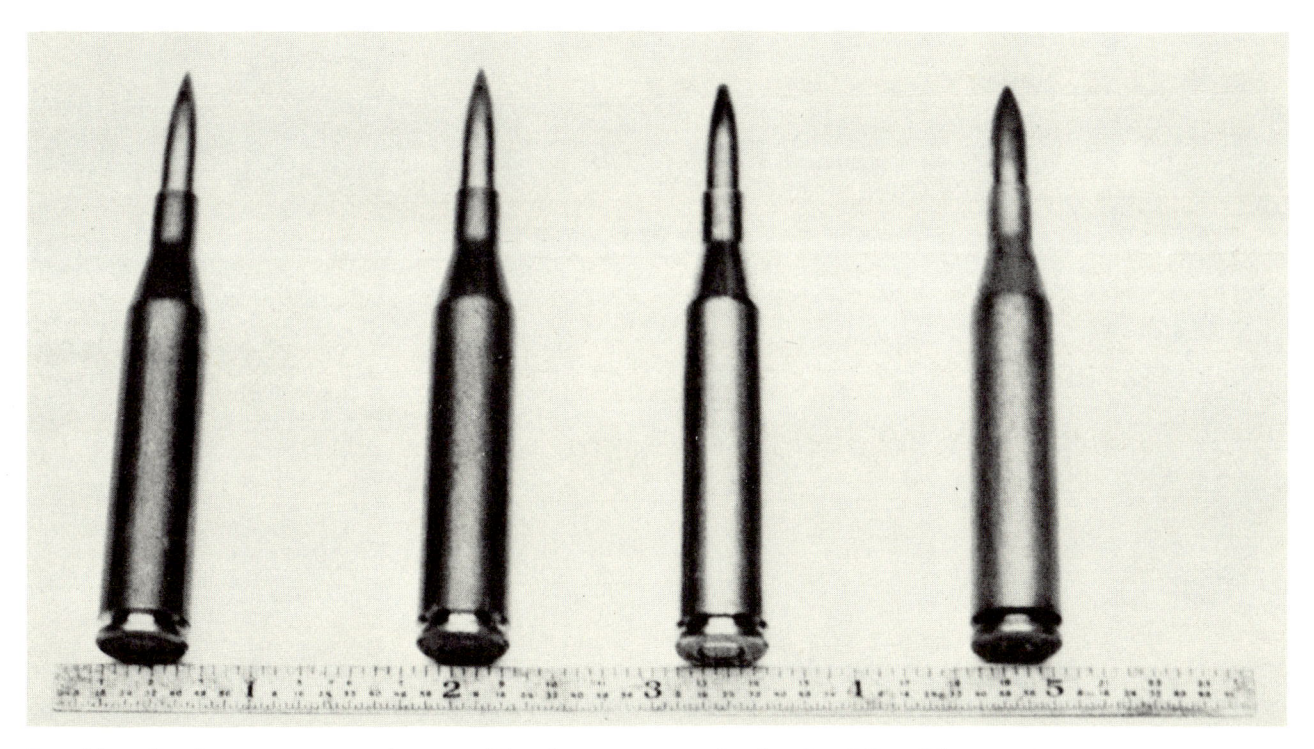

Considered a long-range varmint cartridge by some standards, the 5.6 x 61 vom Hofe cartridges with a 77 grain, 5.6mm diameter bullet, and powder loading of approximately 54 grains gave a muzzle velocity of 3708 fps.

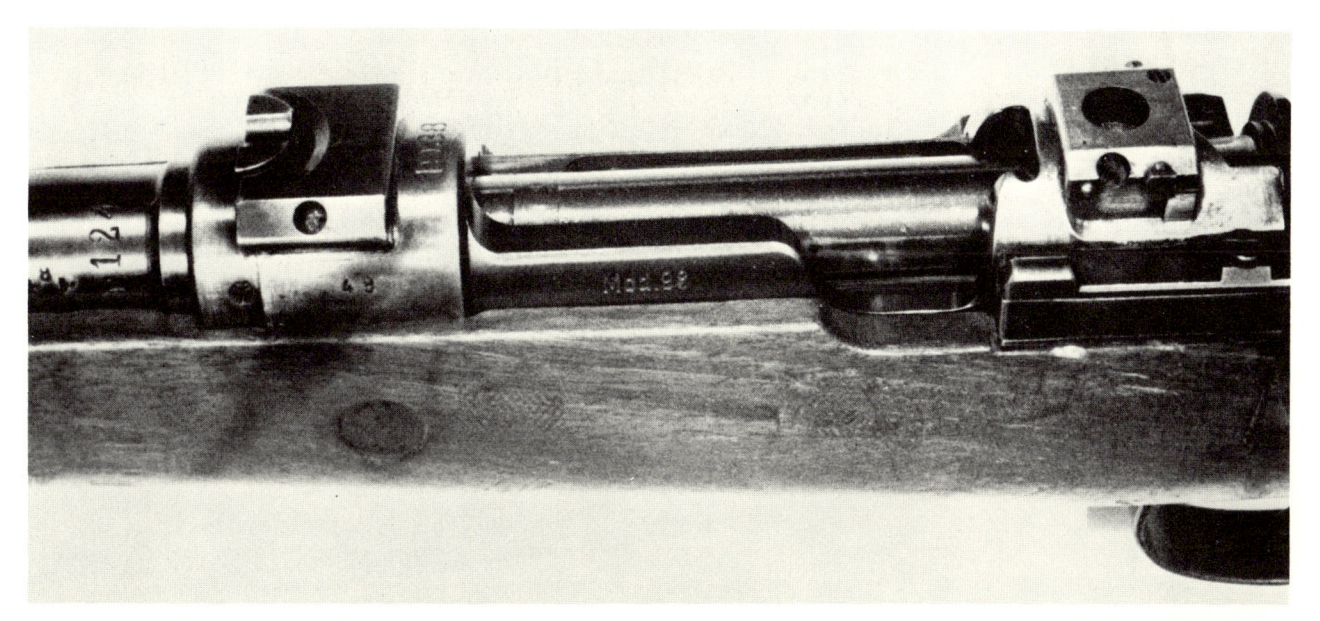

An unexplained, authentic Karabiner 98k rifle with telescope mounting bases identical to those found on the preceding Mauser military sporter. The Berlin-Suhler Waffen und Fahrzeugwerke (BSW 1938) rifle, serial No. 48, bears the number "4" Waffenamt inspection stamp typical on the early BSW 98k variants. Note that the rear base locking lever is missing here.

Right view of BSW manufacture 98k rifle with telescope mounting base. Even though telescopic-sighted 98k rifles were used by the German military before the war, most were not intended for sniping purposes at that time.

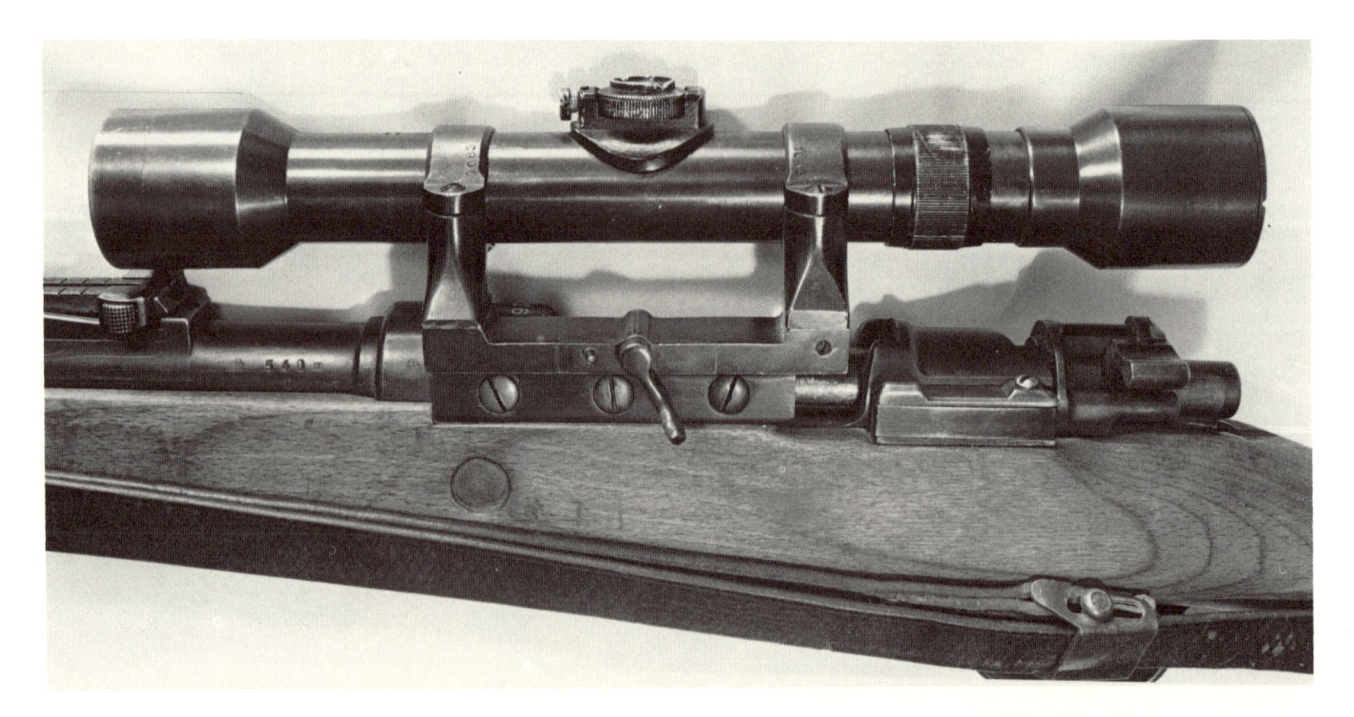

Prewar manufacture 98k rifle by Mauser-Werk, A.-G., Werk Borsigwalde (S/243 1938) with unique side rail telescope mounting and Hensoldt (Dialytan 4X) sight. Note the three base mounting screws and fragile locking lever. A comparison of this upper mount assembly to the short side rail military variant reveals similar characteristics.

SNIPING WEAPONS MISCELLANY

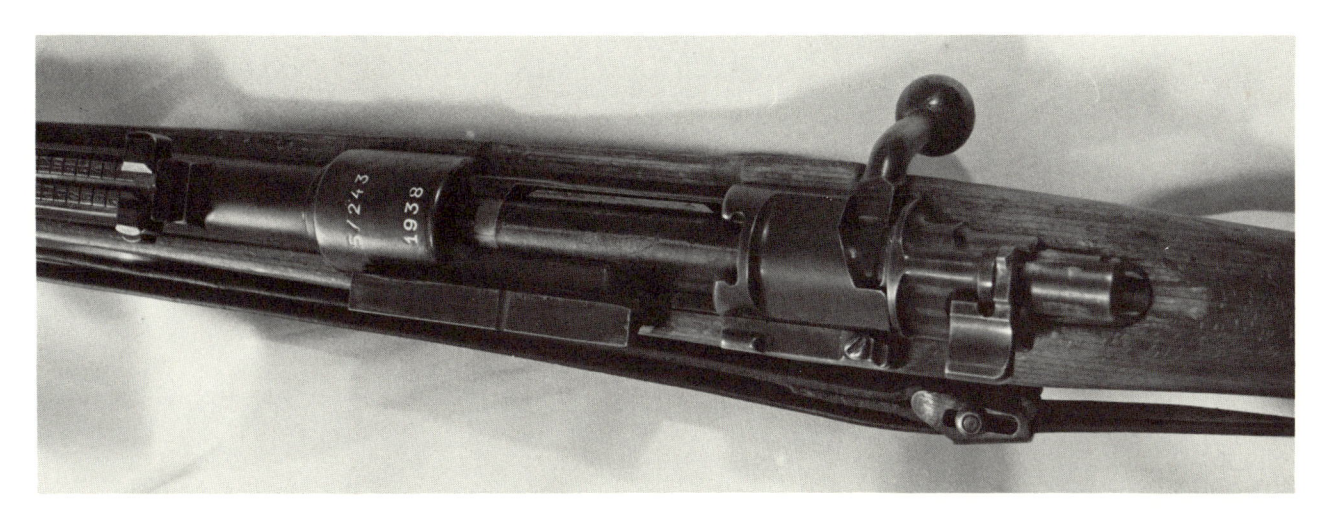

Top view of Mauser (S/243) 98k rifle with telescope removed. Note the circular recess in the base which engaged the mount locking lever.

Right view of side rail mount S/243 98k (280 Waffenamt stamp) with 4-power prewar commercial Hensoldt scope. The knurled ring was used to focus the sight.

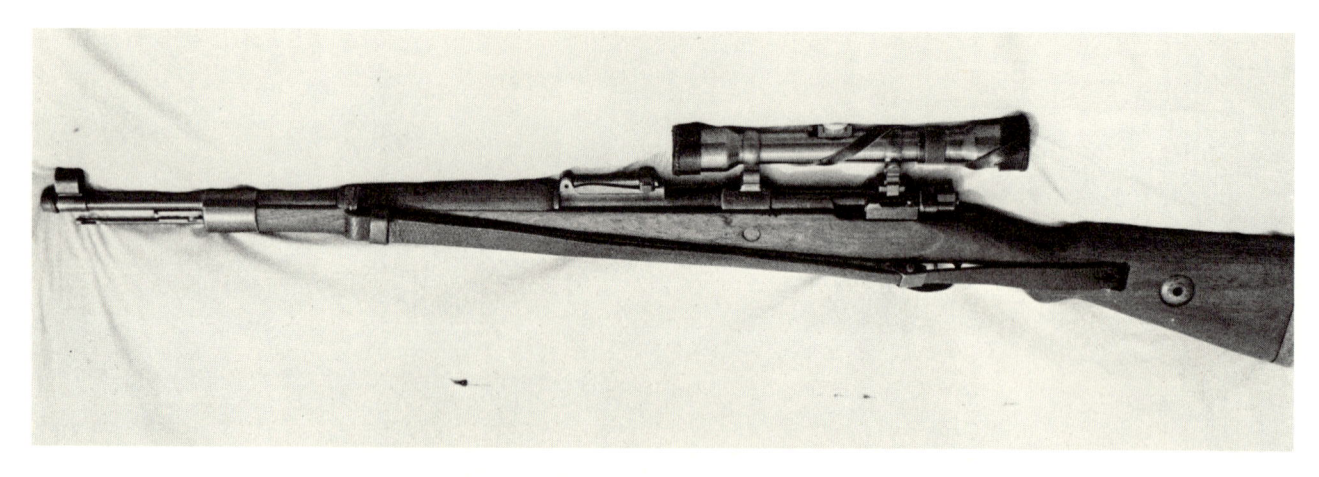

A late war claw mount 98k sniping variant representative of those assembled at the field level using commercial components. While difficult to verify as original, in this case an authentic World War II German sharpshooting rifle was based on a 98k produced by Steyr (bnz 43).

The 98k (bnz 43) spring-loaded rear mount locking latches engaged both sides of the receiver base. Note the manner in which the front base was dovetailed into the receiver ring. Provided with an opening (tunnel) in both mounts, the standard sights could be used with the scope in place.

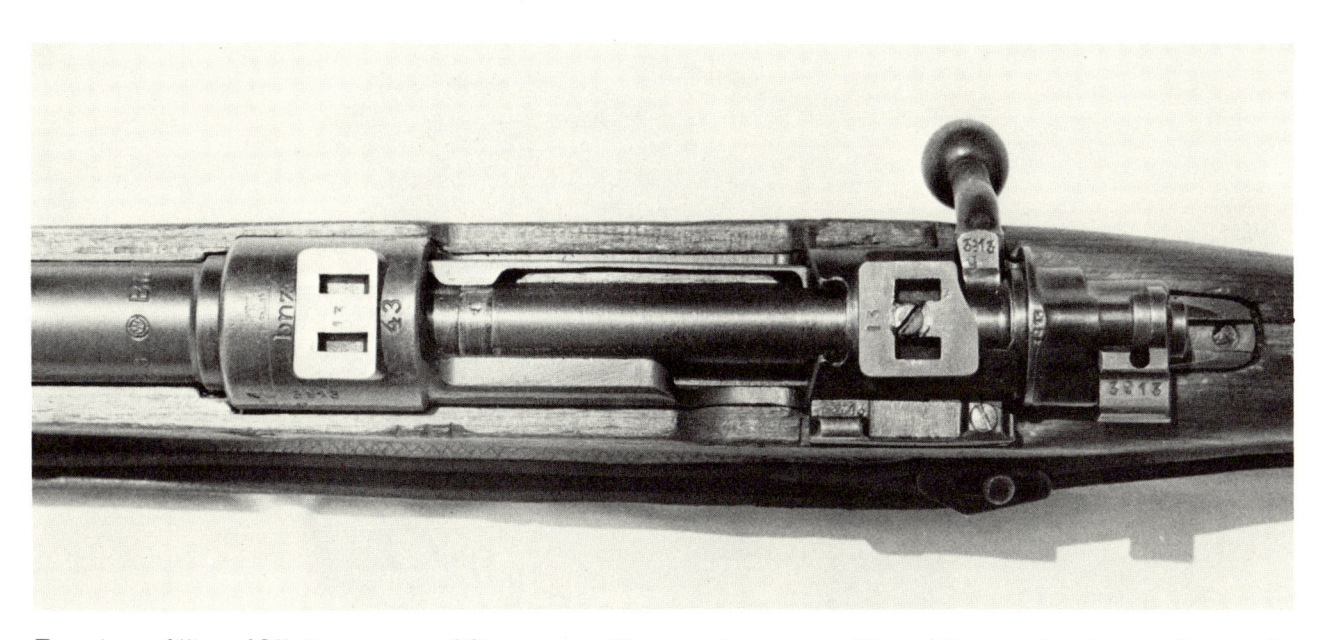

Top view of "bnz 43" claw-mount 98k sniping rifle showing the position of the receiver bases. Originally held by soft-solder, an auxiliary holding screw was added to the rear base at some point. The front claw-mount assembly and dual rear mount projections engaged the openings in both bases.

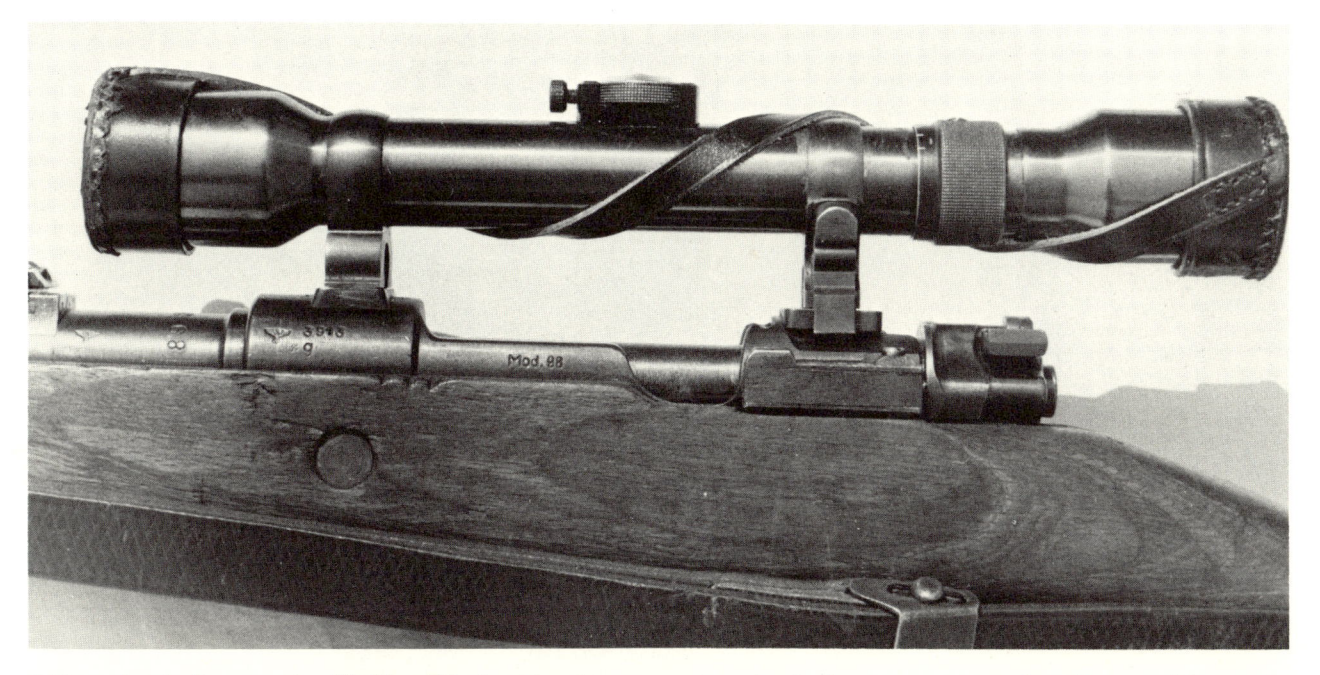

Although similar to the Waffen-SS double-claw mount variant, the scope mounting rings and bases on this rifle are distinctly different. The 4-power scope, one of three known commercial variations made by K. Krahling (Wetzlar) has an aluminum focus ring, a pointed post with side bar reticle pattern, and was finished in a lustrous blue.

Commercial leather scope carrying case typical of those pressed into German military service for sniper use early in the war.

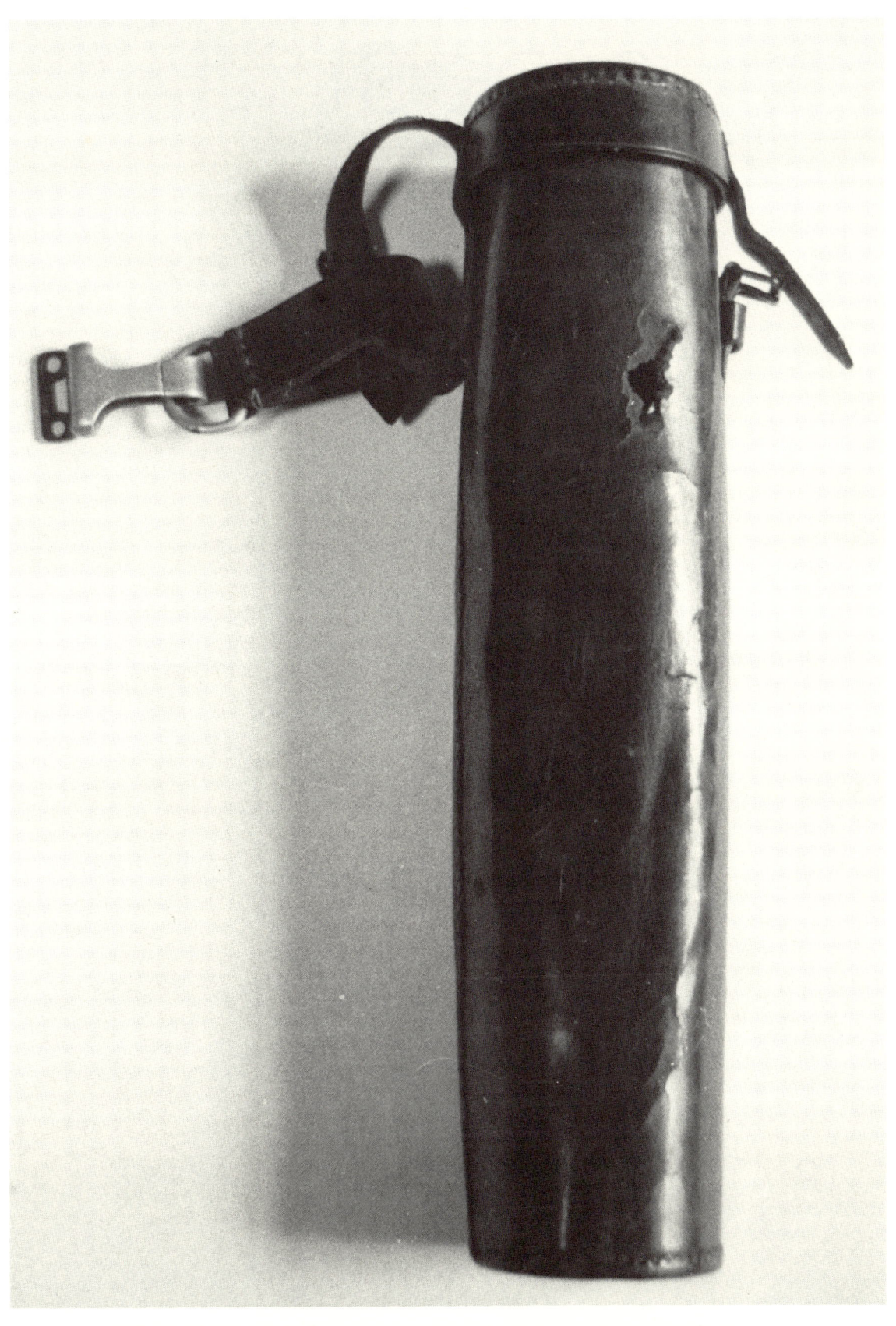

This unusual World War II German scope carrying case held a turret mount assembly when brought back from Europe following the war.

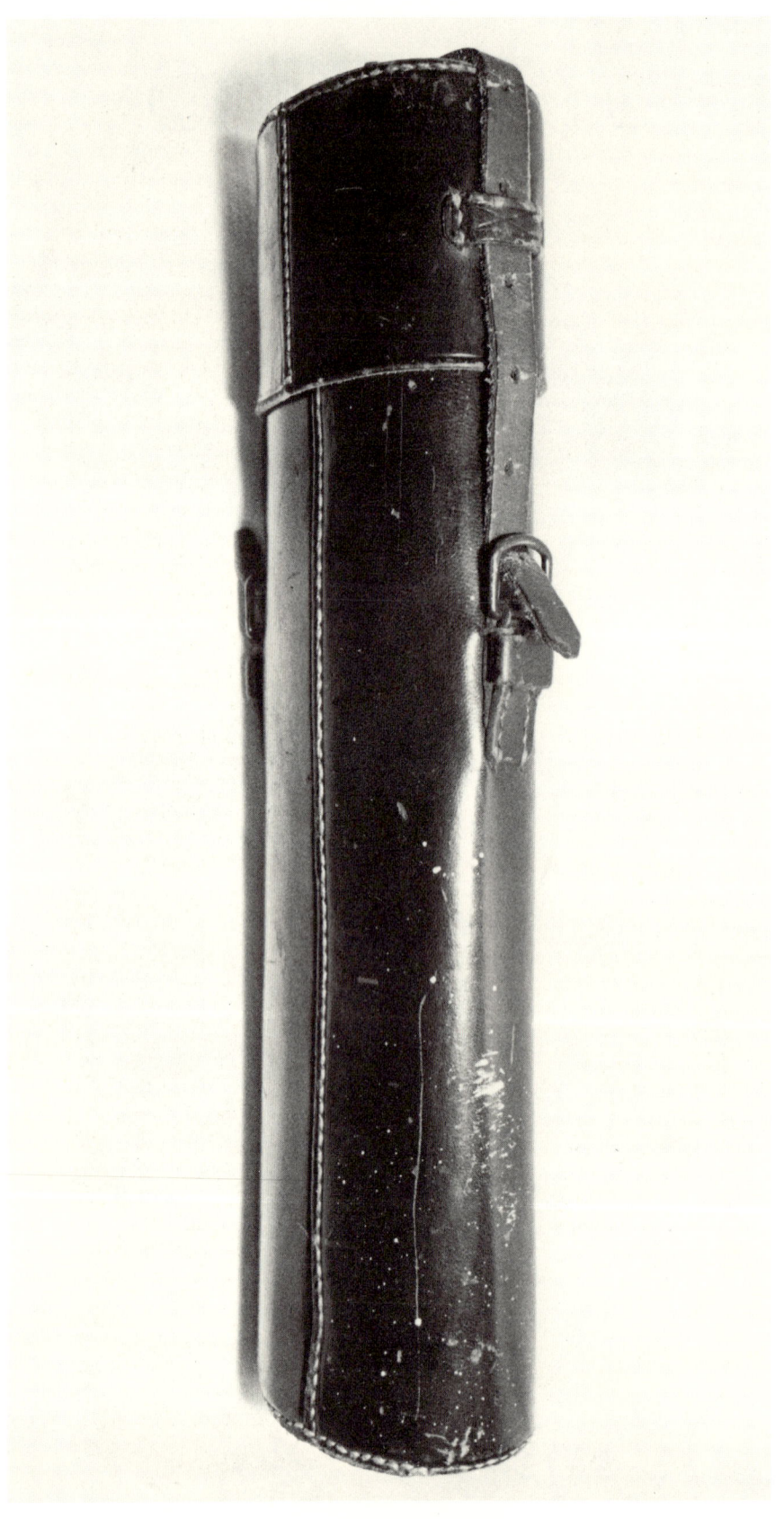

Even though leather scope cases were briefly manufactured for sharpshooting applications, those bearing German military markings are far from common.

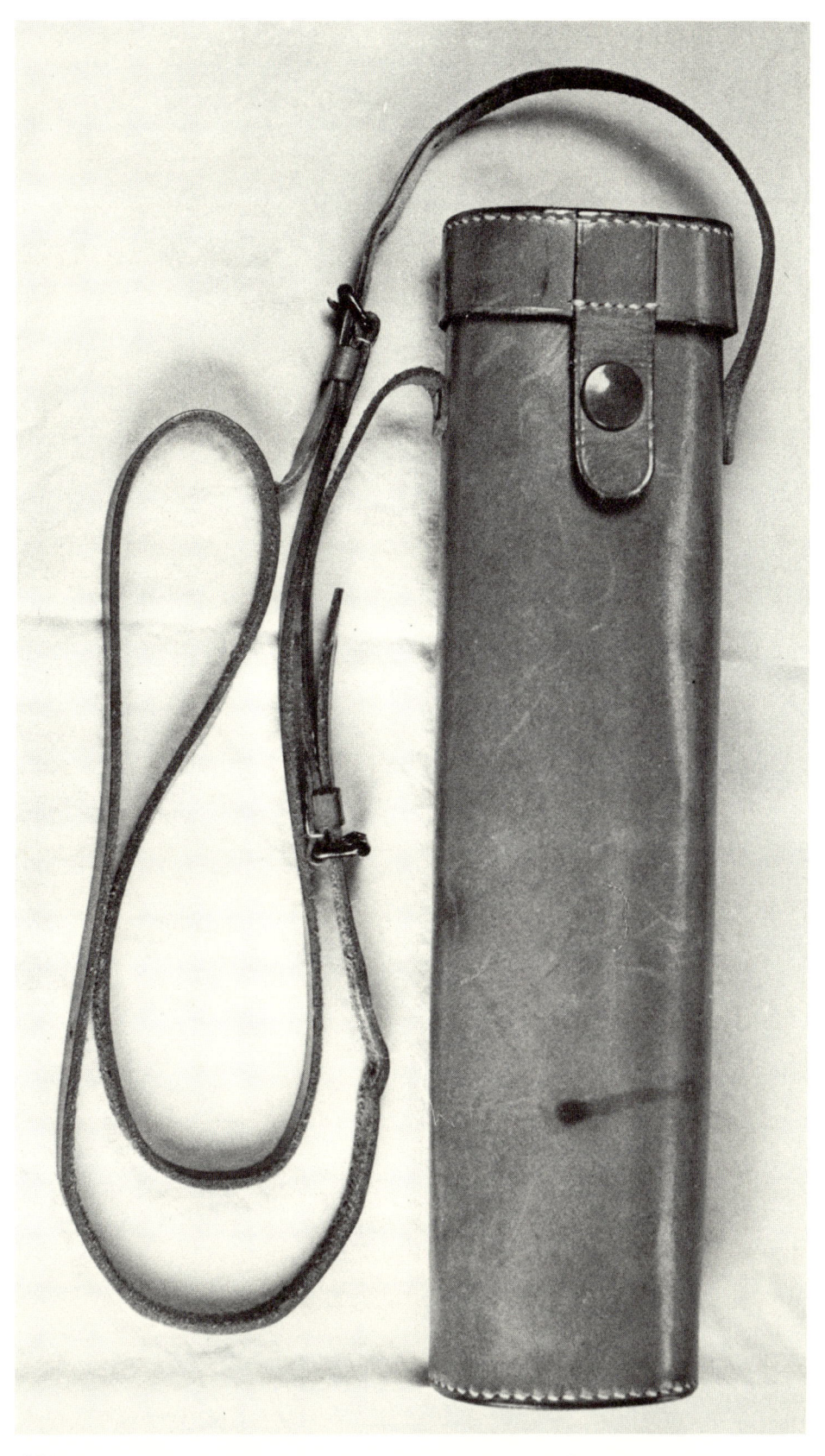

Military marked leather scope carrying case made for sniper use by Carl Weiss, Lederwarenfabrik (cww 43) during the war.

Leather scope case with shoulder strap containing an early 4-power Ajack scope and short side rail mount. Prewar commercial cases fashioned from black or brown leather were readily adapted for military use. Note the rifle number stamped on the lid.

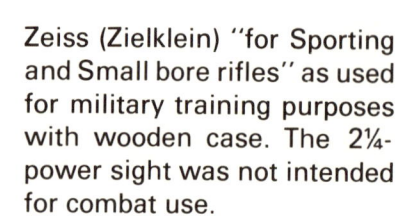

Zeiss (Zielklein) "for Sporting and Small bore rifles" as used for military training purposes with wooden case. The 2¼-power sight was not intended for combat use.

Typical 6 x 30 (bek) and 10 x 50 (rln) World War II German military binoculars with carrying cases as issued for general Wehrmacht use. Manufactured under contract by various optical firms, binoculars were also made by some of the same companies that produced telescopic sights for sniping purposes. Although 6 x 30 binoculars were cited as "regulation issue" for snipers, the 10 x 50 type saw extensive use as well.

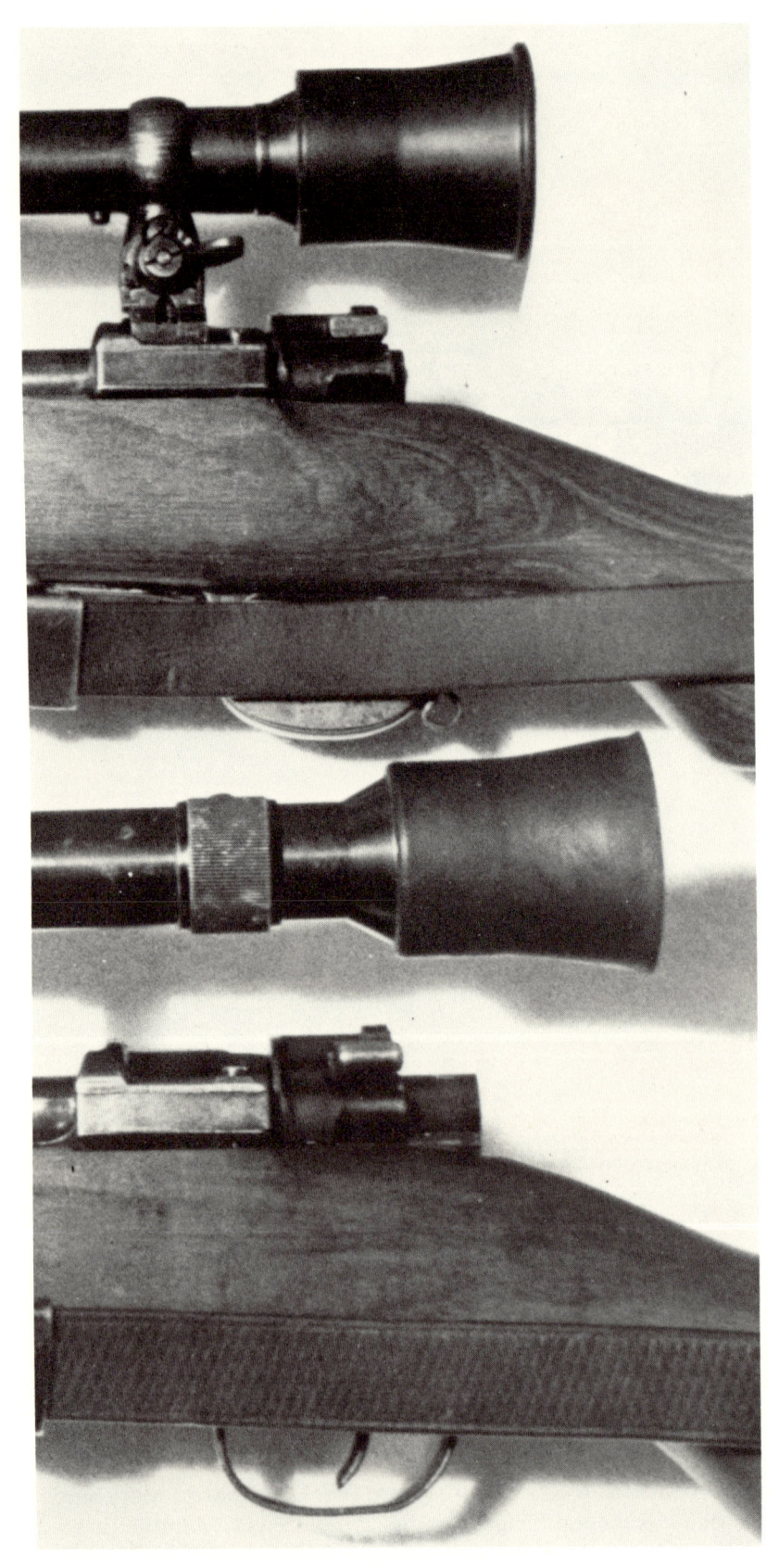

A part of the early sniping issue, surviving rubber eye-guards are far from common. Note the rolled-back edge (top) and the straight variant (below).

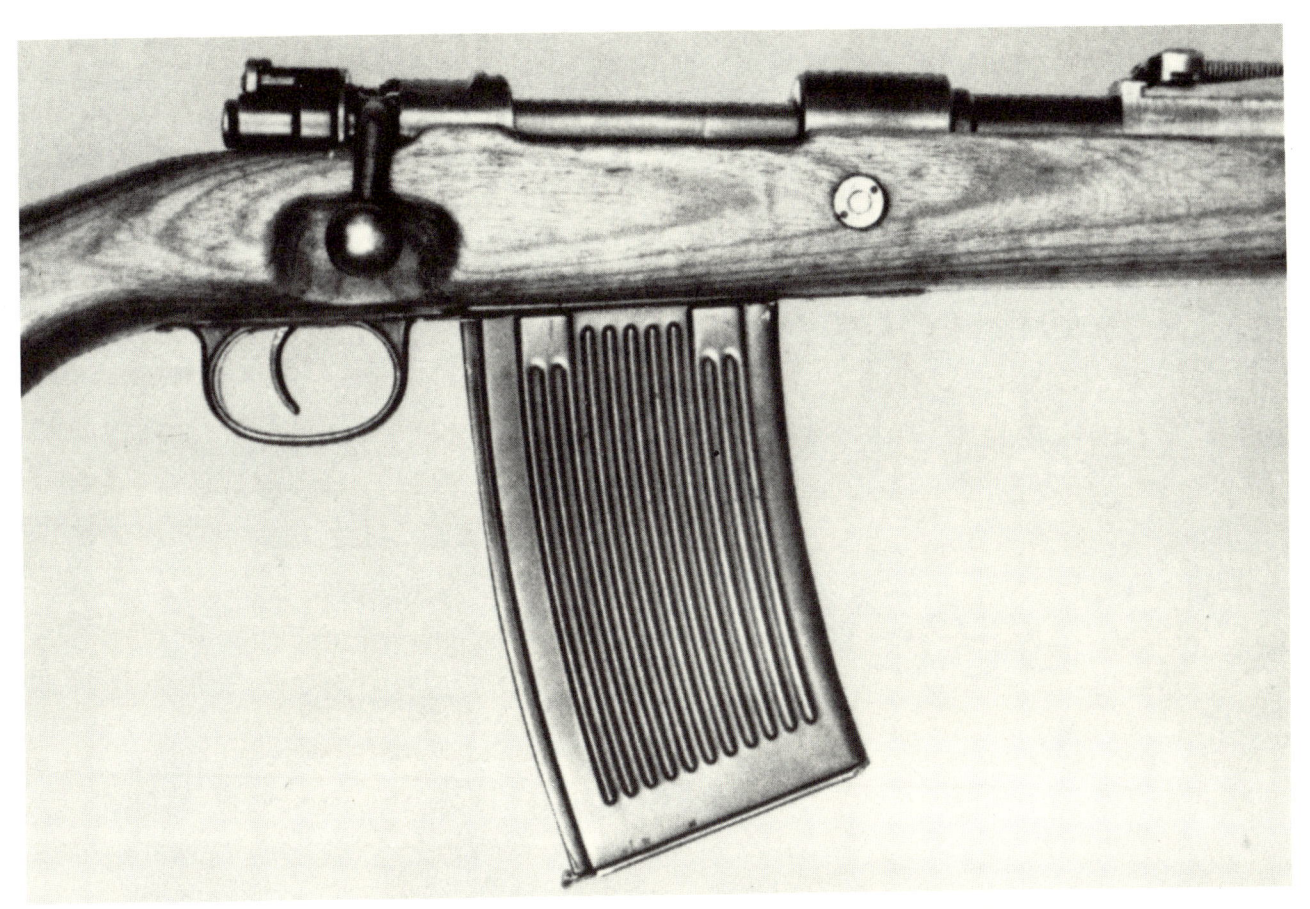

Intended to increase "fire-power," extension magazines were tested with both standard and sniper versions of the 98k. A specially fabricated 10-round magazine for sniping rifles was tested and rejected as late as December 1944. The 25-round variant shown (MG13–Dreyse) is spot-welded to the original housing.

The mount at left is an experimental design by Walther, and its configuration indicates it was intended for use on the 98k. Although the locking mechanism is missing, by comparing the opening with that of a standard ZF 4 mount (right), the similarity is apparent. This mount will not accept a ZF 4 scope and will only fit the left side of the weapon. The design further indicates that the standard military contract 4-power scopes utilized with the 98k were to be employed. The dimensions of the dovetail machining compare with that of the standard Gewehr 43 rail. There are no serial numbers on this mount.

A special sun shield intended for use in arctic conditions reduced the snow glare by about one half. The shield is a metal stamping with an L-shaped groove cut in the side. This groove is aligned with a small screw on the front of the scope tube, pushed on, and turned to the right in familiar bayonet fashion. The shield is shown here on a "bek" code (Hensoldt, Herborn) scope.

Intended for use under subdued light conditions, this rare device has a luminous ramp to assist the shooter when using conventional open sights. Late in the war luminous night sights were manufactured for the K43 and MP44 series weapons in addition to the 98k rifle.

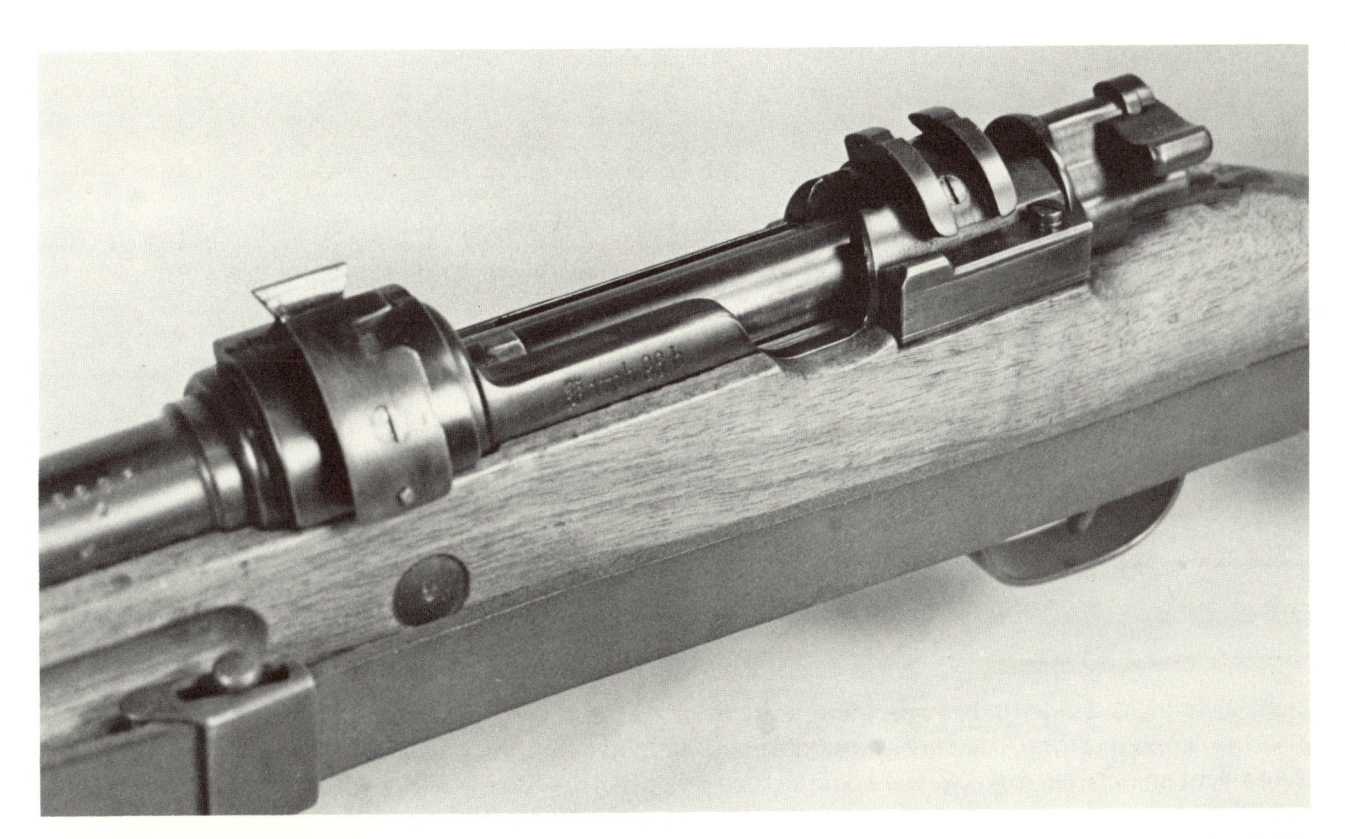

Karabiner 98b (6685e) with highly unusual telescope mounting bases. The only known example of this type.

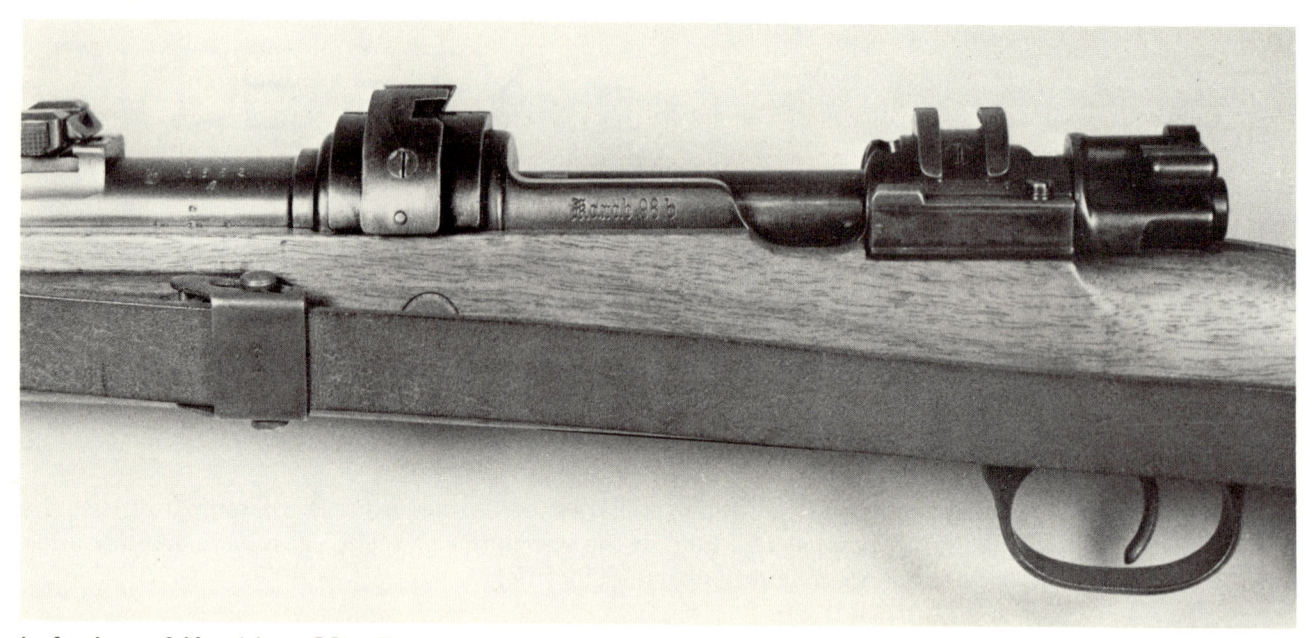

Left view of Karabiner 98b. The front base assembly, a complex piece of machining, is fastened to the receiver ring with three screws. The rear base is held by two screws. The original rifle stock bears Heer (Army) markings.

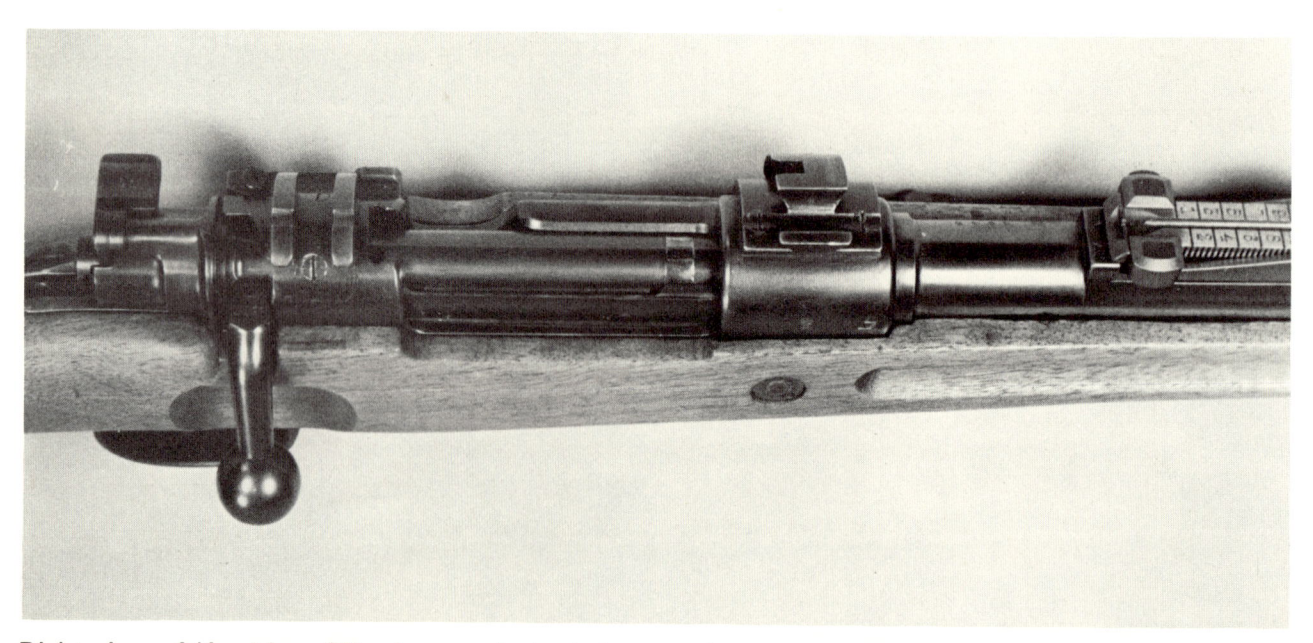

Right view of Karabiner 98b. A recess in both front and rear base permitted use of the standard sights with the scope in place. The configuration of the telescope mounting remains unknown. Although early Wehrmacht snipers did employ telescopic-sighted rifles of this type (Karabiner 98b), bonafide examples are quite rare.

A *current* manufacture ZF 4 telescope mounting produced for the collector trade. The pressure lever markings have since been modified to conform to the original configuration.

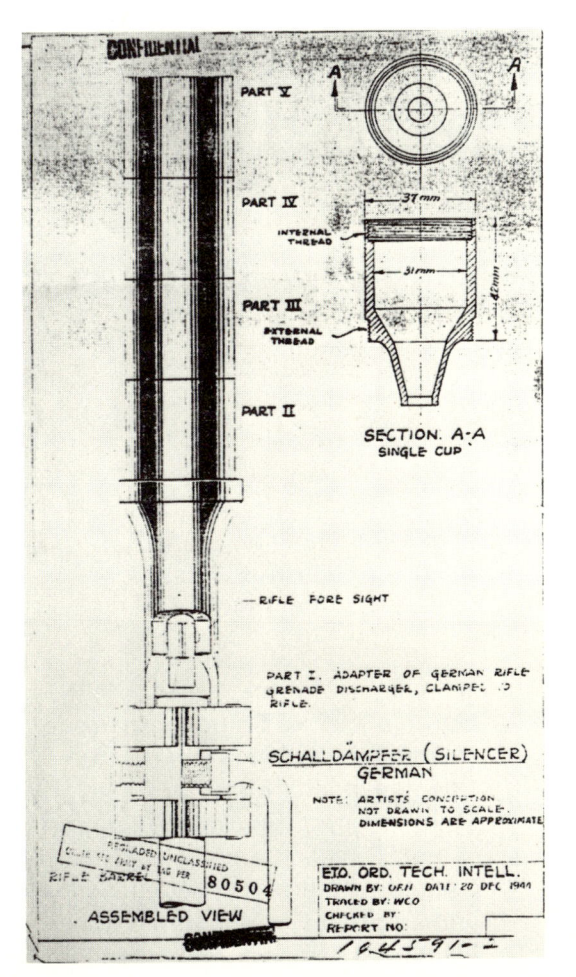

Even though rifle silencers saw limited use by German snipers, the following devices are presented as a matter of reader interest. U.S. Ordnance drawing depicts a typical silencer (Schalldämpfer) described by a German prisoner who served near Berlin sound-testing silencer production destined for combat use.

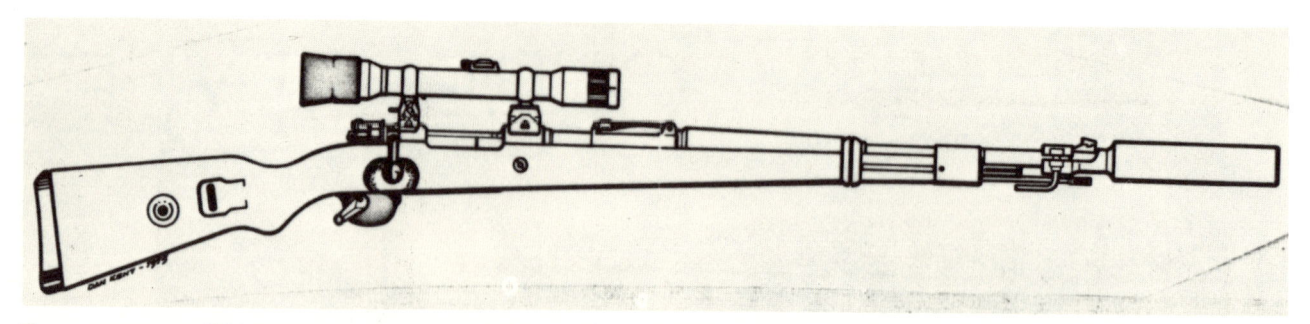

Turret–mount 98k sniping rifle as it would have appeared with a silencer (Schalldampfer). Such devices were not considered practical for sharpshooter use. (Copyright Dan Kent)

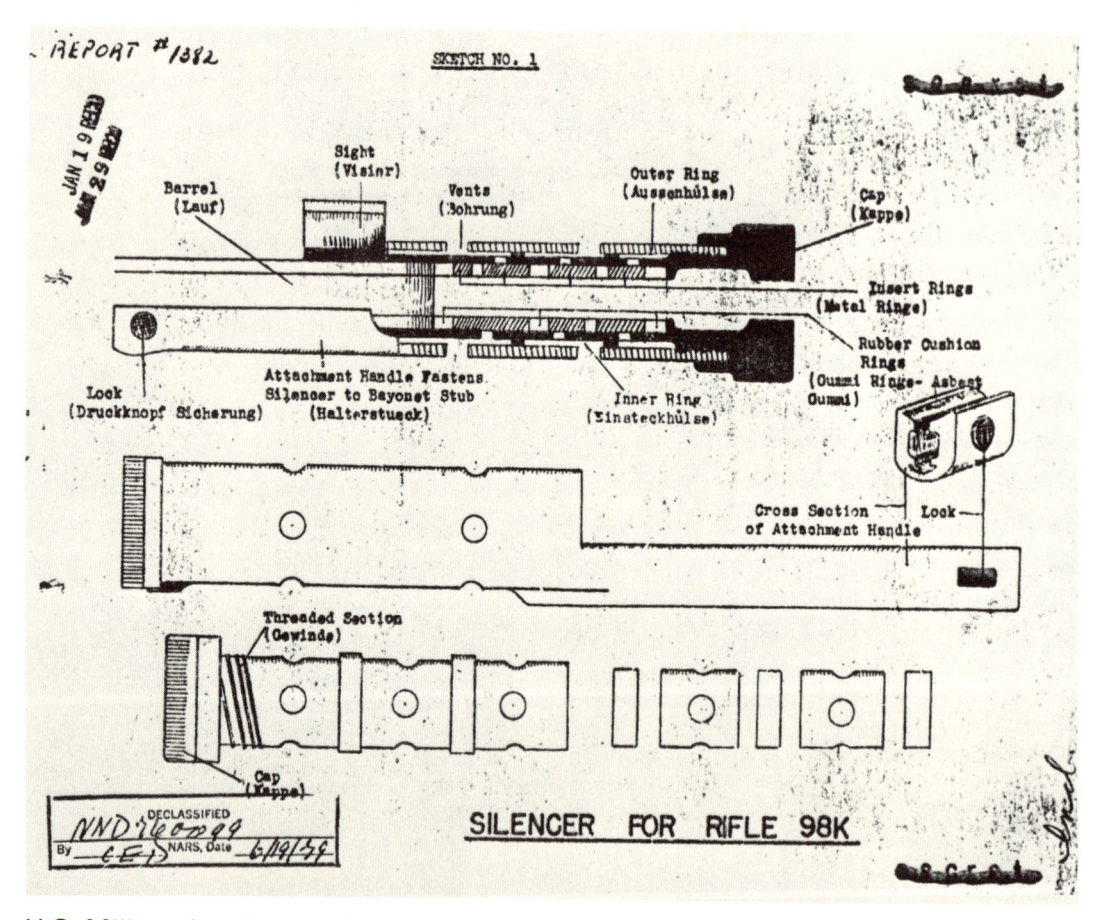

U.S. Military Intelligence Service drawing dated 20 November 1944 showing a German silencer for the 98k rifle. Noted as being in use by regular line troops, this variant easily attached to the bayonet stub. According to the report, "The sound of the shot is like a muffled noise caused by pulling the cork from a bottle."

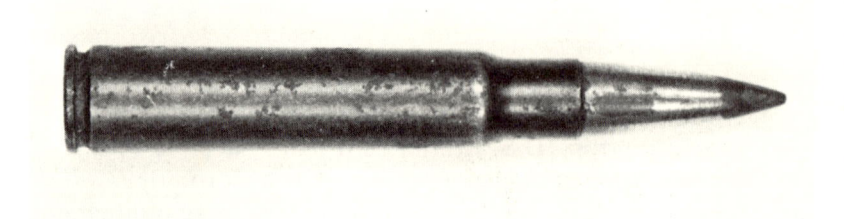

An extremely rare original WW2 German 7.92 x 57 mm Nah-patrone cartridge intended for use with the rifle silencer. Manufactured in 1943 by the Finower Jndustrie G.m.b.H. (cg), the steel case was colored with green opaque lacquer and also had a green primer annulus to set it apart from standard ammunition.

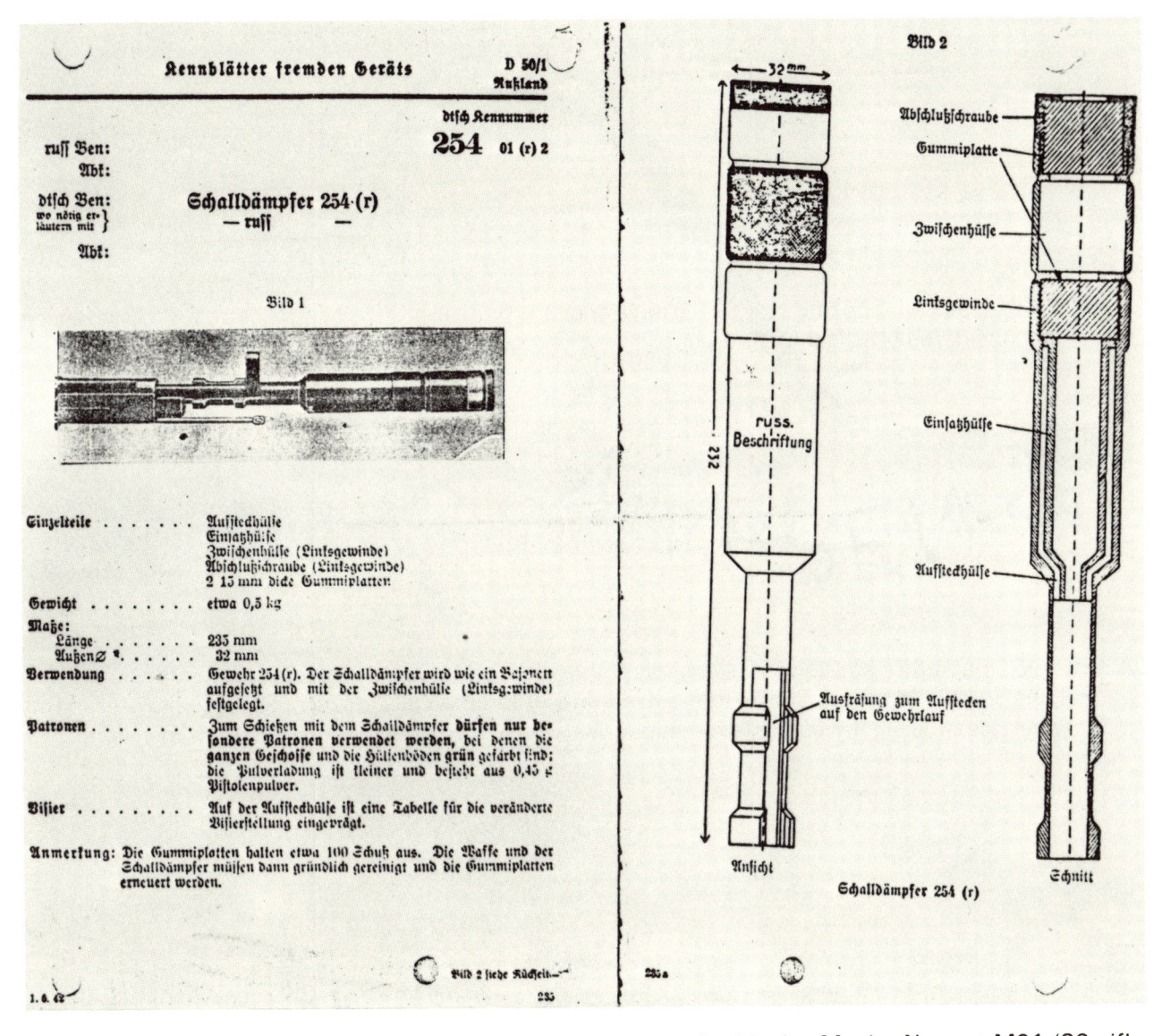

German Ordnance description of the Russian silencer used with the Mosin–Nagant M91/30 rifle. According to reports, these devices were quite reliable.

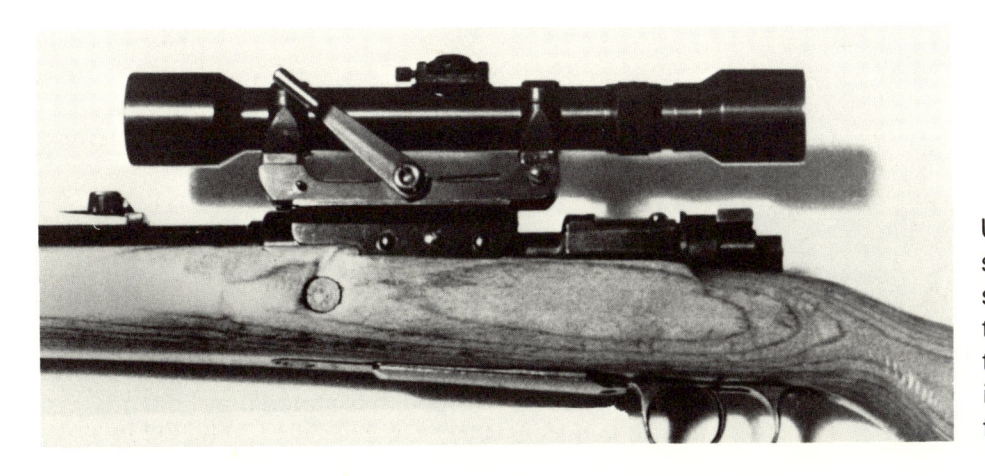

Unconventional 98k long side rail variant with a single large screw and two pins holding the base to the receiver. The scope is an Ajack model with focus ring.

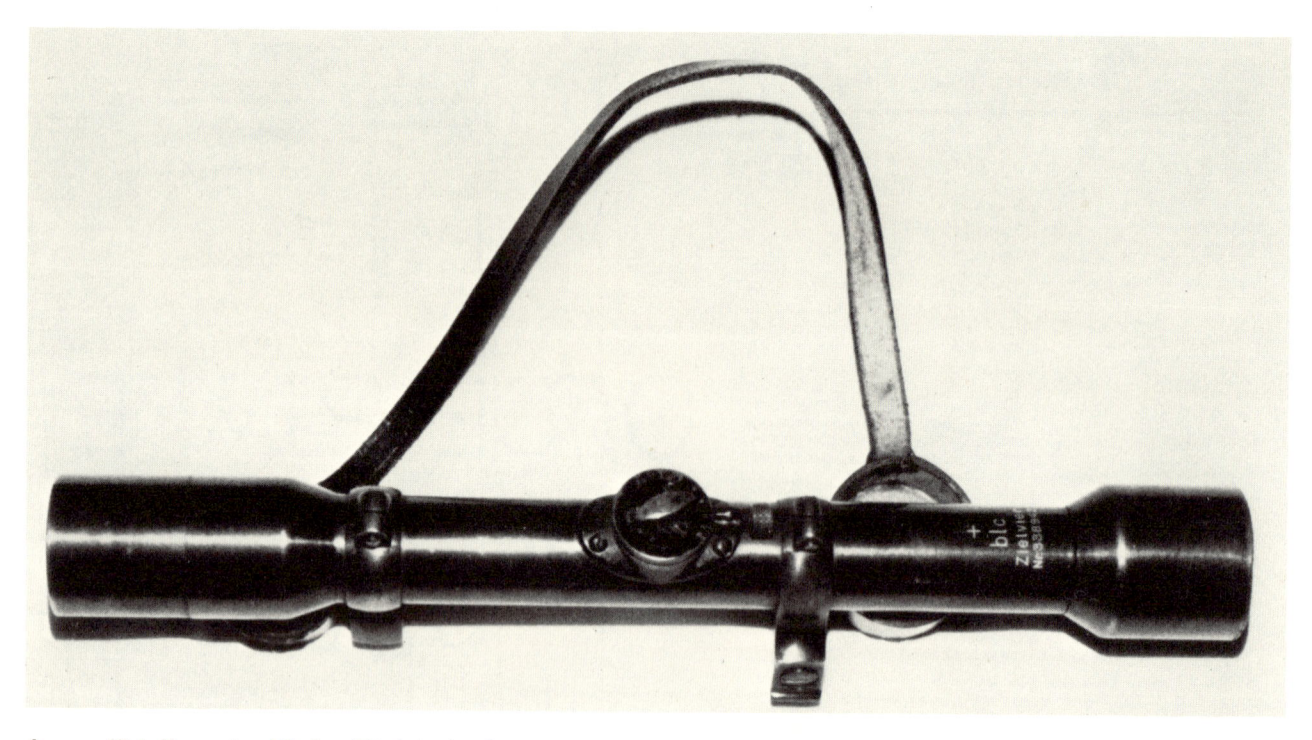

A rare "blc" marked Zeiss (Zielvier) telescopic sight in full military trim with objective extension (shield) and original leather lens caps. Unlike the vast majority of commercial Zeiss scopes that were pressed into military service, this sight was made for sniper use. The mounts were not issued with this sight.

Semi-finished telescope mounts dating from WW2 obtained in *postwar* Germany. Unfortunately, items like this are often used to produce bogus sniping rifles for the collector trade.

Late war manufacture 10 x 50 binoculars bearing the uncommon "rln" Zeiss code. While this mark is variously cited as a Hensoldt code, the Hensoldt firm, now a part of the Zeiss–Gruppe, has stated they "did not" use this code during WW2.

WW2 German Ordnance weapon situation report for the K98k with telescopic sight, dated 1 March 1945. Included with this information are details concerning production previews, acceptance, and ultimate dispersal of 98k sniping arms for the various branches of the Wehrmacht. The quantities are expressed in units of 1,000.

German Ordnance weapon situation report for the K43 with telescopic sight, dated 1 March 1945. While noteworthy, reports of this type often conflict with other official documents of similar content and should be viewed with some reservation.

A 5-round G-K43 magazine of unknown origins. While unverified, experts contend this type was intended to produce a "low weapon profile" for sharpshooting purposes. There are no markings on this magazine.

Adapted for testing purposes with the early selective-fire Gewehr 43 experimental rifles, a 25-round MG 13–Dreyse light machine gun magazine is shown in original form. Even though a 20-round "straight-box" variant similar in form to the standard G-K43 magazine was eventually produced in limited quantity, it was not regular issue.

SNIPING WEAPONS MISCELLANY

A qve 45 K43 rifle with special aluminum ZF 4 scope mount fielded for covert operations in Europe by the Central Intelligence Agency (CIA) after WW2. Made in limited numbers in the United States, the aluminum mount was a simplified version of those produced in Germany during the war.

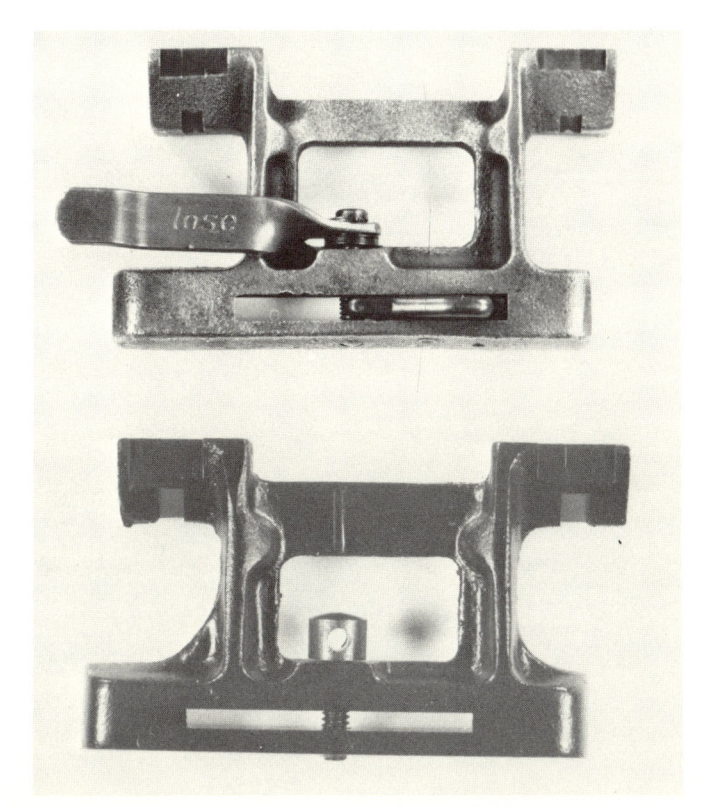

Comparison of a German made ZF 4 mount (top), and the postwar CIA aluminum variant. The large drum-head screw was tightened with a pin to clamp it on the receiver base. Note the absence of an auxiliary locking latch. The aluminum ZF 4 mount is shown in the *CIA Special Weapon Supply Catalog,* dating from that era.

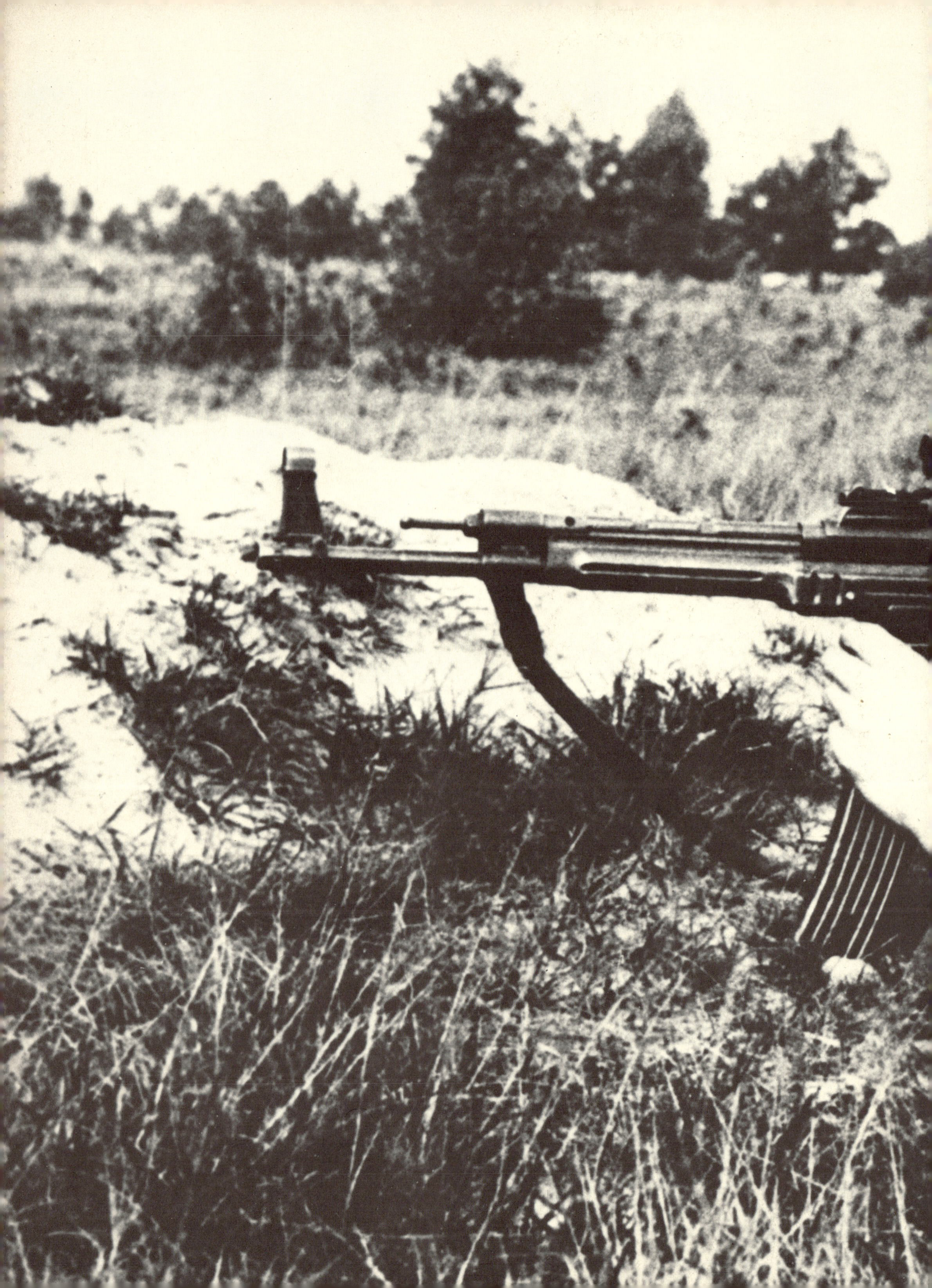